Field-Assisted Sintering

Eugene A. Olevsky • Dina V. Dudina

Field-Assisted Sintering

Science and Applications

Springer

Eugene A. Olevsky
College of Engineering
San Diego State University
San Diego, California, USA

Dina V. Dudina
Lavrentyev Institute of Hydrodynamics, Siberian Branch of the Russian Academy of Sciences
Novosibirsk, Russia

ISBN 978-3-030-09385-3 ISBN 978-3-319-76032-2 (eBook)
https://doi.org/10.1007/978-3-319-76032-2

This Springer imprint is published by the registered company Springer Nature Switzerland AG
The registered company address is: Gewerbestrasse 11, 6330 Cham, Switzerland

Preface

The successful development of computer-aided manufacturing by industry has made it possible to create a multitude of different devices that can be used to process materials using the application of heat, pressure, and an electric current or electromagnetic field simultaneously. These devices and processes belong to the area of so-called field-assisted sintering, which offers fundamentally new materials processing conditions. These sintering techniques enable the creation of materials and components with unique properties. The most widely researched and used representatives of these technologies are microwave, spark plasma, and flash sintering.

Sintering of powder materials is both a widely used technological process and one of the most basic physical phenomena. Indeed, the driving force of sintering is surface tension, or, the minimization of surface energy. Field-assisted sintering, which includes also the impact of electromagnetic factors, is, accordingly, a vivid example of a fundamental and multi-physics process.

Field-assisted sintering technologies thus represent an excellent experimental base for researchers engaged in the discovery of new materials. Simultaneously, field-assisted sintering opens up wide opportunities for modeling the processes of materials consolidation and synthesis while taking into account numerous physical phenomena.

This book is an attempt to unify and compile the diverse field-assisted sintering technologies. This monograph does not offer an exhaustive analysis of all the nuances of technologies and theoretical concepts of field-assisted sintering. However, we tried to introduce the fundamental principles, hardware, and sample applications of both the most popular and less well-known types of field-assisted sintering in a single format.

This monograph should be useful to researchers, industry professionals, and senior undergraduate and graduate students specializing in the field of materials engineering.

San Diego, CA, USA — Eugene A. Olevsky
Novosibirsk, Russia — Dina V. Dudina

Acknowledgments

Eugene Olevsky acknowledges the support of his research group's activities in the area of field-assisted sintering by US Department of Energy, Office of Science, Basic Energy Sciences, Division of Materials Sciences and Engineering.

The authors are grateful to Elena Aleksandrova for her assistance in the preparation of figures.

Contents

About the Authors

Eugene A. Olevsky is the Interim Dean and Distinguished Professor of the College of Engineering at San Diego State University, USA. Dr. Olevsky is the Director of the San Diego State University Powder Technology Laboratory. He has obtained two M.S. degrees in Mechanical Engineering and Applied Mathematics and a Ph.D. degree in Materials Engineering. His primary area of expertise is in experimentation and computational modeling on powder processing, including novel ceramic, metallic, and composite materials synthesis. He is the author of over 500 scientific publications and of more than 150 plenary, keynote, and invited presentations in the area of sintering research. He has supervised scientific sintering studies of more than 100 postdoctoral, graduate, and undergraduate students. Prof. Olevsky is a Fellow of the American Ceramic Society, a Fellow of the American Society of Mechanical Engineers, a Fellow of ASM International, Humboldt Fellow, and is a Full Member of the International Institute of Science of Sintering. Dr. Olevsky's most recent research is focused on field-assisted sintering techniques and sintering-assisted additive manufacturing.

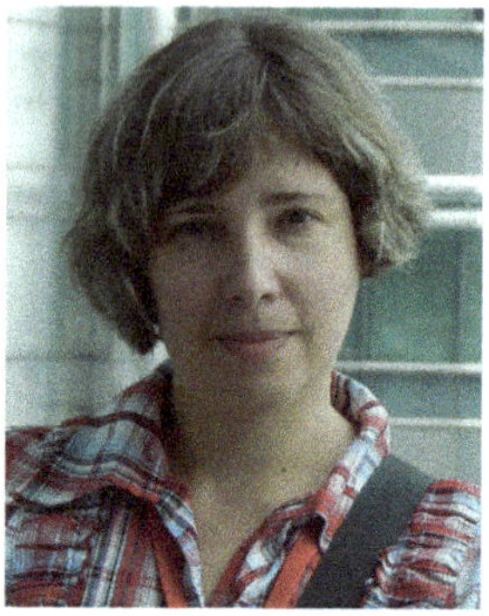

Dina V. Dudina graduated from Siberian State Industrial University, Novokuznetsk, Russia. She obtained a Ph.D. (Candidate of Sciences) degree in Solid State Chemistry in 2004 after completing her postgraduate studies at the Institute of Solid State Chemistry and Mechanochemistry, Siberian Branch of the Russian Academy of Sciences, Novosibirsk, Russia. She conducted her postdoctoral research at the University of California, Davis, USA, and Institut Polytechnique de Grenoble, France. In 2017, she defended a habilitation thesis in Engineering Sciences in Russia. At present, she is a senior scientist with Lavrentyev Institute of Hydrodynamics, Siberian Branch of the Russian Academy of Sciences, Novosibirsk. She teaches Materials Science courses at Novosibirsk State Technical University. Dina Dudina is the author/co-author of more than 90 publications in the area of powder processing, sintering, and composite materials.

Chapter 1
Introduction

1.1 A Brief Historical Overview

Technological advances in many industries depend on the successful performance of parts produced by sintering of powders. More demanding applications of the sintered parts dictate new property requirements, which can be fulfilled only with the use of more efficient sintering techniques. Sintering is a process of consolidating powder particles into a bulk solid, which, in order to proceed, requires thermal activation of diffusion or other mechanisms of mass transport [1]. In traditional thermomechanical methods of powder processing, heat and pressure are the only tools to change the state of a material, and, thus, the possibilities of the conventional powder consolidation methods are limited in influencing the processes occurring in the powder compact during sintering.

Flexibility can be added to the sintering process by the introduction of electric and magnetic fields as sintering tools. A search for heating methods other than conventional furnace heating was stimulated by the need of higher-productivity processes and more uniform heating. Heat transport in initially porous bodies becomes a serious issue as the size of the part to be sintered increases. For conducting materials, a seemingly easy and convenient solution is volume heating by electric current. At present, the field-assisted powder consolidation methods present a matter of great scientific and practical interest, which is indicated by an enormous research activity in this area. The main reasons the field-assisted sintering methods have become so popular in the materials science community are the possibilities of sintering of a wide range of materials to high relative densities in a short time, at temperatures lower that in conventional sintering methods and with a control over grain growth.

Electric current was initially introduced in the field of sintering as a means of rapid heating. Heating filaments of the incandescent lights were introduced by Thomas Edison in the end of the 1870s [2]. The filaments were naturally consolidated during the initial stage of the electric current passage. This idea, however, while inherently present, was not explicitly formulated at that time. The author of the

E. A. Olevsky, D. V. Dudina, *Field-Assisted Sintering*,
https://doi.org/10.1007/978-3-319-76032-2_1

first patent on the application of resistance sintering is W. L. Voelker, who used a DC current to conduct his experiments on the consolidation of filaments of electric light bulbs in 1898 [3]. Resistance sintering was further developed by Sauerwald [4], who conducted the processes under pressure; Taylor [5], whose work had a goal to sinter refractory metals; and Lenel [6]. Soon it was realized that it is not only the presence or absence of current but also its application mode that have to be given particular attention. In the 1960s, patents by Inoue [7, 8] introduced electric discharge sintering, which served as a basis for the development of the spark plasma sintering (SPS) method and equipment by a Japanese company – Sumitomo Coal Mining Ltd. – in the 1990s. Impulse resistance sintering was used by Greenspan [9]. In that process, sintering started at a low pressure to benefit from heating of the contacts between the particles in the beginning of the process; the pressure was further increased as the sintering progressed. Also, microwave sintering and sintering by high-voltage electric pulses have been studied and developed independently for more than half-century, as described in the respective chapters of this book.

A number of review articles have been published since 1993 discussing the advantages and development strategies of field-assisted sintering techniques [10–22]. In these reviews, low-voltage pulsed current-assisted sintering, including SPS, has received particular attention [10–19]. The only monographs on field-assisted sintering are those written by Raichenko [23] and Belyavin et al. [24] published in Russian in 1987 and 1993, respectively. In electric current-assisted sintering, Belyavin et al. [24] saw a solution to the sintering challenges of refractory metals – a solution more promising than the introduction of low-melting-temperature additives. Although the monograph by Raichenko has been cited in research papers a great number of times, it has never been translated into English and remains accessible only to a limited audience. Many years have passed from the publication of this monograph; in these years, the field-assisted sintering has been greatly developed and strengthened by both research results and equipment design. A textbook on the methods of compaction and consolidation of nanostructured materials and products was published in 2008 by Khasanov et al. [25] in Russian, and only one chapter in it was devoted to field-assisted consolidation, namely, SPS.

Multiple possibilities of field application to powder and pre-compacted specimens make field-assisted sintering quite a broad area of solid-state research. This book is an attempt to summarize and critically analyze the up-to-date knowledge on the mechanisms of field-assisted sintering processes operating in different field application schemes. We did not aim at reviewing all successful sintering experiments and trials; many of them have been given credit in published reviews [17, 18]. In order to give the reader an understanding of what drives the sintering process when electromagnetic field is applied to the powder compact, we first briefly describe the phenomena that may be involved in the field-assisted sintering processes. In the following chapters, the physical principles and related equipment of high-voltage consolidation, low-voltage consolidation, flash sintering, induction heating sintering, microwave sintering, magnetic pulse compaction, and some other field-assisted sintering techniques are presented along with the analysis of the relevant theoretical and experimental work elucidating the process mechanisms. At the end of each chapter, selected examples of materials produced by field-assisted sintering are given to show the potential of those sintering methods for practical applications.

The concept of field-assisted sintering allows a wide range of field and pressure application schemes. According to the classification presented by Grasso et al. [18], the processes of pulse electric current-assisted powder sintering can be divided into fast techniques (the current is applied for durations longer than 0.1 s) and ultrafast techniques (the current is applied for durations shorter than 0.1 s). Fast processes are characterized by durations of up to several minutes, current densities reaching 1 kA cm^{-2} and voltages of several volts. The powder sintering facilities using this mode of electric current application include a standard electric pulse generator providing low voltages and high electric currents to the sample. The duration of each pulse ranges from 1 to 300 ms. The pulses are applied to the sample for a total duration of several minutes or tens of minutes. The fast methods are suitable for sintering of both conductive and non-conductive materials. The ultrafast methods usually require relatively high applied pressures (up to several GPa) and very high current densities ($\geq$10 kA cm^{-2}). The voltages vary from several volts to several kilovolts. The discharges with durations of 10^{-5}–10^{-1} s are generated by capacitors. The ultrafast methods are mostly suitable for conductive materials[1].

In the area of sintering assisted by pulsed current, several terms are used to refer to one and the same sintering method. This situation developed as scientists and equipment engineers tried to emphasize different aspects of the developed methods. The same method and equipment can be referred to using different terms in publications by different authors, while one and the same term is often used to refer to methods principally different from each other by the current application scheme. Additional variations of the terms denoting field-assisted sintering techniques appear when those are translated into English from another language.

Belyavin et al. [24] suggested a classification of the field-assisted sintering methods, in which they are grouped by the character of the physical processes involved in heating of the powder material (Fig. 1.1). Direct heating of the powder occurs whenever a current passes through the compact. Indirect heating can be caused by several phenomena associated with electric current. None of the indirect heating methods can be considered versatile, each being suitable for selected applications. The main advantages of induction heating are high heating rates and a possibility of concentrating a high power in a small volume of the material. However, the physical and chemical processes involved in induction heating-based sintering are still rather poorly understood on a microscopic level.

The characteristic values of the parameters of field-assisted sintering techniques are presented in Table 1.1. The following terms are used to refer to high-voltage (mostly ultrafast) processes of sintering that employ electric current applied in a pulsed mode: electric pulse pressing (EPP), electric discharge sintering (EDS),

[1]Material's electric and thermal conductivities may rapidly change during heating.

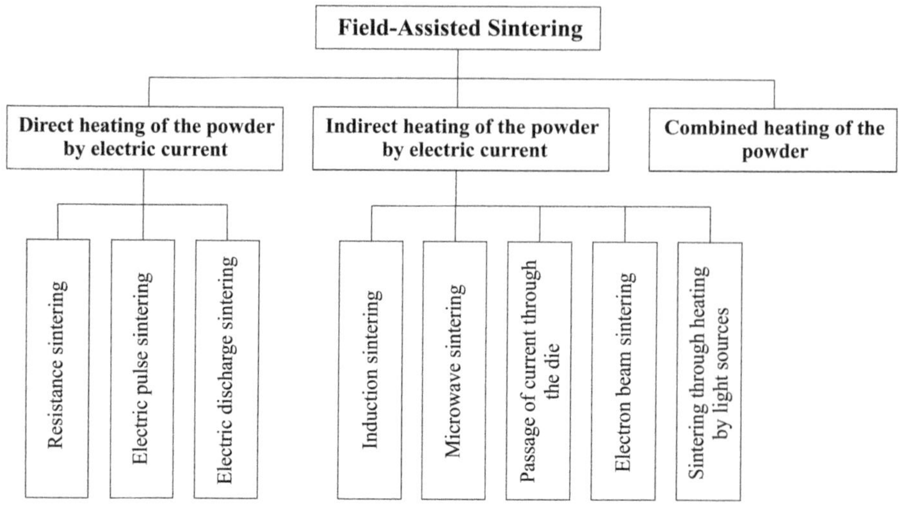

Fig. 1.1 Classification of the field-assisted sintering methods by the heating method of the powder materials. (Drawn using data of Ref. [24])

Table 1.1 Characteristic values of the parameters of field-assisted sintering methods

Field-assisted sintering method	Field application mode	Total process duration	Pressure
Resistance sintering	DC, AC currents	10^{-2}–10^{2} s	10 MPa-1 GPa
High-voltage electric discharge consolidation	Current pulse from a capacitor discharge	<1 s	10 MPa-1 GPa
Flash sintering	DC current	Several seconds	No pressure
Electric discharge sintering, spark plasma sintering (SPS)	Pulsed DC	1–20 min	10 MPa-1 GPa
Sintering in the constant electric field in the noncontact mode	Low DC current or no current through the sample	Several minutes	No pressure
Microwave sintering	Microwave radiation	Several minutes	No pressure
Induction heating sintering	AC	1–20 min	10 MPa-1 GPa
Magnetic pulse compaction (MPC)	Acceleration of the impactor by magnetic field generated by a current pulse from a capacitor discharge	<1 s	Several GPa

electric discharge compaction (EDC), pulsed electric discharge (PED), high-rate electric discharge compaction (HREDC), environmental electric discharge sintering (EEDS), capacitor discharge sintering (CDS), pulse plasma sintering (PPS), and electric pulse sintering (EPS). Low-voltage consolidation is usually referred to as

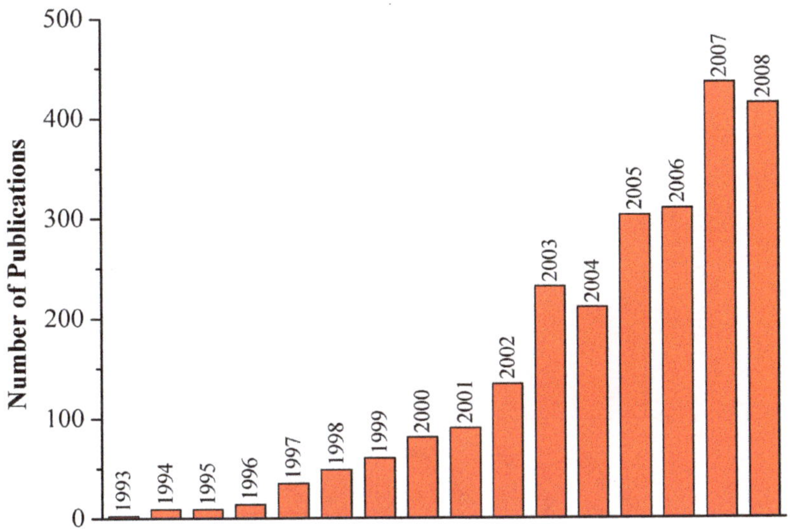

Fig. 1.2 Statistics of publications on spark plasma sintering. (Reprinted from Munir et al. [16], Copyright (2010) with permission of John Wiley & Sons)

Fig. 1.3 Statistics of publications and patents on spark plasma sintering

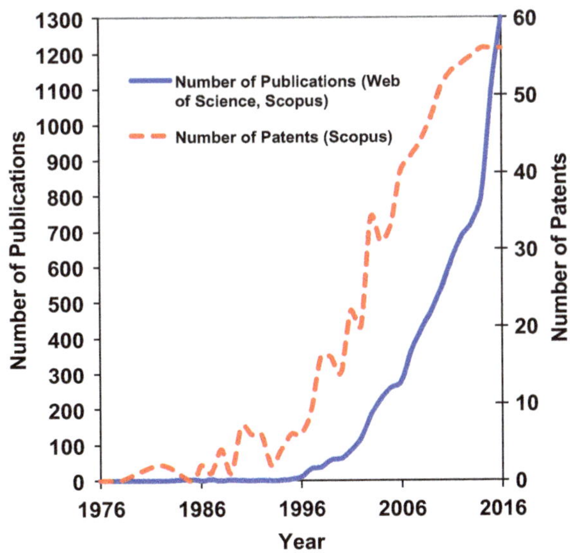

spark plasma sintering (SPS), electric discharge sintering (EDS, especially in the literature published in Russian [23]), pulsed electric current sintering (PECS), or field-assisted sintering technique (FAST). The statistics of publications on SPS from 1993 to 2008 is presented in Fig. 1.2 [16]. The statistics of publications and patents on SPS from 1973 to 2006 was provided in Ref. [26] and extended in Fig. 1.3 to

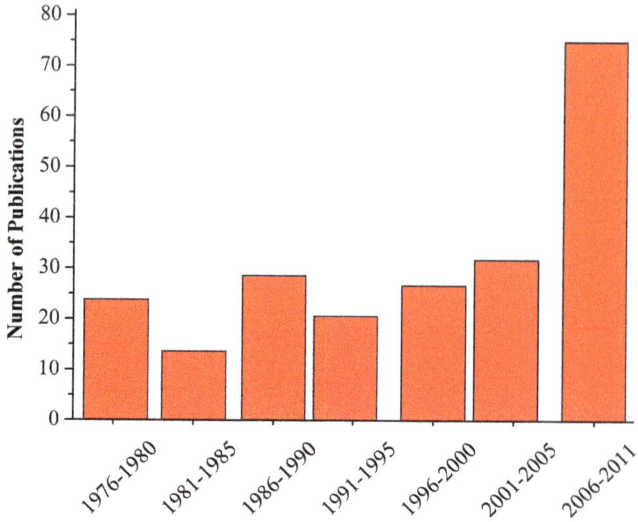

Fig. 1.4 The number of publications on electric pulse sintering (high-voltage techniques). (Reprinted from Yurlova et al. [22], Copyright (2013) with permission of Springer)

2016. The number of publications on electric pulse sintering (high-voltage techniques) over the 1976–2011 period is shown in Fig. 1.4 [22].

Electric discharge sintering (EDS, a low-voltage technique) should, in a general case, include two different stages, the first one favoring the formation of discharges between the particles for efficient heating of the contacts, and the second one favoring densification [23, 27]. Some authors use the term "electric discharge sintering" to refer to a two-stage process, the first stage of which is a high-voltage pulse, while the second stage includes the action of an electric current of high density and lasts several tens of minutes [28]. During the first stage, even in powders having an oxide film on the particle surface, the inter-particle contacts successfully form, promoting efficient sintering at the second stage.

1.2 Thermal and Nonthermal Effects in Field-Assisted Sintering

Most of the studies conducted on field-assisted sintering are based on experimental trial-and-error attempts to consolidate various powder systems. One of the most important questions is which factors provide accelerated densification and, in many cases, limited grain growth in field-assisted sintering compared with regular hot pressing applied to the same material systems. The difference between the outcomes of field-assisted and conventional powder hot consolidation processes can be attributed to various physical factors, which are grouped as phenomena of thermal nature and those of nonthermal nature corresponding to the so-called field effect.

1.2.1 Thermal Effects in Field-Assisted Sintering

The group of factors of thermal nature may include [26]:

1. High heating rates, which enable higher sinterability of the powders
2. High local temperature gradients, which provide conditions for thermal diffusion
3. Highly nonuniform local temperature distributions, which cause local melting within inter-particle contact areas
4. Highly nonuniform macroscopic temperature distributions, which create thermal stresses intensifying dislocation creep

It has been shown that high heating rates increase the sinterability of powder compacts by suppressing surface diffusion at the early stages of sintering. High heating rates also ensure limited grain growth. Experimentally, it has been shown in a number of investigations that an increase in the heating rate considerably increases the consolidation rate of conductive and non-conductive powders during SPS. For an alumina powder, it was shown that switching from a heating rate of 50 $°C·min^{-1}$ to a heating rate of 300 $°C·min^{-1}$ while keeping the same maximum temperature allowed decreasing the sintering time by a factor of six and reaching the same final density [29]. Physically, this was explained by the presence of additional defects in the material directly related to high heating rates and a short duration of the process. Johnson [30] qualitatively showed that high heating rates minimize the effects of surface diffusion processes not contributing to densification; hence, the compact body reaches higher temperatures in a highly sinterable state. From the modeling studies conducted by Olevsky et al. [26] for the SPS process, it was concluded that surface diffusion contributes to the evolution of the pore surface curvature, thereby influencing the intensity of grain-boundary diffusion at the early stages of sintering. The interaction of surface and grain-boundary diffusion is thus one of the major mechanisms enabling the impact of heating rates on the consolidation kinetics. Another heating rate-sensitive contribution to the densification is related to the shrinkage kinetics by power-law creep accelerated by a rapid increase in temperature. Both the calculations and the experiments [26] show an increase in the maximum shrinkage rate for larger particle sizes. This mostly unusual for conventional powder sintering and hot pressing phenomenon is explained under conditions of rapid heating by the delayed pore tip spheroidization and the intensification of grain-boundary diffusion at early stages of sintering for larger grain sizes.

In contrast to conventional powder consolidation techniques, high local temperature gradients caused by nonuniform current distribution in the compacts at the scale of the particle size can instigate additional driving forces for consolidation. Ludwig–Soret effect of thermal diffusion leads to the formation of concentration gradients in initially homogeneous two-component systems subjected to a temperature gradient [31]. For the case of atomic and vacancy diffusion in crystalline solids, this effect was studied by a number of authors; its theoretical interpretation was conducted by Shewmon [32] and Schottky [33]. For the electric current-assisted sintering, the effect of thermal diffusion was analyzed by Kornyushin [34]. Later, for rapid densification, the role of temperature gradients was studied by Searcy [35] and Young and McPherson [36]. Johnson [30], however, argued against the significance of thermal diffusion for the sintering enhancement in microwave sintering in a debate with Young and McPherson [37].

On a microscopic scale, the electric field lines become denser when the current flows around the pores, which leads to a local temperature rise and the formation of large temperature gradients. Gostomelskiy and Krupnova theoretically analyzed the growth and healing processes of isolated pores in a metal subjected to pulses of electric current [38]. They concluded that heating of the metal by electric pulses intensifies the healing (disappearance) of pores in comparison with furnace heating up to the same final temperatures. As the temperature of the pore surface is higher than the volume temperature, the temperature gradient leads to the diffusion of vacancies in the pores resulting in the pore growth. At the same time, the vacancy concentration gradients form due to rapid heating. While a certain time is needed for the vacancy concentration to reach its equilibrium value corresponding to the temperature in the volume of the material, this occurs instantaneously at the pore surface. Consequently, the vacancies diffuse from the vicinity of large pores resulting in the pore healing. To reduce the temperature gradients, the sintering process should consist of short multiple pulses (10^{-5} s) with an interval between them one or two orders of magnitude longer than the duration of the pulse. Large pores are reduced in size at a higher rate, which is related to the vacancy flux being proportional to the surface area of a pore. The reduction of porosity will be successfully achieved if the heating of the material by electric pulses is such that the temperature gradients due to the presence of pores potentially leading to the pore growth are not high, while the vacancy gradients between the pore surface and the volume of the material are sufficient to heal the pores. The healing of large pores can be looked upon as the reaction of the system to the increased intensity of the electric field in the vicinity of pores as the diffusion of vacancies from the regions surrounding large pores in the direction of smaller ones reduces the differences in the intensity of the electric field.

A possible contribution of thermal diffusion to the SPS processes has been discussed by Olevsky and Froyen based on modeling results presented in Ref. [39]. The calculation results indicated that local temperature gradients (concentrated in the area of inter-particle contacts) can exceed the macroscopic temperature gradients by 3–4 orders of magnitude. The intensity of thermal diffusion increases for higher pulse frequencies. Thermal diffusion promotes the separation of atoms and vacancies. At the early stages of sintering, this should lead to the growth of inter-particle necks, which corresponds to the enhancement of sintering. At the final stages of sintering, however, the pores may serve as vacancy sinks under thermal diffusion conditions, which impedes sintering. It is possible that the increased pulse frequencies enhance sintering at the early stages of SPS and hinder sintering at the late stages. In some experimental studies, the pulse frequency was found to have a limited impact on the SPS results [40]; its contributions at early and late stages of SPS could have offset each other.

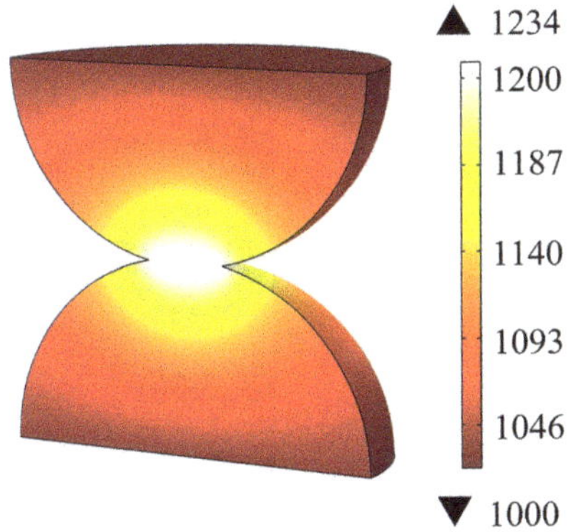

Fig. 1.5 Temperature distribution obtained by modeling of the resistance heating of spherical copper particles 10 μm in diameter

Resistance heating of two spherical particles was modeled by Kuz'mov et al. [41] for spherical particles having a thin oxide film on their surface. In practice, the presence of oxide films on the surface of metallic powders is rather common. The model allowed obtaining the dependences of the temperatures and the temperature gradients on the contact voltage and the relative contact resistance. The contact voltage is the potential difference between the centers of the particles. In the calculations, it was varied from 0.01 V to 0.4 V. The relative contact resistivity k_ρ was calculated as follows:

$$k_\rho = \frac{h}{r} \cdot \frac{\sigma_0}{\sigma}, \tag{1.1}$$

where h is the thickness of the oxide film, r is the particle radius, σ_0 is the electrical conductivity of the material of the particle, and σ is the electrical conductivity of the oxide film. As can be seen in Fig. 1.5 showing the temperature distribution obtained by modeling of the resistance heating of spherical copper particles 10 μm in diameter, the temperature of the inter-particle contacts is substantially different from the average macroscopic temperature of the sample. Figure 1.6 shows the parametric isolines of the average temperature. Unlike temperature, its gradient is inversely proportional to the particle size. If the contact resistivity is constant, the temperature gradient increases with decreasing particle size. The temperature gradients in the contact region between the powder particles with a size of 10 μm exceed 10^6 K·m^{-1}. Such temperature gradients may have a noticeable effect on mass transfer by the mechanisms of thermal diffusion and non-equilibrium distribution of vacancies [36]. If the particle size is varied, while the voltage drop across a macroscopic sample (i.e., macroscopic electric field strength) is kept constant, the contact voltage is proportional to the size of the contacting particles, and the gradient decreases with this size. This is because the inversely proportional relationship between the temperature gradient and the particle size cannot balance the nonlinear (nearly parabolic) dependence on contact voltage. The maximum temperature gradient (Fig. 1.7) differs from the average gradient in that it does not tend to zero as the

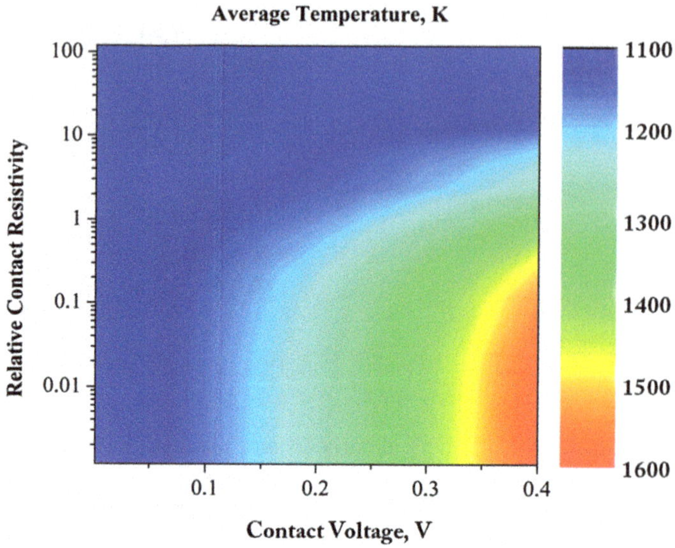

Fig. 1.6 Parametric isolines of the average temperature obtained using modeling of the resistance heating of spherical copper particles 10 μm in diameter. (Drawn using data of Ref. [41])

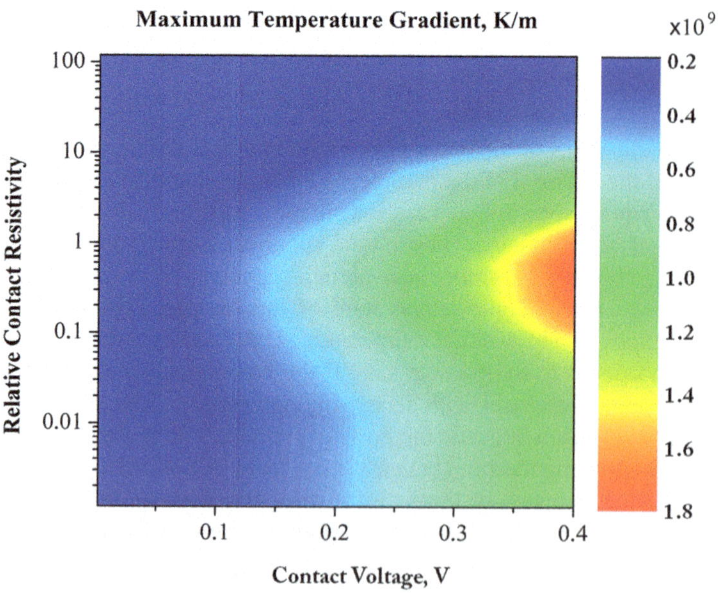

Fig. 1.7 Parametric isolines of the maximum temperature gradient obtained using modeling of the resistance heating of spherical copper particles 10 μm in diameter. (Drawn using data of Ref. [41])

contact resistivity tends to zero. In other respects, the behavior of these quantities is similar, and the above reasoning on the average gradient is true for the maximum gradient. It was shown [36] that the resistance heating of a powder can cause melting in the contact region at very low macroscopic temperature.

The fact that the local temperature at the inter-particle contact region may be much higher than the macroscopic temperature of a consolidating body should be taken into account when shrinkage of the powder compact is described. Since the neck between the particles is also a region of major deformation, especially at the early stages of consolidation, resistance heating causes faster shrinkage than conventional furnace heating at the same temperature and external pressure. The power-law creep equation [42] can be used to prove that even a difference of a few hundred degrees in the temperature causes a significant increase in the shrinkage rate of the powder compact:

$$\dot{\gamma} = A\frac{D_{\mathrm{v}}Gb}{kT}\left(\frac{\sigma_s}{G}\right)^n, \quad D_{\mathrm{v}} = D_{\mathrm{v}0}\exp\left(-\frac{Q_{\mathrm{v}}}{RT}\right) \tag{1.2}$$

In Eq. (1.2) $\dot{\gamma}$ is the shear strain rate, A is a dimensionless constant, D_{v} is the lattice diffusion coefficient, G is the shear modulus, b is the Burgers vector, σ_s is the shear stress intensity, and Q_{v} is the activation energy of diffusion.

While the local temperature gradients lead to thermal diffusion of vacancies, the macroscopic temperature gradients lead to thermal stresses. The thermal stresses, being internal stresses for a porous specimen, cannot independently contribute to the overall macroscopic volume shrinkage. They, however, can impact the processes of the generation and motion of dislocations, which can influence the diffusion rates during sintering. Tuchinskii [43, 44] analyzed thermal stresses in porous compacts originated from a nonisothermal state of the specimen at the beginning of heating. A situation, in which a porous cylinder is heated from its external surface placed in an atmosphere of a constant temperature exceeding that of the initial temperature of the cylinder, was considered. By analyzing the equations for the radial σ_r and peripheral stress σ_θ over the cross section of the cylinder, it was found that the peripheral stress is greater in absolute magnitude than the radial stress. On the cylinder axis, the stresses are equal to each other: $\sigma_r = \sigma_\theta$. During heating of the specimen, the radial stress is tensile throughout the volume of the cylinder increasing from zero on the specimen surface to $\sigma_r = \sigma_\theta$ on the cylinder axis. The peripheral stress in the outer layers is compressive. It reaches its maximum absolute value on the cylinder surface and then decreases reaching zero at a certain distance from the cylinder axis. After that, it changes sign becoming tensile and grows in magnitude.

Tuchinskii emphasizes that thermal stresses alone acting inside the porous body cannot lead to volume shrinkage of the compact as the hydrostatic pressure generated by the thermal stresses averaged throughout the volume is equal to zero. A possible mechanism of activation of sintering under nonisothermal conditions may be related to the generation and motion of dislocations under the action of thermal stresses, which, in turn, accelerate the diffusion processes responsible for mass transport. The thermal stresses can exceed the Laplacian stress. In order to conclude on the ability of thermal stresses to make dislocation sources active, the rms values of thermal stresses, equations for which were derived in Ref. [45], should be compared with the yield stress of the material. Taking into account the dependence of the yield stress on the temperature in his calculations, Tuchinskii showed that thermal stresses generated during rapid heating of a porous nickel compact (at heating rates characteristic of induction heating) are high enough to cause plastic deformation in the outer and inner layers of the compact, actuate Frank–Read sources, and, thus, enhance sintering. In conventional furnace sintering, thermal stresses are likely to reach the value of the yield stress of the material only in the outer layers of the compact [43]. A critical heating rate parameter was introduced in Ref. [44] to conclude on the possibility of plastic relaxation of thermal stresses. At heating rates exceeding the critical value, the dislocation sources become operative, and generation of new dislocations can be expected in the material.

1.2.2 Nonthermal Effects in Field-Assisted Sintering

The nonthermal group of factors contributing to enhanced densification in field-assisted sintering may include [26]:

1. Electromigration and intensified diffusion in ionic conductors
2. Electroplasticity effects
3. Ponderomotive forces
4. Electromagnetic "pinch" effect
5. Dielectric breakdown of oxide films (cleansing effect) and defect generation at grain boundaries
6. Plasma formation

Nonthermal effects of electric current were studied experimentally by Asoka-Kumar et al. [46] and Garay et al. [47]. They used positron annihilation spectroscopy to obtain information on the defect concentrations in solid materials subjected to electric current. Asoka-Kumar et al. showed that an electric current of a density of $8 \cdot 10^4$ A·cm^{-2} passing through the Al-0.5 wt.%Cu alloy induces the dynamic formation of vacancies. The increase in the vacancy concentration was substantially greater than that caused by thermal generation alone (4×10^{18} cm^{-3} vs. 3×10^{17} cm^{-3}). Garay et al. found that a DC current increases the kinetics of annealing of the previously stored defects in Ni_3Ti and attributed the observed effect to the electron wind, although no dependence of the effect on the direction of electric current was observed. A lack of asymmetry could be a consequence of the complex crystalline structure of the Ni_3Ti intermetallic, in which the diffusion of one element is coupled with diffusion of the other because the structure and stoichiometry have to be maintained. These results shed light on the mechanism of the influence of electric current on solid-state chemical reactions involving metals.

Frei et al. [48] showed that DC current has a marked effect on neck growth between copper spheres 3 mm in diameter and copper plates using a SPS apparatus

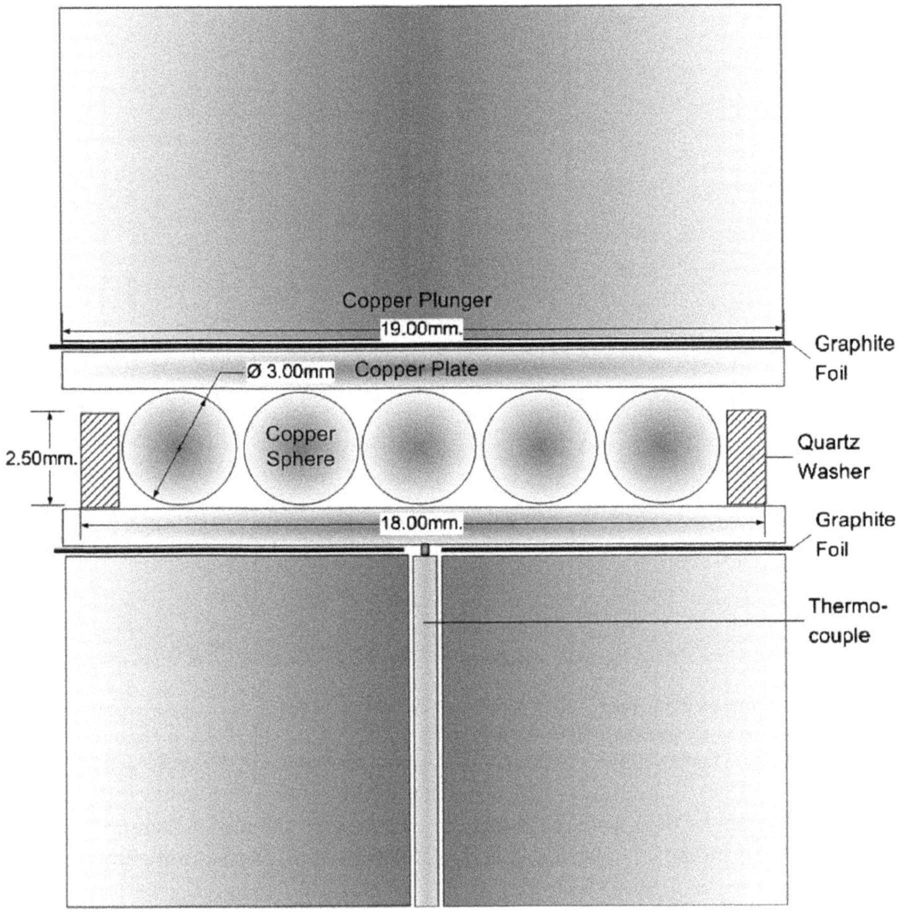

Fig. 1.8 Schematic of the experiment with copper spheres between copper plates to study the effect of current on neck growth in a pulsed electric current sintering apparatus. (Reprinted from Frei et al. [48], Copyright (2007) with permission of AIP Publishing)

and a modified die/punch assembly that allowed separating the influence of the temperature and the current (Fig. 1.8). In a general case of current-assisted sintering, the current and the temperature of the compact are dependent parameters. Since the only source of heat is Joule heating, the power generated to maintain a steady-state temperature is constant. By introducing a stack of alternating copper disks and graphite foils into the SPS die assembly, it was possible to vary the current passing through the sample at a constant electric power (and, consequently, at a sample constant temperature). The resistance of the graphite foil and the contact resistance in this case made the dominant contribution to the resistance of the assembly. The effect of current on neck growth is seen in Fig. 1.9 showing images of the areas on the lower copper plate that were sinter-bonded to copper spheres after their separation.

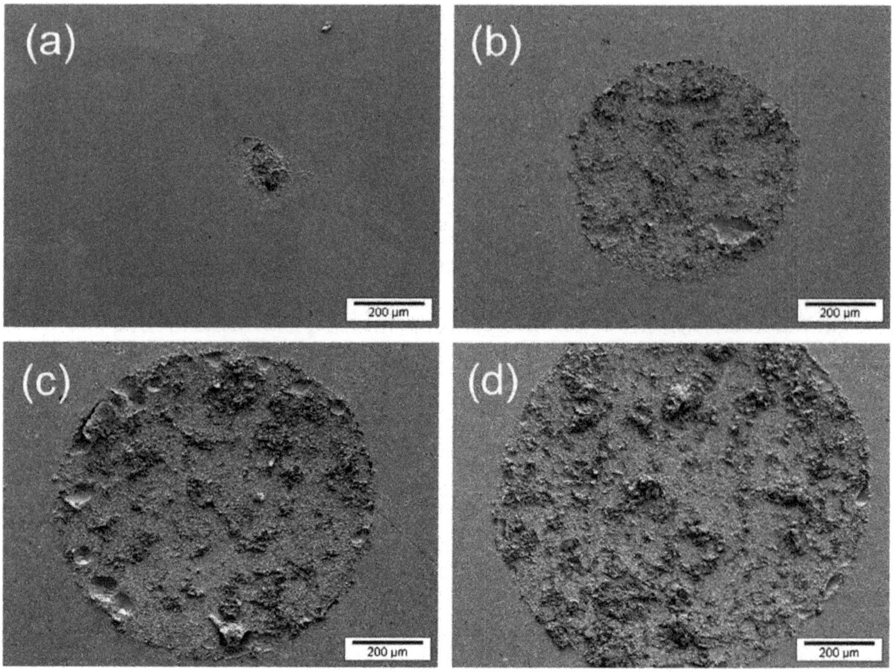

Fig. 1.9 Scanning electron microscopy images of the areas on the lower copper plate that were sinter-bonded to copper spheres at 900 °C for 60 min: (**a**) no current, (**b**) 700 A, (**c**) 850 A, (**d**) 1040 A. (Reprinted from Frei et al. [48], Copyright (2007) with permission of AIP Publishing)

The experimental data were fitted into the equation for the initial stage of sintering for a simplified model of spherical particles sintered at a constant temperature

$$\left(\frac{x}{R}\right)^n = \frac{Bt}{R^m}, \tag{1.3}$$

where B is a constant containing the diffusion coefficient, R is the radius of the sphere, x is the diameter of the neck, t is the time, and n and m are sintering mechanism-dependent constants. Values of sintering exponent n that were found showed that no classic sintering mechanism could be applied to describe the process. An interesting observation was the formation of voids in the neck region, especially at the edge region (Fig. 1.10). The size of the voids increased with the sintering time.

The correlation of sintering enhancement with the level of current demonstrated by Frei et al. [48] was explained by electromigration. The flux J_i of the diffusing species "i" is a result of the momentum transfer from the electron wind effect:

$$J_i = -\frac{D_i C_i}{RT}\left[\frac{RT\partial \ln C_i}{\partial x} + Fz^* E\right] \tag{1.4}$$

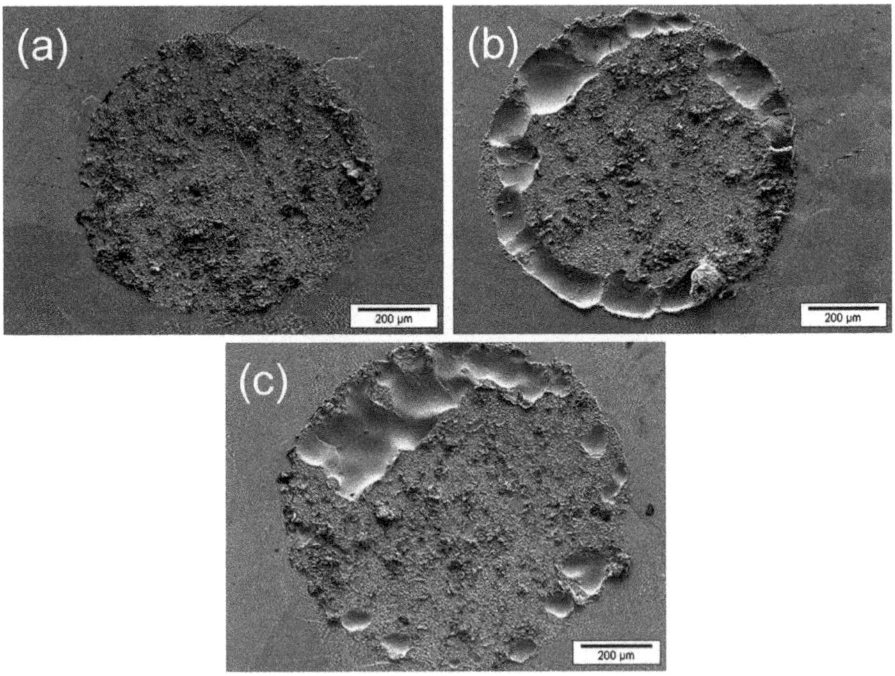

Fig. 1.10 Scanning electron microscopy images of the areas on the lower copper plate, which were sinter-bonded to copper spheres, total current 1040 A, temperature 900 °C: (**a**) 15 min, (**b**) 30 min, (**c**) 60 min. (Reprinted from Frei et al. [48], Copyright (2007) with permission of AIP Publishing)

D_i is the diffusivity of the *i*th species, C_i is the concentration of the species, F is Faraday's constant, z^* is the effective charge on the diffusing species, E is the field, R is the gas constant, and T is the temperature. A large size of the voids observed in these experiments raises a question of the scale of the electromigration-induced effects.

An interesting observation was the presence of ledges on the surface of the copper plate, which became less abundant as the distance from the neck increased. The occurrence of these ledges was related to the current, as they were not observed in experiments conducted in the absence of current. The formation of the ledges was attributed to the effect of surface electromigration on evaporation.

Burenkov et al. [49] provided evidence of the operation of mechanisms other than diffusion during electric current-assisted sintering of porous compacts. With diffusion mechanisms operating, the linear size of the neck between particles X depends on the sintering time t according to a power relationship $X^n \sim t$. For volume diffusion, $n = 5$, while for surface diffusion, $n = 7$. Although this relationship is valid for isothermal conditions, it was considered applicable for the estimation of the neck growth exponent in the electric current-assisted processes. The values of n were found to vary in a wide range – from 0.5 to 5. These estimates, especially the observed cases of very low n, are of great importance for the understanding of the sintering mechanisms, as they point to the operation of very vigorous processes, such as viscous flow and spreading of the liquid metal squeezed out of the contact zone over the solid surfaces. There was no fundamental difference in the interaction between the contacting particles located along the line parallel to the direction of the current and between those located along the line perpendicular to the direction of the current.

The role of melting processes in establishing the inter-particle contacts was addressed by Belyavin et al. [24]. They observed experimentally and described mathematically the processes at inter-particle contacts, the main feature of which was local melting. The behavior of the molten column formed between the particles, namely, shape changes of the molten column, a stability loss, and squeezing of the molten metal from the contact area, depends on the relative influence of the applied mechanical pressure, electromagnetic forces, and surface tension. If mechanical pressure squeezes the molten metal from the inter-particle contact area, the size of the contact decreases and so does its mechanical strength.

A constitutive model of SPS taking into consideration the direct contribution of electric current to grain-boundary diffusion was developed by Olevsky and Froyen [50]. They constructed a densification map for an aluminum powder consisting of three porosity/grain size domains. In each of these domains, one of the three considered driving factors of material transport – externally applied load, surface tension, and electromigration – is dominant (Fig. 1.11). The power-law creep induced by an external stress always dominates for higher porosities. For lower porosity values, the electromigration can become the dominant mechanism, and for smaller particle sizes and low porosity, surface tension is the main driving factor for densification. For very small porosities in the electromigration-dominating zone, the ultimate collapse of voids may require externally applied load as the primary factor.

A significant contribution of high-intensity fields to the materials deformation can be expected thanks to electroplasticity effects [51]. During deformation of aluminum alloy Al7475 in a DC field (field intensity 2 kV·cm^{-1}), the flow stress and the strain hardening exponent of the superplastic deformation were reduced by the electric field, while the strain rate sensitivity exponent was increased. The fracture surface of the specimen tested under the electric field showed the presence of whiskers, whose formation could not be explained by diffusion or dislocation-involving mechanisms because of very high degrees of deformation near the grain boundaries that must occur for such whiskers to form (Fig. 1.12). Therefore, viscous flow of a quasi-liquid was suggested as an explanation of the observed fracture morphology. The application of a DC field led to a number of effects, including dissolution and coalescence of the dispersoids in the alloy and formation of dispersoid-free zones near the grain boundaries.

A reduction in the flow stress during the tensile deformation of ceramics at high homologous temperatures was analyzed by Conrad and Yang for fields <500 V·cm^{-1} [52]. The Joule heating was considered as giving only a minor contribution to the reduction of the flow stress. The main mechanism was a decrease

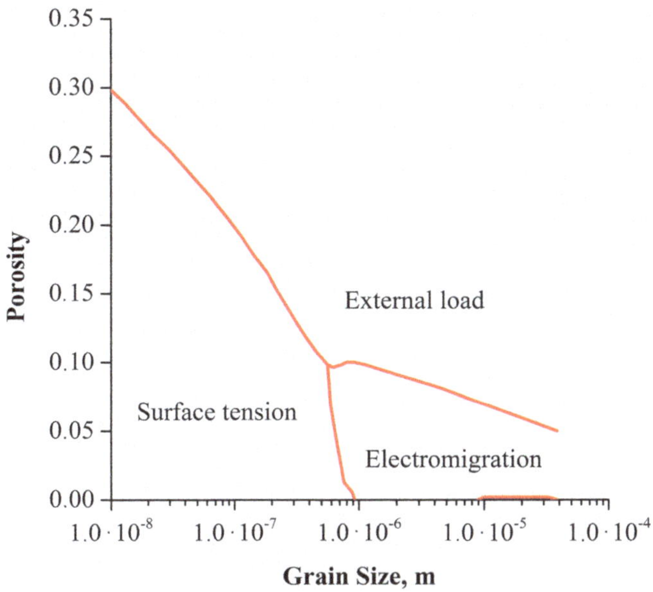

Fig. 1.11 Densification map for Al powder, $\frac{U}{l} = 417\ ^{\mathrm{V}}/_{\mathrm{m}}, T = 673\ K, \overline{\sigma}_x = 283$ MPa. (Reprinted from Olevsky and Froyen [50], Copyright (2006) with permission from Elsevier)

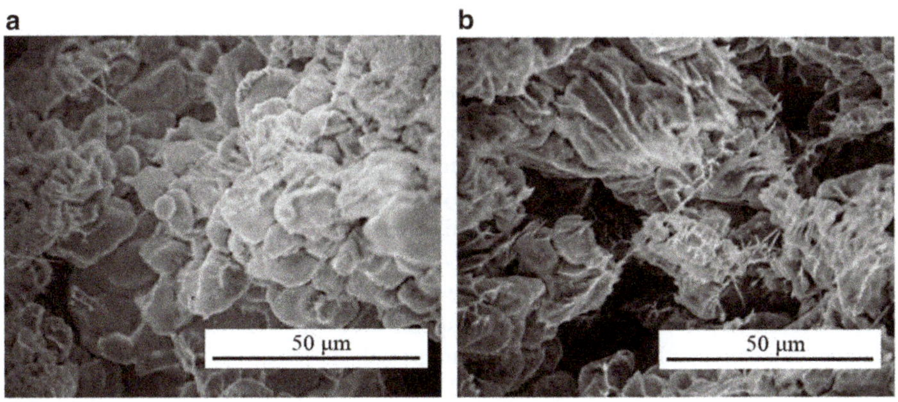

Fig. 1.12 Fracture surface of Al7475 tested without electric field (**a**) and at an applied field with an intensity of 2 kV/cm (**b**) at 516 °C. (Reprinted from Cao et al. [51], Copyright (1990) with permission from Elsevier)

in the potential of the vacancy formation corresponding to the diffusion of the rate-controlling ions in the space-charge region at grain boundaries. It was concluded that, for some oxides, dynamic grain growth is retarded by the electric field influencing grain-boundary mobility within the space-charge zone.

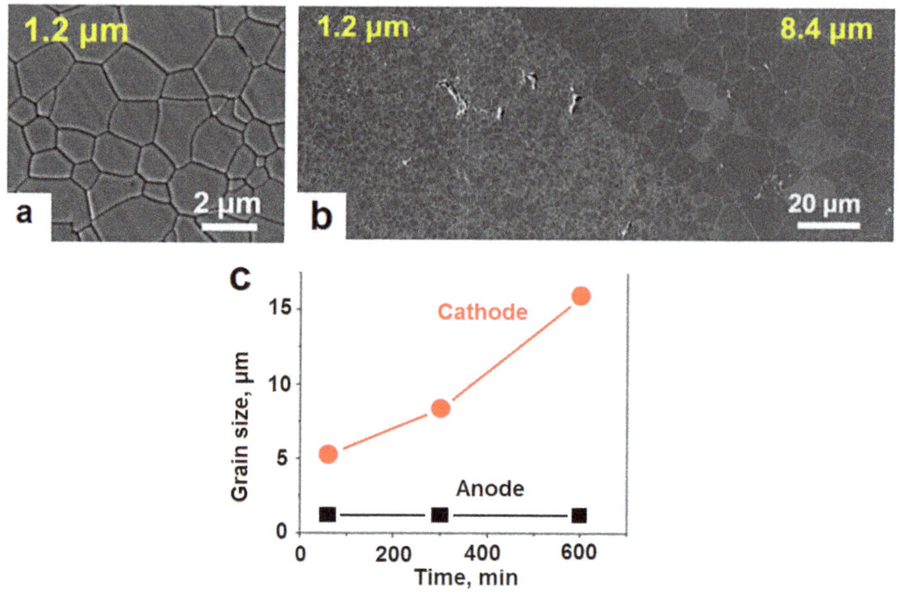

Fig. 1.13 Microstructure of the initial fully dense yttria-stabilized zirconia containing 8 mol.% of Y_2O_3 obtained by hot isostatic pressing, average grain size 1.2 μm (**a**), region of step transition of the grain size in the sample subjected to a current density of 50 A ·cm^{-2} for 300 min at 1340 °C (**b**) and grain size evolution at the anode and cathode sides of the sample with time at 1350 °C and a current density of 50 A ·cm^{-2} (**c**). (Reprinted from Kim et al. [53], Copyright (2011) with permission of John Wiley & Sons)

Kim et al. [53] reported the influence of DC currents on the grain growth of yttria-stabilized zirconia (Fig. 1.13). In their experiments, pre-sintered specimens of yttria-stabilized zirconia were annealed in a furnace with and without the application of a DC current. The microstructure of the fully dense pre-sintered specimen is shown in Fig. 1.13a. The sample temperature was higher than the set furnace temperature due to Joule heating when the current density was greater than 20 A·cm^{-2}. Grain growth in the electrically loaded samples depended on the location. Enhanced grain growth was observed at the cathode side of the specimen (Fig. 1.13b, c). In a fully dense sample subjected to a DC current, the grain growth started at 1150 °C at the cathode side, which is well below the conventional sintering temperature. The lower limit for the current density to trigger grain growth at 1150 °C was 1.5 A·cm^{-2} (from 1.2 to 1.5 μm after 24 h of annealing). Without an electric current, there were no grain growth at 1250 °C and only a slight increase in the grain size from 1.2 to 2.0 μm at 1350 °C after annealing for 20 h. At the anode side, a temperature of 1400 °C was needed to initiate grain growth. The observed effect was attributed to supersaturated oxygen vacancies accumulated at the cathode side leading to reduction of the cations, which, in turn, lowers their migration barrier. When the pre-sintered material was only 70% dense before electric loading, grain growth was initially suppressed, while densification was enhanced by the presence of the electric current. Faster

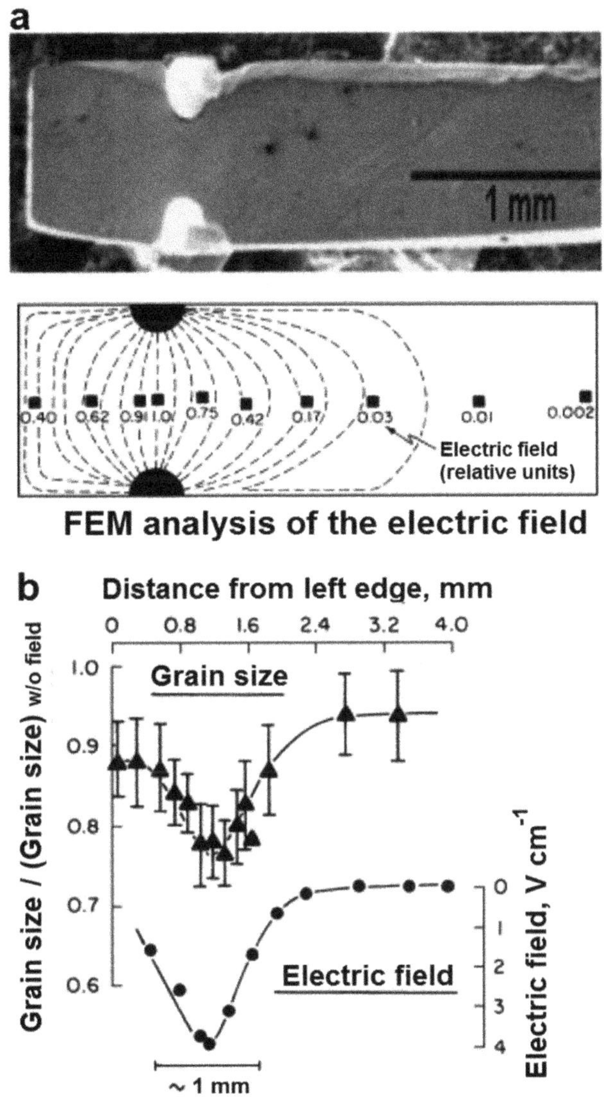

Fig. 1.14 The distribution of the electric field in the specimen obtained by finite element modeling (FEM) (**a**) and the spatial correspondence between the grain size and the magnitude of the applied field for 3 mol % yttria-stabilized zirconia (**b**). (Reprinted from Ghosh et al. [54], Copyright (2009) with permission of John Wiley & Sons)

densification occurred at the cathode side. After full densification was reached, the grain growth started at the cathode side.

Ghosh et al. [54] reported retarded grain growth under applied electric field for dense nanocrystalline 3 mol% yttria-stabilized zirconia prepared by the sinter-forging technique and annealed for 10 h at 1300 °C. The distribution of the electric field in the specimen obtained by the finite element analysis and the spatial correspondence between the grain size and the magnitude of the applied field are shown in Fig. 1.14a–b. The grain size in Fig. 1.14b is normalized with respect to the average grain size obtained without electric field. Among possible underlying mechanisms of retarded grain growth, Ghosh et al. considered a thermodynamic effect of the minimum in the interfacial energy formed due to more intense Joule heating of the boundary having a higher electrical resistance than the grains.

Following the results reported by Ghosh et al., Holland et al. [55] performed a thermodynamic analysis of the grain growth suppression effect in the presence of a thermal gradient. They argued that in mixed or ionically conducting ceramics, no significant contribution of temperature gradients can be expected due to high thermal conductivities and nanometric grain sizes. Along this line, an explanation of the retarded grain growth in the presence of an electric field concerned with an altered grain-boundary mobility and enhanced solute segregation also suggested by Ghosh et al. [54] gains more importance.

The influence of the electric field on grain growth is especially important for nanograined materials, as the densification rate is inversely proportional to the fourth power of the gain size [56]:

$$\dot{\rho} = \frac{Af(\rho)}{Td^4} e^{-\frac{Q_b}{RT}}, \tag{1.5}$$

where A is the material constant, Q_b is the activation energy for self-diffusion at grain boundaries, $f(\rho)$ is a function of the density, T is the temperature, and d is the grain size. As is seen from the above expression, sintering accompanied by grain growth slows down significantly.

Another nonthermal effect of electric field is migration of pores and gas bubbles in ceramics. Such observations were reported by Kim et al. [57] for yttria-stabilized zirconia (8YSZ). Pores and gas bubbles migrated against the field at temperatures, at which the lattice and grain-boundary diffusion was frozen.

The significance of nonthermal effects is currently recognized in the area of microwave sintering. The first studies of the use of microwave radiation for high-temperature processing of materials date back to the 1960s [58]. In the field of microwave sintering of ceramics, the pioneering experimental work was done by Tinga and Voss [59], Berteaud and Badot [60], Meek et al. [61], and Johnson [62, 63]. It was found that microwave sintering has features of high importance for the industry, which include a significant (50–100 °C) decrease in the process temperature and a reduction in the duration of the high-temperature stage of the sintering process. One of the possible reasons for the latter is an inverse distribution of porosity at the intermediate stage of sintering due to volumetric heating, which facilitates densification [64]. The presence of nonthermal effects in microwave sintering was attributed to the action of ponderomotive forces, which reflect the influence of high-frequency electromagnetic fields on mass and charge transport in ionic crystalline solids [65–67]. The nature of ponderomotive forces is the interaction of the microwave field with the effective electric charges of vacancies. As the microwave-induced flux of vacancies oscillates at the microwave frequency, in order to exert influence on mass transport, it needs to be "rectified" through a certain mechanism that transforms the oscillatory motion into a monotonic drift. Such a mechanism has been elaborated by Rybakov et al. [65] and Olevsky et al. [66], who suggested that rectification can be caused by perturbations of the vacancy flow near the pore surfaces of limited permeability for the vacancies.

The pinch effect is self-constriction of the sintered compact in the radial direction by the magnetic field of the current passing through the compact [22]. The distribution of the magnetic pressure over the sample's cross section is of a parabolic character and is more uniform when occurring simultaneously with the skin effect. The pinch effect helps eliminate the problem of friction between the compact and the die wall and maintain the porosity of the compacts as they are pressed out of the die (when the target material is a porous material).

The presence of surface oxides on metallic particles may slow the resistance sintering kinetics or accelerate it depending on the probability of their breakdown. If no breakdown occurs, the films act as barriers for electric current and hinder sintering kinetics. If breakdown of the films does take place, sparking between the particles can accelerate sintering.

Possible plasma effects still remain a controversial and open area of discussion within the field-assisted sintering scientific community. The formation of plasma requires certain levels of vacuum and voltage, which are different in various field-assisted sintering techniques. While high-voltage discharge compaction is routinely carried out in the presence of plasma-related phenomena, low-voltage modes of the field-assisted sintering (such as SPS) in most cases exclude these effects. Some localized plasma effects, depending on the processed material's chemical composition and structure characteristics, are claimed by some authors to be present during low-voltage-assisted processing too [68].

In the following book chapters, we describe in detail the underlying physical phenomena and the respective technical equipment for various known types of the field-assisted sintering techniques. We will demonstrate that the usually high rate of field-assisted processing is based on various transient thermal and nonthermal phenomena, which, if properly managed, provide a unique environment for densification and microstructure retention. In fact, in many cases, a successful field-assisted sintering process utilizes the conditions of *controlled non-equilibrium.*

References

1. Bordia R, Kang SJ, Olevsky EA (2017) Current understanding and future research directions at the onset of the next century of sintering science and technology. J Am Ceram Soc 100:2314–2335
2. Edison TA (1879) Electric Light. US Patent 219628
3. Voelker WL (1898) G.B. Patent 6149
4. Sauerwald F (1922) Apparatus for direct resistance heating to high temperature under pressure
5. Taylor GF (1933) Apparatus for making hard metal compositions. US Patent No. 1896854
6. Lenel FV (1955) Resistance sintering under pressure. JOM 7(1):158–167
7. Inoue K (1962) Electric Discharge sintering. US Patent 3241965
8. Inoue K (1966) Apparatus for electrically sintering discrete bodies. US Patent 3250892
9. Greenspan J (1976) Impulse resistance sintering of tungsten. Army Materials and Mechanics Research Center, USA

10. Tokita M (1993) Trends in advanced SPS spark plasma sintering systems and technology. J Soc Powder Technol Jpn 30(11):790−804
11. Omori M (2000) Basic research and industrial production using the spark plasma system (SPS). Mater Sci Eng A 287(2):183−188
12. Groza JR, Zavaliangos A (2000) Sintering activation by external electrical field. Mater Sci Eng A 287(2):171−177
13. Groza JR, Zavaliangos A (2003) Nanostructured bulk solids by field activated sintering. Rev Adv Mater Sci 5:24−33
14. Mamedov V (2002) Spark plasma sintering as advanced PM sintering method. Powder Metall 45(4):322−328
15. Munir ZA, Anselmi-Tamburini U, Ohyanagi M (2006) The effect of electric field and pressure on the synthesis and consolidation of materials: a review of the spark plasma sintering method. J Mater Sci 41(3):763−777
16. Munir ZA, Quach D, Ohyanagi M (2011) Electric current activation of sintering: a review of the pulsed electric current sintering process. J Am Ceram Soc 94(1):1−19
17. Orrù R, Licheri R, Locci AM, Cincotti A, Cao G (2009) Consolidation/synthesis of materials by electric current activated/assisted sintering. Mater Sci Eng R 63(4–6):127−287
18. Grasso S, Sakka Y, Maizza G (2009) Electric current activated/assisted sintering (ECAS): a review of patents 1906–2008. Sci Technol Adv Mater 10:1–24
19. Dudina DV, Mukherjee AK (2013) Reactive spark plasma sintering: successes and challenges of nanomaterial synthesis. J Nanomater 625218, 12 p
20. Dudina DV, Mukherjee AK (2013) Reactive spark plasma sintering for the production of nanostructured materials. In: Sinha S, Navani NK (eds) Nanotechnology series, vol. 4: nanomaterials and nanostructures. Studium Press LLC, USA, pp 237–264
21. Olevsky EA, Aleksandrova EV, Ilyina AM, Dudina DV, Novoselov AN, Pelve KY, Grigoryev EG (2013) Outside mainstream electronic databases: review of studies conducted in the USSR and post-soviet countries on electric current-assisted consolidation of powder materials. Materials 6:4375–4440
22. Yurlova MS, Demenyuk VD, Lebedeva LY, Dudina DV, Grigoryev EG, Olevsky EA (2014) Electric pulse consolidation: an alternative to spark plasma sintering. J Mater Sci 49:952–985
23. Raichenko AI (1987) Basics of electric current-assisted sintering. Metallurgiya, Moscow, p 128
24. Belyavin KE, Mazyuk VV, Min'ko DV, Sheleg VK (1997) Theory and practice of electric pulse sintering of porous materials. Minsk, Remiko, p 180 (in Russian)
25. Khasanov OL, Dvilis ES, Bikbaeva ZG (2008) Methods of compaction and consolidation of nanostructured materials and products. Textbook, Tomsk Polytechnic University Publishing, Tomsk, 212 p (in Russian)
26. Olevsky EA, Kandukuri S, Froyen L (2007) Consolidation enhancement in spark-plasma sintering: impact of high heating rates. J Appl Phys 102:114913
27. Crivelli IV, Esposito E, Mele G, Siniscalchi A (1973) Formatura per Spark Sintering. Metall Italiana 65(11):611–618
28. Zavodov NN, Kozlov AV, Luzganov SN, Polishchuk VP, Shurupov AV (1999) Sintering of metal powders by a series of heavy current pulses. High Temp 37(1):130–135
29. Zhou Y, Hirao K, Yamauchi Y, Kanzaki S (2003) Effects of heating rate and particle size on pulse electric current sintering of alumina. Scr Mater 48:1631–1636
30. Johnson DL (1990) Comment on "temperature-gradient-driven diffusion in rapid-rate sintering". J Am Ceram Soc 73(8):2576–2578
31. Chipman J (1926) The Soret effect. J Am Chem Soc 48:2577–2589
32. Shewmon P (1958) Thermal diffusion of vacancies in zinc. J Chem Phys 29(5):1032–1036
33. Schottky G (1965) A theory of thermal diffusion based on lattice dynamics of a linear chain. Phys Status Solidi 8(1):357–368
34. Kornyushin YV (1980) Influence of external magnetic and electric-fields on sintering, structure and properties. J Mater Sci 15(3):799–801

35. Searcy AW (1987) Theory for sintering in temperature-gradients-role of long-range mass-transport. J Am Ceram Soc 70(3):C61–C62
36. Young RM, McPherson R (1989) Temperature-gradient-driven diffusion in rapid-rate sintering. J Am Ceram Soc 72(6):1080–1081
37. Young RM, McPherson R (1990) Temperature-gradient-driven diffusion in rapid-rate sintering – reply. J Am Ceram Soc 73(8):2579–2580
38. Gostomelskiy VS, Krupnova LV (1985) Growth and healing of pores in metals under the action of current pulses. Phys Chem Mater Treat (Fizika I Khimiya obrabotki Materialov) 4:82–87 (in Russian)
39. Olevsky E, Froyen L (2009) Influence of thermal diffusion on spark-plasma sintering. J Am Ceram Soc 92:S122–S132
40. Chen W, Anselmi-Tamburini U, Garay JE, Groza JR, Munir ZA (2005) Fundamental investigations on the spark-plasma sintering/synthesis process I. Effect of dc pulsing on reactivity. Mater Sci Eng A 394(1–2):132–138
41. Kuz'mov AV, Olevskii EA, Aleksandrova EV (2013) Effect of micrononuniform heating of powder in field-assisted sintering on shrinkage kinetics. Powder Metall Met Ceram 51 (11–12):657–665
42. Mukherjee AK, Bird JE, Dorn JE (1969) Experimental correlation for high-temperature creep. Trans ASM 62:155–179
43. Tuchinskii LI (1982) Plastic relaxation of thermal stresses during sintering under nonisothermal conditions. Sov Powder Metall Met Ceram 21(11):849–853
44. Tuchinskii LI (1983) Possibility of plastic relaxation of thermal stresses in porous bodies. Sov Powder Metall Met Ceram 22(4):269–273
45. Skorokhod VV, Tuchinskii LI (1978) Condition of plasticity of porous bodies. Sov Powder Metall Met Ceram 17(11):880–883
46. Asoka-Kumar P, O'Brien K, Lynn KG, Simpson PJ, Rodbell KP (1996) Detection of current-induced vacancies in thin aluminum–copper lines using positrons. Appl Phys Lett 68:406
47. Garay JE, Glade SC, Anselmi-Tamburini U, Asoka-Kumar P, Munir ZA (2004) Electric current enhanced defect mobility in Ni_3Ti intermetallics. Appl Phys Lett 85:573
48. Frei JM, Anselmi-Tamburini U, Munir ZA (2007) Current effects on neck growth in the sintering of copper spheres to copper plates by the pulsed electric current method. J Appl Phys 101:114914
49. Burenkov GL, Raichenko AI, Suraeva AM (1987) Dynamics of interparticle reactions in spherical metal powders during electric sintering. Sov Powder Metall Met Ceram 26 (9):709–712
50. Olevsky E, Froyen L (2006) Constitutive modeling of spark-plasma sintering of conductive materials. Scr Mater 55:1175–1178
51. Cao WD, Lu XP, Sprecher AE, Conrad H (1990) Superplastic deformation behavior of 7475 aluminum alloy in an electric field. Mater Sci Eng A 129:157–166
52. Conrad H, Yang D (2010) Influence of an applied dc electric field on the plastic deformation kinetics of oxide ceramics. Philos Mag 90(9):1141–1157
53. Kim SW, Kim SG, Jung JI, Kang S-JL, Chen I-W (2011) Enhanced grain boundary mobility in yttria-stabilized cubic zirconia under an electric current. J Am Ceram Soc 94(12):4231–4238
54. Ghosh S, Chokshi AH, Lee P, Raj R (2009) A huge effect of weak dc electrical fields on grain growth in zirconia. J Am Ceram Soc 92(8):1856–1859
55. Holland TB, Anselmi-Tamburini U, Quach DV, Tran TB, Mukherjee AK (2012) Effects of local joule heating during the field assisted sintering of ionic ceramics. J Eur Ceram Soc 32 (14):3667–3674
56. Wang J, Raj R (1990) Estimate of the activation energies for boundary diffusion from rate-controlled sintering of pure alumina and alumina doped with zirconia or titania. J Am Ceram Soc 73:1172–1175
57. Kim SW, Kang SJL, Chen IW (2013) Ionomigration of pores and gas bubbles in yttria-stabilized cubic zirconia. J Am Ceram Soc 96(4):1090–1098

58. Osepchuk JM (1984) A history of microwave heating applications. IEEE Trans Microw Theory Tech 32(9):1200–1224
59. Tinga WR, Voss WAG (1968) Microwave power engineering. Academic Press, New York, p 1968
60. Berteaud AJ, Badot JC (1976) High temperature microwave heating in refractory materials. J Microw Power 11(4):315–320
61. Meek TT, Holcombe CE, Dykes N (1987) Microwave sintering of some oxide materials using sintering aids. J Mater Sci Lett 6(8):1060–1062
62. Johnson DL (1991) Microwave and plasma sintering of ceramics. Ceram Int 17:295–300
63. Johnson DL (1991) Microwave heating of grain boundaries in ceramics. J Am Ceram Soc 74 (4):849–850
64. Birnboim A, Gershon D, Calame J, Birman A, Carmel Y, Rodgers J, Levush B, Bykov Y, Eremeev A, Holoptsev V, Semenov V, Dadon D, Martin P, Rosen M, Hutcheon R (1998) Comparative study of microwave sintering of zinc oxide at 2.45, 30 and 83 GHz. J Am Ceram Soc 81:1493–1501
65. Rybakov KI, Olevsky EA, Semenov VE (2012) The microwave ponderomotive effect on ceramic sintering. Scr Mater 66:1049–1052
66. Olevsky EA, Maximenko AL, Grigoryev EG (2013) Ponderomotive effects during contact formation in microwave sintering. Model Simul Mater Sci Eng 21:055022
67. Rybakov KI, Olevsky EA, Krikun EV (2013) Microwave sintering: fundamentals and modeling. J Am Ceram Soc 96(4):1003–1020
68. Marder R, Estournès C, Chevallier G, Chaim R (2014) Spark and plasma in spark plasma sintering of rigid ceramic nanoparticles: a model system of YAG. J Eur Ceram Soc 35:211

Chapter 2
Resistance Sintering

2.1 Principle and Physical Mechanisms of Resistance Sintering

Lenel [1] classified resistance sintering under pressure as a type of hot pressing. In resistance sintering under pressure, high currents pass through the powder compact at low applied voltages. In conventional hot pressing, the die is heated by radiation from resistive heaters, by high-frequency induction, or by passing an electric current through it. The material of the die needs, therefore, to be able to withstand temperatures required for high-temperature sintering and possess sufficient mechanical strength. In order to overcome these difficulties, one can heat only the material to be sintered not heating the die directly by passing a high current through the powder compact. In resistance sintering, the heat is generated within the powder and is not conducted from the die. Therefore, only conductive materials can be sintered by resistance sintering. Taylor suggested placing powders in a glass or ceramic tube between the plungers [2]. In a setup described by Lenel [1] (Fig. 2.1), a green compact or a loose powder is placed between wafers, which are made of a material that has a higher electrical resistivity, a lower thermal conductivity, and a higher melting point than the material of the plungers to make the heat distribution in the sample more uniform. If wafers are not used, then the heat is rapidly dissipated through the plungers, which are made of a highly conductive material, resulting in insufficient sintering of regions near the flat ends of the sample. The powder compact is isolated from the die by a ceramic liner.

Resistance sintering is performed when a direct current or an alternating current of low frequency passes through the powder compact. Pressure is applied when the compact is formed by loose powders; pressureless resistance sintering is also possible when electric current is applied to the compact pre-pressed or pre-sintered before the resistance sintering. The sintering time is usually short (a fraction of a second). The resistance sintered material is also cooled rapidly as the current is switched off. The principal requirement to the powder compact for its successful

E. A. Olevsky, D. V. Dudina, *Field-Assisted Sintering*,
https://doi.org/10.1007/978-3-319-76032-2_2

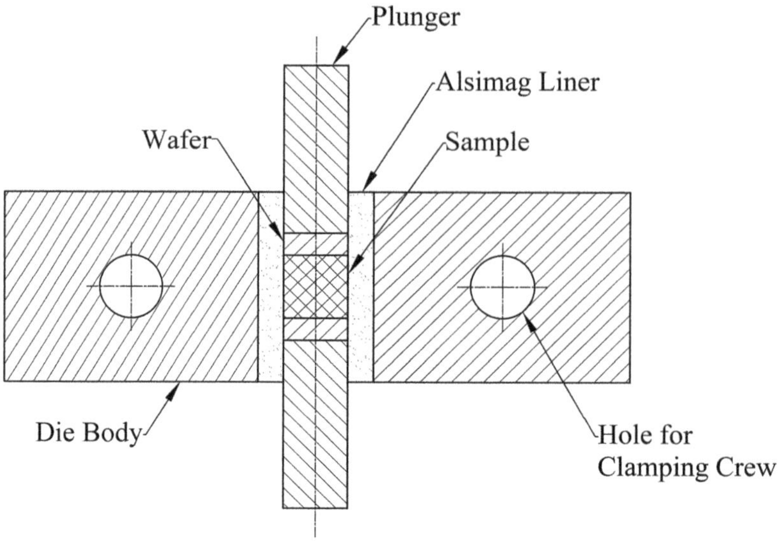

Fig. 2.1 Schematic of the die–plunger setup used for resistance sintering (Adapted from Lenel [1], Copyright (1955) with permission of Springer)

resistance sintering is a rather low resistance. The total resistance of the powder compact is determined by the resistivity of the material itself and contact resistances between the particles. The latter depends on the particle size distribution, the level of applied pressure, and the presence of oxide films on the surface of the particles. The electrical conductivity of the powder compact from carbonyl nickel powders increases with the applied pressure as p^n, where p is applied pressure and $n = 0.85$ [3]. In composite mixtures made of a conducting and an insulating component, the conducting component should form a continuous network within the compact. The sample with 0.1 Ω resistance will not be able to carry enough current for sintering to occur [1]. For materials with a high initial resistance, a two-pulse sintering process can be suggested for a better control of the changes in the sample's resistance during sintering and the microstructure development. Using a two-pulse technique, it is possible to benefit from different parameters of the pulses (time, voltage) chosen for different purposes. The first pulse is intended to break down the resistance, and the second pulse is real sintering. This technique allows avoiding overheating of the compact. Greenspan [4] described so-called impulse resistance sintering – the process designed to fully utilize the heat generated at the inter-particle contacts to bond the particles together. This is done by applying a very low pressure in the beginning of the process (6 MPa) and increasing it when the compact reached a plateau on the shrinkage curve. This method was quite successful in making tungsten compacts of 22% porosity, and despite that, possessing a high flexural strength, which indicated good bonding between the particles. It can be assumed, however, that the early formation of the well-established inter-particle contacts hindered densification of the compacts upon further sintering – this aspect was theoretically

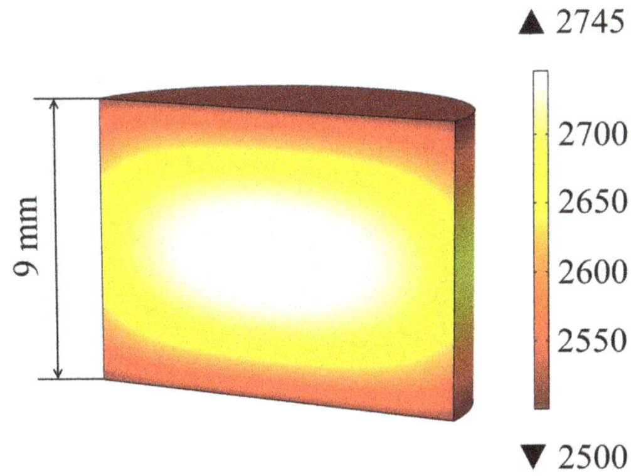

Fig. 2.2 Temperature distribution in an iron compact sintered by resistance sintering (Adapted from Lenel [1], Copyright (1955) with permission of Springer)

elaborated by Olevsky et al. in Ref. [5] – and was the reason for the difficulties of producing fully dense compacts by this technique.

The calculated temperatures resulting from the models can be verified experimentally by placing small pieces of metallic wire with different melting points in the compact [1]. The center of the compact has the maximum temperature (see example in Fig. 2.2 for an iron compact). Lower temperatures of the peripheral regions of the compacts can cause insufficient sintering. In order to achieve a better sintering uniformity, one of the wafers is made concave. A higher current passing through the edges where its path is shorter makes the overall temperature distribution more uniform.

For successful resistance sintering, the ratio of the length to diameter of the compact should not exceed 1. Larger ratios result in poor sintering of the central parts of the compacts. The grain size of the materials resistance sintered under pressure is usually smaller compared with conventionally sintered ones. Although the contact between the particles is established more rapidly during resistance sintering under pressure creating conditions for the grain growth to occur, the short sintering time prevents extensive grain growth. Due to the shortness of the sintering time, the degree of chemical homogenization is also limited in multicomponent mixtures during resistance sintering. The alloy formation can be arrested at a certain stage producing a metastable material from the phase composition and microstructure points of view. Lenel [1] states the shortness of the process of achieving high relative densities and the possibility of producing non-equilibrium microstructures as the main features and advantages of resistance sintering.

As electric current starts passing through the compact upon the application of voltage, the boundaries separating the material with increased electrical conductivity move from the contact surfaces of the compact and the punch (electrode) into the interior of the sample. The current increases dramatically when the two boundaries meet in the central part of the sample. The advantages of two-pulse sintering for compacts of high initial resistance are demonstrated in Fig. 2.3. When current is applied to a low initial resistance compact, it becomes constant after a few cycles

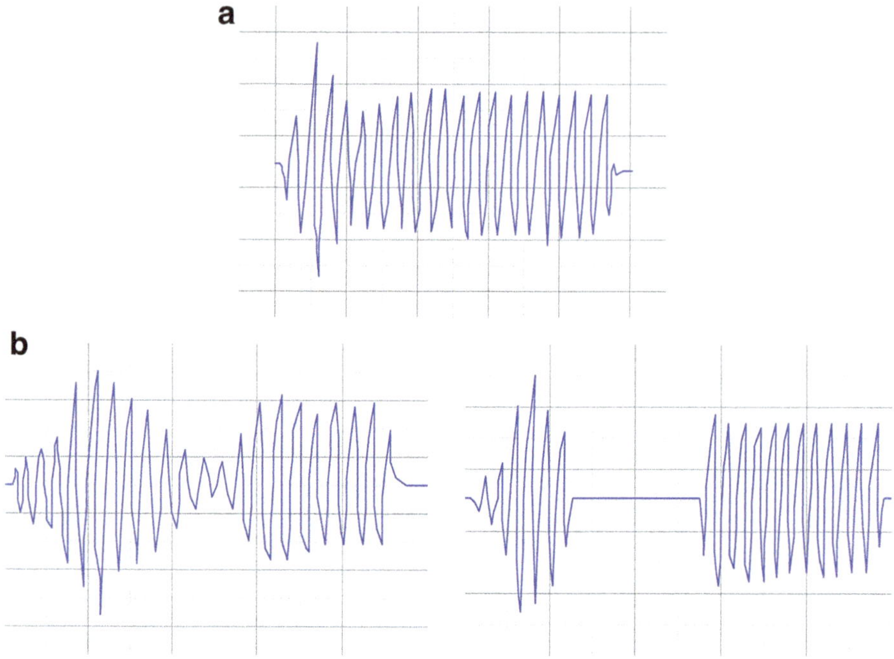

Fig. 2.3 Current traces for the case of a low initial resistance of the compact (**a**) and high initial resistance of the compact, the cases of one- and two-pulse sintering (**b**) (Adapted from Lenel [1], Copyright (1955) with permission of Springer)

because the current regulator can compensate the changes in the resistance. When a high-resistance compact is sintered by a single pulse, the resistance changes so dramatically that the current regulator is unable to follow these changes.

2.2 Resistance Sintering Equipment

Akechi and Hara [6] analyzed the evolution of contributions from the resistance of the powder and that of inter-particle contacts to the total resistance of a titanium compact sintered by resistance sintering using an alternating current of 50 Hz frequency. The analysis was performed based on the measured profiles of electric current and voltage (Fig. 2.4).

As is shown in Fig. 2.5, the following heat-generating elements can be designated: the inter-particle contacts in the powder compact, the powder particles themselves, the contact between the punches and the powder compact punches, the punches, the contact between the punch and the plunger, and internal elements of the setup. At the final stage of sintering, when bonding between the powder particles has been already established, the resistance of inter-particle contacts does not

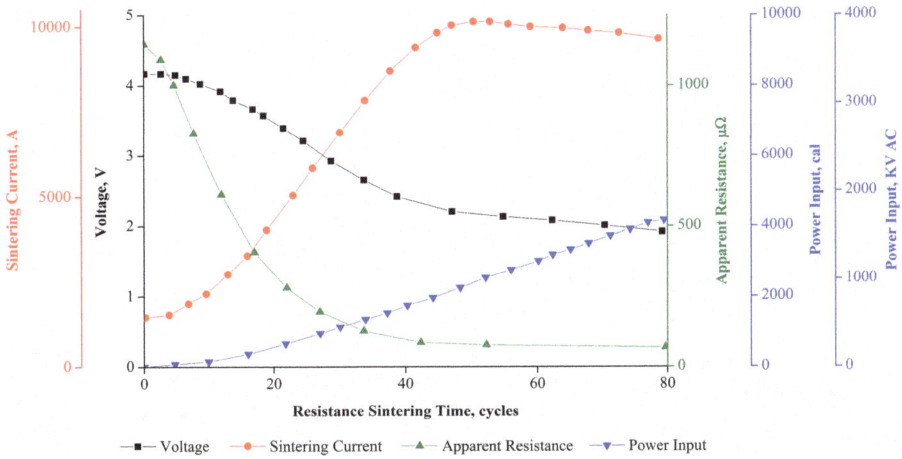

Fig. 2.4 Evolution of current, voltage, apparent electric resistance, and energy input with the sintering time for a resistance sintered titanium powder compact under pressure. Power input is given in calories and kV·A·cycle units (Reprinted from Akechi and Hara [6], Copyright (1977) with permission from Japan Society of Powder and Powder Metallurgy)

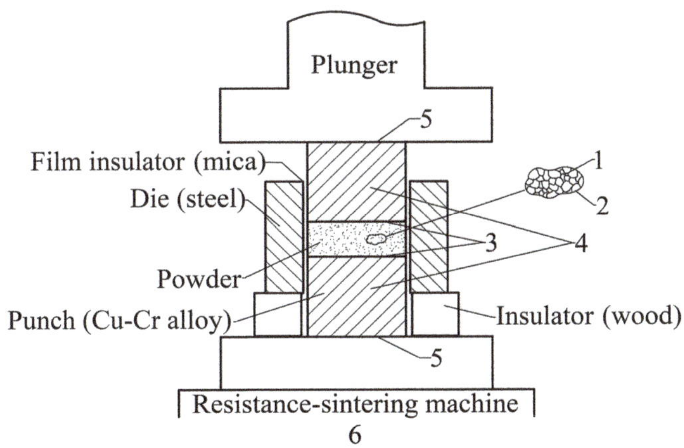

Fig. 2.5 Heat-generating elements of a resistance sintering setup: 1, inter-particle contacts; 2, powder particles; 3, resistance of the contact between the punches and the powder compact; 4, resistance of the punches; 5, resistance of the contact between the punch and the plunger; 6, internal resistance of the setup (Reprinted from Akechi and Hara [6], Copyright (1977) with permission from Japan Society of Powder and Powder Metallurgy)

contribute any longer to the total resistance of the setup. The resistance of the powder compact can be calculated using the resistivity of the bulk metal and dimensions of the compact. The temperature, to which the sample was heated at a certain moment of the sintering process, can be estimated from the measured net power and thus can be taken into account when calculating the resistance of the powder compact and the punches. The sum including the contact resistance between the punches and the

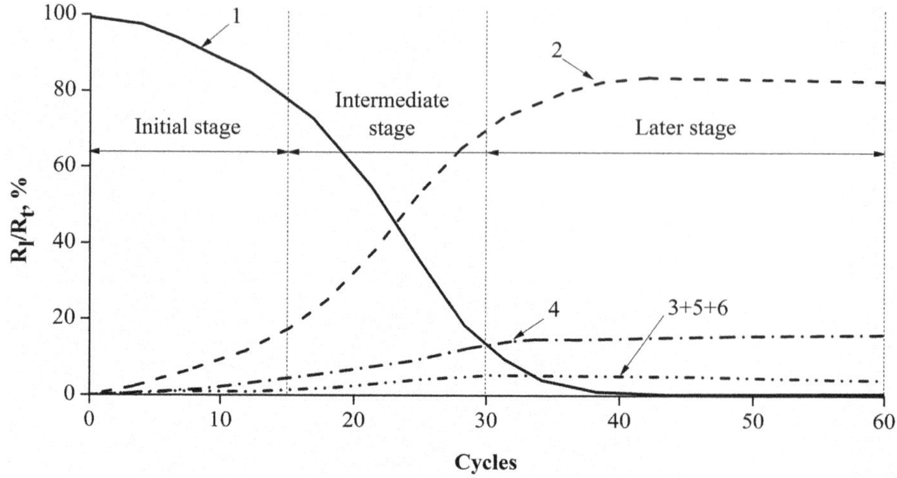

Fig. 2.6 Contributions of resistances of different heat-generating elements of the resistance sintering setup to the total resistance (notations are as in Fig. 2.5) (Reprinted from Akechi and Hara [6], Copyright (1977) with permission from Japan Society of Powder and Powder Metallurgy)

powder compact, the contact resistance between the punch, and the plunger and the resistance of internal elements of the setup can be calculated by subtracting the resistance of the powder compact from the total resistance at the final sintering stage. At the initial sintering stage, the compact is mainly heated due to heat evolution at the inter-particle contacts. At the intermediate stage, the heat is generated both within the particles and the contacts between them. In accordance with these considerations, Fig. 2.6 shows the evolution of the fractions of resistances of different heat-generating elements in the total resistance of the sintering setup.

2.3 Properties of Specimens Processed by Resistance Sintering

In experiments performed by Montes et al. [7], the resistance of a titanium powder compact decreased rapidly during 0.2 s from the beginning of the sintering process; this was accompanied by a certain decrease in the porosity of the compact. Upon further sintering, the resistance did not change, while the porosity continued to decrease (Fig. 2.7). Such behavior was explained by a fast resistivity reduction of oxide films of semiconductor nature with increasing temperature at the initial stage of sintering. Temperatures needed for significant material softening were not reached until the intermediate stage of sintering, during which the densification of the compact continued.

Akechi and Hara [6] emphasized the role of the positive coefficient of resistance in homogenizing the microstructure of the powder compact and stabilizing the

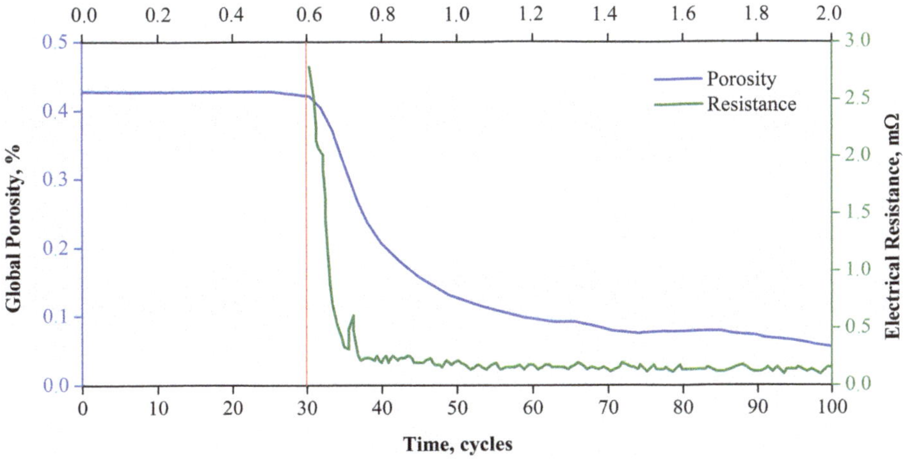

Fig. 2.7 Evolution of porosity and electrical resistance of the compact produced by resistance sintering of a titanium powder (passage of current started at the time equal to 30 cycles) (Reprinted from Montes et al. [7]. Copyright (2011) with permission of Springer)

process of resistance sintering. The regions of the powder compact of low resistance initially carry the most of the current; however, due to Joule heating, the resistivity of the material increases such that redistribution of current occurs. This redistribution causes better sintering of the previously poorly sintered regions of high initial resistance. Rykalin [8] suggests the "self-regulation" term for this phenomenon.

Being a fast sintering technique, resistance sintering does not require a protective atmosphere. However, when reactive metals, such as titanium, are resistance sintered, an increase in the oxygen and nitrogen content in the compact relative to that in the powder can be observed. Montes et al. [7] found that consolidation of a titanium powder with a mean particle size of 24 μm by resistance sintering (total duration of sintering was 1.4 s) results in an order of magnitude increase in the oxygen content, while the content of nitrogen doubles.

Since during the resistance sintering both the electric current and the sample's resistance change with time, in order to assess the contribution of the Joule heat in comparative studies of the microstructure and properties of the compacts sintered from the same powder under variable conditions of resistance sintering (current, number of cycles of current), the calculation of the thermal energy generated per unit mass of the powder (specific thermal energy) due to Joule effect is helpful:

$$\eta = \frac{1}{M}\int_{0}^{t} I^2(\tau)R(\tau)d\tau, \tag{2.1}$$

where $I(\tau)$ is the current, $R(\tau)$ is the electrical resistance, and M is the powder mass.

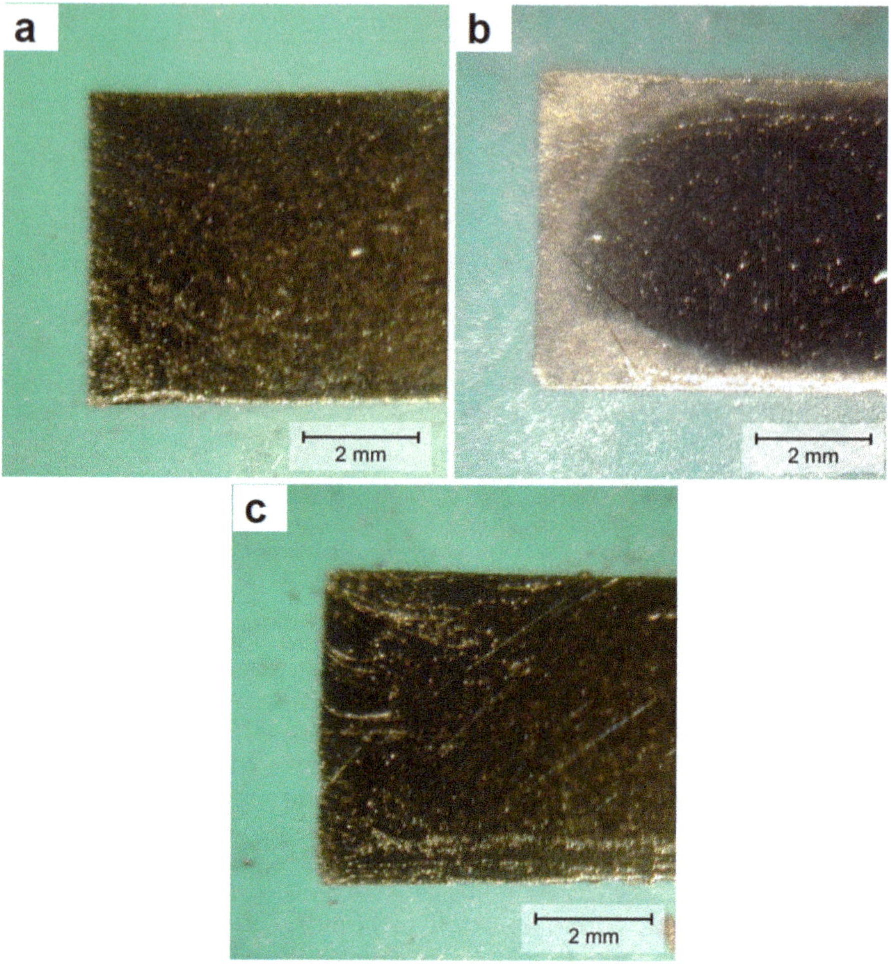

Fig. 2.8 Macrographs of diametrical sections of halves of compacts: (**a**) conventionally sintered (porosity 2%), (**b**) resistance sintered compact (4.0 kA, 50 cycles, porosity 15.8%), (**c**) resistance sintered (6.0 kA, 80 cycles, porosity 3.3%). A higher value of η for sample (**c**) resulted in a more uniform distribution of the porosity (Reprinted from Montes et al. [7]. Copyright (2011) with permission of Springer)

A higher value of η results in a more uniform distribution of the porosity in the resistance sintered compacts (Fig. 2.8). The highest temperature is reached in the center of the sample making it denser than the peripheral regions adjacent to the electrodes cooled during the process and walls of the die. These effects are more pronounced at lower η.

Montes et al. distinguish three types of pores in resistance sintered compacts. The first type is the porosity in poorly sintered regions; these pores are comparable in size with the initial powder particles (Fig. 2.9a). The second type of pores is the residual

Fig. 2.9 Three types of pores in resistance sintered compacts: (**a**) pores in poorly sintered regions, (**b**) residual pores in well-sintered regions, (**c**) pores formed during pressing (shear cracks) (Reprinted from Montes et al. [7]. Copyright (2011) with permission of Springer)

pores in well-sintered regions (Fig. 2.9b). Pores of the third type are large pores that have the greater axis perpendicular to the pressing direction and can be considered as shear cracks introduced during pressing (Fig. 2.9c). These pores can be found only in samples sintered at high currents, as they form in a well-sintered material that does not allow any particle rearrangement in response to shear forces.

Thanks to short processing times and high cooling rates in resistance sintering, the resistance sintered compacts may show microstructures different from those of the conventionally sintered ones. In resistance sintered compacts heated above the temperature of the $\alpha \rightarrow \beta$ transition of titanium, lamellar grains of the α-phase form during fast cooling of the β-phase [7]. In conventionally sintered titanium during slow furnace cooling, large equiaxed grains of the secondary α-phase form.

According to Belyavin et al. [9], resistance sintering by DC current in the case of long process durations (longer than 1 h) can lead to the differences in the microstructure and grain size of the cathode and anode ends of the specimen and nonuniform distribution of porosity.

Lagos et al. [10] and Schubert et al. [11] presented a modified method of resistance sintering based on the action of a current of high density. In their processing, WC–Co powder was filled in a ceramic die between two copper electrodes. Sintering was performed in air with a holding time of only 500 ms

[10]. The current was produced by a low-voltage transformer (around 10 V). The sintered samples consisted of a dense core surrounded by a porous surface layer. The authors pointed out that, in contrast to spark plasma sintering, in which the processing time is in the range of minutes and a controlled atmosphere is needed, a fast single-pulse technique allows conducting sintering in air without oxidizing the material. This is a very important advantage, as it reduced the cost of the sintering equipment. Schubert et al. [11] compared the microstructure and properties of WC–Co materials consolidated by the fast resistance sintering and hot isostatic pressing and found that the former possessed finer microstructures and higher hardness.

2.4 Summary

In resistance sintering, the heat is generated within the powder and is not conducted from the die. Therefore, only conductive materials can be sintered by resistance sintering. During resistance sintering, temperature distribution within the sample is spatially nonuniform, as shown by experiments and modeling. During resistance sintering, the center of the compact has the maximum temperature, while peripheral regions of the compacts are prone to insufficient sintering. At the same time, the short sintering time in resistance sintering prevents extensive grain growth. In powder mixtures, it allows compositional homogenization to proceed only to a limited extent. A high potential of a recently suggested modification of resistance sintering – sintering by a current pulse of high density – should be noted. This variation of resistance sintering allows working in air and still avoiding oxidation of the material due to a very short duration of the pulse.

References

1. Lenel FV (1955) Resistance sintering under pressure. JOM 7(1):158–167
2. Taylor GF (1933) Apparatus for making hard metal compositions. US Patent No. 1896854
3. Euler KJ, Fromm R, Sperlich B (1980) Electric-conductivity of compressed nickel powder. Planseeber Pulvermet 3:127–134
4. Greenspan J (1976) Impulse resistance sintering of tungsten. Army Materials and Mechanics Research Center
5. Olevsky E, Bogachev I, Maximenko A (2013) Spark-plasma sintering efficiency control by inter-particle contact area growth: a viewpoint. Scr Mater 69(2):112–116
6. Akechi K, Hara Z (1978) Analysis of electric resistance-sintering process of titanium powder. J Jpn Soc Powder Powder Metall 9:193–197 (in Japanese)
7. Montes JM, Rodríguez JA, Cuevas FG, Cintas J (2011) Consolidation by electrical resistance sintering of Ti powder. J Mater Sci 46:5197–5207
8. Rykalin NN (1959) Thermal processes in contact welding. Moscow, 277 p (in Russian)
9. Belyavin KE, Mazyuk VV, Min'ko DV, Sheleg VK (1997) Theory and practice of electric pulse sintering of porous materials. Minsk, Remiko, 180 p (in Russian)

10. Lagos MA, Agote I, Gallardo JM, Montes JM, Schubert T, Weissgaerber T, Prakash L, Andreouli C, Oikonomou V, Lopez D, Calero JA (2016) Development of the ERS process for the fabrication of hardmetal parts. Proc. World PM2016, Congress and Exhibition, World PM2016– HM – Processing I
11. Schubert T, Weißgärber T, Lagos MA, Agote I, Gallardo JM, Montes JM, Prakash L, Oikonomou P, Andreouli C (2016) Comparison of hard metals fabricated by Electrical Resistance Sintering (ERS) and Sinter-HIP. Proc. World PM2016, Congress and Exhibition, World PM2016– HM – Processing I

Chapter 3
Sintering by High-Voltage Electric Pulses

3.1 Principle and Physical Mechanisms of High-Voltage Consolidation

Consolidation of powders by high-voltage electric pulses generally attracts attention as a method of very fast consolidation, in which most of the heat is released at the inter-particle contacts. High-voltage electric pulse sintering is a processing of choice when refractory metals are to be consolidated [1–3]. The principle of this technique is the passage of a high-voltage current pulse through the powder sample under applied external pressure. In the most widely used variation of high-voltage consolidation, the current is produced by discharging a capacitor bank. The described sintering methods utilize electric discharges of several kV, electric current densities exceeding 10 $kA{\cdot}cm^{-2}$, and pressures of up to 10 GPa. In most cases, the powder to be consolidated is subjected to a single electric pulse with duration shorter than 0.1 s. Consequently, electric pulse methods allow consolidating near-net-shape powder compacts much faster than in the conventional methods of densification, such as pressureless sintering, hot pressing, or hot isostatic pressing. Fast sintering in high-voltage techniques can be thought of as a possibility to conduct sintering with minimal microstructure changes, if those are undesirable in the consolidated material. Belyavin et al. [1] approached electric current-assisted sintering pursuing the sintering enhancement possibilities of refractory metals. The activation of sintering by current was suggested as more efficient than the activation by the introduction of low-melting-temperature additives.

Pulse consolidation processes exist in several variations: high-voltage electric discharge consolidation (HVEDC) [4–10], high-energy high-rate consolidation (HEHR) [11–16], pulse plasma sintering (PPS) [17, 18], and capacitor discharge sintering (CDS) [19–23]. A setup for HVEDC uses pulsed current generated by discharging a capacitor bank to rapidly heat the powder sample, to which external pressure is simultaneously applied. The main parameters of the HVEDC process are the external pressure and the electric current generated by the discharge. In HEHR

E. A. Olevsky, D. V. Dudina, *Field-Assisted Sintering*,
https://doi.org/10.1007/978-3-319-76032-2_3

consolidation, a homopolar generator is used, which transforms the rotation energy into the electrical energy as a result of Faraday's effect. Although the HEHR method uses a voltage below 100 V, it is often discussed together with high-voltage consolidation due to its highly dynamic nature: consolidation by means of a single pulse of a very high electric current (100–500 kA). In PPS, a current source generates periodical pulses of current. Sintering is conducted in vacuum, which differs this method from other methods based on the application of pulsed current. The CDS method uses two circuits coupled by mutual inductance instead of a single RLC circuit. This configuration allows applying low voltages to the powder compact, thus reducing the possibilities of discharges, breakdown, and local plasma formation during the process. The CDS normally produces nearly fully dense compacts, the porosity being present only in the surface layer and uniformly distributed. The CDS method is based on the storage of high-voltage electrical energy in a capacitor bank inserted in a freely oscillating system composed of a primary circuit and a mutually coupled secondary circuit. The secondary circuit works in conjunction with a mechanical press, which is controlled by a programmed logical controller. Once the desired pressure from the press is reached, the switch closes the circuit, and the electromagnetic energy is transferred to the secondary circuit by means of a transformer that enables conversion from high voltage and low current in the primary circuit to low voltage and high current in the secondary circuit. A comprehensive review by Yurlova et al. [1] summarizes the current knowledge of the process mechanisms and tracks the historical development of the high-voltage methods. The advantages of HVEDC can be exploited only through the optimization of the consolidation parameters since excessive energy dissipation during this type of processing can lead to the instability of the compaction process, the formation of an undesirable material structure, and even to the destruction of the sintered specimens and of the equipment used. The time dependence of the associated thermal processes at the inter-particle contacts plays a key role in electric pulse powder consolidation.

A schematic of a HVEDC setup, an equivalent electric diagram, and HVEDC tooling are shown in Fig. 3.1. The behavior of the electric circuit consisting of a capacitor, a consolidation setup, and a powder column is described by the following equation [10, 24]:

$$\frac{d^2U}{d\tau^2} + \frac{R(\tau)}{L}\frac{dU}{d\tau} + \frac{U}{LC} = 0, \tag{3.1}$$

where U is the voltage, τ is the time, L is the inductance, and C is the capacitance. The principal difficulty in describing the process of HVEDC mathematically is the changing resistance of the powder compact during the process. If $R^2(\tau) > 4LC$, i.e., the resistance of the powder compact exceeds the reactive component of the circuit, the discharge has an aperiodic character. If $R^2(\tau) < 4LC$, the discharge is a damped oscillation process. Electric discharge consolidation facilities usually use a Rogowski coil to measure the current and oscillographs to record the current waveform. The typical current waveform during HVEDC is a damped sinusoid (Fig. 3.2)

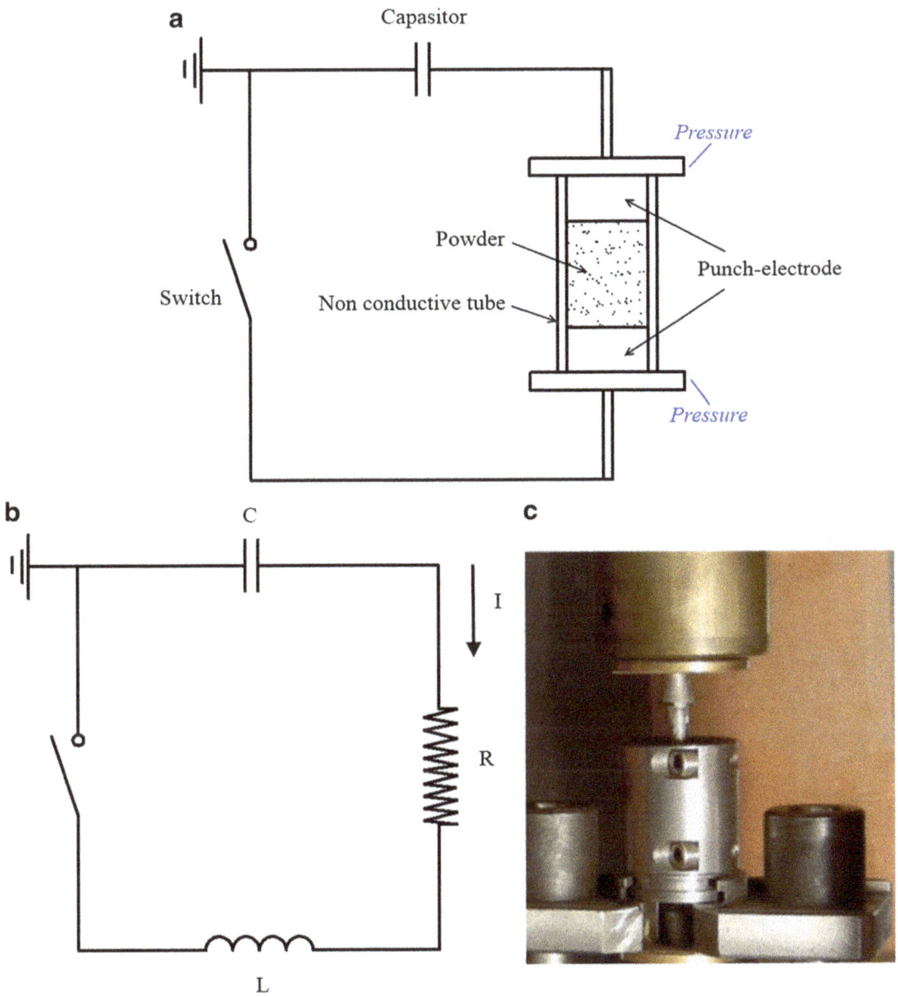

Fig. 3.1 Schematic of a HVEDC setup (**a**), equivalent electric diagram (**b**), and HVEDC tooling (**c**)

[25]. A change from a damped oscillation discharge to an aperiodic discharge (Fig. 3.3) can be caused by an increase in the compact resistance due to temperature rise due to Joule heating [24].

Al-Hassani et al. [26] suggested using the electric circuit theory to describe the electric pulse sintering. The mathematical model is based on the assumption that the inductance and capacitance remain constant, while the resistance varies with temperature. The variation of resistance with time during electric discharge sintering is expressed by a two-term exponential form:

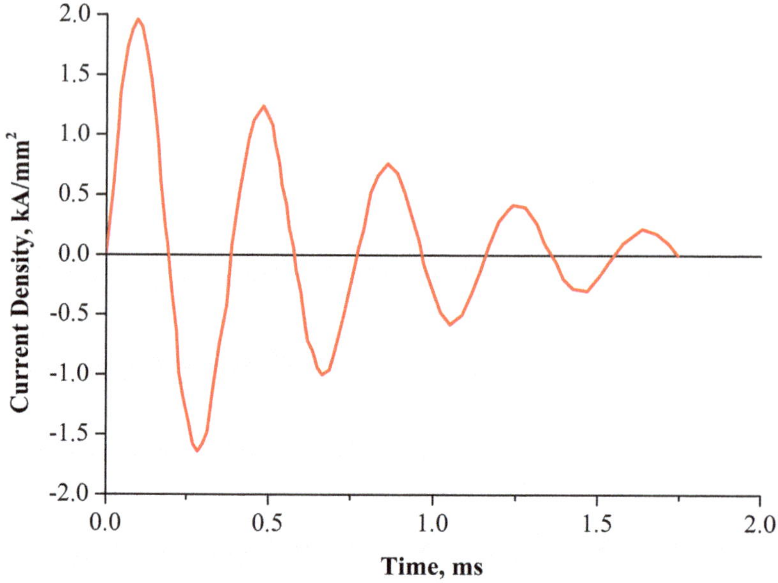

Fig. 3.2 Typical pulse current waveform during HVEDC (Rogowski coil) [25]

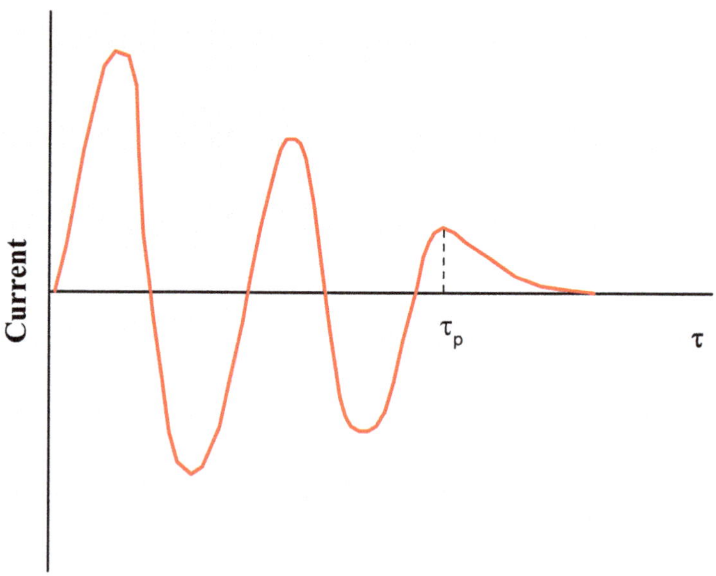

Fig. 3.3 A change from a damped oscillation discharge to an aperiodic discharge [24]

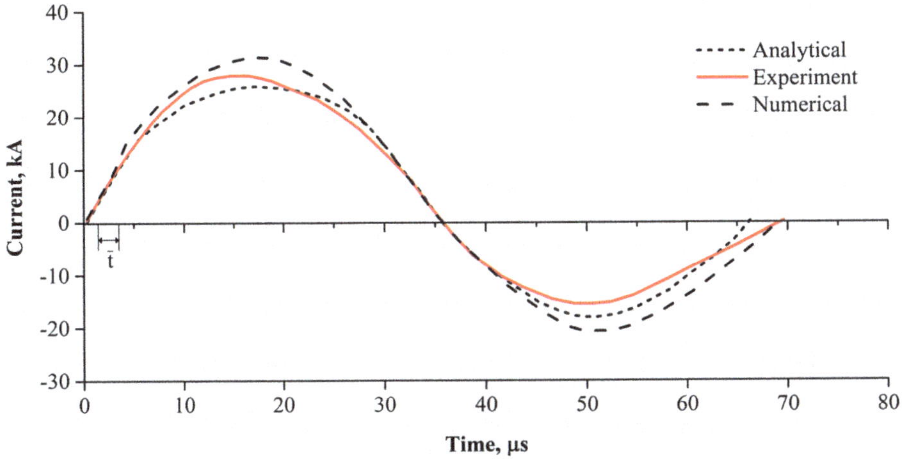

Fig. 3.4 Current waveforms of a commercial steel powder sample. (Reprinted from Al-Hassani et al. [26], Copyright (1986) with permission from Elsevier)

$$R(t) = R_1(t) + R_2(t) \text{ where } R_1 = A_1 e^{-\alpha_1 t} \text{ and } R_2 = A_2 e^{-\alpha_2 t} \text{ here } A_1 \gg A_2, \alpha_1 \gg \alpha_{21} \tag{3.2}$$

Assumption (3.2) implies that when t is small, the first term $R_1(t)$ dominates, and when t is large, the second term does. A mathematical analysis of the equations was carried out to determine the time dependence of the electric current. As is seen in Fig. 3.4, the calculations are in good agreement with the experimental results [26].

The experimental results on the instantaneous resistance of three steel powders oxidized for different periods of time are presented in Fig. 3.5 [27]. The resistance of the powder column shows an initial sharp reduction, which is consistent with the breakdown of the insulating oxide layers on the particle surfaces. In the subsequent stage of consolidation (after the breakdown of the oxide layers), the powder resistance slightly increased, apparently due to an increase in temperature. It can be seen that the instantaneous resistance of the powder column depends on the thickness of the oxide film. As can be concluded from Figs. 3.5 and 3.6, the resistance of the compacts is dramatically reduced by the time the current reaches its maximum.

In order to exclude heating of the sample throughout its volume, the electric pulse duration should be shorter than the time needed to heat the materials through heat conduction. Short pulses lead to the skin effect and the formation of thin skin layers. In order to ensure the uniform heating of the sample through its cross section, the thickness of the skin layer should be greater than the characteristic size of the sample. The pulse duration should be shorter than the time needed to fully heat the particles; otherwise complete melting of the sample might take place. The pulse duration should satisfy the relationship:

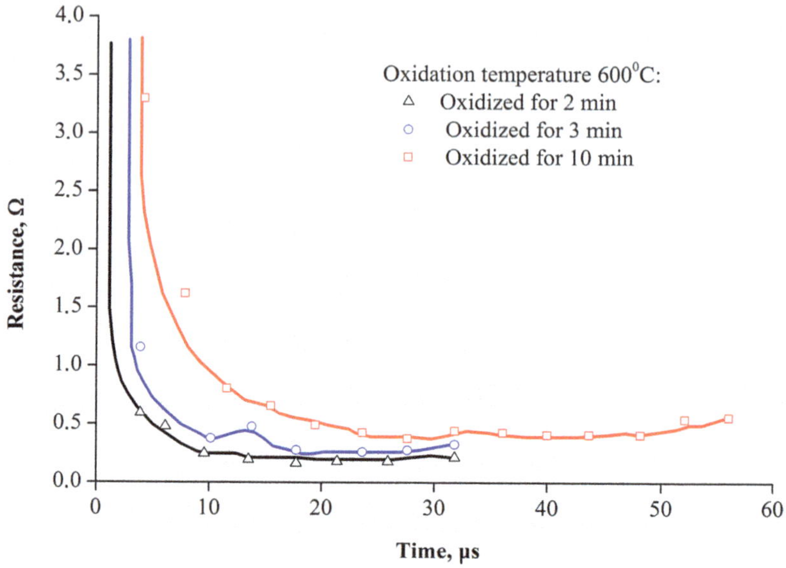

Fig. 3.5 Evolution of resistance of oxidized steel powders during electric pulse consolidation. (Reprinted from Alitavoli and Darvizeh [27], Copyright (2008) with permission form Elsevier)

$$\tau_M < \tau_P \ll \tau_T, \tag{3.3}$$

where τ_P is the pulse duration, τ_M is the time, during which the magnetic field penetrates the sample determined by the skin effect, and τ_T is the time needed to heat the powder particles [29]. These two characteristic times depend on the material properties, the sample dimensions, and the particle size of the powder. The sample electrical resistance during electric pulse sintering usually changes according to the following scenario: during the first 10–20 μs, the electrical resistance rapidly drops, then it decreases more slowly, and after 100–200 μs from the process start, the resistance stops decreasing and may as well increase due to heating of the sample. The faster the discharge, the shorter the time, during which the electrical resistance levels off. Zavodov et al. shortened the discharge duration by using exploding wires [29]. It was shown that sintering can occur within a time as short as ~2 μs. The sintered material did not, however, possess the required strength.

The characteristic cooling time in electric pulse sintering is several seconds. The duration of the densification process is several milliseconds [30]. Based on this, densification can be assumed to occur at a constant temperature. The analysis of the electric pulse sintering parameters reveals the hierarchy of the characteristic times in the electric pulse sintering process. The energy input into the powder material is characterized by the pulse duration $\tau_0 < 10^{-3}$ s. The time, during which the compact is formed from the powder, depends on the mechanical loading system and lies in the $2 \cdot 10^{-3} < \tau_1 < 2 \cdot 10^{-2}$ s range. The cooling time of the sintered sample τ_2 is a

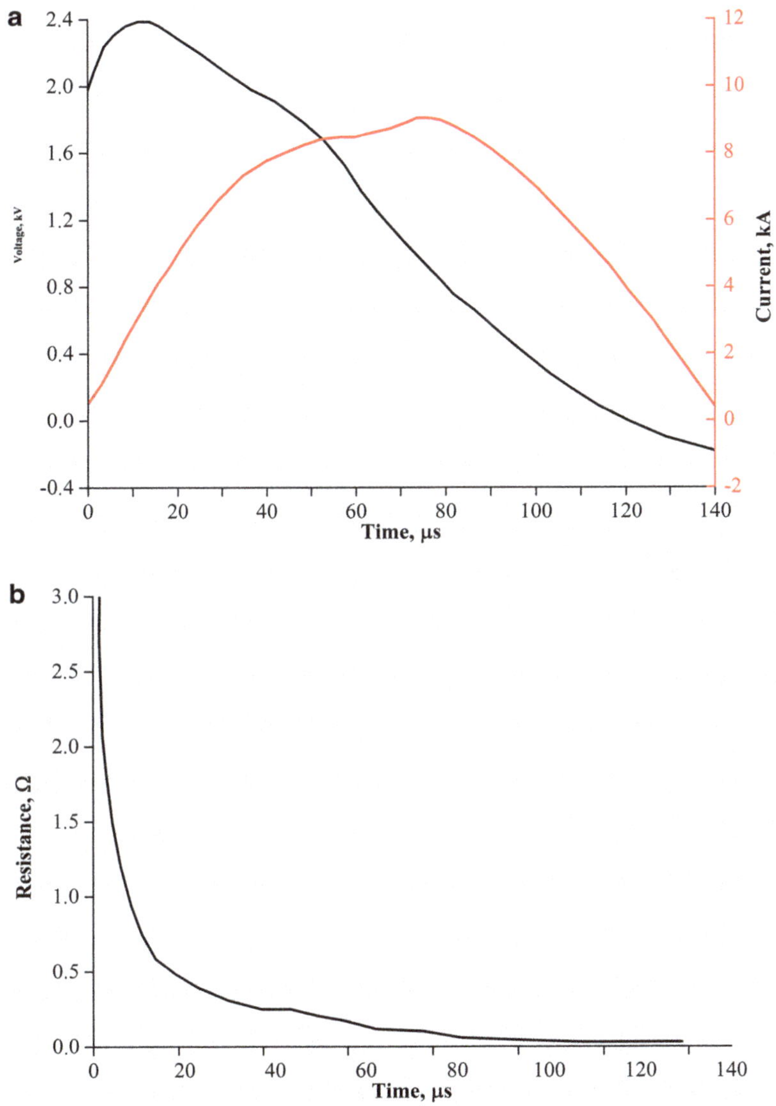

Fig. 3.6 Voltage and current (**a**) and resistance (**b**) of a compact consolidated from an oxidized nickel powder by an electric discharge. (Reprinted from Kim et al. [28], Copyright (1988) with permission from Elsevier)

function of thermal conductivity of the material and the characteristic sample size; $\tau_2 \approx 2.5$ s. The following relationship is observed: $\tau_0 < \tau_1 \leq \tau_2$ [5, 30].

The pinch effect causing constriction of the powder sample in the radial direction is observed only at certain current values passing through the sample during the discharge. When the electric current is uniformly distributed over the cross section of

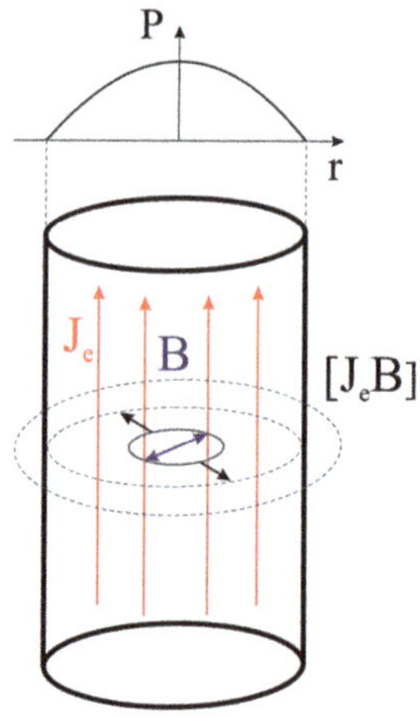

Fig. 3.7 The pinch effect in a powder sample during electric pulse sintering and magnetic pressure p profiles

a cylindrical powder sample of radius r_0, the magnetic pressure can be calculated using equation [31]:

$$p_m(r) = -\frac{\mu_0 \left(r_0^2 - r^2\right) I^2}{4\pi^2 r_0^4} \tag{3.4}$$

where μ_0 is the vacuum permeability, r is the radial distance, and I is the current. The action of the pinch effect is schematically shown in Fig. 3.7. The distribution of the magnetic pressure over the sample cross section is of a parabolic character with a maximum in the center of the sample.

The action of magnetic pressure can produce a nonuniform pore distribution in the porous powder material. However, if

$$p_l \geq p_m, \tag{3.5}$$

where p_l is the pressure on the lateral surface of the compact caused by the compressive force applied to the electrode, this will not take place [10]. If $p_l = p_m$, the friction force will be zero between the powder compacts and the die wall.

The distribution of the magnetic pressure is more uniform when a strong skin effect is observed. The pinch effect makes it easier to take the sintered sample out the die; the repeated use of the die is possible reducing the processing costs. The greater is the length of the sample, the lower is the current, and the lower is the pressure caused by the pinch effect. This consideration explains a reduction in the sintered density of samples with increasing their length.

Kim et al. [28] reported microscopic evidence of heavy deformation of nickel as a result of electric discharge consolidation carried out without external pressure and suggested that this deformation was caused by electromagnetic forces generated during the discharge. The indications of heavy deformation were observed microscopically as cells with small-angle boundaries.

3.2 Stages of High-Voltage Consolidation

As was already mentioned, an important feature of high-voltage electric pulse sintering is the concentration of the released energy in the contact zone between the particles. Therefore, the initial state of the powder particles (the thickness and structure of oxide films, the presence of second-phase inclusions, etc.), the shape and size of the particles, as well as the applied pressure greatly influence the physical processes during the consolidation, which are spatially inhomogeneous and time-dependent.

Al-Hassani et al. distinguish five sintering stages [32]:

- Destruction of the surface contaminating layers between the adjacent particles in the axial direction upon the critical applied stress depending on the nature of the powder material and geometry of the sample; this is accompanied by an instantaneous drop in the sample resistance and an increase in the current passing through the paths of least resistance.
- Inter-particle sintering leading to the formation of inter-particle necks parallel to the direction of electric current; the inter-particle necks grow, and this is accompanied by a further reduction in electrical resistance; however, an increase in the resistivity with temperature can be significant, and it is possible that the total resistance of the powder sample will increase.
- At the third stage, the conductive inter-particle necks are self-constricted as a result of the pinch effect; the surface layers of the particles are destroyed in the radial and azimuthal directions, which lead to sintering of the particles to each other in these directions.
- The electric current passes through the continuous metal using the previously formed electrical paths.
- Disintegration and stability loss or electric explosion.

In powder compacts, the density fluctuations, which can often be present due to powder agglomeration, can cause fluctuations in electrical conductivity within the compact. The paths of least resistance can be overheated. If overheating results in the formation of a molten metallic channel, a short circuit forms leaving the rest of the sample poorly sintered. Working with a mixture of steel and polyethylene powders, Alp et al. [33] determined a critical voltage, above which the compact disintegrates in a similar way a metallic wire explodes under a high current.

Fundamental parameters of electric pulse sintering are the specific energy input (SEI) and applied pressure. According to Ervin [34], SEI is the result of the real part of the current and real part of the voltage acting on the sample:

$$\mathrm{SEI} = \lim_{t\to+\infty} \frac{E_{\mathrm{j}}(t)}{w} \approx \frac{1}{w}\int_0^{t^*} i(\tilde{t}) \cdot v(\tilde{t})d\tilde{t}, \tag{3.6}$$

где $E_{\mathrm{j}}(t)$ is the energy used to heat the powder (Joule effect); w is the mass of the powder placed in the die; t^* is an approximate discharge duration; $i(t)$ and $v(t)$ are

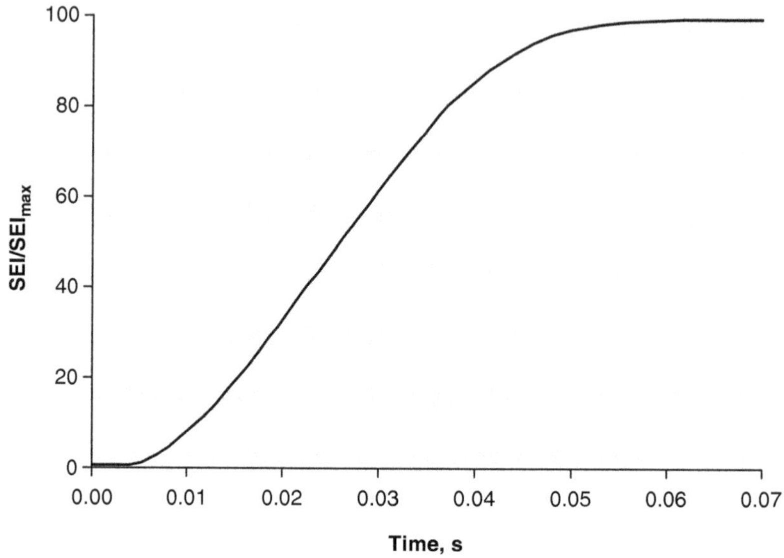

Fig. 3.8 Variation in time of the energy dissipated during an electric pulse in CDS due to Joule effect. (Reprinted from Fais [21], Copyright (2010) with permission from Elsevier)

electric current and voltage during the discharge. For instance, the typical pulse duration in capacitor discharge sintering is 30 ms, and 90% of the energy is released in less than 20 ms (Fig. 3.8). Higher values of SEI can lead to higher compact densities [21]. The thermal cycle lasts about 1 s [20].

In order to characterize an electric pulse sintering process, Anisimov and Mali [35] used the integral of current:

$$J = \int_0^t j^2 dt \tag{3.7}$$

where j is the current density in the sample. Sintering occurs when the integral of current exceeds a critical value, which is lower than required to fully melt the material.

3.3 Processes at Inter-particle Contacts During High-Voltage Consolidation

When no electric current is involved in the sintering process, the densification rate will increase with increasing applied pressure; however, in electric pulse sintering, as the level of pressure influences the resistance of the powder compact, an opposite influence may be expected. As the pressure increases, the contact resistance between the particles decreases reducing Joule heating. Al-Hassani has found that the

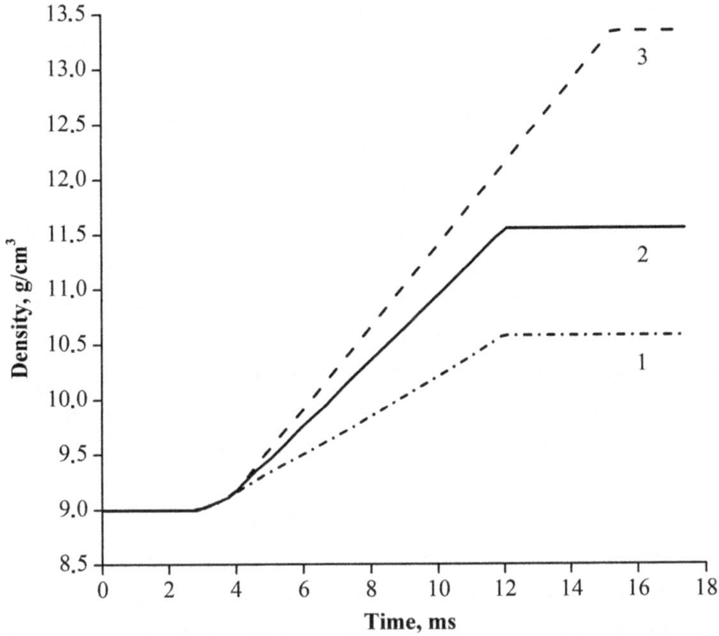

Fig. 3.9 Variation of the sample density with time during electric pulse sintering of WC–Co powders: 1 – 75 kA cm^{-2}, 2 – 85 kA cm^{-2}, and 3 – 97 kA cm^{-2} at a constant applied pressure of 150 MPa. (Reprinted from Grigoryev [6], distributed under Creative Commons Attribution 3.0 Unported License (CC BY 3.0), https://creativecommons.org/licenses/by/3.0/)

resistance of the powder volume is inversely proportional to $P^{2/3}$; here P is the axial pressure applied to the powder sample [26]. The yield strength of the material decreases with temperature; therefore, the lower is the temperature of the material, the higher is its yield strength, and, consequently, the lower is its densification rate. At a constant initial resistance of the powder compact ensured by the same applied pressure, the temperature of the inter-particle contacts increases with increasing current density and the powder consolidation rate increases, as shown in Fig. 3.9 for WC–Co compacts [6]. At a constant pulsed current, the densification rate is determined by the temperature dependence of the yield strength of the material. Grigoryev [4, 6] proposed using a dimensionless parameter $\beta = \sigma(T)/P$, where $\sigma(T)$ is the yield strength of the material, to analyze the change in the densification rate with applied pressure. Indeed, the pressure determines the initial resistance of the powder column and the amount of heat released in the powder material as well as its temperature. With an increase in the pressure, the resistance of the powder column decreases, and the heating of the material is reduced. Normally, a higher external pressure leads to a higher consolidation rate. However, a direct pressure-associated increase in the consolidation rate is leveled by the temperature dependence of the yield strength of the powder material in the case of consolidation assisted by a current pulse, as the sample's temperature does not remain constant (Fig. 3.10).

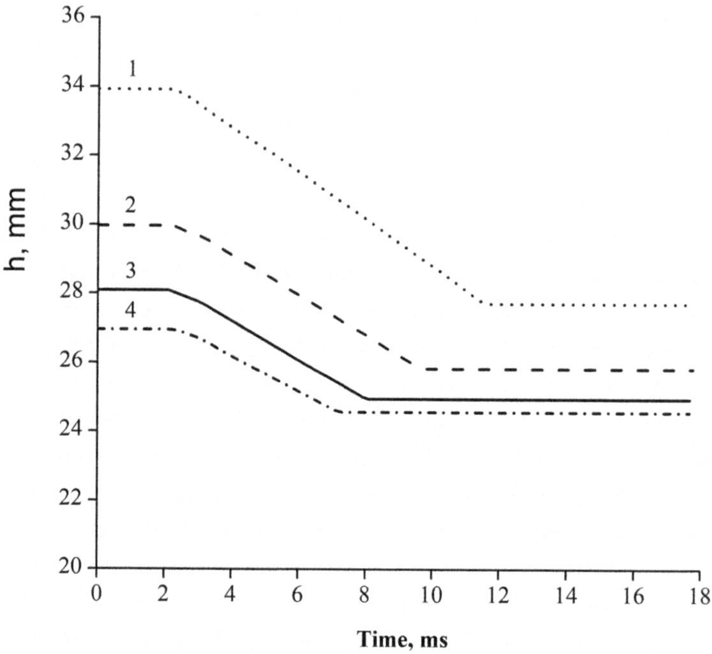

Fig. 3.10 Variation of the sample height with time during electric pulse sintering of an iron powder for a fixed current density (156 kA cm^{-2}) and different pressures: (1) 106 MPa, (2) 176 MPA, (3) 247 MPa, (4) 282 MPa. (Reprinted from Grigor'ev [4], Copyright (2008) with permission of Springer)

The pressure distribution along the vertical axis of the sample is shown in Fig. 3.11 [1]. Such a character of the pressure distribution is caused by the friction between the powder and the walls of the die. The difference in the applied pressure causes differences in the resistance, and, consequently, heat evolved in the compact. For parts of significant lengths, a multistep process has been suggested, in which new "portions" of a powder are "added" to the rod by means of sintering.

Belyavin et al. [1] consider the behavior of molten columns (bridges) that form between the particles during electric pulse sintering. Under the combined action of the applied mechanical pressure, electromagnetic forces, and surface tension, the bridges can change their shape, lose their stability, and move away from the contact area. The dependences of the size of the inter-particle contacts and porosity and strength of titanium compacts are shown in Fig. 3.12. When a certain pressure is reached, the molten bridges lose their stability, and some molten metal flows away from the compact. The capillary and electromagnetic forces are responsible for the stability of molten bridges, while the internal hydrostatic pressure of the liquid column can induce squeezing of the melt from the inter-particle contact. When this happens, the contacts are reduced in size and the mechanical strength of the sintered compact decreases. Figure 3.12b shows that there is an optimal pressure that should

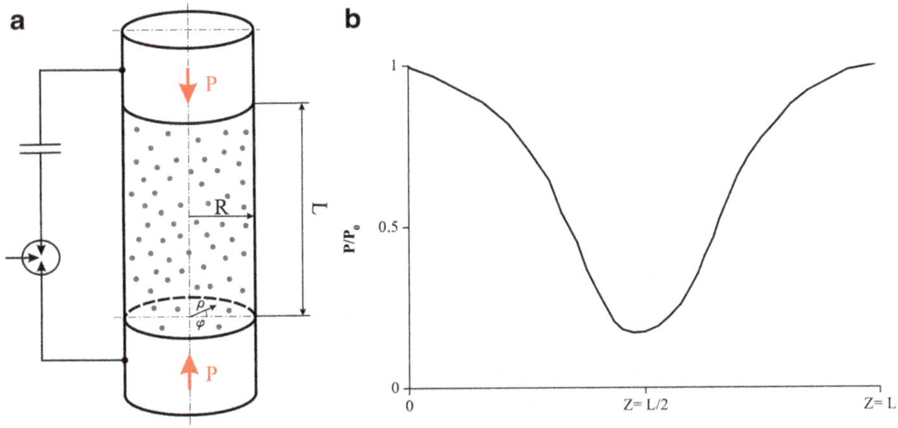

Fig. 3.11 Pressure distribution along the sample length [1]

Fig. 3.12 Relative size of the inter-particle contacts (**a**) and porosity (1) and strength (2) of the compacts (**b**) vs. applied mechanical pressure for titanium compacts. (Drawn using data of Ref. [1])

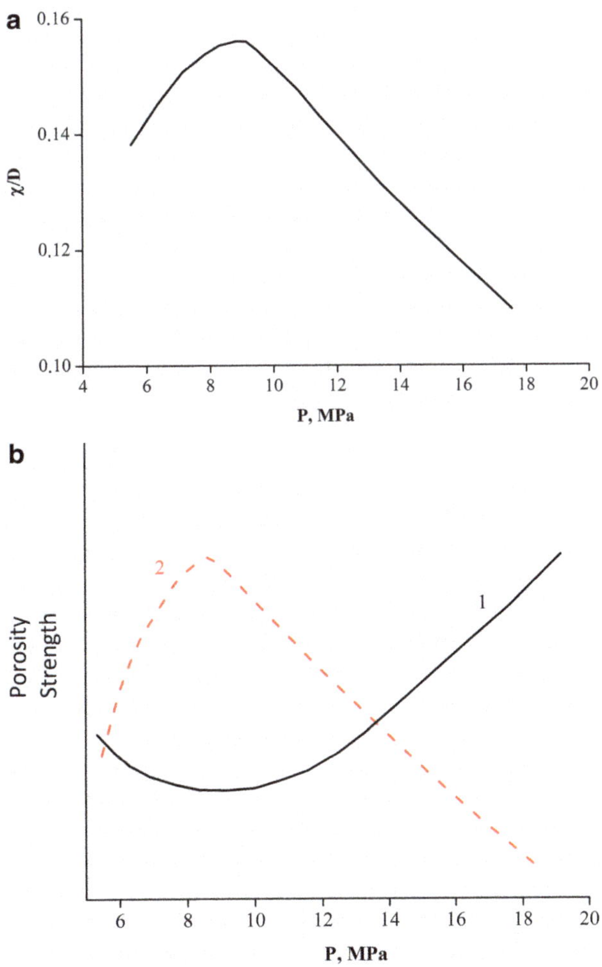

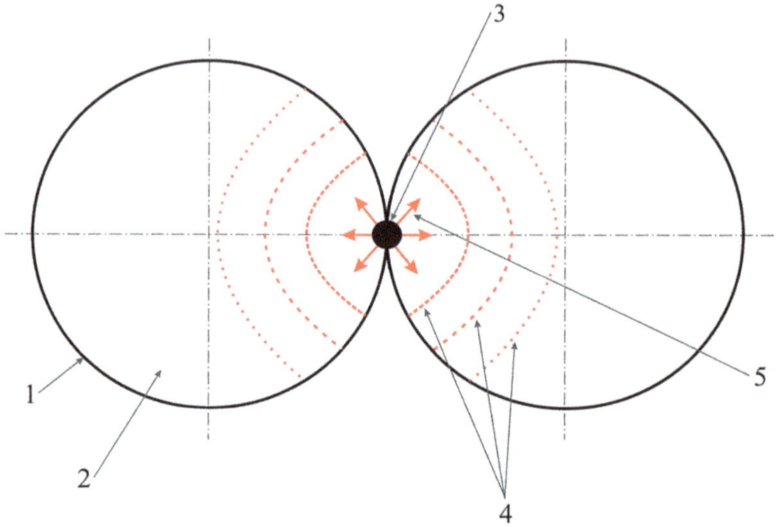

Fig. 3.13 The heat exchange scheme between the heated layers of the contacting powder particles (1, oxide film; 2, powder particle; 3, contact zone; 4, isotherms; 5, heat dissipation direction). (Reprinted from Belyavin et al. [10], Copyright (2004) with permission of Springer)

be applied to maximize the compact strength and minimize the porosity. If no pressure is applied, the electric pulse-sintered compacts are only 40–70% dense.

An important sintering factor is the temperature of the sample. The neck size between the particles was shown to increase with the heat generated in the sample during the pulse [36]. Belyavin et al. have found a relationship between the discharge voltage and the temperature of the powder particles [10]. The following assumptions have been made:

- The heat in the contact zones of all powder particles evolves uniformly; the heat exchange processes occur independently of each other.
- Due to short discharge durations, there is no heat exchange with the surroundings.
- The particles are heated from the heat sources, which are the contact zones.
- The contact zone is a semisphere.
- The dissipation of heat inside the particle occurs according to the scheme of Fig. 3.13; the heat is distributed between the particles equally.
- Heat exchange between the inner layers of the particles is governed by Fourier's law.

Figure 3.13 shows the scheme of the heat exchange between the heated layers of the powder particles. The temperature in the contact zone is determined by [10]:

$$T_q = \frac{3q - 2.25b\gamma\pi d^3}{4(c_1 + c_2)\gamma\pi d^3}, \tag{3.8}$$

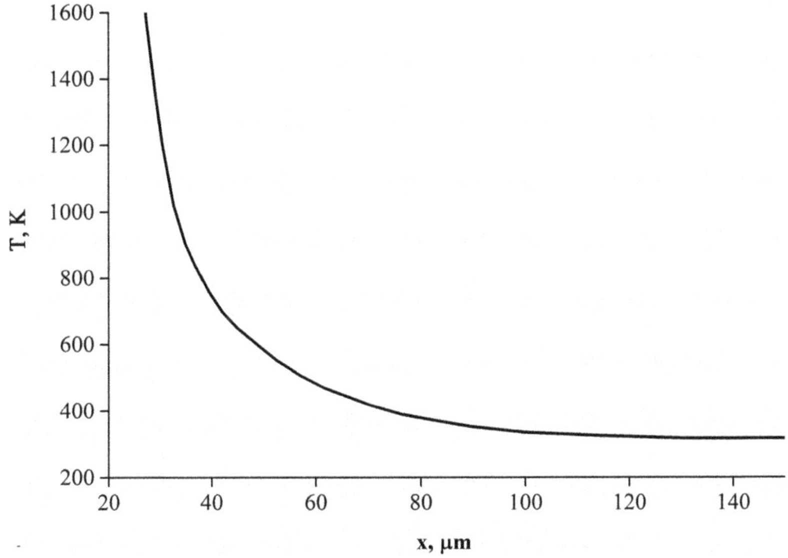

Fig. 3.14 The temperature inside a powder particle vs. the distance from its surface. (Reprinted from Belyavin et al. [10], Copyright (2004) with permission of Springer)

where T_q is the temperature in the contact zone, b is the specific melting heat, c_1 and c_2 are the heat capacities of the solid and liquid state of the material, d is the diameter of the contact zone, γ is the density of the particle, and q is the specific discharge energy. The characteristic temperature variation inside a powder particle with distance from the particle surface is shown in Fig. 3.14. An advantage of electric pulse sintering is the evolution of heat in the contact zone between the particles, which leaves the central regions of the particles relatively cold.

The temperature at the inter-particle contacts depends on the angle between the contact zone and the direction of electric current. In the contact zones normal to the direction of electric current, the temperature is the highest, while in the contact zones parallel to the direction of electric current, the temperature may not increase at all.

The temperature of the contact zone between the particles can be calculated by incorporating current distribution into the random packing of particles [37]. The following assumptions are made: the particles are incompressible, the Joule heat is not dissipated by radiation or conduction, and the particle surface becomes clean of the surface oxide before the peak current is reached (the surface cleaning time is much shorter than the discharge time). The temperature of each contact zone can be calculated using the following equation [38]:

$$\Delta T_i = \int_0^{t_\mathrm{p}} [j \cos(\theta_i)]^2 \rho / (dc) dt \tag{3.9}$$

where ΔT_i is the temperature in i contact zone, t_p is the pulse duration, j is the microscopic current density, θ_i is the solid angle between the normal to the contact

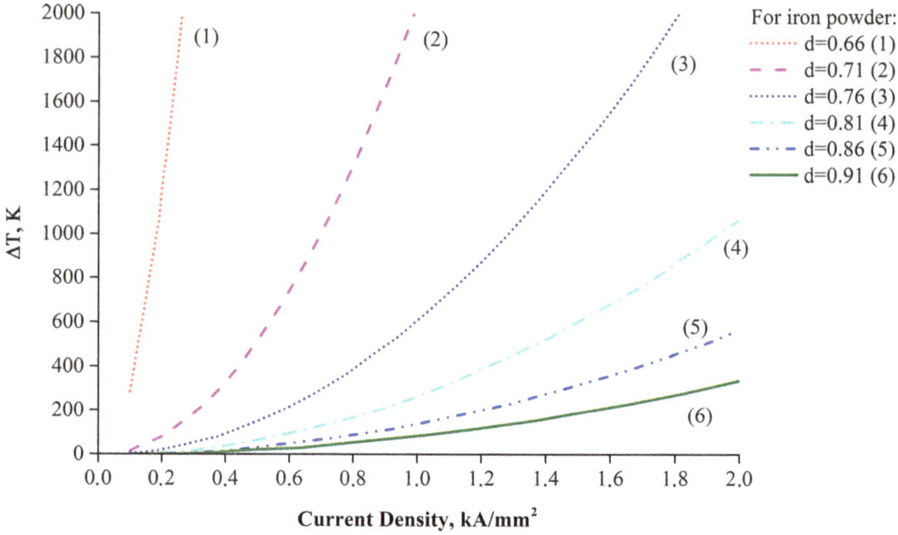

Fig. 3.15 Temperatures at the inter-particle contacts calculated by Wu and Guo [25] for iron compacts of different relative density. (Reprinted from Wu and Guo [25], Copyright (2007) with permission of Springer)

surface and the direction of the current, ρ is the resistivity, d is the density of the bulk material, and c is the heat capacity.

At the contact area perpendicular to the current direction, the temperature is the highest. Wu and Guo [25] calculated the average temperatures of the inter-particle contacts using the following equation:

$$\Delta T = \overline{\Delta T_i} = \sum\nolimits_1^{Z/2} (\cos\,\theta_i)^2 \int_0^{t_p} j^2\rho/(dc)dt, \qquad (3.10)$$

where Z is the average coordination number in the pressing limit, which, as was shown by Artz [37], depends on the relative density D:

$$\begin{aligned} Z(D) &= Z_0 + 9.5(D - D_0) \text{ for } D < 0.85, \\ Z(D) &= Z_0 + 2 + 9.5(D - 0.85) + 881(D - 0.85)^3 \text{ for } D > 0.85, \end{aligned} \qquad (3.11)$$

where $Z_0 = 7.3$ is the initial coordination number. The calculation results for iron compacts of different relative density (66–91%) are presented in Fig. 3.15. The temperature is sensitive to the relative density and increases with increasing current density dramatically at low relative densities. Calculations of the average temperatures at the contact regions allow determining whether a substantial role in densification will be played by liquid-phase sintering mechanisms. The role of liquid phases in the densification of tungsten heavy alloys by electric discharge is demonstrated in Fig. 3.16, from which it can be concluded that as the content of tungsten in

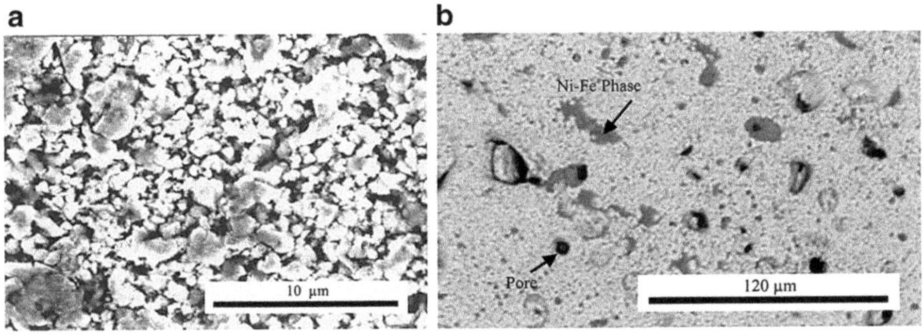

Fig. 3.16 Microstructure of W–Ni–Fe alloys consolidated under the same current density (2.5 kA mm^{-2}): (**a**) 98W–Ni–Fe, (**b**) 90W–Ni–Fe. (Reprinted from Wu and Guo [25], Copyright (2007) with permission of Springer)

the alloys decreases, the densification becomes more efficient and the evidence of the liquid-phase sintering is seen as areas in the microstructure corresponding to the Ni–Fe phase.

The nature of thermal processes in the inter-particle contacts has a significant influence on the temperature distribution throughout the volume of the consolidated material. Experimental results show that generation of an insufficient local Joule heat in the inter-particle contacts leads to weak intergranular bonding, a low final density, and a low mechanical strength of the consolidated specimen. On the other hand, during HVEDC, there is an upper critical level for the local Joule heating of the inter-particle contacts, beyond which the powder material disintegrates like an exploding wire [33]. Therefore, the optimum electric pulse current amplitude and pulse time length are necessary to generate a sufficient amount of heat for the stable formation of strong inter-particle bonds.

Grigoryev and Olevsky [5] developed a mathematical model of the physical processes occurring under HVEDC, taking into account the processes in the particle contact zones, which is described below. The system of equations describing the processes under HVEDC was based on the mass, momentum, and energy conservation laws and on the electrodynamic equations for the consolidated powder materials. Simulation of the HVEDC confirmed two different time scales for the thermal processes occurring during the processing: the first stage – the stage of the energy input into the powder, and the second (final) stage (the stage of cooling) of the consolidated material. The numerical results indicated the possibility of the localization of heat in the inter-particle contacts for certain parameter values of the high-voltage pulse electric current. The simulation of the thermal processes in the inter-particle contacts has identified an upper critical level for the high-voltage pulse current amplitude, beyond which the inter-particle contacts in the powder material disintegrate via an electrothermal explosion:

$$j_o = \sqrt{\frac{2\xi\sigma}{\rho h}} T_b, \tag{3.12}$$

where $\xi \leq 1$, σ is the Stefan–Boltzmann constant, T_b is the boiling point of the material, ρ is the electrical resistivity of the contact spot, and h is the thickness of the contact area. Equation (3.12) was obtained from the analysis of the heat balance at the initial period of time when the electric current density rapidly increases during the pulse (at its leading edge). The heat balance assumes the equivalence of the Joule heat generation rate and the heat dissipation by heat transfer through radiation.

Belyavin et al. [1] calculated the depth of the molten region at the inter-particle contacts and cooling rates of the molten material after the completion of the pulse. The depth increases with increasing energy of the discharge and decreases with increasing pulse duration (Fig. 3.17). Figure 3.18 shows the cooling rate variation with time for different distances from the particle surface for a titanium powder. It can be seen that immediately after the completion of the process (the pulse duration was 50 μs), the cooling rates as high as 10^8 K s^{-1} are achieved and remain at a level of 10^6 K s^{-1} for about 10 μs. Such cooling rates favor the formation of a fine-grained structure or amorphous zones at the inter-particle contacts.

High energy density in the contact zones can bring about changes of the state of the material (solid into liquid and partially into a dense low-temperature plasma). Vityaz' et al. [39] studied the distribution of the alloying elements across the contact formed between the spherical particles of a Ti alloy, in which α and β phases were present, stabilized by Al and Mo alloying elements, respectively. It was found that during conventional vacuum sintering, a uniform distribution of the alloying elements was established, while in electric pulse sintered compacts, concentration "jumps" were observed when moving from one particle to another indicating a far-from-equilibrium state of the material in the neck region formed in the conditions of highly localized and rapid heat release.

The microstructural evidence of higher temperatures of the particle outer layers relative to the interior and the occurrence of melting/rapid solidification processes was obtained by Cho et al. [40], who used a metallic glass powder, which crystallized with increasing temperature. A spherical metallic glass powder of the $Cu_{54}Zr_{22}Ti_{22}Ni_6$ composition with particles ranging from 5 to 45 μm was sintered by a single capacitor discharge, and the microstructure of the porous compact was studied. While morphological changes of the particle surfaces are usually accepted as the evidence of melting and solidification in the systems of crystalline metals, a grainy structure was to indicate the thermally induced crystallization processes in the sintered metallic glass powder. The crystallized outer layer of the particles 30–45 μm in diameter observed along with an unchanged amorphous core indicated higher temperatures of the particle surfaces. Smaller particles crystallized completely. When the energy of the pulse was sufficient to fully crystallize the material, there was still a difference in the microstructure of the outer regions of the particles and that of the core, the grains being smaller in the former due to faster cooling.

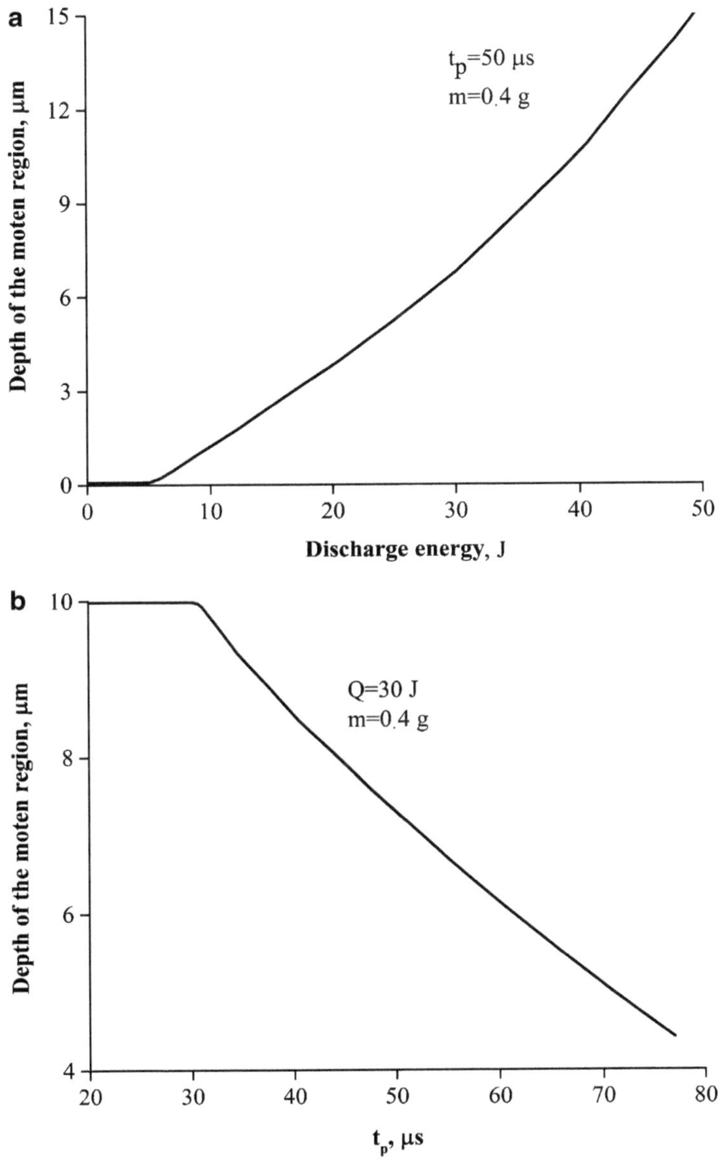

Fig. 3.17 Depth of the molten region at the inter-particle contacts vs. discharge energy (**a**) and duration of the pulse (**b**) for a titanium powder. (Drawn using data of Ref. [1])

The role of diffusion during electric pulse sintering is not fully understood. Fais argues that atomic diffusion plays the major role, but the interaction of thermal, electromagnetic, and mechanical fields makes it challenging to analyze the processes [21]. However, Wu and Guo point out that diffusion and viscous flow are negligible during electric pulse sintering owing to extremely short sintering durations [25].

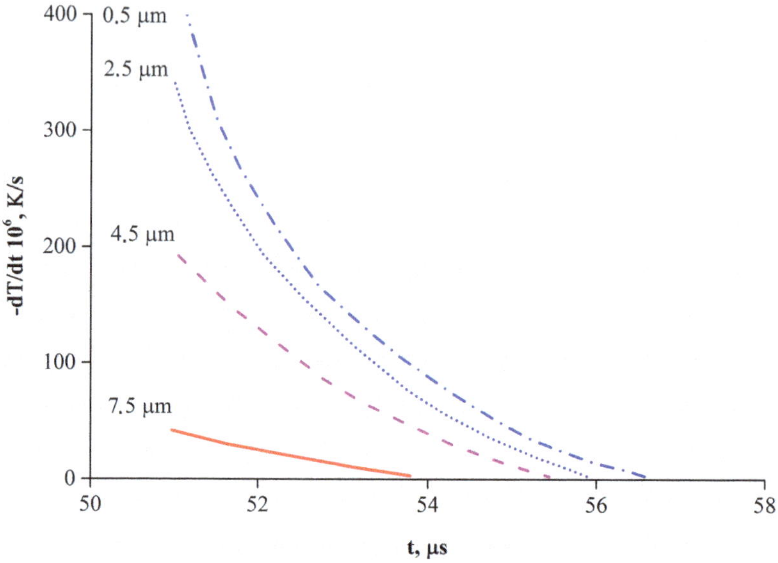

Fig. 3.18 Calculated cooling rates vs. time for different distances from the particle surface for a titanium powder. (Drawn using data of Ref. [1])

The possibility of improving the quality of the sintered compacts by applying several high-voltage pulses has been a matter of discussion. Belyavin found that a porous compact forms during the first discharge [10]. Zavodov et al. studied the effect of the voltage application mode and of the number of pulses on the quality of the sintered samples [29]. The voltage was increased with each subsequent pulse. It was found that when the voltage was gradually increased, the final resistance of the materials was 2–3 times lower than in the case of a single pulse. The neck diameter was 30–50% of the particle diameter. The strength of the material sintered using this step-by-step voltage increase approach was higher than the strength of the material sintered by a single pulse.

The efficiency of electric pulse sintering is the ratio of the thermal energy released in the powder compact during the discharge to the electrical energy stored by the capacitor bank. In order to increase the efficiency of electric pulse sintering facilities that can be described by a simple oscillatory RLC circuit, Belyavin et al. [10] derived a relationship, which, if observed, guarantees that during the first period of oscillations, the major fraction of the stored energy of the capacitor (95%) is released as heat:

$$\sqrt{\frac{L}{C}} \leq R < \sqrt{\frac{2L}{C}} \tag{3.13}$$

where R is the resistance, L is the circuit inductor, and C is the capacitance of the capacitor bank.

Applying modeling methods, the circuit parameters can be found such that the efficiency calculated using Eq. (3.13) would be the highest:

$$\eta = (1 - \delta)\frac{R_1}{\sqrt{R_1^2 + \left(\frac{1}{\omega C} - \omega L\right)^2}}, \tag{3.14}$$

where R_1 is the resistance of the powder column, ω is the frequency of the damped oscillations, η is the efficiency, and $\delta = 0.05$ is the damping factor. In order to achieve a reasonable efficiency of electric pulse sintering, a RLC circuit should possess a certain resistance.

Assuming that the particles have spherical shape, the diameter of the contact area between the particles is much smaller than the particle diameter and the particles deform elastically under a compressive force applied to the punches, Belyavin et al. analytically derived the following equation to calculate the initial resistance of the powder compact pre-pressed prior to sintering [10]:

$$R = \rho\frac{4D_0}{\pi(1 - 2\varepsilon)^2 n_V^{1/3}}\left(\frac{E}{3FD_0(1 - \nu^2)(1 - \sigma)}\right)^{2/3}\frac{h}{S}, \tag{3.15}$$

where ρ is the resistivity of the material of the powder particles, D_0 is the particle diameter, ε is ratio of the thickness of the oxide film to the linear deformation, n_V is the volume concentration of inter-particle contacts in the compact, F is the compressive force of the electrode, E is the Poisson ratio, σ is the friction coefficient, E is the Young's modulus, h is the height, and S is the cross-sectional area of the powder compact.

The discharge efficiency is proportional to the ratio of the powder surface resistance to the sum of the circuit resistance and that of the powder surface. Rock et al. [41] suggested mixing Nb–Al mechanically milled powders produced by low- and high-energy mechanical milling to achieve a certain resistance of the compact to be sintered by electric discharge sintering. The XRD patterns of the compacts of different initial resistances are shown in Fig. 3.19. Slightly narrower peaks in the XRD patterns of higher initial resistance compacts indicate a small degree of grain growth caused by heating during the discharge. However, it is normally accepted that the duration of an electric discharge is sufficient for densification of the powders but too short for any noticeable grain growth during electric pulse sintering [1, 23].

The size of the powder particles is also important for electric pulse sintering. Zavodov et al. [29] have observed fine particles to be thrown out of the die due to intensive gas evolution during a high-voltage electric pulse [29]. The electrical conductivity of a fine powder is lower than that of a coarse powder. However, the coarse powders are consolidated into compacts of higher densities, which is related to the discharge energy dissipating between a smaller number of contacts so that each contact receives more energy at a constant discharge energy [42].

Alitavoli and Darvizeh [27] discussed the existence of the optimal thickness of the oxide film on metallic powders required to produce sufficient heat for the neck

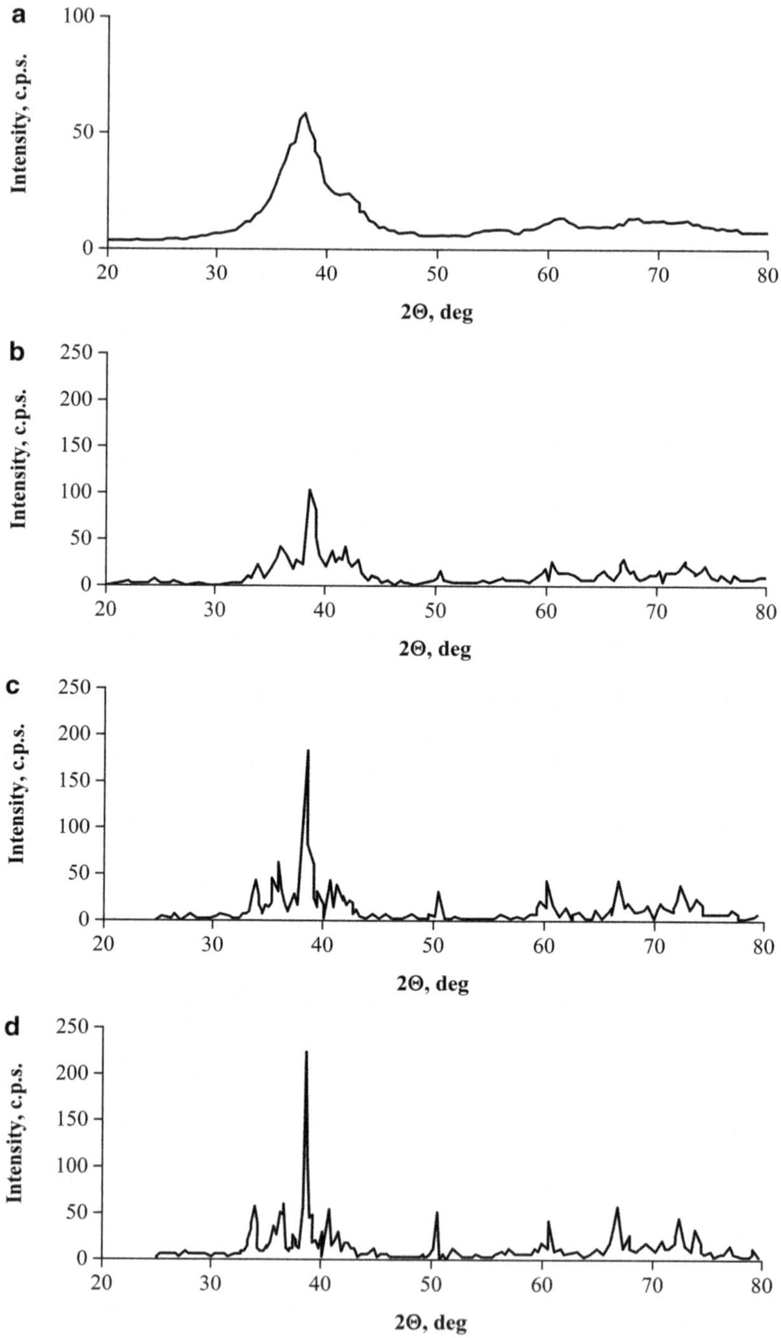

Fig. 3.19 XRD patterns of the Nb–Al powder and Nb-Al compacts of different initial resistances consolidated using the same input energy of 0.5 kJ g^{-1}: (**a**) mechanically milled mixture of Nb and Al, (**b**) 80 mΩ, (**c**) 100 mΩ, (**d**) 130 mΩ. (Reprinted from Rock et al. [41], Copyright (1998) with permission of Springer)

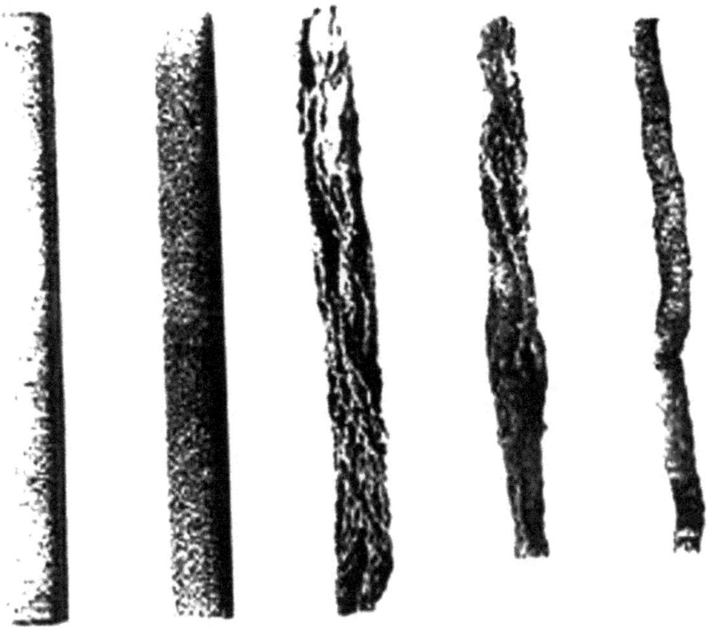

(a) (b) (c) (d) (e)

Fig. 3.20 The effect of preliminary oxidation of the steel powder on the structure of the compact: non-oxidized powder (**a**), oxidized at 500 °C for 15 min (**b**), 30 min (**c**), 40 min (**d**), and 60 min (**e**). (Reprinted from Alitavoli and Darvizeh [27], Copyright (2008) with permission from Elsevier)

formation during electric pulse consolidation. The compacts produced by hydrogen-annealed steel powders showed lower bending strength compared with those produced from the as-received powders. When the oxide film was completely removed, insufficient amounts of heat were generated, which led to weak inter-particle bonding and low mechanical strength. On the other hand, when the thickness of the oxide layer increases and the critical value is reached, the current distribution in the powder volume becomes very nonuniform such that uniform densification cannot be achieved even at the expense of an increased voltage or capacitance of the capacitor bank. Figure 3.20 shows the effect of preliminary oxidation of steel powders on the structure of the compact [27].

Figure 3.21 shows the contact areas of spherical Mo powder particles formed by the action of a constant compressive pressure of 80 MPa and of heat sources of various intensities [5]. The action of the heat sources of lower power locally enhances the plasticity of the material and causes the subsequent intense deformation of the contact region (Fig. 3.21a). A more powerful heat source leads to the local melting of inter-particle contacts and to a high rate of crystallization refining the grain structure of the material of the contact region (Fig. 3.21b).

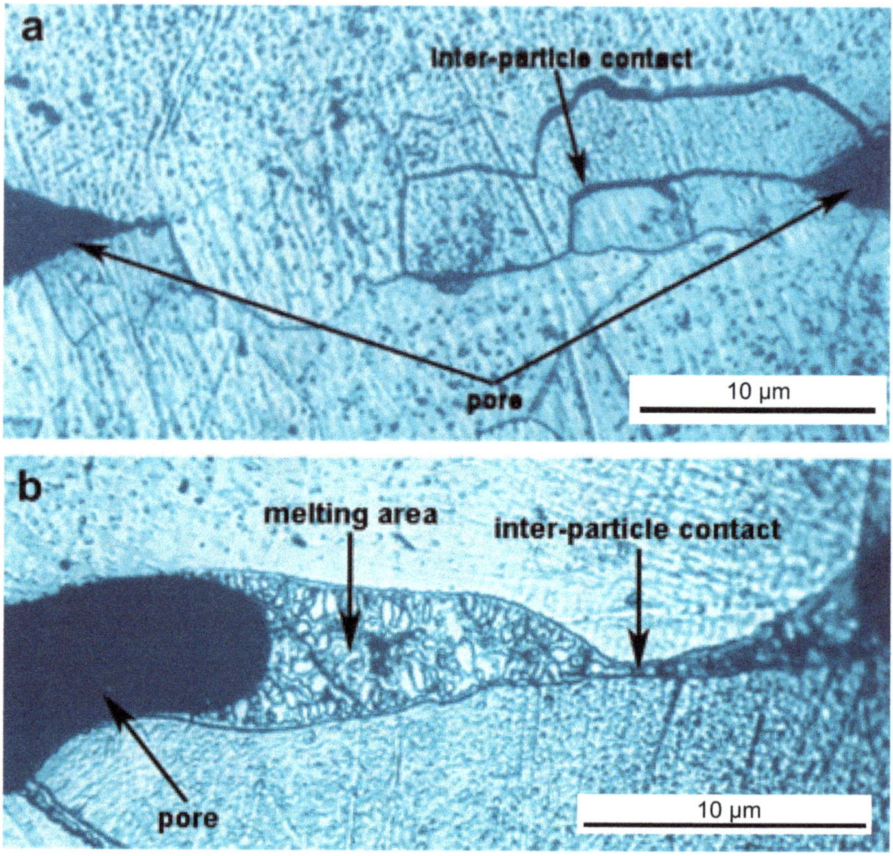

Fig. 3.21 Plastic deformation (**a**) and melting (**b**) of the contact region between Mo particles. (Reprinted from Grigoryev and Olevsky [5], Copyright (2012) with permission from Elsevier)

Electric pulse sintering is promising for making not only dense compacts but also porous materials and parts, such as filters, getters, acoustic and radio absorbers, and implants. Thus, when energy exceeding 1 kJ is evolved in the backfill of a titanium powder, a dense core forms in the center, while the adjacent material remains porous [43]. The core diameter depends on the energy at a constant charge of the capacitor bank. This effect is observed due to the action of magnetic field during the discharge (the pinch effect), which contracts the sample in the radial direction and reduces the sample diameter [44]. The porosity of the sample depends on the sintering parameters and can be varied according to the aim of the experiment. Belyavin suggests that manufacturing of porous refractory metals by electric pulse sintering will allow reducing energy consumption, make the technology more simple, and ensure the desired structural properties (porosity, pore size, specific surface area), hydrodynamic properties (permeability coefficient, local permeability), and physical and mechanical properties (strength, electrical conductivity) [45].

Investigations comparing the sintering results achieved by different methods using one and the same powder material are quite rare. Due to well-known difficulties of reliably measuring the sintering parameters, it is nearly impossible to make a direct comparison of the sintering conditions and conclude on the dominant role of a certain experimental parameter. However, by sintering the same powder by different techniques, it is still possible to conduct comparative studies. Scardi et al. [22] and Fais et al. [23] have made an attempt to find certain characteristics of the microstructure development and the crystalline structure evolution typical to CDS and spark plasma sintering (SPS) producing dense samples from the same powder. They found that due to a short sintering time in CDS, the dislocations present in the mechanically milled powder remain quite uniformly distributed in the crystalline lattice of the sintered material, while in the SPS-produced samples, the dislocations tend to interact, form pile-ups, and reach lower-energy configurations. The coherently scattering domains are smaller in the CDS samples than in the SPS ones of the same composition and having close values of density [23]. In a Fe-based alloy (Fe–1.5 wt.% Mo) sintered by CDS, austenite was found as a minor phase, while it was not present in the SPS samples of the same composition, the latter consisting of the ferritic phase. The authors conclude that in terms of preserving the nanocrystalline state of metals, the CDS method appears to be more successful.

3.4 High-Voltage Electric Discharge Consolidation (HVEDC) Apparatus

A single electric pulse passing through the powder sample of high resistance carries energy that is enough to sinter many materials. Therefore, electric pulse sintering can have its applications and advantages over the other electric current-assisted sintering techniques, such as spark plasma sintering (SPS) [1]. However, the best use of the advantages of the high-voltage electric pulse sintering can be achieved only through the optimization of the consolidation parameters. A variety of apparatus have been developed to conduct electric pulse sintering. Many facilities are unique such that their only users are their developers. Commercial manufacturing of the facilities is also rapidly developing making them available for research purposes.

The high-voltage electric discharge consolidation (HVEDC) [4, 9] setup developed in Moscow Engineering Physics Institute (MEPhI) is shown in Fig. 3.22. The setup includes an energy storage unit, a switchboard, a pressing unit, and measuring systems of electric pulse parameters, sample temperature, and densification kinetics. The energy storage unit contains a capacitor bank with a stored energy of 75 kJ, which during the discharge ensures a powerful energy release in the powder, and a charging unit. The capacitor bank contains 30 capacitors each with a capacitance of 200 μF; voltages as high as 6 kV can be achieved. The switch board in the setup is a vacuum discharger (trigatron switch), which is used to complete the circuit. The trigatron switch allows current pulses of up to 10^6 A. The setup uses pulsed current

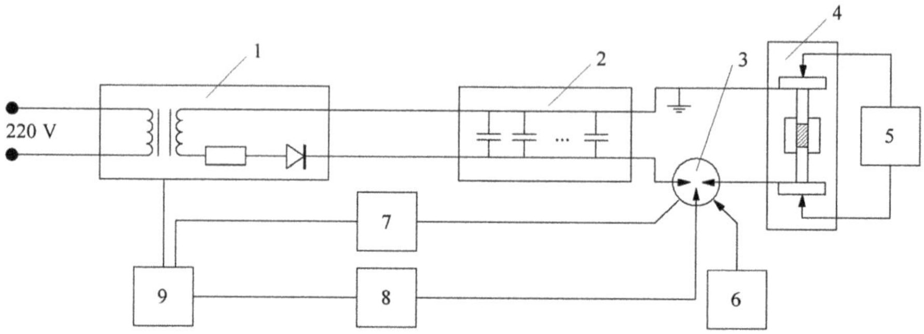

Fig. 3.22 HVEDC setup developed in Moscow Engineering Physics Institute (MEPhI): (1) charging unit, (2) capacitor bank, (3) trigatron, (4) press form, (5) pressure system, (6) current pulse recorder, (7) forevacuum pump, (8) ignition unit, (9) control unit [4, 9]

generated with the capacitor bank to rapidly heat the powder sample, to which external pressure is simultaneously applied during the sintering process. The main parameters of HVEDC are the external pressure and the electric current generated by the discharge. The powder is poured into a nonconducting ceramic die and pressed by two molybdenum punches also serving as electrodes. The powder volume ("column") is a cylinder ~10 mm in diameter and 10–15 mm high. A modification of this configuration exists featuring a ceramic tube, in which the powder is placed. In order to prevent it from damage, the tube is then placed into a metallic die. The punches, which are at the same time the electrodes, transfer pressure to the powder with the help of a pneumatic system and serve as current leads bringing electric pulses to the powder sample. The applied pressure reaching 400 MPa is created by an air-operated press or an air cylinder developing a force of 50 kN. The high-voltage capacitor bank is discharged through the powder.

The measuring system records the current amplitude and the pulse waveform in the discharge circuit of the setup. The discharge current is measured using a Rogowski coil, which is placed around the powder column. The pulse duration for this setup does not exceed 600 μs. This value determines the time, during which the energy is transferred to the powder. The temperature on the powder surface is measured by thermocouples. The weighed powder is placed inside the tooling consisting of a ceramic die and molybdenum punches, which is then set in the pressing equipment. The pressure generated by the air cylinder is controlled by a pressure gauge. Using this pressure, the powder is pre-compacted. Then the capacitor bank is charged up to a selected voltage. The trigatron receives a triggering pulse, the electric circuit closes, and the capacitor bank discharges through the pre-pressed powder material. The pressure is then released allowing the consolidated part to be taken out of the pressing equipment. The HVEDC process does not use any protective atmosphere or vacuum. Rogowski coils are usually used to measure the current, and oscillographs are used to record the current waveform. Typical temperature variations of the side surface of the powder column and the outer surface of the die during HVEDC at a constant pressure are shown in Fig. 3.23.

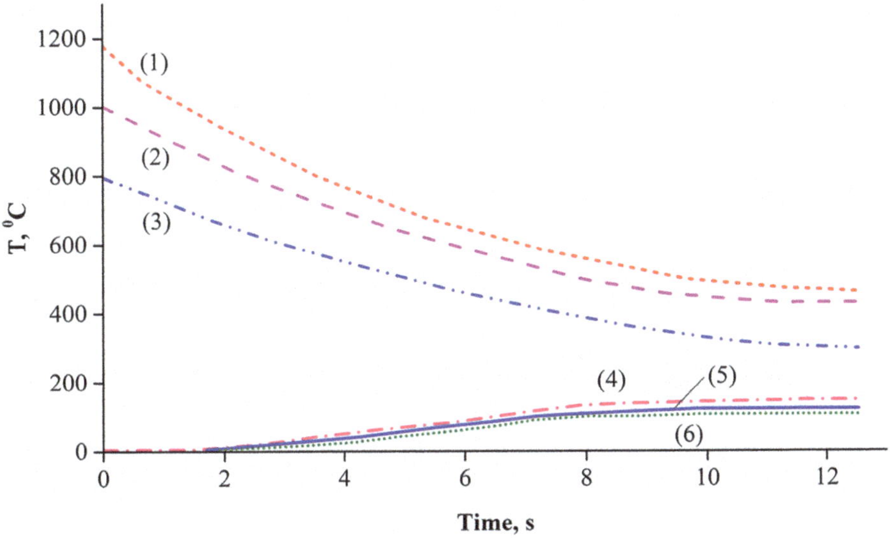

Fig. 3.23 Typical temperature variations of the side surface of the powder column (1–3) and the outer surface of the die – insulating tube (4–6) during HVEDC at a constant pressure (thermocouple method). Curves (1) and (4) were obtained at an amplitude of current density of 234 kA cm^{-2}, (2) and (5) at 195 kA cm^{-2}, (3) and (6) at 156 kA cm^{-2}. (Drawn using data of Ref. [9])

A different variation of the electric pulse sintering setup has been developed by Anisimov and Mali [35]. The powder material is pressed between two punches in a die made of fabric-based laminate. A force of 20 kN is applied by a screw press. Sintering is performed as a result of the discharge of a capacitor bank with a capacitance of 3.4 mF and a voltage of 5 kV. The discharge consists of two half-periods of current; the current amplitude in the second half-period is three times lower than that in the first one. The sample diameter can be varied from 5 to 10 mm and its thickness from 2 to 3 mm.

Electric discharge consolidation is also possible to implement in a setup, in which the direction of current is normal to the pressure application axis. Such a design was suggested by Anisimov and Mali (Fig. 3.24) [35]. The current density in the sample was sufficient to sinter compacts with dimensions of $2 \times 20 \times 30$ mm^3. In order to increase the pressure during sintering, an electromagnetic inductor was used utilizing the same current that passes through the sample or working from an additional capacitor bank. The magnetic pressure reaches 500 MPa at an inductor current of 300 kA. The inductor experiences significant damage and survives only a single experiment. The values of the current integral required for sintering measured for this geometry are close to those obtained for a conventional loading scheme.

Sintering of a powder by electric current passing directly through the sample is only possible when the powder material is electrically conductive. It is known that upon heating semiconductor materials can become conductive. An important characteristic affecting the sintering process is the sample resistance [46].

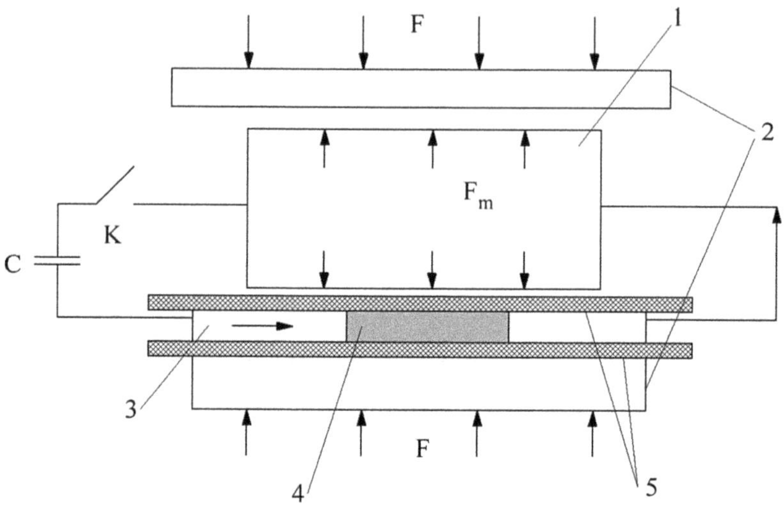

Fig. 3.24 Electric discharge sintering setup: 1, electromagnetic inductor; 2, plunger; 3, electrode; 4, powder sample; 5, insulator; C, energy storage capacitor; K, switch. (Reprinted from Anisimov and Mali [35], Copyright (2010) with permission of Springer)

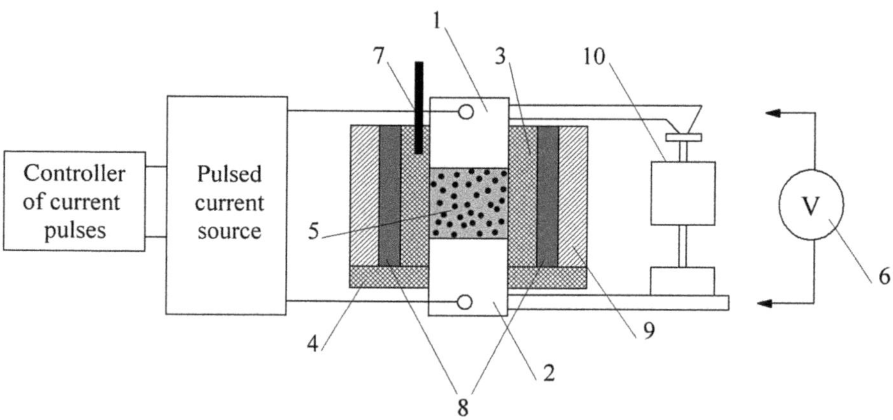

Fig. 3.25 Schematic of the electric pulse sintering setup designed to produce SiC-based ceramics: 1, 2, electrodes; 3, die; 4, insulator; 5, powder; 6, voltmeter; 7, thermocouple; 8, heating element; 9, steel binding; 10, pressure measuring unit. (Based on Ref. [47])

Bilalov et al. [47] tried to sinter silicon carbide and aluminum nitride powders by electric current using a setup shown in Fig. 3.25. The consolidation was performed in insulating dies made of sapphire; the punches served as current leads. The powders of SiC and AlN were loosely poured into the die. A high-power pulse generator was the current source. Unlike other electric pulse sintering facilities, this one includes a heating element to heat the powder prior to sintering.

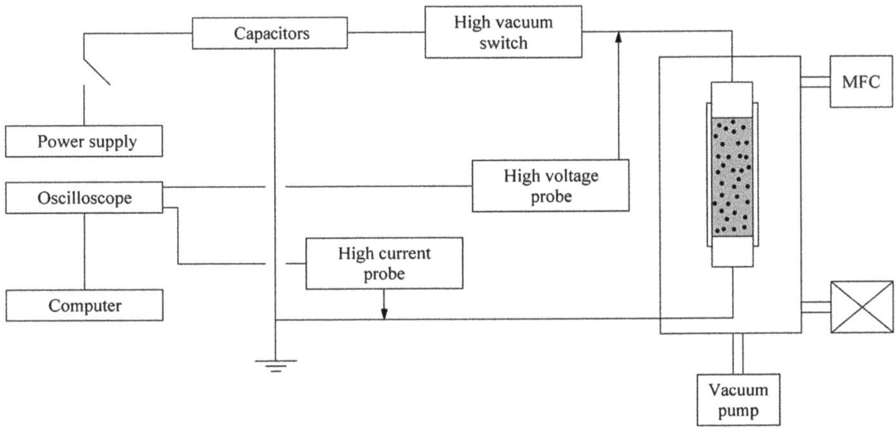

Fig. 3.26 Environmental electro discharge sintering (EEDS) setup. (Reprinted from An et al. [49], Copyright (2005) with permission from Elsevier)

Jung et al. [48] have patented a method of making porous titanium implants by electric discharge consolidation. Pyrex dies of different shapes including complex ones with cavities and arcs were used. The die is connected to copper or brass electrodes. The power supply of 100 or 200 V is used in this setup. By means of a transformer, the voltage is increased up to 1–5 kV (the preferable value is 2.5 kV). The current of a high voltage passes through the first switch and charges the capacitor. After that the capacitor is instantaneously discharged through the powder column. The consolidated samples are dense in the center and porous in the outer regions.

High chemical reactivity of titanium causes the formation of an oxide film on the surface of the Ti powder particles. To tackle this issue, a modified electric discharge consolidation setup was developed [43, 49, 50, 51]. The alterations included the addition of a vacuum system, automatic insertion of electrodes, a heat sink system, and a high-vacuum switch (Fig. 3.26). This modified consolidation method was named "environmental electro discharge sintering" (EEDS).

The vacuum system allows creating residual pressures of $2 \cdot 10^{-3}$ torr. After pumping, the discharge chamber is filled with argon up to a pressure of 1000 torr to create an inert atmosphere. The EEDS method opens up new opportunities to modify the surface of the consolidated materials. An et al. [49] used nitrogen as the gaseous environment and observed the formation of titanium nitride on the surface of the compact. The setup uses a quartz die of 4 mm diameter, tungsten cathodes, and copper heat sink elements. The upper electrode has an automatic drive. The powder is poured into the die and subjected to vibrational packing. No external pressure is applied during the sintering process.

Alp et al. have suggested a hybrid consolidation method – electro impact compaction (EIC), in which the material is subjected to a DC current prior to electric discharge consolidation [52]. This combination of the two processes resulted in

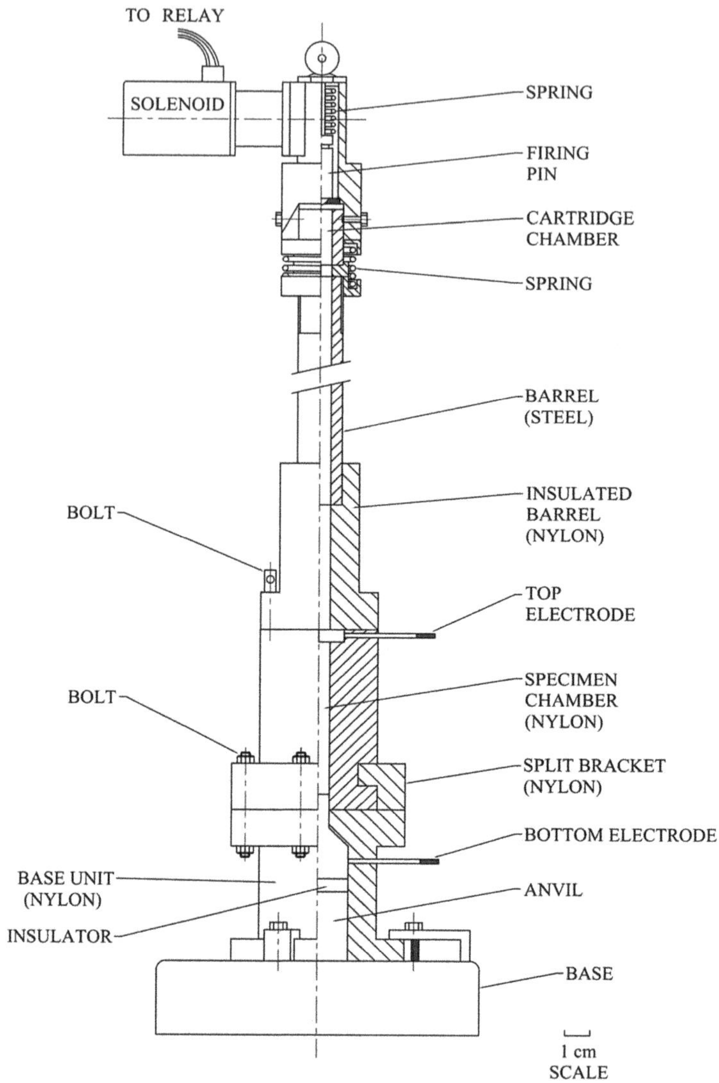

Fig. 3.27 Schematic of an electro impact compaction (EIC) apparatus. (Reprinted from Alp et al. [52], Copyright (1993) by permission of Taylor & Francis Ltd., www.tandfonline.com)

higher relative densities compared with consolidation by the electric discharge method only. An EIC apparatus shown in Fig. 3.27 consists of an assembly of a punch and a hammer, a pulse coil, and a steel casing. An insulating material separates the casing from the cylindrical high-voltage electrode. The low-voltage electrode possesses an insulating cylinder. The powder is placed into a nonconducting die. The apparatus is fixed on a steel plate. A high-voltage capacitor bank can be charged up to 20 kV and is used together with an impact device. The coil of a mechanical relay receives energy from a DC current generator. The pulse coil makes the hammer held by a spring to move. The current is measured by a Rogowski coil. The capacitor discharge induces the electromotive force in the toroid proportional to the current. The current is recorded by a digital oscilloscope. The hammer velocity is measured with the help of two optical fiber cables. The displacement of the hammer during the discharge is measured by a capacitor sensor.

3.5 High-Energy High-Rate (HEHR) Consolidation Setup

High-energy high-rate (HEHR) consolidation was developed in the USA and belongs to a group of processes, in which the electric energy evolves in the powder leading to its heating. The power levels pertaining to this method are of the order of 1 MJ s^{-1} [11]. In this facility, a homopolar generator is used, which transforms the energy of rotation into the electrical energy as a result of Faraday's effect (Fig. 3.28). The energy stored by the generator can reach 10 MJ. The homopolar generator can create current pulses of 100–500 kA, which are used to sinter the powder materials.

During the discharge, the current passing through the powder reaches maximum and then rapidly decreases to zero. An electric current waveform typical to the HEHR method is shown in Fig. 3.29. The voltage applied to the powder is <100 V. The pressure from a hydraulic press is applied for several minutes to provide time for the heat to be dissipated from the electrodes to the copper plates. The external pressure can reach 0.5 GPa. In Ref. [16], during consolidation of an aluminum alloy powder, the maximum current reached 100 kA after 200 ms. At the onset of the pulse, a pressure of 104 MPa was applied to ensure sufficient electrical conductivity of the powder. As the current pulse decayed, the pressure was stepped up to 207 MPa and held constant for another 5 min.

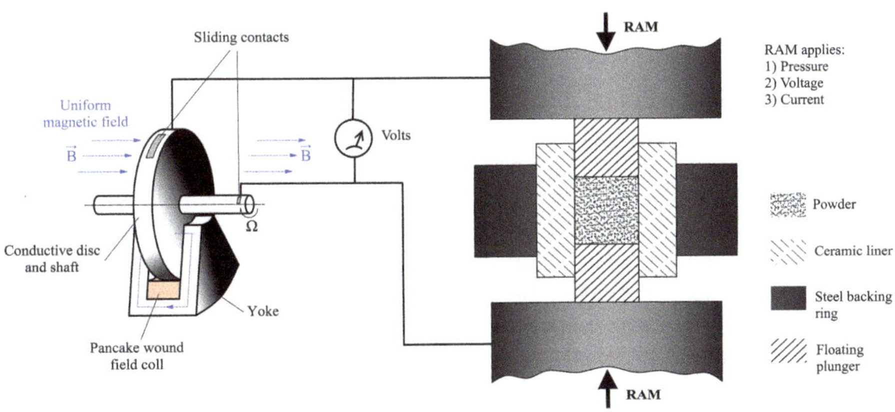

Fig. 3.28 Schematic of a HEHR setup. (Drawn using Refs. [11, 12])

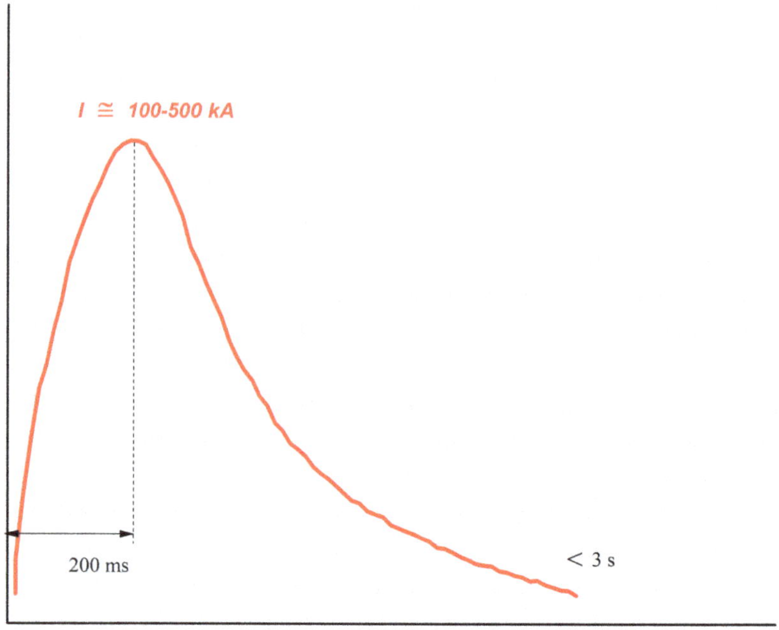

Fig. 3.29 Current waveform in HEHR. (Drawn using Ref. [11])

The die, electrodes, and contact surfaces of supporting rings are coated with boron nitride to prevent the formation of electric arcs during the process of consolidation. The die is a seamless tube made of corrosion-resistant steel. The electrodes are also made of steel. Another possibility is to work with an alumina die and protect it by a steel tube to prevent it from damage during sintering [14, 15].

3.6 Capacitor Discharge Sintering (CDS) Setup

Capacitor discharge sintering (CDS) was first used for consolidation of powders by Knoess and Schlemmer to produce dense compacts [19]. It was further studied and improved by Fais, who used it for making nearly fully dense metallic materials and metal–ceramic composites [21]. The CDS method uses two circuits coupled by mutual inductance instead of a single RLC circuit. This configuration allows applying low voltages to the powder compact thus reducing the possibilities of discharges, breakdown, and local plasma formation during the process. The CDS normally produces nearly fully dense compacts, the porosity being present only in the surface layer and uniformly distributed.

The CDS method is based on the storage of high-voltage electrical energy in a capacitor bank inserted in a freely oscillating system composed of a primary circuit and a mutually coupled secondary circuit (Fig. 3.30). The secondary circuit works in

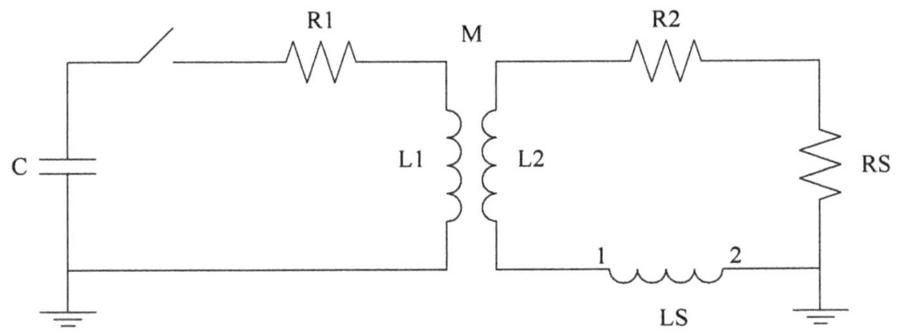

Fig. 3.30 Equivalent electric diagram of the CDS setup [21]

conjunction with a mechanical press, which is controlled by a programmed logical controller. Once the desired pressure from the press is reached, the switch closes the circuit, and the electromagnetic energy is transferred to the secondary circuit by means of a transformer that enables conversion from high voltage and low current in the primary circuit to low voltage and high current in the secondary circuit.

Knoess and Schlemmer [19] utilized a ceramic die as well as a conducting die with an insulating coating. The punches were made of molybdenum. The applied pressure was 1 GPa, while the voltage was less than 50 V. The powder was subjected to 1–2 pulses of current (100 kA) during 2–3 ms.

The main advantage of CDS is the reliability of the electric discharge system, which uses solid-state devices instead of ignitrons. The service life of the electronic components of the primary circle is more than 10^7cycles, which ensures reliable and reproducible discharges. When high energies are transferred to the powder compact, the applied voltage can be reduced by a transformer. In the primary circuit, the capacitors are charged up to voltages in the 1.5–3.5 kV range, whereas the current passing through the powder has a voltage of 5–50 V. A high voltage is transformed into a lower voltage during 2–3 ms, while the current pulse is 10–40 ms long. The current density can reach 100 MA mm^{-2}. The CDS does not normally use a protective atmosphere or vacuum. A schematic of a CDS setup and variation of the process parameters with time are presented in Fig. 3.31a.

In order to consolidate the powder, a certain amount of energy should be supplied to the material. In order for the powder to receive an optimal amount of energy during the discharge, the other elements of the circuit should possess an optimal electrical conductivity. This concerns the capacitors, transformers, cables, and electrodes. It has been proved that electrical conductivity can be improved by pre-pressing the powder compact before sintering. In order to decrease resistance, copper electrodes may be used. As pressure has to be applied during sintering, copper has to be replaced by an alloy. Fais uses electrodes made of the Cu–1% Co–0.5% Be alloy [21]. Schutte et al. [42] showed that the best results were obtained when the electrodes made of an oxide particle-strengthened alloy were used (ODSC3/11). Egan and Melody [53] suggested Cu–W protection for the electrodes.

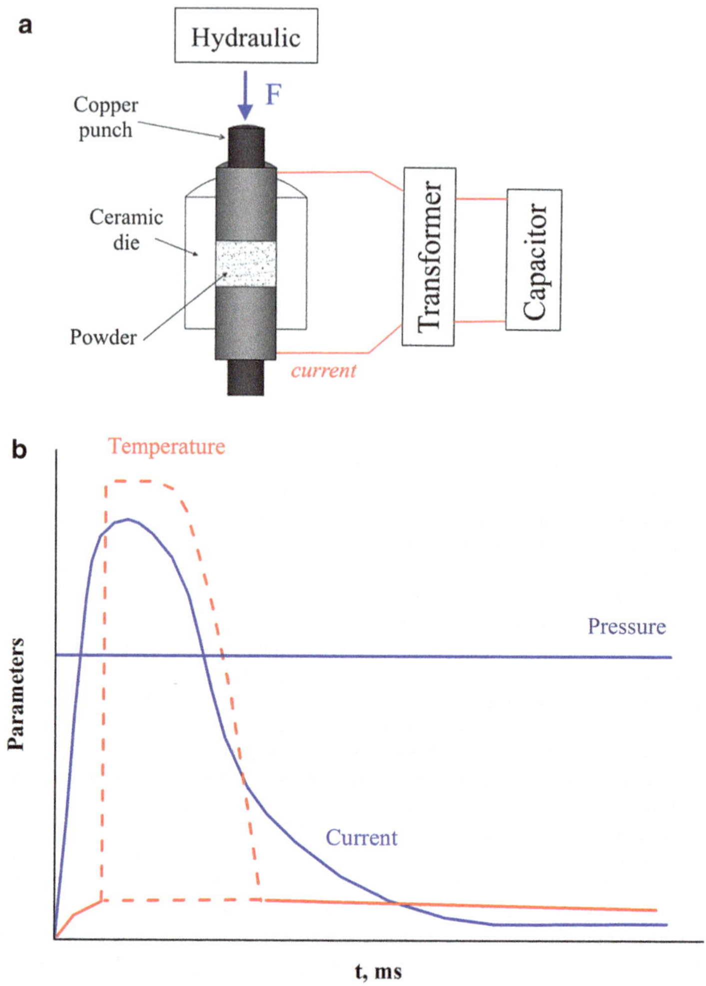

Fig. 3.31 Schematic of a CDS setup (**a**) and sintering parameters (**b**)

The temperature is measured by a pyrometer connected to a digital oscilloscope. The maximum temperature is reached in 10–30 ms, which indicates the end of the electric discharge (Fig. 3.31b). Cooling starts immediately after the discharge is finished and lasts for about 1 s. During sintering, the surface layers of particles melt at the microscopic level, while the integral temperature of the sample remains relatively low. The pressure applied to the sample is provided by a hydraulic press and reaches 1000 MPa. Schutte et al. used a hydraulic press with a possibility of increasing load stepwise (with a step of 1 kN) [42].

3.7 Pulse Plasma Sintering (PPS) Setup

Pulse plasma sintering (PPS) was developed at the Faculty of Materials Science and Engineering of Warsaw University of Technology, Poland [17, 18, 54–56]. The sintering technique used by Michalski and other researchers was initially called pulse electric discharge and was a predecessor of PPS. The consolidation process was carried out in two steps. The first step consisted of shaping the material at pressures up to 1 GPa at room temperature. At the second step, the pre-pressed sample 10 mm high and 10 mm in diameter was placed inside a setup, and a pressure of 1.2 GPa was applied [17]. Sintering was carried out in argon. The discharge of the capacitors (capacitance 200 μF) produces current pulses, which heat the sample. The pulse duration was 100 μs, while the pulse frequency was 0.5 Hz. The capacitors could be charged up to 7 kV. The punches (electrodes) carrying high current densities (of the order of $\approx$10 MA/m^2) were made of titanium and aluminum. The current during the pulse reached several kA.

PPS uses pulses of electric current passing through the sample with a simultaneous application of pressure. The capacitor used in the setup has a capacitance of 300 μF with the maximum charging voltage of 10 kV. The powder is poured into a graphite die placed in between two graphite punches. The diameter of the die is 20 mm. The height of the samples can reach 15 mm. The pulse duration is usually several hundred of microseconds, while the current amplitude can reach several kA. The current source generates periodical pulses of current. The current pulse has an oscillating character similar to the other electric pulse sintering techniques. Arc discharges occur between the particles cleaning their surfaces and accelerating diffusion. The PPS process has high thermal efficiency. As the pulse duration (500 μs) is short in comparison to the interval between the pulses (1 s), the measured temperature is lower than that of the powder particle experiencing the action of the passing current, the latter possibly reaching several thousand degrees (Fig. 3.32). As a result, the process acquires a quasi-adiabatic character [54, 55]. Prior to sintering, the sample can be heated up to a selected temperature. PPS allows sintering of powders into compacts of relative densities exceeding 99%. Sintering is conducted in vacuum (Fig. 3.33), which differs this method from other methods based on the application of pulsed current. The samples are cooled in vacuum under applied pressure.

3.8 Briquetting by Electric Pulse Sintering

Pulsed current was suggested as a means to briquette dispersed metallic materials (Fig. 3.34) [57, 58]. In this case, pulsed current solves the problem of reclamation of metallic chips and other waste products of metallurgy. Materials for filters, getters,

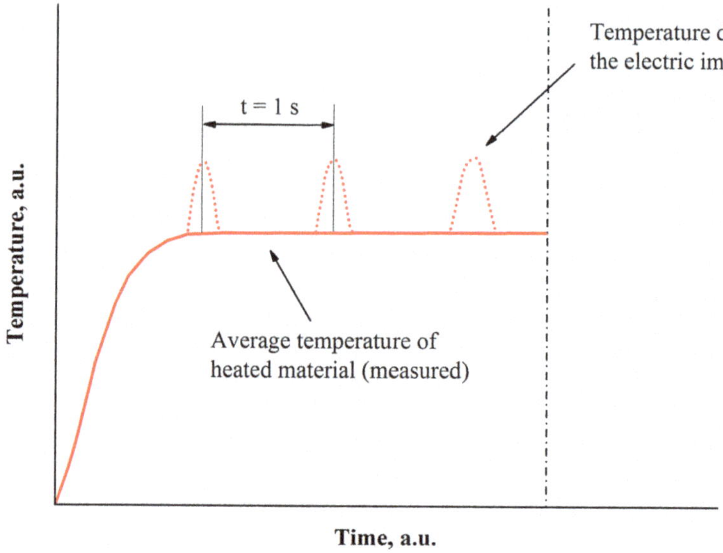

Fig. 3.32 Temperature variation during sintering by PPS. (Reprinted from Jaroszewicz and Michalski [54], Copyright (2005) with permission from Elsevier)

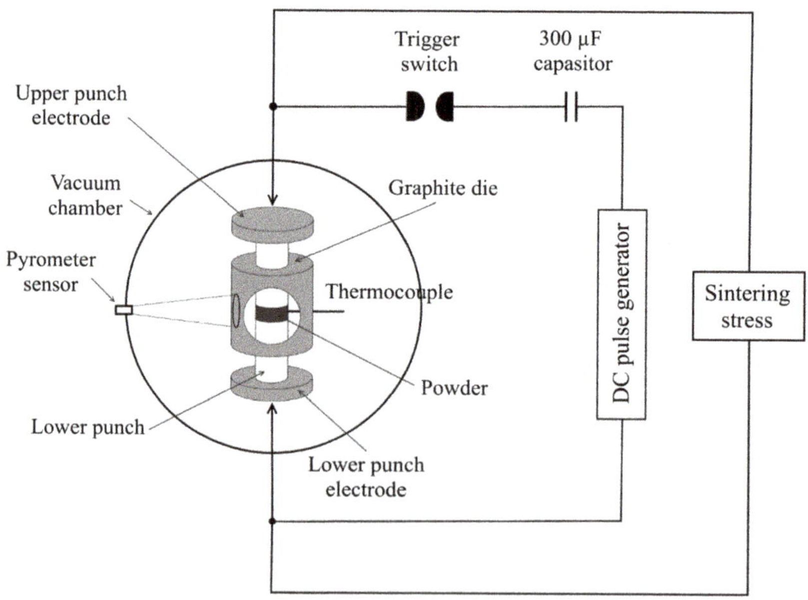

Fig. 3.33 Schematic of an apparatus for PPS. (Reprinted from Rosinski et al. [55], Copyright (2007) with permission from Elsevier)

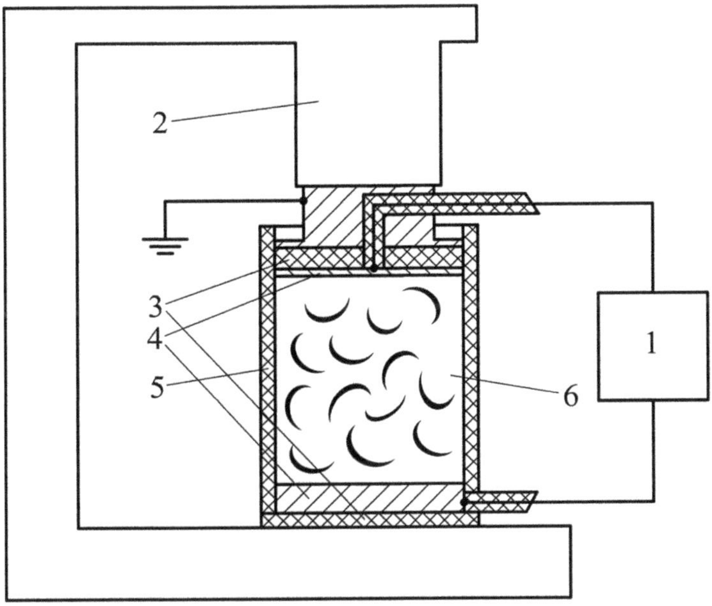

Fig. 3.34 A briquetting setup: 1, electric pulse generator; 2, press; 3, insulators; 4, punches (electrodes); 5, insulating die; 6, chips [58]

and acoustic and radio absorbers can be produced using the same technique. Composite briquettes containing inclusions of a different material can be manufactured; thus, briquettes with dielectric and electrically conducting inclusions can be made. The proposed pulsed current-based technology has been tried on chips of metals and alloys of varying quality; briquettes of various shapes have been obtained. The proposed technique includes pressing of the chips under a low pressure of 0.7–6 MPa in a nonconducting die. Keeping the pressure on the sample, electric current of high density is allowed to pass through it to form a briquette. The capacitor bank has a total capacity of 15 mF. The stored energy is 200 kJ, and the voltage is 5 kV. The current amplitude is 500 kA; the pulse duration is 250 μs. Briquettes with a diameter of 50 mm and heights ranging from 100 to 140 mm were obtained [57].

Metallic chips are pressed under relatively low pressures (up to 50 MPa, the remaining porosity is about 50%); then, keeping the pressure on the material, an electric pulse is allowed to pass through it resulting in the formation of a briquette. The material to be consolidated can be the waste consisting of metallic chips, powders, granules, or flakes including those coated with paint. Being pre-pressed only, the material remains weak and does not retain its shape after the pressure has been released. It is only with the application of electric current that the compact acquires strength and the ability to hold its shape.

3.9 Pulsed Current-Assisted Shock Consolidation

An interesting use of the capability of electric current to enhance consolidation has been found by Shvetsov et al. [59], who combined shock consolidation with a passage of a current pulse coming from a discharging capacitor connected to the consolidation ampoule. In Fig. 3.35, the setup for field-assisted shock compaction is shown for the consolidation with a central rod, which is common in shock wave technology. The powder was poured into a gap between the central rod and the inner wall of the ampoule. The ampoule was placed along the axis of a cylindrical charge of an explosive. Before the explosion, the central rod and the steel ampoule were connected to a high-voltage supply. The powder compact was insulated from the ampoule by a polyethylene film. After the capacitor was charged, the assembly was in the waiting mode. As the shock wave disrupted the insulation, an electric current passed between the ampoule and the central rod through the shock compressed powder.

The effect of the current pulse on the microstructure of the shock consolidated compact was significant: the part of the compact that was located closer to the capacitor was denser than parts located farther from the capacitor. The application of electric current has thus helped improve densification of what is known in the shock consolidation technology as a "cold" layer – a layer adjacent to the central rod and experiencing only moderate heating during shock consolidation. Evidence of melting of the central rod has also been found. The developed scheme offers a possibility to improve bonding between the central rod and the powder compact consolidated by shock waves.

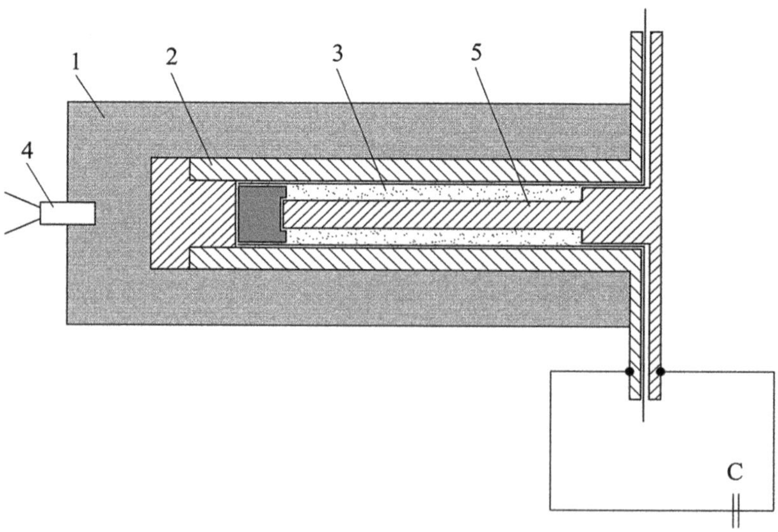

Fig. 3.35 Field-assisted shock consolidation setup: 1, explosive; 2, steel ampoule; 3, powder to be consolidated; 4, detonator; 5, central rod. (Drawn using data of Ref. [59])

3.10 Densification Kinetics Imposed by HVEDC

Since the governing parameters of the HVEDC technology are pressure and voltage, the densification map describing the density dependence on these two quantities can be built, as pictured in Fig. 3.36 for a ZrN powder [60]. The density of the specimens increases with the load applied in the pre-pressing stage. This trend appears, nevertheless, to have an upper bound. Given a value of electric current, a densification enhancement is observed up to a certain value of the cold-pressing load, after which any further increase affects density only slightly. Figure 3.36 also shows that the dependence of the relative density on the voltage is linear. The electric power is proportional to the squared voltage. If the voltage is applied for more than a few seconds, the zirconium nitride specimens can reach a thermodynamically stable structure. In this context, we expect the ZrN relative density to depend on the squared voltage. However, the extremely short duration of HVEDC processes leads to kinetically trapped structures, which renders the relative density linearly dependent on the applied voltage. A raise in the electric current density that flows through the powder compact leads to an increase in the density of the compact [60]. However, beyond a certain critical value, the powders will release a significant amount of voltage through the matrix. Quantitatively, the dependence between the energy and the voltage is complex, since the material resistance depends on the

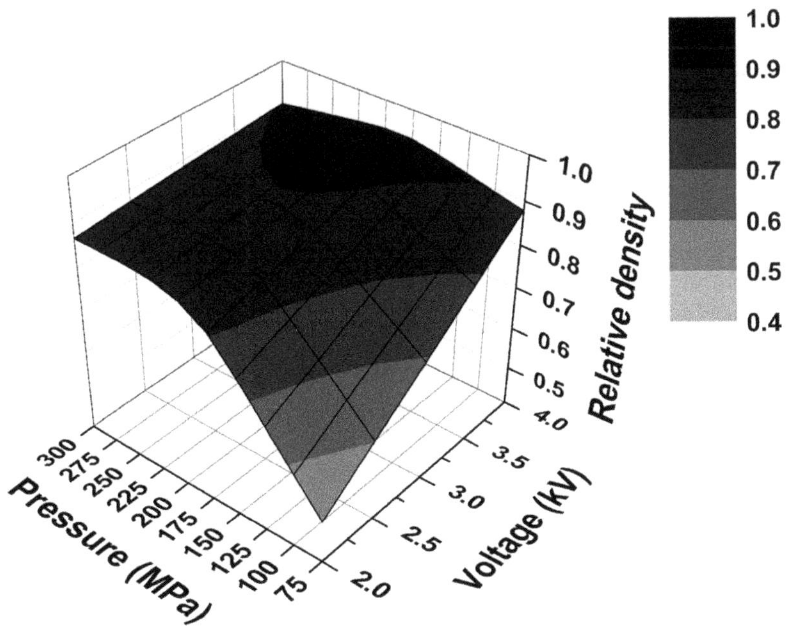

Fig. 3.36 The densification map of ZrN consolidated by high-voltage electric discharge consolidation. The relative density of the specimens is given as a function of pressure and voltage. The legend indicates the relative density. (Reprinted from Lee et al. [60], Copyright (2015) with permission from Elsevier)

relative density of the material, which evolves during the processing. In general, the dependences between the process energy and the instant values of system internal parameters, such as density, can be described by rather complex relationships, including exponential/logarithmic terms [60]. Therefore, it may be expected that, when explored over wider ranges of applied voltages, the relationship between the relative density and the voltage may become nonlinear.

It should be noted also that for the completeness of the conducted analysis, it is highly desirable to know the temperature of specimens according to the applied voltages. However, in its current apparatus' configuration, the temperature of specimens could not be measured during HVEDC process due to very short pulse duration.

The constitutive equation of the continuum theory of sintering [61] developed for hot pressing can be used to analyze the effect of electric currents on densification kinetics during HVEDC processes [60]. The constitutive equation for a nonlinear viscous material is [60, 61]:

$$\sigma_z = A_m \sqrt{\frac{2}{3}\frac{(1-\theta)}{\theta}(1-\theta)\left[\sqrt{\frac{2}{3}\frac{(1-\theta)}{\theta}}\left|\frac{\dot{\theta}}{(1-\theta)}\right|\right]^m} \tag{3.16}$$

where σ_z is the load applied (Pa), θ is porosity (dimensionless), $\dot{\theta}$ is the porosity change with time (1/s), and m is the strain rate sensitivity exponent. The material constant A_m (Pa $\cdot$ s^m), which is a function of microstructure (principally reflecting the influences of the grain size, sub-grain size, and dislocation density), is expressed by an Arrhenius-type relationship [60]:

$$A_m = A_{m0} T^m \exp\left(\frac{m \Delta H_{SD}}{RT}\right) \tag{3.17}$$

where A_{m0} is a material's constant (Pa·(s/K)^m), T the absolute temperature (K), ΔH_{SD} is the activation energy for self-diffusion (J/mol), and R is the gas constant (J/mol•K).

The sintering mechanism is related to the strain rate sensitivity m. For HVEDC, a significant aspect to be taken into account for the modeling of the process is the extreme rapidity characterizing this technique, which allows powders to be consolidated within a few seconds. A "zero-order approximation" of such technology based on its peculiarly short processing time can disregard the influence of heat on densification and, in view of the applied high pressures, can describe HVEDC as a cold-pressing process.

From the constitutive equation for the cold-pressing case, the following relationship can be derived [61]:

$$\sigma_z = \sigma_y \sqrt{\frac{2}{3}\frac{(1-\theta)^3}{\theta}} \tag{3.18}$$

where σ_y is the yield stress of the bulk material.

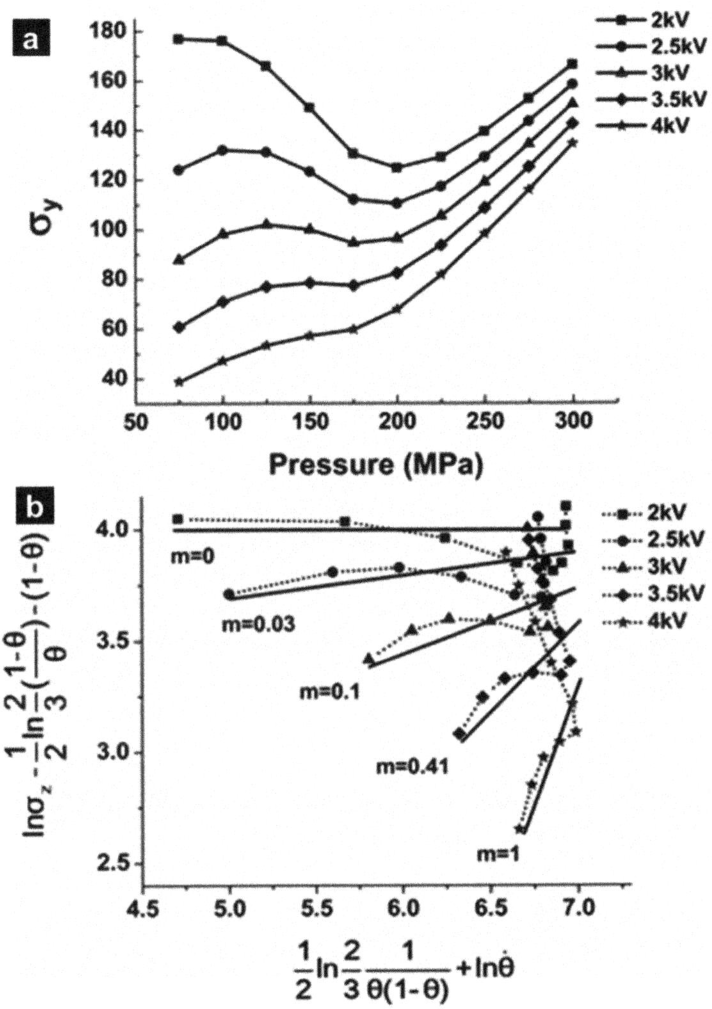

Fig. 3.37 Constitutive equation fitting of high-voltage electric discharge consolidation results. (**a**) yield stress variation with pressure and (**b**) m variation with voltage. (Reprinted from Lee et al. [60], Copyright (2015) with permission from Elsevier)

By substituting the experimental data on pressure and final porosity into Eq. (3.18), the yield stress for each condition can be calculated, as shown in Fig. 3.37a. It appears that when a low voltage is applied, Eq. (3.18) provides good correlations with experimental data for a constant value of the yield stress of the bulk material. Indeed, in low-voltage processes the short sintering time does not allow enough time for the effect of heating to be significant, and the adoption of a cold-pressing model can be considered adequate. When the voltage is increased, the high electrical currents are involved, and the consequent increased heat and/or field-related phenomena compromise the validity of the purely plastic material assumption, inducing a pressure-dependent behavior on the "apparent yield stress."

The constitutive equation for a nonlinear viscous material, Eq. (3.16), was applied to the HVEDC results to determine the densification mechanism (m value). Taking the natural logarithm of Eq. (3.16) and rearranging,

$$\ln \sigma_z - \frac{1}{2}\ln\frac{2}{3}\left(\frac{1-\theta}{\theta}\right) - (1-\theta) = m\left\{\frac{1}{2}\ln\frac{2}{3\theta}\frac{1}{(1-\theta)} + \ln\dot{\theta}\right\} + \ln A_m \quad (3.19)$$

After inserting the average porosity rate by calculating slope and pressure given in Fig. 3.36 into Eq. (3.19), a plot can be generated to determine m, as shown in Fig. 3.37b. In the conventional sintering theory, the densification of a powder by means of hot pressing provides 1 m value that is associated with a specific sintering mechanism. However, Fig. 3.37b shows mostly an increase of the slopes (m value) with increased voltage. Under low voltage, $m = 0$, which confirms that HVEDC acts like quasi-cold pressing. Increasing the voltage leads to a change of the strain rate sensitivity value m from 0 to 1, which is attributed to the increasing heat generation.

It appears that the combination of high pressure and voltage during HVEDC induces a different behavior in the material with respect to those used for modeling of hot pressing. The obtained first-order approximation provides an impetus for future work on refining the mechanisms underlying the novel and still rare technology of HVEDC.

3.11 Selected Examples of Materials Processed by High-Voltage Electric Pulse Consolidation

WC–Co composites were successfully sintered by HVEDC, as reported in Refs. [6, 7, 9]. Figure 3.38 shows the dependence of the relative density of the sintered material on the peak current density. The WC–Co composite has the maximum density when the peak current density reaches 95 kA cm^{-2}. When the current density exceeds 100 kA cm^{-2}, the density of the compact dramatically decreases.

Fais [21] showed that 99% dense WC–Co composites can be produced by CDS starting from a powder mixture containing 88% WC with a particle size of 120 nm and 12% Co and using a discharge power of 30 kJ. Siemiazko et al. [62] used PPS to conduct self-propagating high-temperature synthesis in the W–C–Co mixtures. The synthesized material showed a 30% increase in hardness relative to the material sintered from the mixtures of tungsten carbide and cobalt. Sintering was performed at 1500 K. Figure 3.39 shows the evolution of hardness, relative density, and grain size of the WC–Co composites with the sintering time; Fig. 3.40 demonstrates the variation of fracture toughness of the composites sintered for different times. The highest fracture toughness was achieved in the samples sintered for 10 min. With increasing sintering time, the fracture toughness degraded.

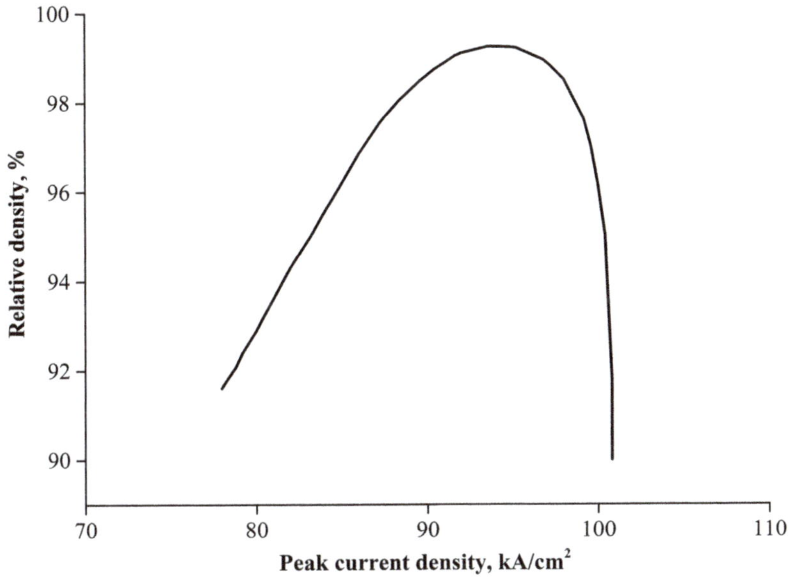

Fig. 3.38 Relative density of the sample vs. current density for WC-Co composites processed by HVEDC (applied pressure 200 MPa). (Reprinted from Grigoriev and Rosliakov [7], Copyright (2007) with permission from Elsevier)

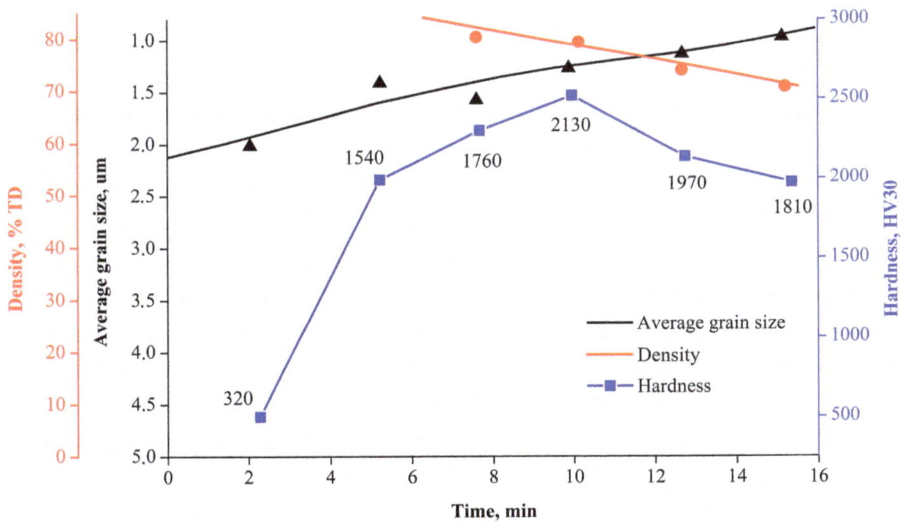

Fig. 3.39 Hardness, relative density, and grain size of WC–Co composites obtained by PPS vs. sintering time. (Drawn using data of Ref. [62])

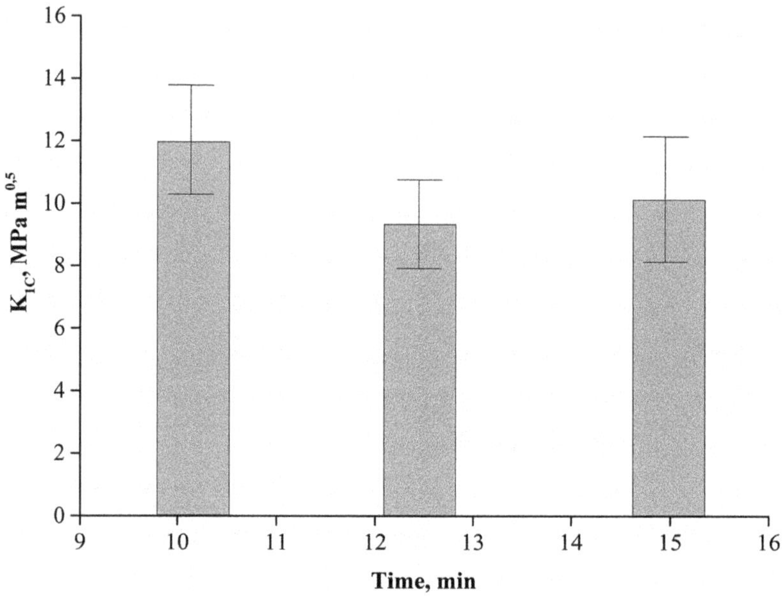

Fig. 3.40 Fracture toughness K_{IC} of WC–Co composites sintered by PPS for different times. (Drawn using data of Ref. [62])

Pulsed electric current can assist in the preparation of complex composite materials, which are challenging to produce by other methods. Thus, cBN/WC–Co composites with high strength and hardness were successfully produced by PPS [63]. The composites were obtained by sintering of WC + 0.5 wt.%VC + Co + cBN in vacuum at a temperature of 1100°C under an applied pressure of 100 MPa for 5 min.

The composites were not fully dense having densities of 92.6% and 98.6%, while the density of the material sintered without any addition of cubic BN was much higher (99.5%). The fracture surface of the sintered composites revealed transgranular fracture of the cBN particles (Fig. 3.41), which indicated strong bonding at the cBN/WC–Co interface.

Another example of sintering of a multiphase composite is consolidation of diamond/WC6Co by PPS [56] from a mixture of WC–6 wt.% Co (average particle size 0.8 μm) and diamond particles ranging from 40 from 60 μm comprising 30 wt.% of the composite. The hardness of the sintered material was 23 GPa.

PPS was successfully applied to join a W–Cu composite material to a tungsten substrate [64]. The composite–substrate interface was free of cracks and pores; no delamination effects have been observed (Fig. 3.42). Technologies based on pulsed electric current are promising for the production of Cu–W composite stacks [12]. A copper foil and a tungsten grid were used as the initial components. The stack was pressed at room temperature and in ambient air. Keeping the pressure

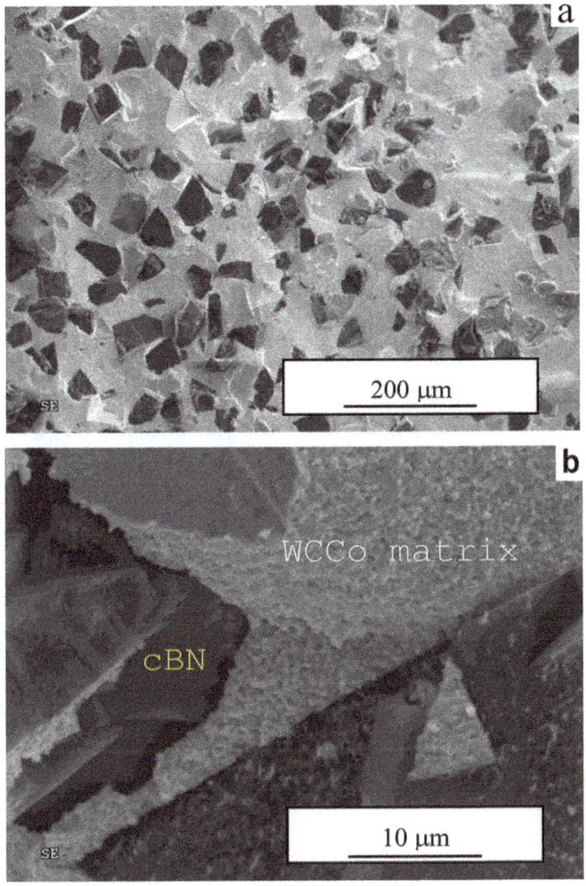

Fig. 3.41 Fracture surface of the 30 vol% cBN-(WC-0.5wt%VC-6wt%Co) composite ((**a**) microstructure, (**b**) interface between BN and the matrix). (Reprinted from Michalski et al. [63], distributed under Creative Commons Attribution (CC BY) license, https://creativecommons.org/licenses/by/3.0/)

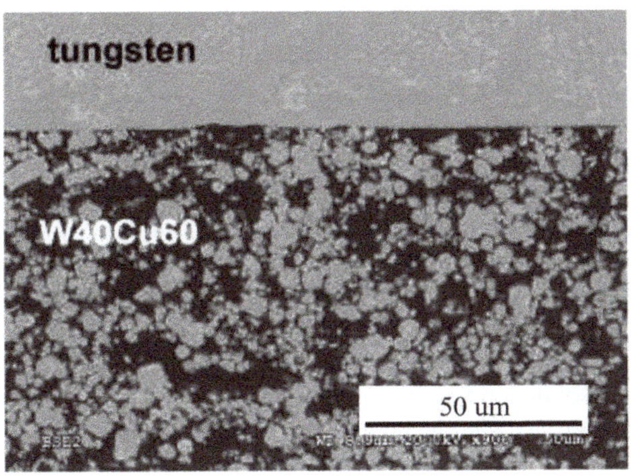

Fig. 3.42 W–Cu composite joined to a W substrate [64]

applied to the stack, a current pulse was allowed to pass through it. The resultant composite was 95% dense.

PPS was shown to be suitable for the sintering of Cu–diamond composites containing 50% of diamond [65]. Commercially available diamond powders were mixed with pure copper and a Cu–Cr alloy containing 0.8 wt.%Cr. High relative density and thermal conductivity of the Cu0.8Cr–diamond composite can be explained by the formation of chromium carbide at the metal–diamond interface. High thermal conductivity of the Cu0.8Cr–diamond composite makes it attractive for heat sink management in high-power circuits. Although sintering was carried out at a temperature, at which diamond is thermodynamically unstable, graphite was not found in the sintered samples.

The preparation of dense copper by CDS has been described in Ref. [23]. A copper powder with the particle size of 25 nm was used. Sintered by a discharge of 6 kJ, the compacts were 94–100% dense with a grain size of 26 nm and microhardness of 183 HVB.

Electric pulse sintering by a single discharge was used by Anisimov and Mali [35] to consolidate copper powders at current densities $\geq 10^2$ A mm^{-2} and pulse durations $\leq 10^{-4}$ s. Sintering was successful when the current integral (Eq. 3.7) exceeded its critical value $J^* = 1.4 \times 10^{15}$ A^{2}s m^{-4}, which is significantly lower than that required to cause complete melting of copper ($J^* = 1.05 \times 10^{17}$ A^{2}s m^{-4}). Nearly fully dense Cu monoliths were obtained. The same method was used to sinter Cu–40 vol.% TiB_2 composites and mixtures of Ti–B–Cu containing Ti and B in the amounts to form 40 vol.% TiB_2–Cu composites. Porous composites were obtained as a result of electric pulse sintering containing particles of TiB_2 20–40 nm in size distributed in the copper matrix. The critical value of the current integral was lower in the case of reactive sintering (Ti–B–Cu) due to an exothermic reaction between the components of the mixture ($Ti+2B=TiB_2$). Although the sintered composites maintained some porosity, their electric erosion resistance was four times greater compared with monolithic copper. These experiments have proved that electric pulse sintering is suitable for both reacting and non-reacting powder systems.

High-voltage electric discharges have been used to synthesize nanocrystalline intermetallics Nb_2Al and Nb_3Al from a mechanically alloyed mixture of Nb (77 wt.%) and Al (23%) [41]. The grain size of the mechanically alloyed material was 5–8 nm. The pre-pressed samples of 75–85% relative density (the external pressure was 450 MPa) were subjected to an electric discharge with a specific energy of 1 kJ g^{-1}. The two-phase material was 98% dense having grains from 26 to 36 nm and hardness ranging from 14.7 to 17.6 GPa.

An et al. [49] employed EEDS to produce implant materials. Sintered parts having a dense core and porous peripheral regions were obtained in a single step. They are suitable for the successful intergrowth with the bone tissue. The implants were sintered from an atomized titanium powder having particles 150–200 μm in size. An electric energy of 1.5 kJ was evolved in the 0.7 g of the Ti powder. The consolidation was performed in vacuum of about ~$2 \cdot 10^{-2}$ Torr. Figure 3.43 shows the surface and the cross-section of the sintered sample [49].

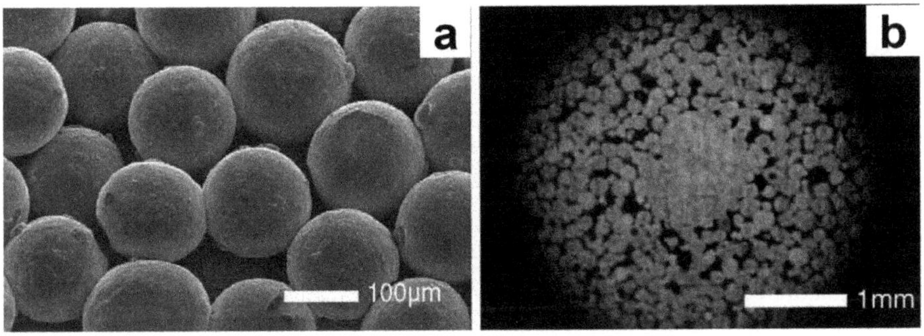

Fig. 3.43 Porous surface (**a**) and cross section (**b**) of the sintered Ti sample. (Reprinted from An et al. [49], Copyright (2005) with permission from Elsevier)

A review of the structure and properties of porous metallic materials produced using high-voltage consolidation was presented by Minko and Belyavin [66]. The electrical resistivity of consolidated metals was suggested as a convenient parameter to control the quality of inter-particle bonding after consolidation. The dependences of the electrical resistivity of the samples on the pressing pressure for titanium, niobium, and tantalum were similar. As the pressure increased, a drop in the resistivity of the samples was observed due to destruction of oxide films and the formation of good metallic contacts between the particles of the powder. After reaching a minimum, the resistivity started to increase with a further increase in the pressing pressure. This effect was explained by insufficient heating of the samples (at pressures greater than 20 MPa) during the consolidation (see Sect. 3.3).

3.12 Summary

High-voltage electric pulse sintering techniques have many characteristics common to other consolidation methods that use electromagnetic fields; however, several features pertaining to electric pulse sintering make it a self-standing area of powder metallurgy. High-voltage electric pulse sintering has potential to provide technological and economical benefits, which is possible due to the following process characteristics:

- Short process duration
- Fewer processing steps in comparison with other sintering methods (a possibility of combining shaping and sintering steps)
- Often palliated requirements for the sintering conditions (protective atmosphere or vacuum may not be necessary)
- A possibility of making parts of the desired shape

One of the main challenges of high-voltage electric pulse sintering is achieving a uniform microstructure of the material, which is a prerequisite for the uniformity of the properties at a macro-level. In each particular case, however, optimal parameters of consolidation should be found, which is an exigent task. The relative density of the sintered material can vary in a wide range depending on the consolidation parameters.

High-voltage electric pulse sintering is a unique technology suitable for sintering of materials with a wide range of possible applications (from space industry to medicine), including:

- Materials based on iron, copper, nickel, refractory metals, metals of high reactivity that are easily oxidized, complex mechanically alloyed materials.
- Electric pulse sintering is promising for consolidation of Al- and Ti-based rapidly solidified alloys, which have oxide films on the surface.
- Dissimilar materials joined together (metal–ceramic composites and steel, composites with non-conductive inclusions).
- Materials for specific applications (metallic glasses, superconducting composites, high-strength materials, materials for filters, getters, acoustic absorbers), etc.

However, the list of materials that can be processed by electric discharge consolidation is limited as the electric current should pass directly through the powder sample. Another drawback of the method is the necessity of high-voltage equipment.

Overall, high-voltage electric pulse sintering is promising to improve the properties of structural materials by forming nanostructures. This is possible due to short sintering durations implying that only slight changes in the microstructure of the material will occur during the process. However, very high sintering rates make it difficult to control the process parameters. Indeed, due to the transient nature of many associated phenomena, the *control of non-equilibrium* concept is especially important for high-voltage electric pulse sintering compared with many other field-assisted sintering techniques.

References

1. Belyavin KE, Mazyuk VV, Min'ko DV, Sheleg VK (1997) Theory and practice of electric pulse sintering of porous materials. Minsk, Remiko, p 180 (in Russian)
2. Yurlova MS, Demenyuk VD, Dudina D, Lebedeva LY, Grigoryev EG, Olevsky EA (2014) Review: electric pulse consolidation: an alternative to spark plasma sintering. J Mater Sci 49:952–985
3. Olevsky EA, Aleksandrova EV, Ilyina AM, Dudina DV, Novoselov AN, Pelve KY, Grigoryev EG (2013) Outside mainstream electronic databases: review of studies conducted in the USSR and post-soviet countries on electric current-assisted consolidation of powder materials. Materials 6:4375–4440
4. Grigor'ev EG (2008) Kinetics of the consolidation processes in dispersed materials under electric-pulse effect. Bull Rus Acad Sci: Phys 72(9):1210–1212
5. Grigoryev EG, Olevsky EA (2012) Thermal processes during high voltage electric discharge consolidation of powder materials. Scr Mater 66:662–665
6. Grigoryev EG (2011) High voltage electric discharge consolidation of tungsten carbide-cobalt powder. In: Cuppoletti J (ed) Nanocomposites with unique properties and applications in

medicine and industry. InTech, Rijeka, pp 345–360 ISBN: 978-953-307-351-4. Available from: http://www.intechopen.com/books/nanocomposites-withunique-properties-and-applications-in-medicine-and-industry/high-voltage-electric-discharge-consolidation-oftungsten-carbide-cobalt-powder
7. Grigoriev EG, Rosliakov AV (2007) Electro–discharge compaction of WC–Co and W–Ni–Fe–Co composite materials. J Mater Process Technol 191:182–184
8. Popov VP, Grigor'ev EG, Novikov SV, Baidenko AA, Goucharov SV (1996) Mathematical modeling of the densification process in the electrical discharge sintering of copper-tin powder. Powder Metall Metal Ceram 35:32–35
9. Grigoryev EG (2009) Kinetics of densification processes of powder materials under electro pulse sintering. Arab J Sci Eng 34(1):29–33
10. Belyavin KE, Min'ko DV, Kuznechik OO (2004) Modeling of the process of the electric-discharge sintering of metal powder. J Eng Phys Thermophys 77(3):628–637
11. Marcus HL, Weldon WF, Persad C, Eliezer Z, Bourell D (1990) Controlling fundamentals in high–energy high–rate pulsed power materials processing of powdered tungsten, titanium aluminides and copper–graphite composites. In: Final technical report, Center for Materials Science and Engineering, Texas University, Austin
12. Persad C, Peterson DR, Zowarka RC (1989) Composite solid armature consolidation by pulse power processing: a novel homopolar generator application in EML technology. IEEE Trans Magn 25(1):429–432
13. Orth JE, Wheat HG (1997) Corrosion behavior of high energy high rate consolidated graphite/ copper metal matrix composites in chloride media. Appl Compos Mater 4:305–320
14. Wang MJ, Persad C, Eliezer Z, Weldon WF (1987) High-energy/high-rate consolidation of copper-graphite composite brushes for high-speed, high-current applications. In: Gully JH (ed) Proc 3rd international conference on current collectors, paper 20, Austin, TX, USA
15. Eliezer Z, Wang MJ, Persad C, Gully J (1988) A novel processing technique for metal–ceramic composites. Mater Sci Forum 34-36:505–509
16. Elkabir G, Rabenberg L, Persad C, Marcus HL (1986) Microstructural evaluation of a high-energy high-rate P/M processed aluminum alloy. Scr Metall 20:1411–1416
17. Oleszak D, Jaroszewicz J, Rosinski M, Michalski A (2002) Structure of NiAl–TiC composite fabricated by mechanical milling and pulse electric discharge sintering. Rudy Metal 47:432–434
18. Rosinski M, Kruszewski M, Michalski A, Fortuna-Zalesna E (2011) W/steel joint fabrication using the pulse plasma sintering (PPS) method. Fusion Eng Des 86:2573–2576
19. Knoess W, Schlemmer M (1996) US Patent No 5529746
20. Fais A, Maizza G (2008) Densification of AISI M2 high speed steel by means of capacitor discharge sintering (CDS). J Mater Process Technol 202:70–75
21. Fais A (2010) Processing characteristics and parameters in capacitor discharge sintering. J Mater Process Technol 210:2223–2230
22. Scardi P, D'Incau M, Leoni M, Fais A (2010) Dislocation configurations in nanocrystalline FeMo sintered components. Metall Mater Trans A 41:1196–1201
23. Fais A, Leoni M, Scardi P (2011) Fast sintering of nanocrystalline copper. Metall Mater Trans A 27:1517–1521
24. Raichenko AI (1987) Basics of electric current-assisted sintering. Metallurgiya, Moscow, 128 p (in Russian)
25. Wu X, Guo J (2007) Effect of liquid phase on densification in electric–discharge compaction. J Mater Sci 42:7787–7793
26. Al-Hassani STS, Can M, Watson EJ (1986) A second order approximation to nonlinear circuit equations as applied to high energy electrical discharge processes. J Comput Appl Math 15:175–189
27. Alitavoli M, Darvizeh A (2009) High rate electrical discharge compaction of powders under controlled oxidation. J Mater Process Technol 209:3542–3549

28. Kim DK, Pak HR, Okazaki K (1988) Electro discharge compaction of nickel powders. Mater Sci Eng A 104:191–200
29. Zavodov NN, Kozlov AV, Luzganov SN, Polishchuk VP, Shurupov AV (1999) Sintering of metal powders by a series of strong current pulses. High Temp 37(1):130–135
30. Grigoryev EG (2009) Modelling of the macroscopic processes in powder medium under the powerful electric-pulse effect. Bulletin of the MSPU 1:52–56 (in Russian)
31. Clyens S, Al-Hassani STS (1976) Compaction of powder metallurgy bars using high voltage electrical discharges. Int J Mech Sci 18(1):37–40
32. Al-Hassani STS (1979) Consolidation of powder metallurgy bars by direct electrical discharge and rotary swaging. Wire Ind 46:809–816
33. Alp T, Darvizeh AF, Al-Hassani STS (1988) Preforming of metal-polymer composites by electric discharge compaction of powders. Powder Metall 3:173–177
34. Ervin DR, Bourell DL, Persad C, Rabenberg L (1988) Structure and properties of high energy, high rate consolidated molybdenum alloy TZM. Mater Sci Eng A 102(1):25–30
35. Anisimov AG, Mali VI (2010) Possibility of electric–pulse sintering of powder nanostructural composites. Comb Expl Shock Waves 46(2):237–241
36. Kim YH, Cho YJ, Lee CM, Kim SJ, Lee NS, Kim KB, Jeon EC, Sok JH, Park JS, Kwon H, Lee KB, Lee WH (2007) Self–assembled microporous Ti–6Al–4V implant compacts induced by electro–discharge sintering. Scr Mater 56:449–451
37. Arzt E (1982) The influence of an increasing particle coordination on the densification of spherical powders. Acta Metall 30(10):1883–1890
38. Sprecher AF, Mannan SL, Conrad H (1983) On the temperature rise associated with the electroplastic effect in titanium. Scr Metall 17(6):769–772
39. Vityaz' PA, Kaptsevich VM, Belyavin KE, Prezhina TE, Kerzhentseva LF, Govorov VG (1990) Contact formation during the electric-pulse sintering of a titanium alloy powder. Soviet Powder Metall Metal Ceram 29(7):527–529
40. Cho JY, Song GA, Choi HS, Kim YH, Kim TS, Lee MH, Lee HS, Kim HJ, Lee JK, Fleury E, Seo Y, Kim KB (2012) Necking mechanisms on porous metallic glass and W compacts using electro-discharge sintering. J Alloys Compd 536:S78–S82
41. Rock C, Qiu J, Okazaki K (1998) Electro-discharge consolidation of nanocrystalline Nb–Al powders produced by mechanical alloying. J Mater Sci 33:241–246
42. Schütte P, Moll H, Theisen W (2010) In: Proceedings of the PM2010 powder metallurgy world congress. European Powder Metallurgy Association, Florence
43. An YB, Oh NH, Chun YW, Kim YH, Kim DK, Park JS, Kwon JJ, Choi KO, Eom TG, Byun TH, Kim JY, Reucroft PJ, Kim KJ, Lee WH (2004) Mechanical properties of environmental electro discharge sintered porous Ti implants. Mater Lett 59:2178–2182
44. Grigoryev EG, Mitrofanov AV, Rosliakov AV (2000) In: Proceedings of the scientific session of MEPhI, part 9. Moscow, Russia (in Russian)
45. Belyavin KE (2000) Theoretical and technological bases of electric pulse sintering of refractory metal powders and application of the technology in industrial manufacturing of porous metallic parts. Synopsis of thesis, Research Institute of Powder Metallurgy with pilot production, Minsk (in Russian)
46. Bilalov BA, Kardashova GD, Magomedova EK, Ahmedov RR (2010) In: Proceedings of the international scientific and technical conference INTERMATIC–2010, part 2, Moscow, Russia (in Russian)
47. Bilalov BA, Gikitikchev MA, Magomedova EK, Dallaeva DS, Bilalov AB (2010) Process investigation of silicon carbide ceramic obtaining by electro pulse sintering. In the World of Scientific Discoveries (V mire nauchnyh otkrytiy) 6:191–193 (in Russian)
48. Jung J, Kim K, Lee W (2001) US Patent No 7347967
49. An YB, Oh NH, Chun YW, Kim DK, Park JS, Choi KO, Eom TG, Byun TH, Kim JY, Byun CS, Hyun CY, Reucroft PJ, Lee WH (2006) One–step process for the fabrication of Ti porous compact and its surface modification by environmental electro–discharge sintering of spherical Ti powders. Surf Coat Technol 200(14–15):4300–4304

50. An YB, Oh NH, Chun YW, Kim YH, Park JS, Choi KO, Eom TG, Byun TH, Kim JY, Hyun CY, Kim DK, Byun CS, Sok JH, Kwon JJ, Lee WH (2005) Surface characteristics of porous titanium implants fabricated by environmental electro–discharge sintering of spherical Ti powders in a vacuum atmosphere. Scr Mater 53:905–908
51. Lee WH, Hyun CY (2006) XPS study of porous dental implants fabricated by electro–discharge sintering of spherical Ti–6Al–4V powders in a vacuum atmosphere. Appl Surf Sci 252:4250–4256
52. Alp T, Can M, Al-Hassani STS (1993) The electroimpact compaction of powders: mechanism, structure and properties. Mater Manuf Process 8:285–289
53. Egan D, Melody S (2009) EDS as a method of manufacturing diamond tools. Met Powder Rep 64(6):10–13
54. Jaroszewicz J, Michalski A (2006) Preparation of a TiB_2 composite with a nickel matrix by pulse plasma sintering with combustion synthesis. Je Eur Ceram Soc 26:2427–2430
55. Rosinski M, Fortuna E, Michalski A, Pakiela Z, Kurzydlowski KJ (2007) W/Cu composites produced by pulse plasma sintering technique (PPS). Fusion Eng Des 82:2621–2626
56. Michalski A, Rosinski M (2008) Sintering diamond/cemented carbides by the pulse plasma sintering method. J Am Ceram Soc 91:3560–3565
57. Abramova KB, Bocharov YN, Samuilov SD, Shcherbakov IP (2001) Molding loose metal particles into briquettes with the use of short pulses of high density current. Tech Phys 46 (4):484–489
58. Samuilov SD (2002) Electrophysical method of briquetting of metallic scrap, Synopsis of thesis, Saint-Petersburg, Russia, 17 p. (in Russian)
59. Shvetsov GA, Mali VI, Bashkatov YL, Anisimov AG, Matrosov AD, Teslenko TS (2005) Effect of magnetic fields on explosive welding of metals and explosive compaction of powders. In: Schneider-Muntau HJ, Wada H (eds) Materials processing in magnetic fields, Proceedings of the international workshop on materials analysis and processing in magnetic fields, March 2004 in Tallahassee, Florida, World Scientific Publishing Co. Pte. Ltd., pp 360–370
60. Lee G, Yurlova MS, Giuntini D, Grigoryev EG, Khasanov OL, Izhvanov O, Back C, McKittrick J, Olevsky EA (2015) Densification of zirconium nitride by spark plasma sintering and high voltage electric discharge consolidation: a comparative analysis. Ceram Int 41:14973–14987
61. Olevsky E (1998) Theory of sintering: from discrete to continuum. Mater Sci Eng R 23:41–100
62. Siemiazko D, Michalki A, Rosinski M (2008) In: Skorokhod V (ed) Proceedings of symposium I "Functional and Structural Ceramic and Ceramic Matrix Composites (CCMC)", Warsaw, Poland
63. Michalski A, Rosiński M, Płocińska M, Szawłowski J (2011) Synthesis and characterization of cBN/WCCo composites obtained by the pulse plasma sintering (PPS) method. IOP Conf Ser Mater Sci Eng 18:202016 4 pages
64. Fortuna E, Rosiński M, Michalski A, Lisowski W (2007) FGM W-Cu composites and W-Cu/W joints fabrication route based on pulse plasma sintering (PPS) method. In: Annual report IPPLM, Warsaw University of Technology
65. Ciupiński L, Siemiaszko D, Rosiński M (2009) Heat sink materials processing by pulse plasma sintering. Adv Mater Res 59:120–124
66. Minko D, Belyavin K (2016) A porous materials production with an electric discharge sintering. Int J Refr Metals Hard Mater 59:67–77

Chapter 4
Sintering by Low-Voltage Electric Pulses (Including Spark Plasma Sintering (SPS))

4.1 Principle and Physical Mechanisms of Low-Voltage Electric Pulse Sintering

Consolidation methods of powder materials that use low-voltage electric pulses are referred to in the literature as spark plasma sintering (SPS) and electric discharge sintering (EDS) (should not be mixed up with "high-voltage electric discharge sintering" or "capacitor discharge sintering"). These methods are based on the principle introduced by Inoue [1, 2] in the 1960s. Different types of equipment and a wide range of experimental parameters have been used since; many sintering studies were conducted under an applied pressure. Raichenko and his team at Frantsevich Institute for Problems of Materials Science, National Academy of Sciences of Ukraine, started developing the method of EDS in the beginning of the 1970s. The majority of studies performed until 2007 in this institute and other research organizations were conducted using a facility developed and manufactured by a research team led by Raichenko in 1976 [3, 4] (Fig. 4.1 (top)). EDS was also the subject of research at Tomsk State Polytechnic University, Russian Federation [5, 6]. Figure 4.1 (bottom) shows a schematic and tooling of a modern SPS facility.

EDS was also conducted using ERAN 2/1 facility of Frantsevich Institute for Problems of Materials Science, National Academy of Sciences of Ukraine [7–12], shown in Fig. 4.2. This facility allows heating samples 8 mm in diameter up to 1700 °C in 90–120 s. Samples were sintered under the following conditions: an alternating current of 0.3 kA, a direct current of 1.1 kA, and a pressure of 80 MPa. Electric current was applied to the powder compact using graphite tooling. When graphite tooling is heated up to 1100 °C, it can react with oxygen to form CO, which serves as a protective atmosphere for the consolidated samples.

According to Raichenko [4], the low-voltage current-assisted sintering (EDS), in its optimized regime implementation, should be characterized by two different stages

E. A. Olevsky, D. V. Dudina, *Field-Assisted Sintering*,
https://doi.org/10.1007/978-3-319-76032-2_4

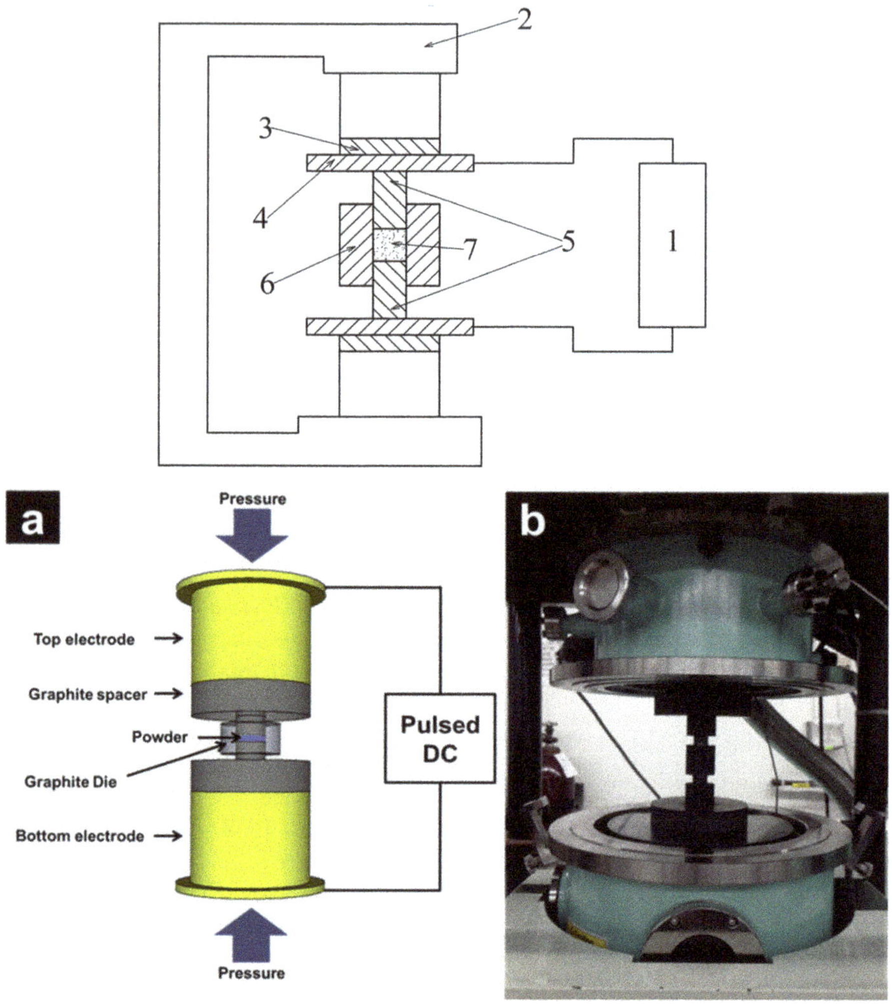

Fig. 4.1 Top: Schematic of an electric discharge sintering (EDS) setup: 1, current source; 2, press; 3, insulating plates; 4, current-carrying plates; 5, punches (electrodes); 6, die; and 7, powder compact. (Reprinted from Raichenko et al. [3], Copyright (1977) with permission of Springer). Bottom: SPS schematic (**a**), chamber and tooling (**b**) (Dr. Sinter SPSS-515, Fuji Electronic Industrial Co., Ltd., Japan)

of the consolidation process. The first stage is called a "sparking" stage because conditions favoring sparking should be created to form contacts between previously loose powder particles. The powder compact experiences a relatively low pressure during the first stage ($\leq$10 MPa) and a superposition of AC and DC currents. Due to the sparking effect, heating of the inter-particle contacts is more significant than in

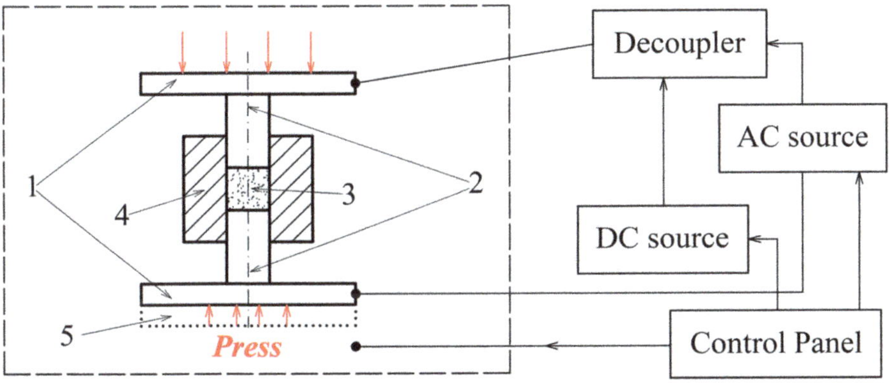

Fig. 4.2 ERAN 2/1 installation for electric discharge sintering (EDS): 1, current-carrying plates of the hydraulic press; 2, graphite punches; 3, powder mixture; 4, graphite die; 5, press. (Reprinted from Petukhov et al. [8], Copyright (2007) with permission of Springer)

the case of conventional Joule heating in the absence of sparks. The second stage is the "consolidation" stage and should include the simultaneous impact of more powerful current pulses (in some cases together with a DC current) and high mechanical pressure applied to the sample. For successful processing, the right choice of the applied pressure and of the pulse current parameters (amplitude, duration, and the interval between pulses) is of paramount importance. Crivelli et al. [13] showed that an optimal regime for spark sintering of an Al alloy powder consisted of two stages, during the first ("sparking") stage that lasted 1 min, the current density was 11.25 A cm^{-2}, and during the second ("densification") stage, the current density was 382 A cm^{-2}.

Raichenko emphasizes the importance of the load application schedule for EDS and uses a term "sintering with programmed loading" [4]. At the first stage, low pressures result in poor contacts between the particles, which favor conditions for actual discharges in the corresponding gaps. The inter-particle contacts experiencing the discharge action are heated to a greater extent than they would have been in the case of conventional resistance heating. The inter-particle contacts thus formed are more established than in the absence of discharges. If the inter-particle contacts become too strong, they will hinder further densification by making it difficult for the particles to rearrange and slide relative to each other [14]. However, if the second stage of EDS immediately follows, the material of the necks may be still hot enough to enable a high degree of deformation.

It should be noted that although the term "electric discharge sintering" has been widely used since its introduction into the community and accepted by different scientific schools, one should bear in mind that the conditions for the discharge formation are created only when certain, optimized sintering schedules are applied. In many experiments reported in the literature, the sintering had been conducted through a single-stage process and at a constant pressure. The term itself, nevertheless, is used, very often without a comment on its direct meaning. So, the term

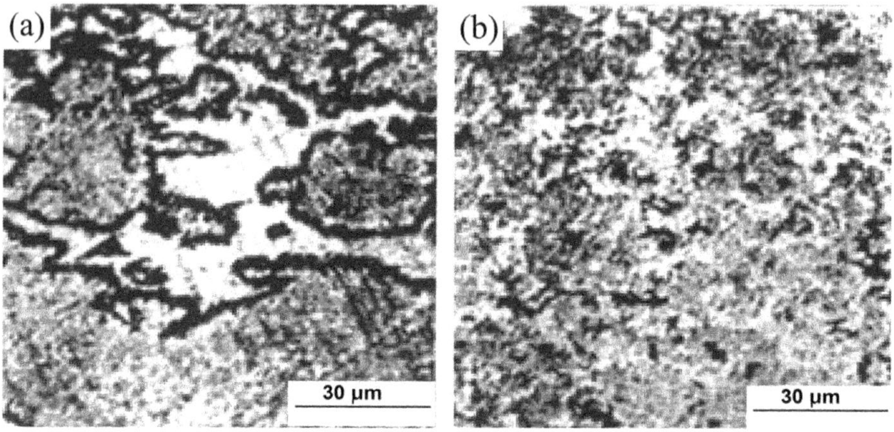

Fig. 4.3 Microstructure of Cu–Sn specimens produced at pre-pressing pressures (pressures applied during the first stage of EDS) of 4.17 MPa (**a**) and 12.50 MPa (**b**), magnification ×400. (Reprinted from Sukhov et al. [15], Copyright (1987) with permission of Springer)

emphasizes that discharges between the particles are possible, but they do not necessarily accompany all sintering processes. Sukhov et al. [15] demonstrated a more intense shrinkage of a Cu–Sn alloy powder at a pressure of 4.17 MPa than at a pressure of 12.50 MPa applied during the first stage of EDS and explained the observed effect by the formation of discharges between the particles at a low pressure. Indeed, the microstructure of the material sintered at 4.17 MPa shows evidence of melting of a significant portion of the material, which caused faster densification of the powder (Fig. 4.3). Raichenko [4] discusses the applicability of the EDS facilities for conducting processes that are more close to hot pressing. This can be done by using relative high pressures at the first sintering stage (or through the while sintering cycle), thus avoiding the formation of discharges.

A nonmonotonic dependence of the porosity of the sintered compact on the pressure applied during the sparking (first) stage of EDS has been experimentally found by Raichenko et al. [16] for Al and Ni powders. Figure 4.4 shows the dependence of the porosity of the Ni compacts produced from Ni powders differing by the oxygen content on the pressure applied during the sparking stage. At low pressures below the critical value, the application of current causes the formation of a single channel, along which the material melts. It is clear that this is not a pathway to efficient densification. As the pressure increases, the number of contacts between the particles that are suitable for the sparking events to occur increases leading to a uniform distribution of the sparking points in the sample volume. An increase in the porosity with increasing pressure above the critical value corresponding to the minimal porosity can be explained by the formation of inter-particle contacts that do not favor sparking as they are too well established with oxide films present on the particle surface possibly disrupted under the applied pressure. As it is more difficult to establish contacts suitable for sparking between particles with a thicker oxide, a higher critical pressure was found for the Ni powder with a higher oxygen.

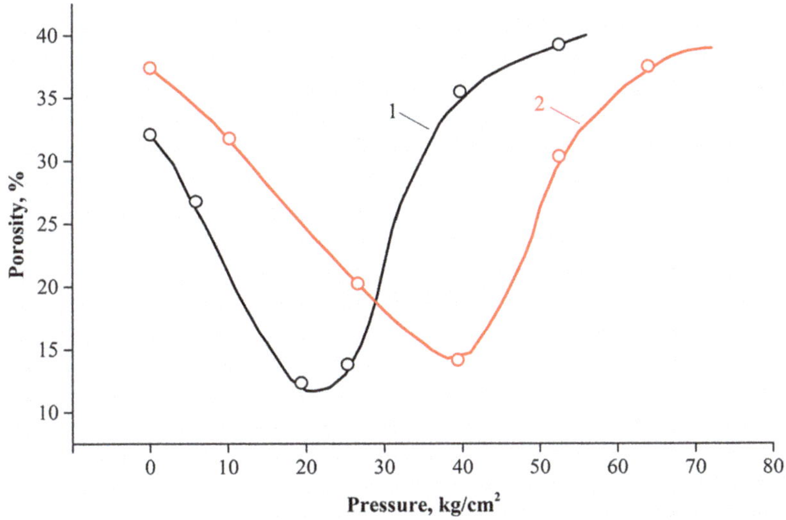

Fig. 4.4 Porosity of Ni compacts produced from powders differing by oxygen content on the pressure applied during the sparking (first) stage (1–0.06% of oxygen; 2–0.12% of oxygen). (Reprinted from Raichenko et al. [16], Copyright (1977) with permission of Springer)

The situation for Al powders was more complicated, as can be seen in Fig. 4.5. Finer Al powders could be densified more efficiently by EDS due to a greater number of inter-particle contacts and the availability of a higher number of points of potential sparking.

As was proved by Raichenko et al. [16], EDS is more efficient when DC and AC currents are superimposed. Unlike compacts sintered by a DC current only, which possess a dense core but a porous peripheral part of nonuniform microstructure, compacts sintered by superimposed AC and DC currents have a uniform microstructure throughout the cross-section. The role of the AC current in sintering intensification of the peripheral part due to the skin effect was confirmed by calculating the current frequency (assuming the penetration depth is equal to the radius of the compacts) and experimentally finding that at a frequency close to the calculated, the compact had the lowest porosity and the highest strength. In the absence of the AC components, the peripheral parts of the compact are densified to a lower extent compared with the core due to heat losses.

Mechanical pressure applied during the second stage of EDS inhibits contact heat sources and facilitates volumetric heating. Raichenko et al. [17] have shown that a pressure applied in a pulsed mode during the concluding stage of sintering intensifies densification. A periodically repeating load of a constant sign was applied during the second stage of EDS of copper. The duration of the pressure pulses was 1.5–4.6 s. The compacts obtained with a pulsating pressure were denser, the effect becoming more significant as the duration of the second sintering stage increased.

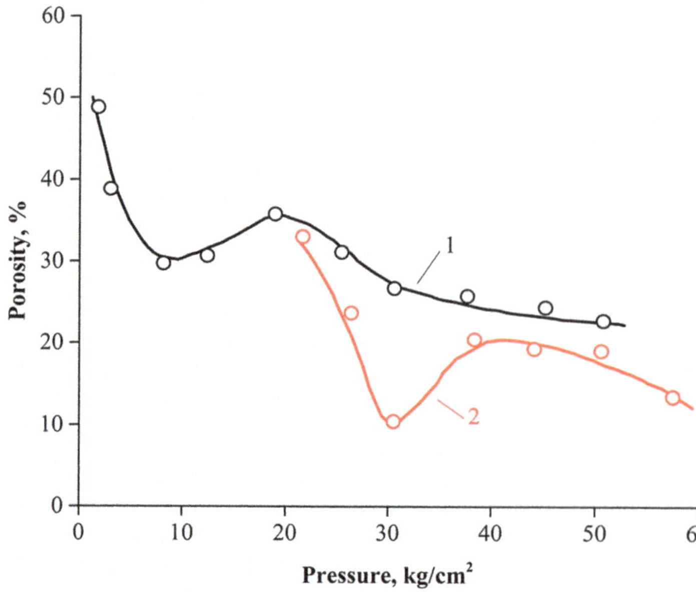

Fig. 4.5 Porosity of Al compacts produced from 250–315 μm powders (1) and 63–160 μm powders (2). (Reprinted from Raichenko et al. [16], Copyright (1977) with permission of Springer)

4.2 Low-Voltage Electric Pulse Sintering Equipment

Low-voltage pulsed electric current is the major feature of SPS, which is currently regarded as the most advanced and promising powder consolidation method. Indeed, in the past two decades, this sintering technique has been thoroughly elaborated. The details of the process have been described by Tokita [18, 19], Omori [20], Groza and Zavaliangos [21, 22], Munir et al. [23, 24], and Garay [25]. Reactive processes in the SPS have been addressed by Orrù et al. [26] and Dudina and Mukherjee [27, 28]. SPS consists essentially of conjoint application of high temperature, axial pressure, low voltage, and high current. Terminologically, the term "pressing" better fits the process character, since for sintering the free surface area reduction should be the dominant driving factor. The field component is associated with electric current passing through a powder specimen. The electric field generates Joule heat, which provides the conditions of hot compaction; the current also enhances densification and can improve the final grain structure in quite a remarkable way. This approach significantly shortens the processing and enhances the performance time and quality wise. In particular, it carries the potential of maintaining nano- and submicrometer structures in the materials after consolidation. A technique using AC instead of pulsed DC has many similarities with SPS and is called field-activated pressure-

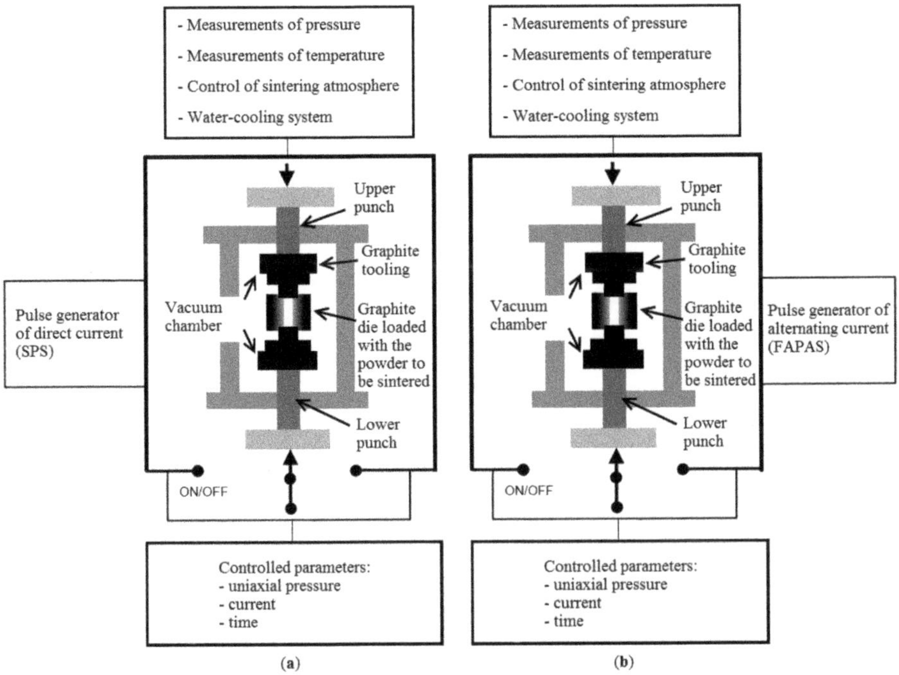

Fig. 4.6 Schematics of spark plasma sintering (SPS) (**a**) and field-activated pressure-assisted synthesis (FAPAS) (**b**) facilities

assisted synthesis (FAPAS) [29, 30]. Figures 4.6 and 4.7 show the schematics of the SPS and FAPAS facilities and the corresponding current profiles. The pulse sequence in the SPS facilities can be varied, but normally 12:2 sequence is used (12 pulses ON and 2 pulses OFF, which means that after 12 pulses, the DC current is switched off for the duration equal to the length of 2 pulses). The duration is each pulse is normally 3 ms. The term "FAPAS" is, however, sometimes used to refer to sintering by a DC current [31]. Groza and Zavaliangos [21, 22] use the term "field-assisted sintering technique (FAST)" to refer to sintering processes by means of pulsed DC or combined processes with the first stage using pulsed DC (sintering activation) and the second one using resistance heating by DC.

The SPS equipment was introduced into the market by Sumitomo Coal Mining Co., Ltd. (Japan) in 1989. At present, the SPS equipment is developed and manufactured by Sinter Land Inc. (Japan), Fuji Electronic Industrial Co., Ltd. (Japan), FCT Systeme GmbH (Germany), and Thermal Technology LLC. (USA). The two basic types of DC pulse generators used in the SPS apparatus are the thyristor-type and the inverter-type pulse generators. An inverter-type power generator with pulse width modulation has a lower power consumption and space-saving advantages reducing the production costs of materials [19]. Sintering is usually conducted under vacuum; however, the system allows operating in an atmosphere

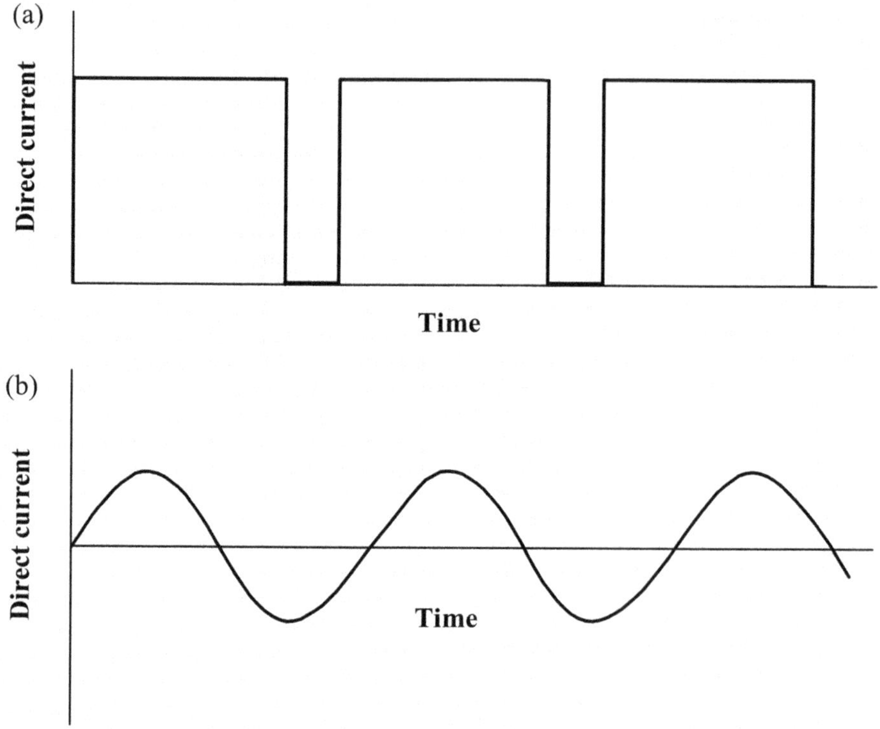

Fig. 4.7 Current profiles in SPS (**a**) and in FAPAS (**b**) facilities

of (inert) gas – argon or nitrogen can be used. Usually, a residual pressure of 5–6 Pa in the chamber is achieved after 15 min of pumping (from the atmospheric pressure). Industrial manufacturing with the use of SPS is possible thanks to the development of multi-head, large-size batch-type, tunnel-type, rotary-type, and shuttle-type SPS systems [19]. In addition, automatic handling and powder stacking equipment has been developed.

4.3 Macroscopic Temperature Gradients in SPS

Anselmi-Tamburini et al. [32] performed numerical modeling of the current and temperature distribution in a die/punches/graphite spacer assembly using a finite volume approach. The analyses were carried out to investigate the current and temperature distributions under typical SPS conditions utilizing two materials dramatically differing in electrical conductivity (alumina and copper). The current distribution was evaluated in the radial and axial directions (Figs. 4.8 and 4.9). Radial current density gradients result in significant differences in the temperature between the surface of the graphite die and its center. Both the simulation and

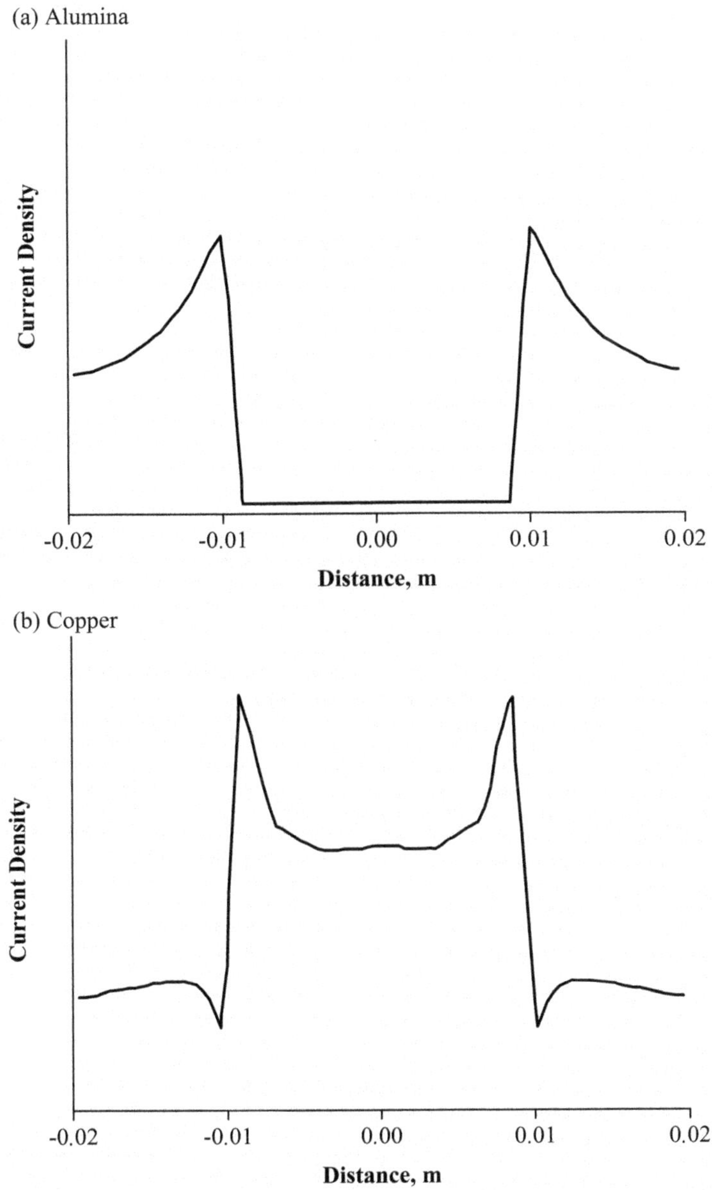

Fig. 4.8 Radial distribution of the current density: (**a**) alumina, (**b**) copper. Applied voltage is 5 V. The sample thickness is 3 mm; the sample diameter is 19 mm. (Reprinted from Anselmi-Tamburini et al. [32], Copyright (2004) with permission from Elsevier)

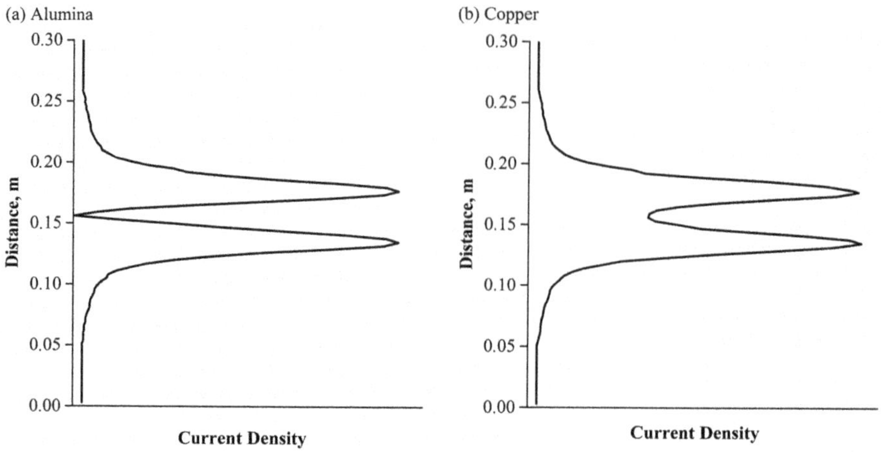

Fig. 4.9 Distribution of the current density along the vertical axis of the assembly: (**a**) alumina, (**b**) copper. The highest current density in both cases corresponds to the exposed portion of the plungers. The sample thickness is 3 mm; the sample diameter is 19 mm. (Reprinted from Anselmi-Tamburini et al. [32], Copyright (2004) with permission from Elsevier))

the experimental results show a considerable difference between the temperatures of the sample and the external surface of the die, which is due to current distribution and radiation heat losses. Figure 4.10 shows the calculated radial temperature distribution for alumina and copper compacts sintered using a constant applied current of 1000 A. The temperature difference, however, disappears when the current is turned off, which implies that the central role in maintaining the gradient is played by the electric current, and, thus, higher currents will induce that larger temperature gradients.

Vanmeensel et al. [33] developed a finite element model to gain a better understanding of the temperature distributions in the die–punch–sample assemblies of the SPS facilities. The model utilized experimentally measured material properties and thermal/electrical parameters during real sintering cycles. It was shown both theoretically and experimentally that the temperature distribution inside the tooling and the sample in the die is not homogeneous during the sintering cycle. The electrical properties of the sample were shown to be crucial for the character of the temperature distributions inside the tooling and the sample. Much higher thermal gradients were observed in an electrically conductive samples compared with samples of insulating materials under the same sintering conditions (Fig. 4.11). In the case of zirconia (ZrO_2), the current flows only through the punches and the die; consequently, the heat losses due to the radiation from the tooling surfaces are compensated. The case of titanium nitride (TiN), which is a good electric conductor, presents an opposite situation: the current mainly flows through the sample. As a result, the radiation heat

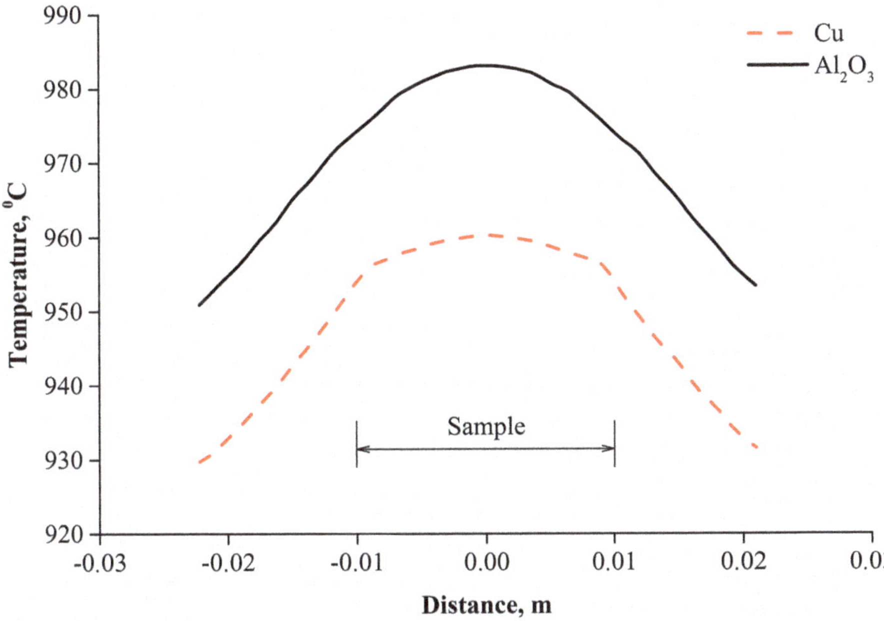

Fig. 4.10 Calculated radial temperature distribution for nonconducting (alumina) and conducting (copper) sintered samples (the sample thickness was 3 mm; the applied current was kept constant at 1000 A). The sample diameter is 19 mm. (Reprinted from Anselmi-Tamburini et al. [32], Copyright (2004) with permission from Elsevier)

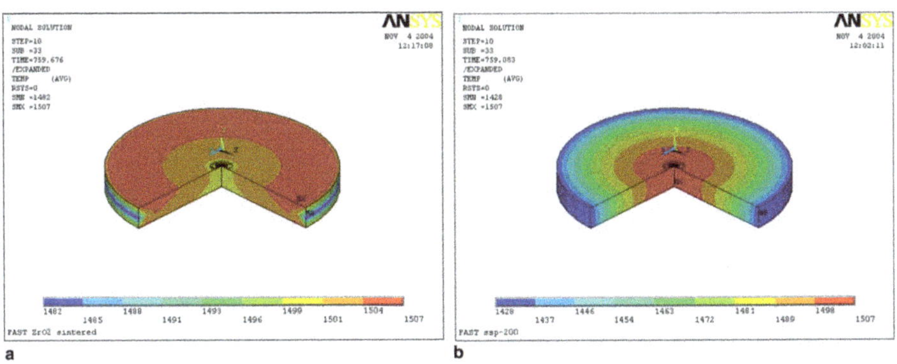

Fig. 4.11 Temperature distribution inside the ZrO_2 (**a**) and TiN (**b**) fully dense specimens during the dwell at 1500 °C (the steady-state temperature distribution reached 15 s after the full load was applied). (Reprinted from Vanmeensel et al. [33], Copyright (2005) with permission from Elsevier)

losses are not compensated, and the temperature distribution in the sample becomes very inhomogeneous. The authors emphasize the importance of surrounding the die with a thermal insulation material to bring the heat losses to a minimum and reduce the radial temperature gradients.

4.4 Temperature Measurements and Heat Dissipation in SPS/FAST Facilities

Langer et al. [34] emphasize the importance of the location of the measurement point in drawing conclusions on a lowered temperature of field-assisted sintering for non-conductive materials. They compared the results of the sintering procedure in terms of the real temperature of the sample conducted in the SPS and FAST equipment keeping the pressure, the heating rate, and the pulse on/off ratio constant. Although the FAST and SPS apparatus are based on the same principle, direct comparison between them is accompanied by certain difficulties due to the difference in the temperature measurement approaches. Figure 4.12 shows a schematic indicating the location of the temperature measurement points and methods in the SPS (Japan) and FAST (Germany) pulsed current sintering facilities. Alumina and zinc oxide showed higher densities when processed by SPS than when processed with FAST under the same apparent processing parameters. Densification curves for alumina are shown in Fig. 4.13 for the SPS and FAST processing alternatives. It was also found that the spark plasma sintered samples had larger grains. The observed differences are due to the differences in the temperature measurements in the two sintering systems. In the SPS process, the same relative density is reached at a lower apparent temperature than in the FAST. Langer et al. [34] performed calibration by the copper-melting experiments and showed that the radial temperature measurement in the SPS leads to an underestimation of the real temperature of the specimen of at least 100 °C. Therefore, it can be concluded that the measurement position plays a significant role. The measured temperature at the bottom of the upper punch in a FAST facility is closer to the real sample temperature in comparison with the temperature radially measured on/in the die wall during the SPS. These findings imply that the lower sintering temperatures reported for the SPS technique can be partially due to the underestimation of the actual temperature within the sample.

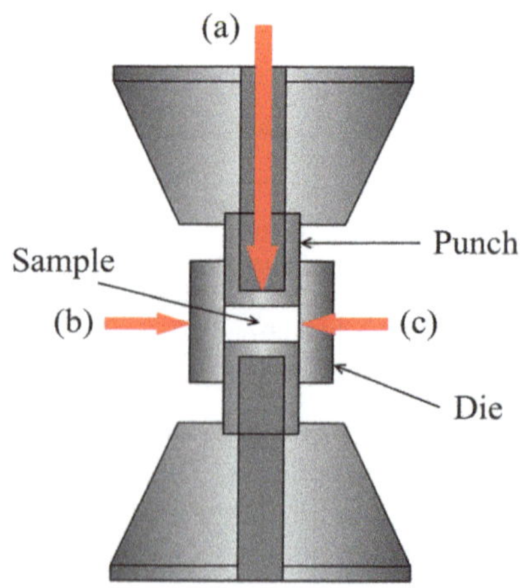

Fig. 4.12 Schematic showing the positions of the temperature measurements: (**a**) pyrometer in FAST, (**b**) pyrometer in the SPS, (**c**) thermocouple in the SPS. (Reprinted from Langer et al. [34], Copyright (2010) with permission of John Wiley & Sons)

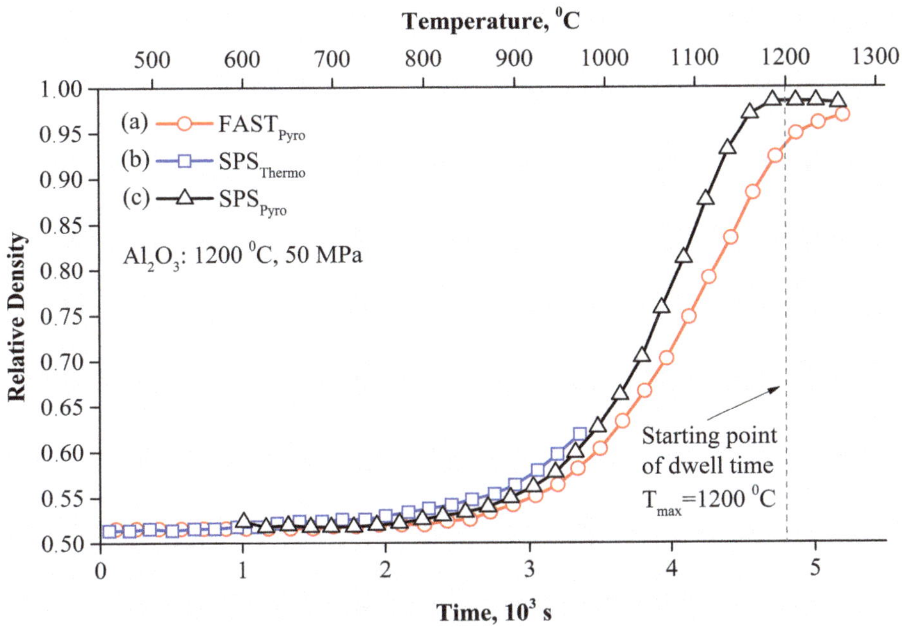

Fig. 4.13 Densification curves for Al_2O_3 as a function of time: (**a**) FAST (pyrometer), (**b**) SPS (thermocouple), (**c**) SPS (pyrometer), heating rate 10 K/min, applied pressure 50 MPa, and dwell time 15 min. (Reprinted from Langer et al. [34], Copyright (2010) with permission of John Wiley & Sons)

In a series of publications, Langer et al. have directly compared the results of hot pressing and FAST using Al_2O_3 [35], 8YSZ [36], and ZnO [37] powders. The main conclusion from these studies was the significance of the temperature control below 450 °C, which is often done by means of a pyrometer. In the absence of the regulation loop, the power input is limited to 35% of the nominal power before the system reached 450 °C, as is recommended by the manufactures of the FAST equipment.

This leads to transient overheating of the sample in the beginning of the sintering cycle (Fig. 4.14). As a result, FAST-processed materials have bigger grains. However, when the temperature is controlled carefully by means of a thermocouple measuring the temperature at a distance of 5 mm from the sample and the transient overshoot is thus avoided, the grain sizes of the FAST and hot-pressed specimens are close to each other. Rapid densification of ZnO in FAST with a pyrometer control (Fig. 4.14b) is due to transient overheating; such a densification curve is not suitable for any analysis of the densification mechanisms.

Lee et al. [38] conducted SPS experiments with ZrN powder, during which the temperature was measured by a thermocouple inserted in a hole in the die, a pyrometer focused on a hole in the die and a thermocouple inserted directly in the sample (sacrificial thermocouple), as is shown in Fig. 4.15a. The pyrometer measured the temperature until the final temperature was reached, and the sacrificial

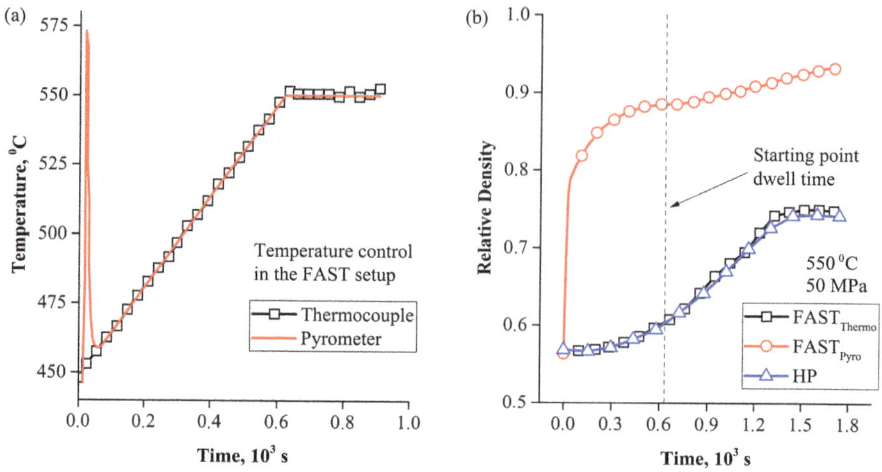

Fig. 4.14 Temperature overshooting in a FAST facility (**a**) and densification curves of ZnO nanopowder (20–50 nm initial particle size) hot-pressed and processed by FAST with a pyrometer and thermocouple temperature control, 50 MPa, 550 °C (**b**) Rapid densification of ZnO in FAST with a pyrometer control is due to transient overheating. (Reprinted from Langer et al. [37], Copyright (2011) with permission of John Wiley & Sons)

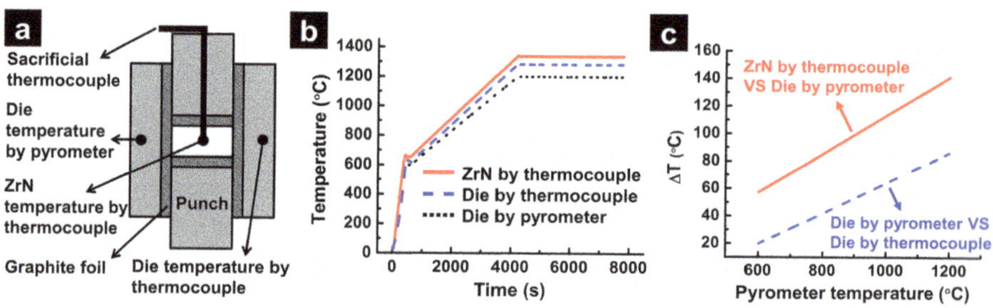

Fig. 4.15 (**a**) Schematics of a SPS setup with temperature measurement points, (**b**) temperature evolution of different measurement spots, (**c**) differences (ΔT) between ZrN temperature measured by thermocouple and die temperature measured by pyrometer and between die temperature measured by pyrometer and die temperature measured by thermocouple (blue dash line). (Reprinted from Lee et al. [38], Copyright (2017) with permission from Elsevier)

k-type thermocouple embedded in ZrN powder was melted and finally broken at 1250 °C. Temperatures >1250 °C were estimated based on the calibration experiments ($T > 600$ °C). Compared with the die temperature readings from the pyrometer, the die and powder measured by thermocouple showed higher temperatures during the SPS cycle (Fig. 4.15b, c). For example, at 1200 °C, ΔT between the ZrN sample and the die is 141 °C (Fig. 4.15c). This temperature difference is due to heat dissipation from the die surface. Indeed, the location of the sample in the SPS die set is more favorable to heat accumulation compared with the die surface.

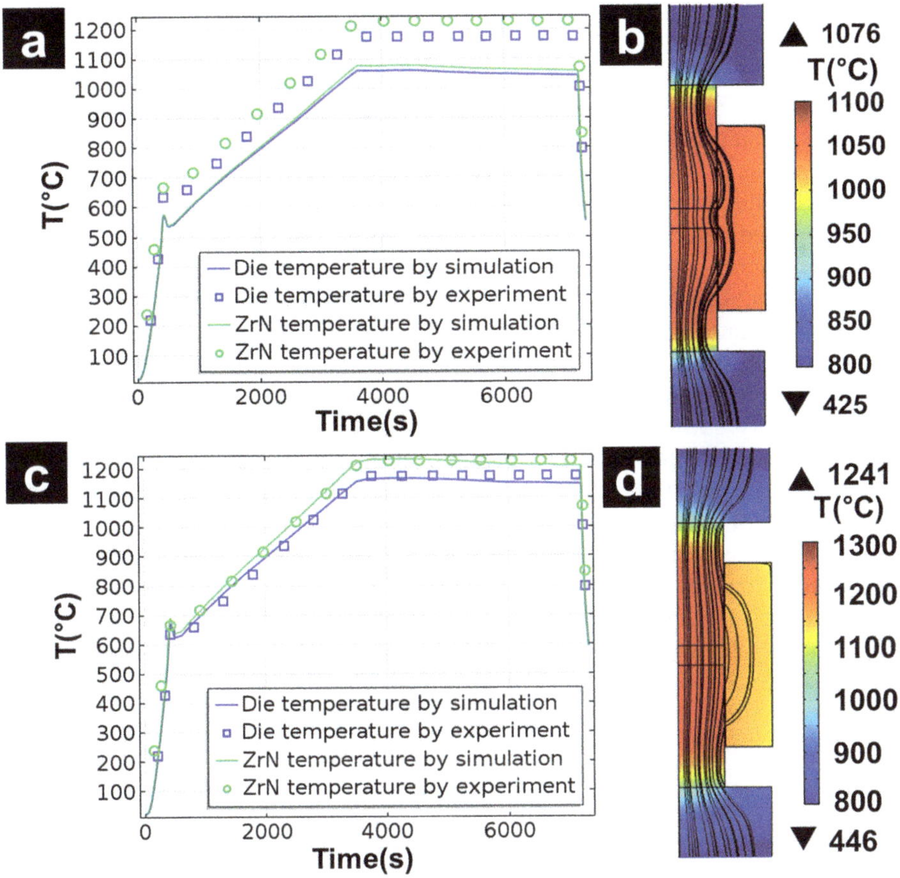

Fig. 4.16 Experimentally measured and finite element modeling-simulated temperatures of ZrN powder and graphite die during a SPS thermal cycle (**a**) without contact resistance and (**c**) with electrical and thermal contact resistance. Simulation map of the temperature and electric current flow at the end of the holding time (**b**) without contact resistance and (**d**) with electrical and thermal contact resistance. SPS at 1100 °C under 60 MPa and 10 °C·min was used. (Reprinted from Lee et al. [38], Copyright (2017) with permission from Elsevier)

Contact resistances of SPS die/punch/sample assemblies were studied in Refs. [39, 40]. It was shown that the vertical contact resistance has an important impact on the magnitude of the fraction of the electric current passing through the conductive powder [38]. Without the consideration of the vertical electrical and thermal contact resistances, the simulated temperatures for the die and sample during the holding stage are 138 and 172 °C lower than the experimentally measured temperatures (Fig. 4.16a), which means that the electrical and thermal contact resistances cause heat dissipation. The major fraction of the electric current is flowing through the die (Fig. 4.16b). However, the results of the modeling considering the vertical electric and thermal contact resistances are close to the experimentally measured temperatures (Fig. 4.16c). The electric current from the top punch was split, and the high portion of electric current passed through the powder during the holding stage (Fig. 4.16d). A nonhomogeneous temperature distribution between the die and the central column of the tooling set (punches and powder) was observed due to the presence of the contact resistance.

Possibilities of rapid heating in the SPS facilities are well appreciated and widely discussed in the sintering community; however, the cooling rates of the samples are rarely reported, intentionally varied, or adjusted. Usually, the specimen is cooled after the sintering procedure is over using a water cooling system; alternatively, argon flooding can be used. Zhang et al. [41] suggested a way to overcome this limitation by integrating a gas-quenching system into a SPS facility of standard design, which would allow for rapid cooling of sintered specimens right after the sintering step. Operations, such as sinter hardening, are, therefore, made possible in the SPS facility, saving the energy and costs associated with re-heating of the sample necessary for conventional sinter hardening, which is performed separately from the sintering step itself. It was suggested to use high-velocity nitrogen flow to provide rapid cooling of the sintered specimens. Natural cooling occurred at 1.6 $°Cs^{-1}$; with the quenching system, a cooling rate of 6.9 $°Cs^{-1}$ was reached. SPS with an integrated quenching system was successfully applied to Ti6Al4V alloys, in which gas quenching resulted in microstructural changes (precipitation of nanocrystalline Al_2Ti and lamellar morphology of α-phase opposed to globular morphology of $\alpha + \beta$ phases observed in the naturally cooled samples), an increase in hardness, an increase in ultimate compressive strength, and an increase in ductility.

4.5 Proportional–Integral–Derivative (PID) Control of Temperature During SPS and Regulation Quality Improvement

The origin of difficulties in the proportional–integral–derivative (PID) control of the temperature during the SPS was analyzed by Manière et al. [42] using finite element modeling. It was confirmed that PID regulation is strongly influenced by the heating response and the lag between the area of the main heat generation and the area of heating control. This lag is a function of tooling dimensions and the thermal contact resistance (TCR), which decreases the heat flux between the heat generation and control areas. The authors used a coupled electrothermal finite element model of the regular SPS setup (Fig. 4.17). The model includes the Joule heating equations and a set of temperature-dependent electric and thermal contacts. The verification experiment was performed; in both the experiment and the simulation, the sample was a fully dense silver pellet.

The configuration shown in Fig. 4.17 uses a high-pressure enabling thick die (inner diameter 10 mm, external diameter 30 mm) and a conductive sample. In these conditions, the heat generation is mainly located in the punches, and the die heating

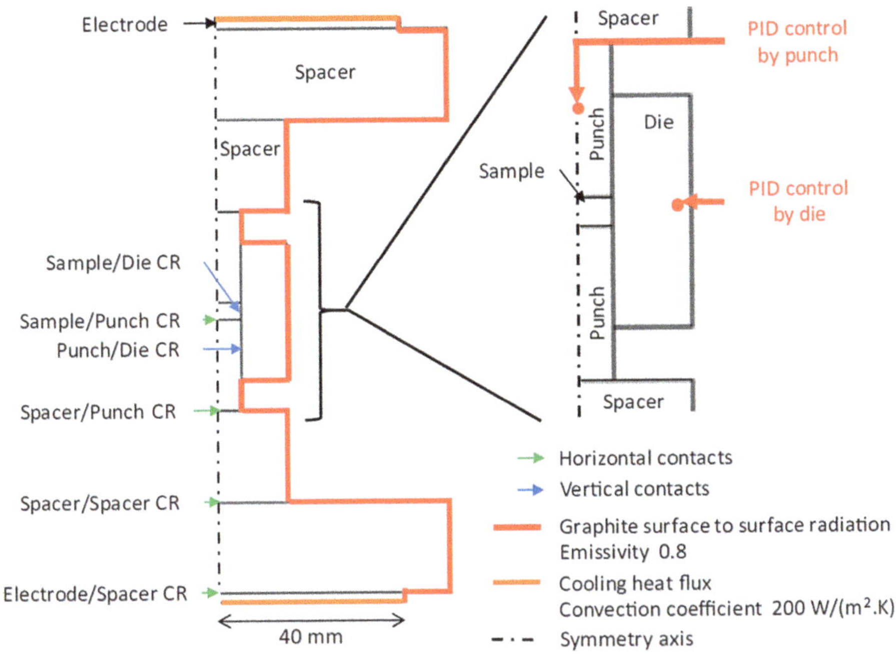

Fig. 4.17 Cross-section of a SPS assembly showing the location of each electric/thermal horizontal/vertical contact resistance taken into account in the model, the main thermal boundary conditions, and the two different temperature control locations. (Reprinted from Manière et al. [42]. This work is licensed under a Creative Commons Attribution License, http://creativecommons.org/licenses/by/4.0/)

occurs with a time lag (Fig. 4.18a). In this case, the PID regulation shows a lot of oscillations of the electric current and temperature (Fig. 4.19a). When the electric current increases, for instance, at 180 s (Fig. 4.18a), the maximum temperature is in the middle of the punches, and the die heating takes place with a time lag. During the electric current decreasing stage (220 s), the heat is mainly evacuated by thermal conduction in the punches in contact with the colder spacer. Under these conditions, the die cooling has a time lag. This lag of the die heating and cooling is explained by the presence of a high TCR at the vertical punch/die contact. During the heating stage, the TCR decreases the heat exchanges from the punches to the die. During the cooling stage, the heat flux that passes across this interface is reduced by the TCR, inducing the time lag in the cooling at the die surface. This impact of the vertical interface TCR on the temperature regulation was verified. If the TCR is multiplied by two, the punch/die temperature difference increases from 100 K to 200 K (Fig. 4.18a–b), and both the temperature and electric current oscillations' magnitude increase (Fig. 4.19a). If the die thickness is increased, the mass of the die material to be heated by the punches is increased, and the punch/die temperature difference is also about 200 K (Fig. 4.18c) causing high regulation oscillations (Fig. 4.19b). The heat generation by punches was reduced by reducing their heights in order to

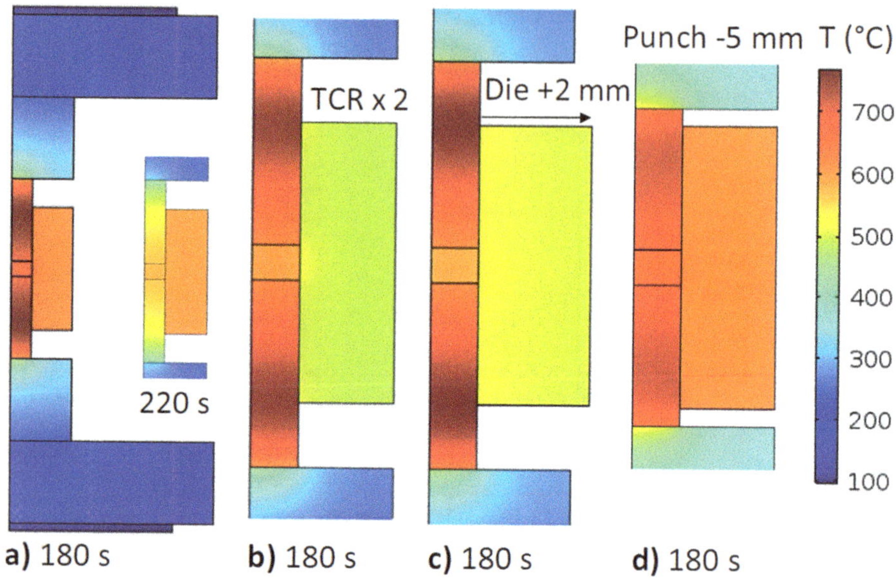

Fig. 4.18 Temperature fields determined for the die-controlled regulation: (**a**) at 180 and 220 s, (**b**) with vertical thermal contact resistance (TCR) multiplied by 2, (**c**) with a thicker die, (**d**) with smaller punches. (Reprinted from Manière et al. [42]. This work is licensed under a Creative Commons Attribution License, http://creativecommons.org/licenses/by/4.0/)

homogenize the temperature and stabilize the PID regulation. In this case, the punch/die temperature difference is reduced down to 80 K, but the PID regulation still included oscillations (Fig. 4.19b).

The finite element simulation allowed determining an area within the SPS setup having the highest heating responsiveness [42]. In order to reveal such areas, finding the heating rate field generated by the electric current impulse is a key issue. It was found that the areas of maximum responsiveness are located not in the die but at the mid-height of the punches. As a result, the punches are better candidates for the PID regulation compared with the die, which is heated with a time lag.

4.6 "Plasma" Issue in SPS

In a study by Hulbert et al. [43], two experimental methods were employed in an attempt to determine the presence or absence of plasma during SPS: in situ atomic emission spectroscopy (AES) (Figs. 4.20, 4.21, and 4.22) and ultrafast in situ voltage measurements (Fig. 4.23). With the use of these methods, no evidence of plasma was found during the SPS processes. This result was confirmed using several different powders across a wide spectrum of SPS conditions. The equipment consisted of a high-temperature fiber-optic flame probe connected to a fiber-optic patch cable. The

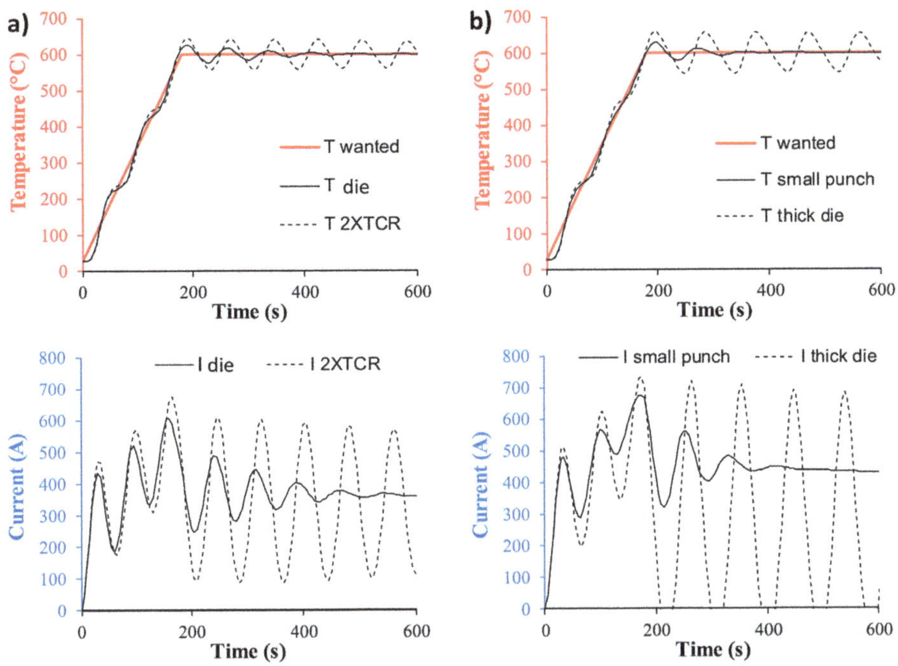

Fig. 4.19 PID regulated temperature and electric current under different conditions: (**a**) die temperature control regulation with vertical TCR multiplied by 2, (**b**) die temperature control regulation for small punch and thick die configurations. (Reprinted from Manière et al. [42]. This work is licensed under a Creative Commons Attribution License, http://creativecommons.org/licenses/by/4.0/)

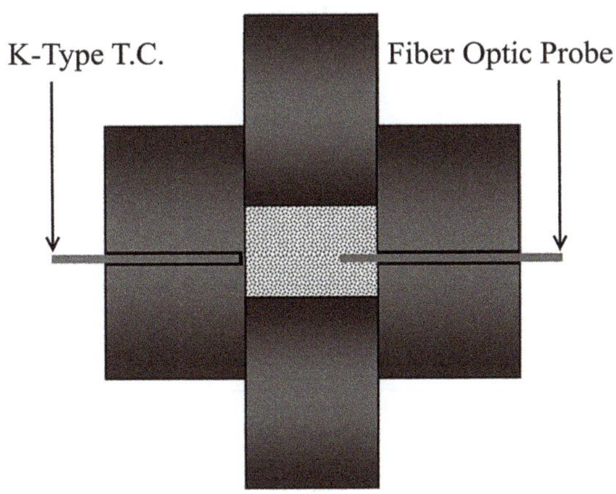

Fig. 4.20 The experimental setup used for in situ atomic emission spectroscopy studies of the phenomena involved in the SPS. (Reprinted from Hulbert et al. [43], Copyright (2008) with permission of AIP Publishing)

fiber-optic flame probe was placed in a borosilicate glass capillary to prevent damage to the probe during sintering. The glass sleeve lined a through-hole located at the middle plane of the die and was butted up against the powder compact (Fig. 4.20). In

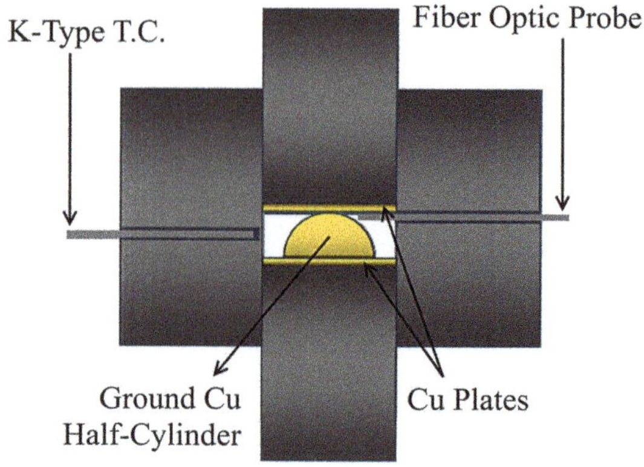

Fig. 4.21 The configuration of the Cu plate/Cu half-cylinder experiments. The sides of the punches were coated with boron nitride to force the current through the plate/cylinder assembly. (Reprinted from Hulbert et al. [43], Copyright (2008) with permission of AIP Publishing)

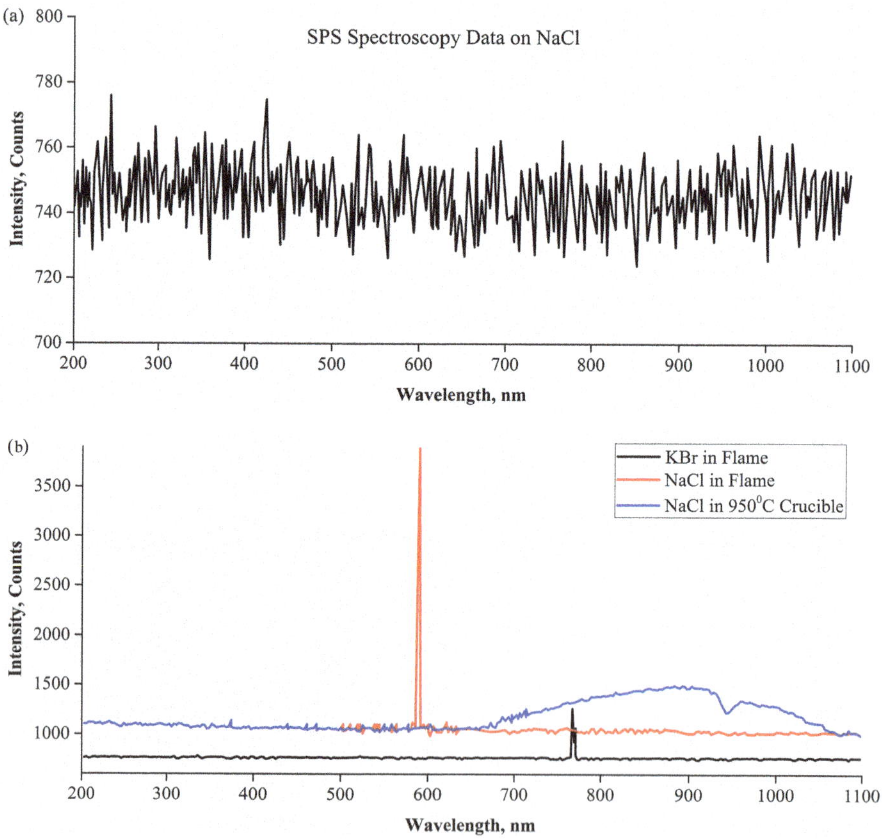

Fig. 4.22 (**a**) AES data for NaCl sintered by SPS, the plot is representative of all AES spectra collected in different configurations of the experiment and (**b**) flame spectroscopy data of calibration experiments of the sensitivity and resolution of the equipment. (Reprinted from Hulbert et al. [43], Copyright (2008) with permission of AIP Publishing)

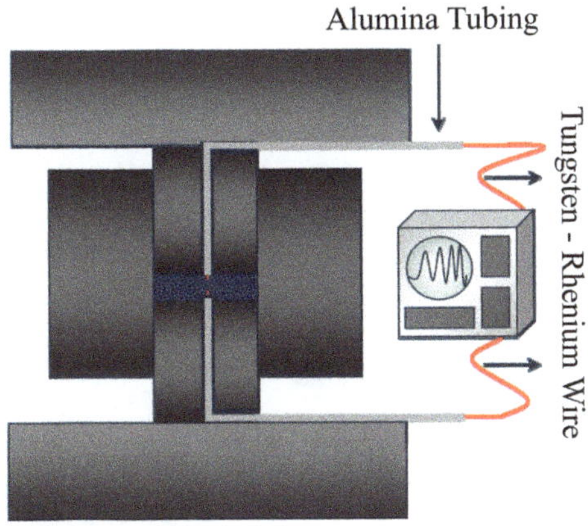

Fig. 4.23 A diagram of the setup for the ultrafast in situ voltage measurements. (Reprinted from Hulbert et al. [43], Copyright (2008) with permission of AIP Publishing)

the case of metallic powdered specimens, the probe, along with the glass sleeve, was actually pressed into the powder charge, with the tip of the probe located at the center of the specimen. Another fiber-optic patch cable was connected to a spectrometer with a high photonic sensitivity. In order to try and force a sparking event, a special experiment was designed using a standard SPS die and punch. Two copper plates were polished; a copper half-cylinder was sandwiched between the two copper plates and the two punches (Fig. 4.21).

The ultrafast in situ voltage measurements between the upper and lower punches were made using a digital storage oscilloscope (Fig. 4.23) that is capable of detecting events down to approximately 1 ns. Such time resolution is desirable if one is trying to detect highly transient voltage anomalies associated with plasma generation, sparking, or arcing events. The wires were embedded in the powder charges of Al, Cu, and Al_2O_3 powders and were approximately 1–3 mm apart.

The spectrum shown in Fig. 4.22a is for NaCl sintered at 650 °C for 3 min. This spectrum is representative of all the SPS spectra taken during this investigation including the control experiment with no powder charge. The results of the flame spectroscopy experiment with characteristic photons clearly detected for NaCl and KBr are shown in Fig. 4.22b. Those were used to sensitivity and resolution calibration of the equipment. In the SPS spectroscopy experiments with powders, no characteristic photons were detected. They were not detected during the experiment involving the copper plates and the partially ground copper cylinder either. Hulbert et al. [43] admit that, although the AES results do not speak favorably of the presence of plasma in the SPS, it is not possible to completely dismiss the possibility of plasma based solely on the AES results. When the results of the in situ voltage measurements were analyzed (Fig. 4.23), no voltage anomalies associated with sparking, arcing, or plasma generation were found. With a temporal resolution of approximately 1 ns, the technique would have detected any voltage transients necessarily associated with any type of plasma generation, arcing, or sparking events. Hulbert et al. concluded that the SPS voltage is approximately one order of magnitude less than the voltage required to generate plasma during the process.

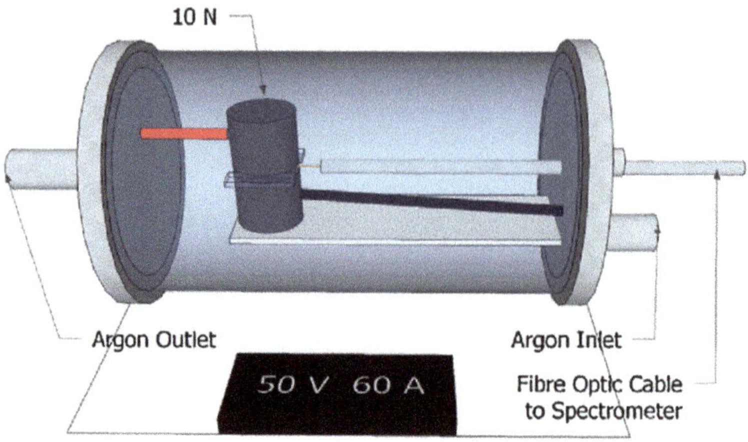

Fig. 4.24 Experimental setup for investigating the formation of plasma states in electric current-assisted sintering (quartz tube). (Reprinted from Saunders et al. [44], Copyright (2014) with permission from Elsevier)

In a study conducted by Saunders et al. [44], AES was used to detect possible transitions of the material to the plasma state during electric current-assisted sintering; in their experiments, conductive materials – zirconium diboride (ZrB_2) and tungsten – were used. The authors arrived at the same conclusion: SPS under typical conditions (<10 V applied across the sample) does not generate plasma states. However, when a voltage of 50 V was applied across a 1-mm-thick specimen of a loosely packed tungsten powder under pressureless conditions in a flowing argon atmosphere (in a setup specially designed for the "plasma search" (Fig. 4.24)), evidence of plasma was obtained in the form of spectral lines of oxygen, tungsten, and argon.

It is reported that bright flashes were visible with the naked eye during the first 2–3 s of treatment. The spectral emission data were acquired during the first 600 ms of discharge (Fig. 4.25). The starting time *t* was taken as the time, at which a profile with zero emission was observed. The temperature of the particles was calculated by interpreting the black body radiation curve captured by the spectrometer. After 100 ms and 300 ms, sharp spectral emission lines are detected, while at 200, 500, and 600 ms, temperatures of up to 3800 K were detected. The presence of tungsten lines indicated that some of the tungsten was sputtered from the surface of the powder particles. Such observations were not reported for experiments in conventional SPS facilities. The presence of oxygen lines showed that the formation of plasma was accompanied by surface cleaning – removal of oxides from the particles' surfaces. Microstructural evidence of local melting was observed in tungsten and ZrB_2 samples subjected to treatment at 50 V for 5 s.

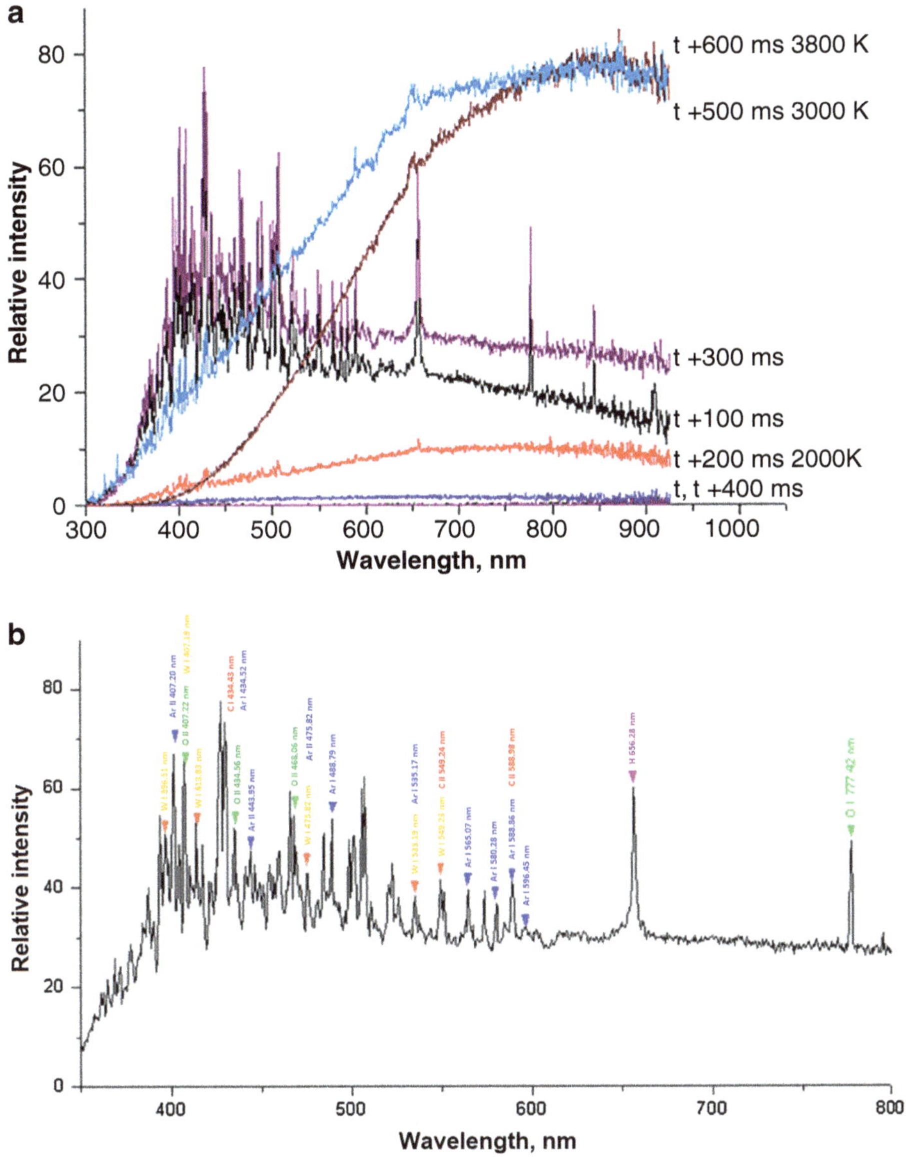

Fig. 4.25 Spectra obtained in the case of sintering of a tungsten powder in the setup shown in Fig. 4.24: black body radiation and spectral emission lines (**a**), identification of the spectral lines of the ionized species (**b**). (Reprinted from Saunders et al. [44], Copyright (2014) with permission from Elsevier)

4.7 Processes at the Inter-particle Contacts in SPS

The influence of an increase in electrical resistivity with temperature on the process of sintering has been addressed by Song et al. [45], who suggested that SPS of metallic powders proceeds through a "self-adjusting" mechanism. The temperature profiles were calculated by considering the geometrical features of the contact of two spheres. The electrical resistivity was assumed uniform throughout the particle. Heat conduction during current pulses was fully excluded from consideration; the heat produced in a microunit (a cylinder of a small thickness taken in the vicinity of the contact between the particles) as a result of Joule heating was used to increase the temperature of this microunit. Redistribution of current between the two particles is determined by the competition between two factors: the growing cross-sectional area of the neck causing a reduction in the resistance of the neck region and the increasing resistivity of the material with temperature due to Joule heating. The calculated temperature distribution for a pair of spherical copper particles within a compact sintered by SPS for a sequence of 12 pulses 3.3 ms long and a current of 1000 A is shown in Fig. 4.26.

The fracture surface of the compacts formed from a spherical copper powder reveals the neck zones with a structure evolving with the SPS temperature (Fig. 4.27). The fracture of a single inter-particle neck shows fine grains, which, according to the authors, formed due to local melting and rapid cooling of the inter-particle contacts.

Aman et al. [46] reported unconventional neck morphology between copper particles formed in pressureless SPS, which suggested an ejection mechanism of mass transfer operating during sintering. Figure 4.28 compares neck morphologies found in the spark plasma sintered (in a pressureless process) and conventionally sintered compacts. In the latter, the necks were smooth, which pointed to surface diffusion-enabled bonding. An explanation for the observed neck morphologies in the compacts produced by SPS was suggested by the authors based on the possible local field enhancement and ion-enhanced field emission. The unconventional necks thus formed under conditions favoring local thermal breakdown and local sparks/plasma. It was suggested that local plasma generated between the two copper particles involved enough matter and energy to bond the particles by an ejection mechanism.

Yanagisawa et al. [47] showed that sparking events between copper particles are more likely to occur at low pressures and high current densities. The contact region in the case of sparking caused by a single discharge (Fig. 4.29) has a morphology shown in Fig. 4.30. A scheme describing the contact formation through mass transfer assisted by melting is presented in Fig. 4.31 (mass transfer is shown by arrows).

The processes occurring at the inter-particle contacts are also of great importance for nonmetallic materials. According to Vasiliev et al. [48], the strength of porous monoliths produced by SPS from a zeolite powder was due to strong inter-particle bonding formed during SPS. The formation of strong bonds between the particles was associated with a structural collapse of the contacting zeolite grains, the

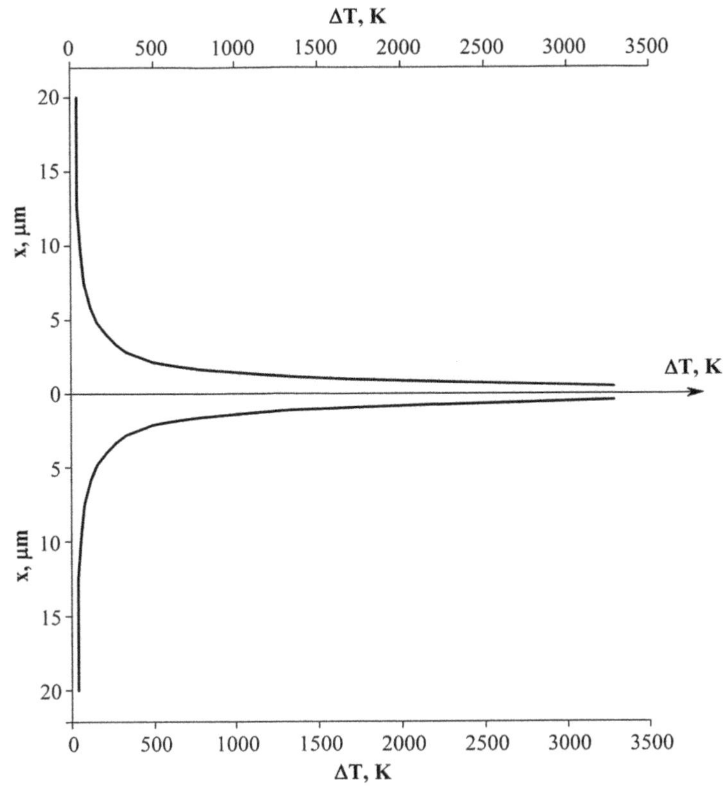

Fig. 4.26 The calculated temperature distribution at the contact between two spherical copper particles within a compact sintered by SPS. (Reprinted from Song et al. [45], Copyright (2005) with permission of John Wiley & Sons)

breakage and rearrangement of chemical bonds, and the formation of amorphous regions in the proximity of the contact between the particles (Fig. 4.32).

Dudina et al. [49] observed melting of the copper matrix during SPS of Ti_3SiC_2–Cu composite particles produced by mechanical milling. An interfacial reaction occurring preferentially in the vicinity of contacts between the powder particles was observed. During sintering of the 5 vol. % Ti_3SiC_2–Cu powders, a smooth surface of the powder agglomerates, their close-to-spherical shape and well-established contacts between them helped avoiding melting of the copper matrix, which would otherwise have taken place as a result of a local increase in the temperature at the particle contacts under pulsed electric current. In the 18 vol.% Ti_3SiC_2–Cu composite (Fig. 4.33), the agglomerates had a rough surface and an irregular shape, which prevented the formation of intimate contacts between them. In that case, SPS resulted in partial melting of the Cu matrix, the evidence of which is seen in both the cross-sectional view and fracture surface of the composite (Fig. 4.34a–b). The XRD phase analysis (Fig. 4.34c) shows that the

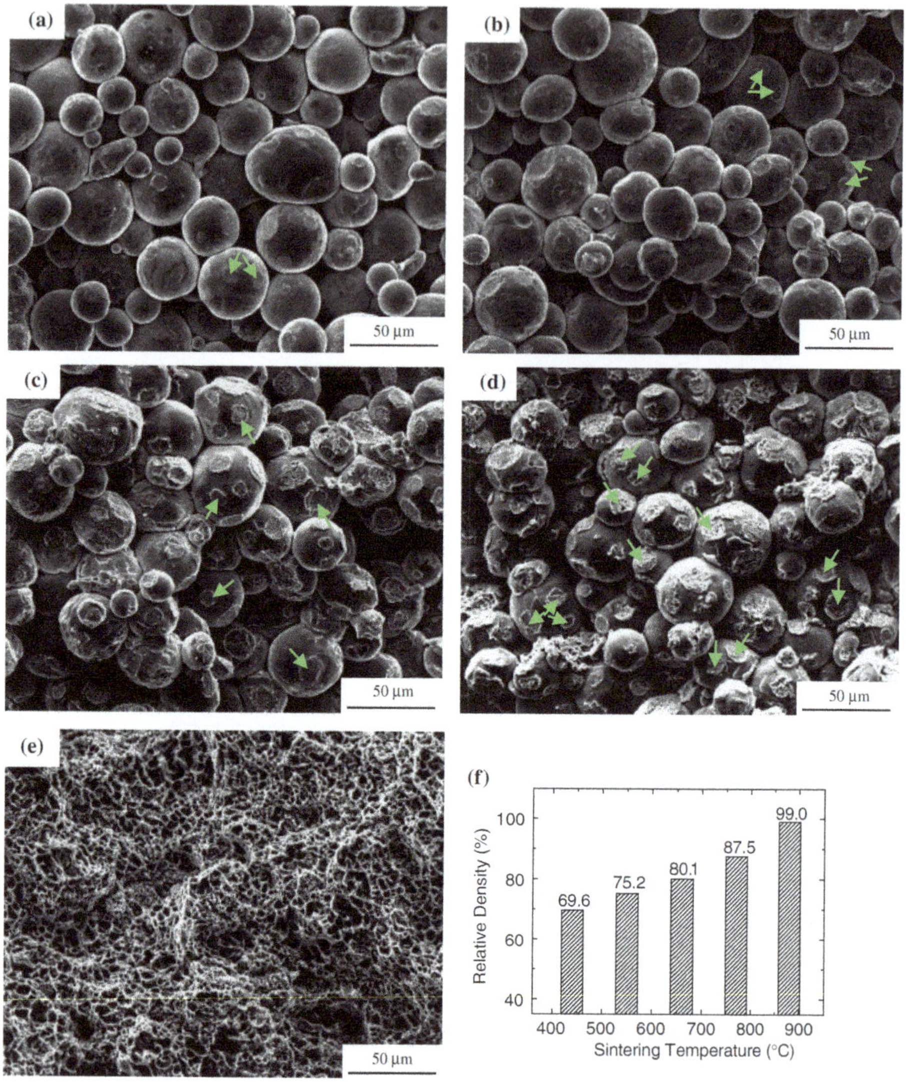

Fig. 4.27 Fracture surfaces of the copper compacts spark plasma sintered at different temperatures ((**a**) 440 °C, (**b**) 550 °C, (**c**) 660 °C, (**d**) 770 °C, (**e**) 880 °C) and (**f**) the relative density of the compacts (the arrows indicate the neck zones). (Reprinted from Song et al. [45], Copyright (2005) with permission of John Wiley & Sons)

local melting of copper results in its fast interaction with Ti_3SiC_2 leading to the formation of TiC_x separated from the Cu–Si melt during the SPS. The energy-dispersive spectroscopy analysis confirmed the absence of Ti and the presence of Cu and Si in the re-solidified areas. This result demonstrated that local melting during the SPS can destabilize the Ti_3SiC_2 phase leading to de-intercalation of Si,

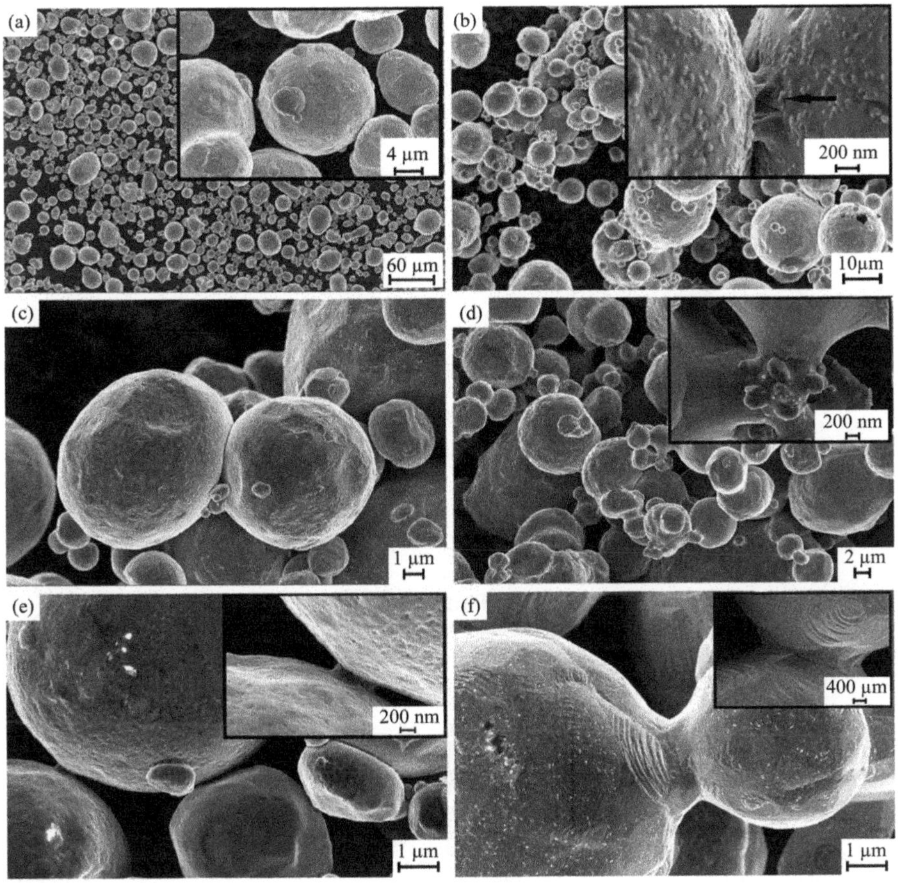

Fig. 4.28 Initial Cu powder (**a**), pressureless SPS at 565 °C (**b**, **c**), pressureless SPS at 782 °C (**d**), conventionally sintered at 600 °C (**e**), and conventionally sintered at 800 °C (**f**). (Reprinted from Aman et al. [46], Copyright (2012) with permission of Springer)

which dissolves in metallic Cu. Fast cooling of inter-particle contacts, which are the locations of the melt formation during the SPS, enables supersaturated Cu (Si) solid solution to crystallize. Partial melting of the Cu matrix compromises the phase stability and uniformity of the microstructure of the Ti_3SiC_2–Cu composites and thus cannot be suggested as a pathway to enhanced densification.

Very few studies have been conducted until now to compare the SPS behavior of powders having the same composition but different microstructures and phase distribution and, as a result, different contact conditions of the particles, the importance of which during the initial stages of the SPS process cannot be exaggerated. Worth noting, however, is a study conducted by Zhao et al. [50], showing that the structure of the composite powder has a significant effect on the final microstructure, density, and mechanical properties of bulk composites obtained by SPS. One of the mixtures prepared by Zhao et al. was obtained by mixing a submicron W powder

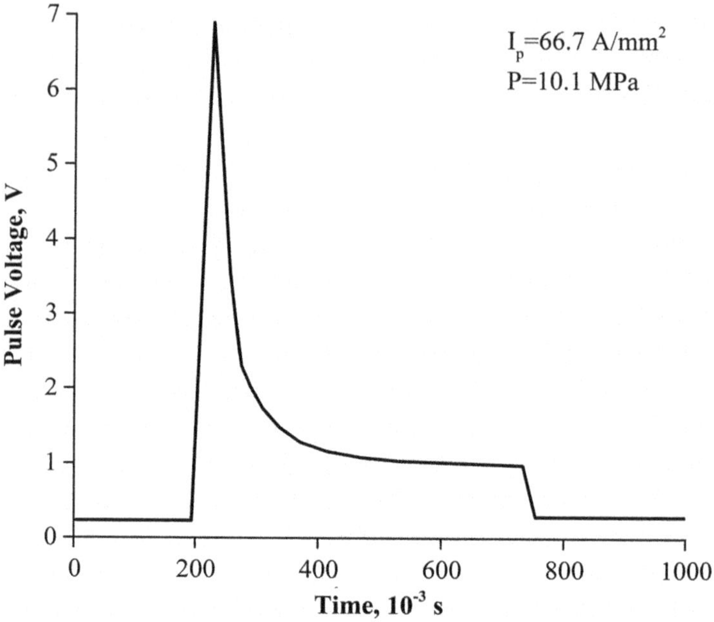

Fig. 4.29 Voltage evolution with time during a pulse discharge. (Reprinted from Yanagisawa et al. [47], Copyright (2002) with permission form Elsevier)

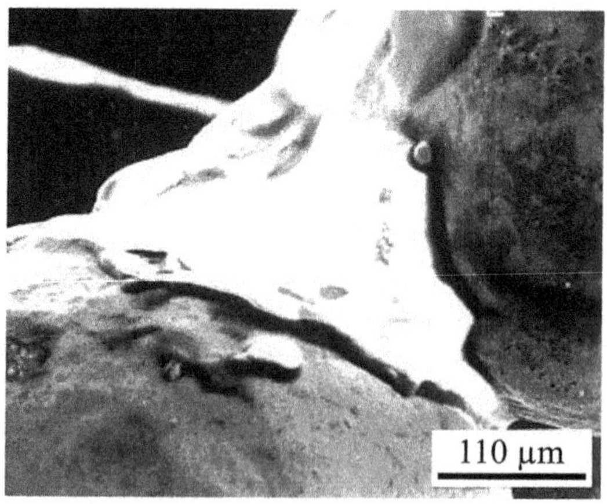

Fig. 4.30 Contact between copper particles in the case of sparking by a single discharge. (Reprinted from Yanagisawa et al. [47], Copyright (2002) with permission form Elsevier)

with a Co nanopowder by wet milling; the other was produced by dry milling of the same powder W powder and a Co powder with particles of micrometer size (Fig. 4.35). In the latter, Co thin films coated the WC particles. Because of the good contact condition of the Co coating and the WC cores and a high electrical conductibility of cobalt, most of the current passes through the cobalt thin films during the sintering process.

This contact condition was very favorable for the current distribution in the SPS process (Fig. 4.36). As a result of self-Joule heating of the Co film-coated WC

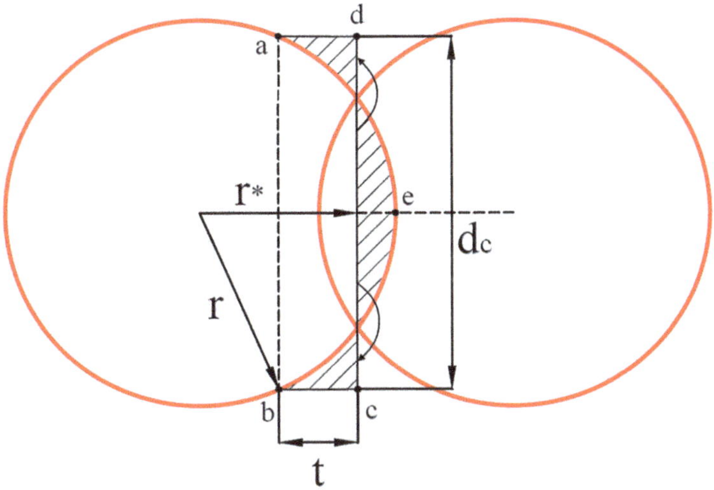

Fig. 4.31 A scheme of the neck formation via mass transfer by melting. (Reprinted from Yanagisawa et al. [47], Copyright (2002) with permission form Elsevier)

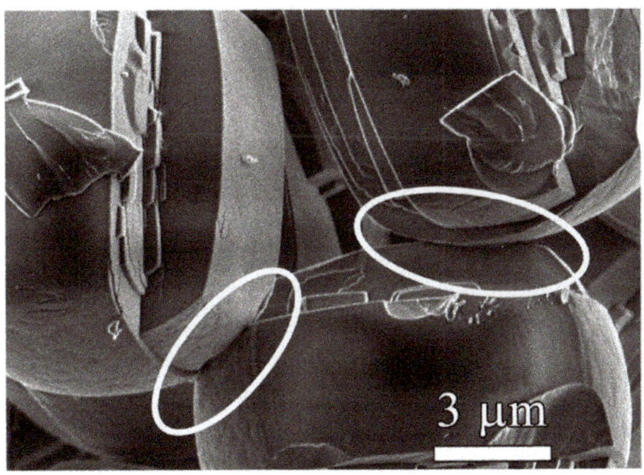

Fig. 4.32 Contact zones between the particles of zeolite spark plasma sintered to produce a porous structure. (Reprinted from Vasiliev et al. [48], Copyright (2010) American Chemical Society)

particles, a homogeneous temperature distribution in the sample was achieved, leading to a rapid densification progress, a homogeneous microstructure, and an excellent combination of mechanical properties of the sintered material (Fig. 4.37).

Holland et al. [51] reported in situ transmission electron microscopy (TEM) observations of the stages of pressureless field-assisted sintering of nanosized nickel particles. Scanning tunneling microscopy (STM) inside a TEM was used to apply an electrical current directly to the powder particles. An isolated Ni particle with a diameter of 100 nm located on the STM tip was brought into mechanical contact with a powder agglomerate on the TEM sample consisting of grains 100–200 nm in size. To initiate sintering through electrical current, a negative DC bias (up to 10 V)

Fig. 4.33 Morphology of the 18 vol.% Ti_3SiC_2–Cu composite powder used for SPS experiments in Ref. [49]

was applied to the STM tip (Fig. 4.38). The consolidation occurred in the absence of any external heat source through multistage sintering. At the first sintering stage, relative particle rotation was observed. After an annealing time of approximately 145 s at 10 V, rapid consolidation of the particle agglomerate occurred. The intermediate stage of sintering was very short. The result of the third sintering stage was a polycrystalline agglomerate with grain boundaries and no apparent interconnected porosity. The current evolution measured at a bias of −10 V (Fig. 4.39) shows instances of decreased current, which imply increased resistance due to the early neck formation and involvement of the new grains.

Matsuno et al. [52] used TEM to study the neck formation between nickel particles having a layer of NiO on their surfaces during the early stages of sintering. They found that the surface oxides remain during the early stages of sintering and the neck formation is limited by self-diffusion of nickel through the oxide layer. The presence of carbon, however, facilitates the neck formation, as it lowers the reduction temperature of the nickel oxide. Bonifacio et al. [53, 54] showed that by applying bias to agglomerated Ni nanoparticles in situ in a TEM, it is possible to directly observe the dielectric breakdown effect. Figure 4.40 shows the initial contact between Ni nanoparticles before the bias was applied as well as the evolution of this contact as successive runs of applied voltage were applied. After the third run, the oxide layer on the Ni nanoparticles became discontinuous, and after the fourth run, a metallic neck continued to grow. Using the electron energy loss spectroscopy (EELS), spatial distribution of the O/Ni atomic ratio was studied, which showed that under the applied field, oxygen migrated from the contact.

The application of these findings directly to the SPS is not possible, as the field strength in the experiments was very high (or the order of 10^8 V m^{-1}). However, considering a possibility of field amplification at the contact area of dielectric particles, the results of these experiments are of interest for discussing the sintering mechanisms involved in different field-assisted consolidation techniques.

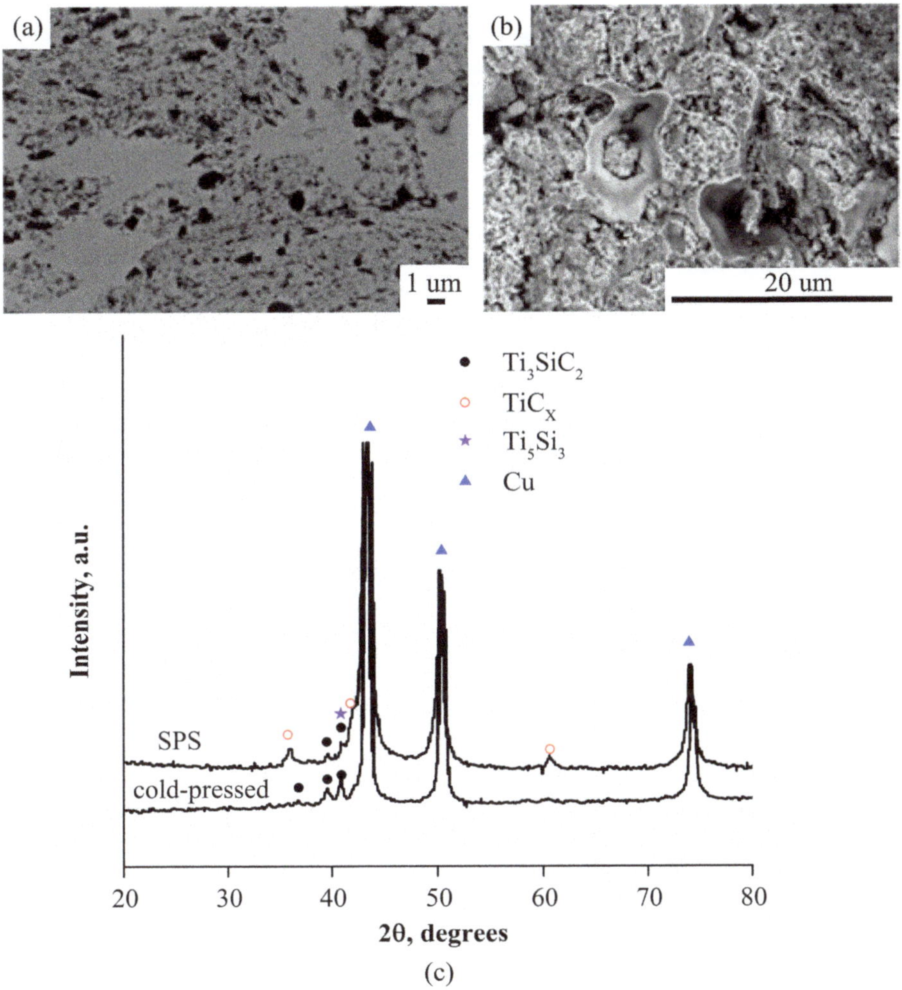

Fig. 4.34 Microstructure of the spark plasma sintered composite ((**a**) cross-section and (**b**) fracture surface) and XRD patterns of the cold-pressed and sintered compacts (**c**). (Reprinted from Dudina et al. [49], Copyright (2013) with permission of Springer)

A schematic describing the process of breakdown through the vacancy formation is discussed by Bonifacio et al. [54] (Fig. 4.41). The applied electrical field polarizes the metal oxide layer on the surface of the particles during stage I, which causes the formation of oxygen vacancies at stage II. The oxygen vacancies act as electron trapping sites. Once a critical number of charge traps are reached (according to the percolation model), the local field enhancement causes an electrostatic field across the dielectric layer that is equivalent to the breakdown voltage of the oxide (stage

Fig. 4.35 Morphology of WC–Co composite powders: (**a**) submicron WC powder mixed with Co nanopowder (by wet milling), (**b**) submicron WC powder dry-milled with Co micron particles (Co layers coat WC particles). (Reprinted from Zhao et al. [50], Copyright (2007) with permission from Elsevier)

III). The electrothermal depletion of the surface oxides occurs by means of oxygen migration away from the inter-particle contact constituting stage IV.

4.8 The Effect of High Heating Rates: Experimental Studies

A high heating rate is one of the characteristic features of pulsed current-assisted sintering process and a necessary prerequisite of a successful consolidation by SPS. Heating rates up to 2000 °C/min can be achieved in the SPS. For non-conductive powders, the high heating rate is the main distinction between SPS and traditional hot pressing. Zhou et al. [55] reported that the same final density was obtained upon an increase of heating rate from 50 to 300 °C/min with the same maximum temperature, which corresponded to a sixfold decrease in the allowed sintering time. This means the average shrinkage rates are directly proportional to the heating rates. Similar observations were reported by Shen et al. [56], Chu et al. [57], and Stanciu et al. [58].

Shen et al. [56] reported experimentally measured shrinkage rates for spark plasma sintered alumina and suggested an explanation of a shift in the shrinkage rate maximum to lower temperatures with increasing heating rate by enhancement of the grain-boundary diffusion with increasing current, by which higher heating rates are achieved. With heating rates lower than 350 °C/min, fully dense samples were obtained, but at higher heating rates, the porosity of the compacts increased, as can be seen in Fig. 4.42 (the experiments were conducted at a constant pressure of 50 MPa, without holding time at maximum temperatures of 1300° or 1400 °C). Munir et al. [24] point out that the interpretation of the experimental data on the effect of the heating rate on densification may be complicated by the differences in the effective thermal and electrical conductivities of the powder compact, contact

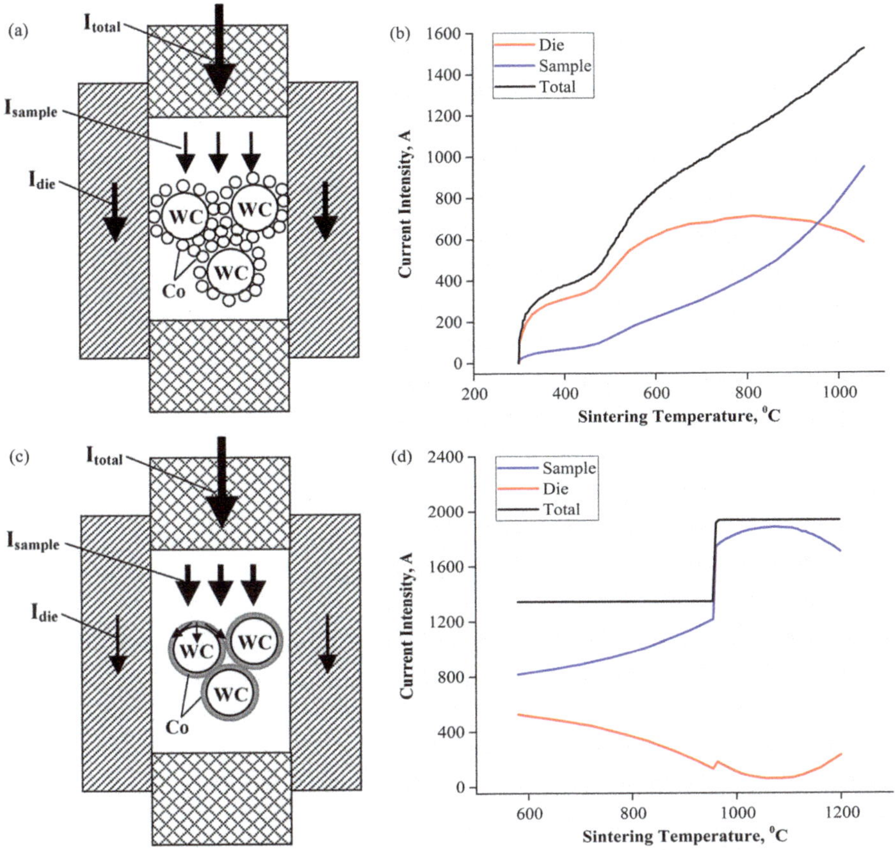

Fig. 4.36 Schematic of the SPS of WC–Co powders of two different structures (**a**, **c**) and estimated current distribution between the die and the sample (**b**, **d**). Submicron WC powder mixed with Co nanopowder (**a, b**) and Co-coated submicron WC particles (**c, d**). (Reprinted from Zhao et al. [50], Copyright (2007) with permission from Elsevier)

resistances between the sample and the die and between parts of the die assembly, and the pressure application schedule. However, the effect of the heating rate on the grain size of the sintered compacts is rather clear, as is seen in Fig. 4.43 [56]. The observed decrease in the grain size with increasing heating rate is related to bypassed grain coarsening processes: at higher heating rates the sample has a shorter dwell time, and grain growth should be anticipated in the material to a lower extent.

Higher densification rates at higher heating rates can be explained by the presence of the additional defects stored in the material during processing. These defects can be initial "biographic" defects resulting from the processes of thr powder synthesis, as was suggested by Ivensen [59], or defects in the boundaries between particles, as was suggested by Dabhade et al. [60]. It has been found that long high-temperature treatments substantially change the internal structure of grain boundaries: the fraction of high-angle random boundaries with high diffusivity decreases with time

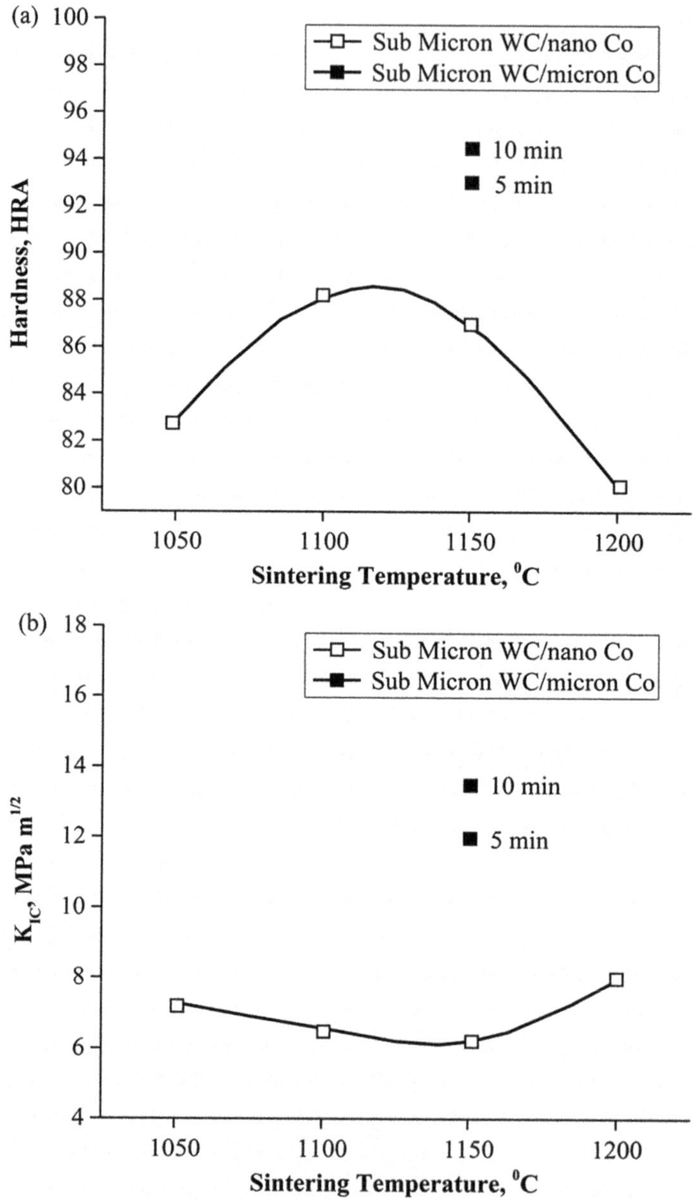

Fig. 4.37 Hardness (**a**) and fracture toughness (**b**) of WC–Co spark plasma sintered materials obtained from a submicron WC powder mixed with a Co nanopowder (by wet milling) and a submicron WC powder dry-milled with Co micron particles (Co layers coat the WC particles). The holding time for the mixture of the submicron WC powder with the Co nanopowder was 10 min. The applied pressure in all experients was 60 MPa. (Reprinted from Zhao et al. [50], Copyright (2007) with permission from Elsevier)

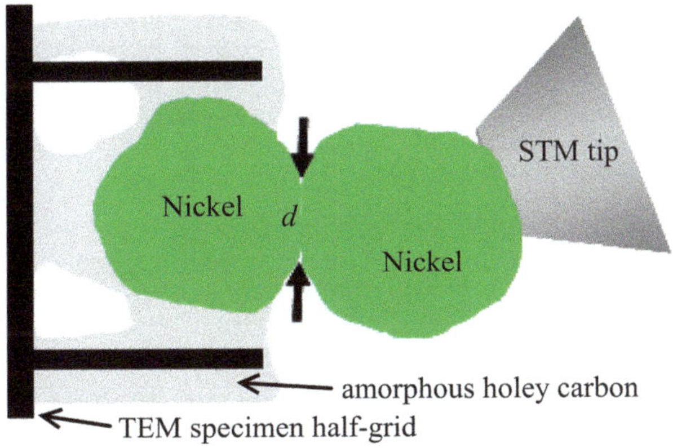

Fig. 4.38 Schematic of the in situ electric field-assisted sintering experiments in a transmission electron microscope. (Reprinted from Holland et al. [51], Copyright (2010) with permission of AIP publishing)

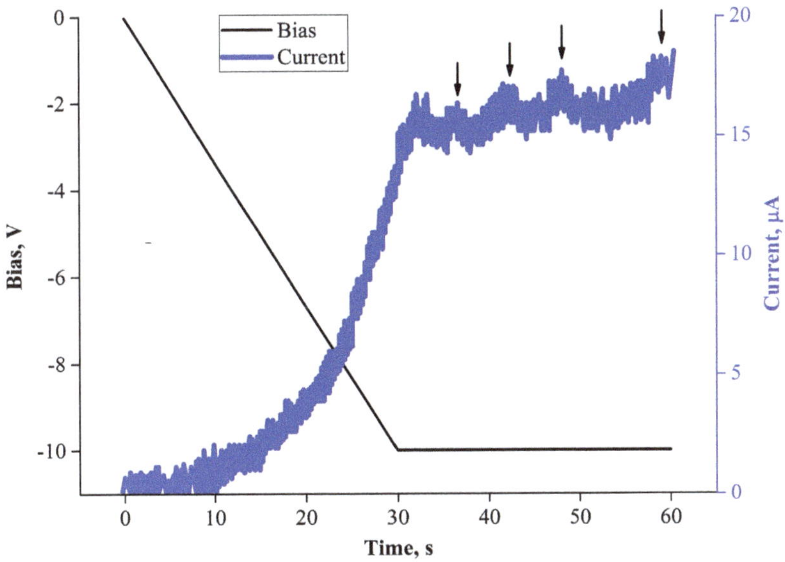

Fig. 4.39 Current and voltage during the first annealing of Ni nanoparticles at −10V in situ in electric field-assisted sintering experiments in a transmission electron microscope. (Reprinted from Holland et al. [51], Copyright (2010) with permission of AIP publishing)

[61]. High heating rates and short consolidation times could thereby preserve high diffusivity during sintering. The high heating rate can influence the densification rate through its impact on the phase transformations, such as glass crystallization, which changes the effective theoretical density of the solid phase and, therefore, affects densification [62, 63].

Experiments by Gillia and Bouvard [64] indicate the dependence of the densification rate on the average heating rate but no dependence on the temperature history. These experimental results can be used as a justification of why conventional theories of sintering and hot pressing do not explicitly include the heating rate as a

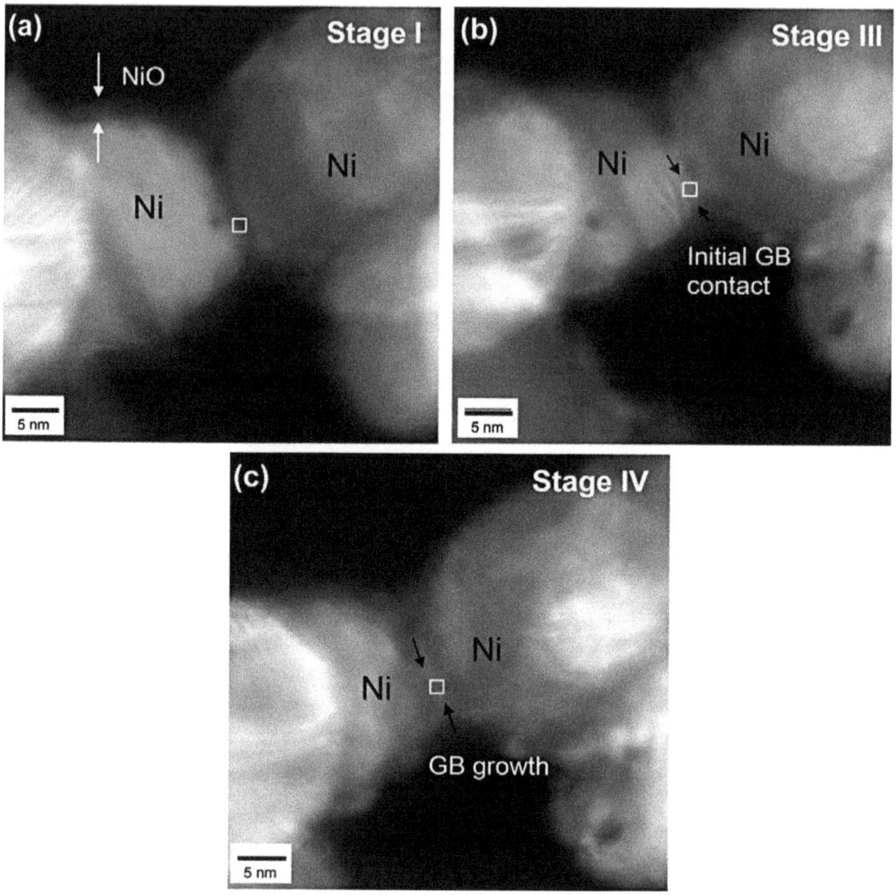

Fig. 4.40 High-angle annular dark-field images (scanning transmission electron microscopy) of two contacting nickel particles before the application of an electrical bias (**a**, stage I). The particles are covered with continuous NiO layers. (**b**) and (**c**) show the progression of surface oxide removal and the subsequent neck formation. (Reprinted from Bonifacio et al. [54], Copyright (2013) with permission from Elsevier)

parameter in kinetic relationships. Yet, the ability of conventional theories to implicitly incorporate heating rates was earlier demonstrated by Johnson [65] who qualitatively showed that high heating rates minimize the effects of surface diffusion processes not contributing to densification; hence, the compact body reaches higher temperatures in a highly sinterable state. The latter essentially means that rapid heating should accelerate densification during sintering due to the early activation of diffusion mechanisms responsible for consolidation (grain-boundary diffusion and lattice diffusion) and – in the case of SPS – also due to power-law dislocation creep and due to the reduction of low-temperature stage of sintering when surface diffusion dominates. Furthermore, short consolidation times preserve a fine-grained

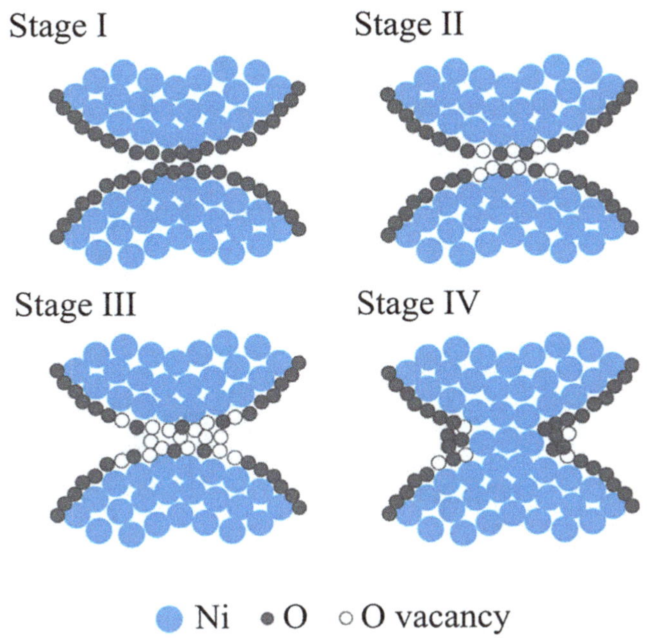

Fig. 4.41 Stages of surface cleaning of Ni nanoparticles during field-induced dielectric breakdown. The process involves the formation of oxygen vacancies and oxygen migration. (Reprinted from Bonifacio et al. [54], Copyright (2013) with permission from Elsevier)

structure in materials, which, in turn, also promotes high densification rates. The implicit impact of the heating rate on the densification kinetics during sintering has also been demonstrated by Roura et al. [66], who argued that the concept of a greater shrinkage rate at rapid heating ramps is valid for any transport mechanism (they assumed the vacancy diffusivity to be independent of any other internal parameter beside temperature). Following Johnson [65], the interaction between surface and grain-boundary diffusion has been noted by various authors as an important factor determining sintering consolidation rate. Coupled surface-grain-boundary diffusion models of sintering have been developed both numerically [67–72] and analytically [73–75]. Most of these studies did not include the direct analysis of the heating rate influence on the possibility of sintering acceleration; in most cases, they did not take into account the presence of external loads.

An interesting explanation of the effect of high heating rates on the densification of ZnO nanopowder was suggested by Schwartz et al. [76], who observed the formation of 95% dense samples as a result of sintering by the FAST technique with a maximum temperature of 400 °C. When lower heating rates were used, more porous samples were obtained, although the same maximum temperature was reached (Fig. 4.44). While studying the differences in the densification of two ZnO powder samples (Fig. 4.45), one of which was kept in a humid and the other in a dry environment before sintering, it was suggested that residual water is

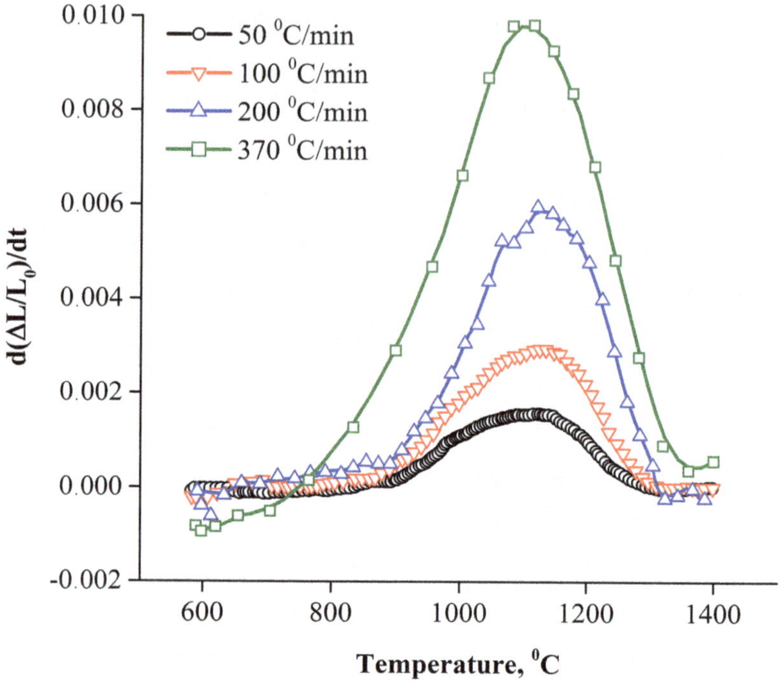

Fig. 4.42 Shrinkage rates of spark plasma sintered alumina at different heating rates. (Reprinted from Shen et al. [56], Copyright (2002) with permission of John Wiley & Sons)

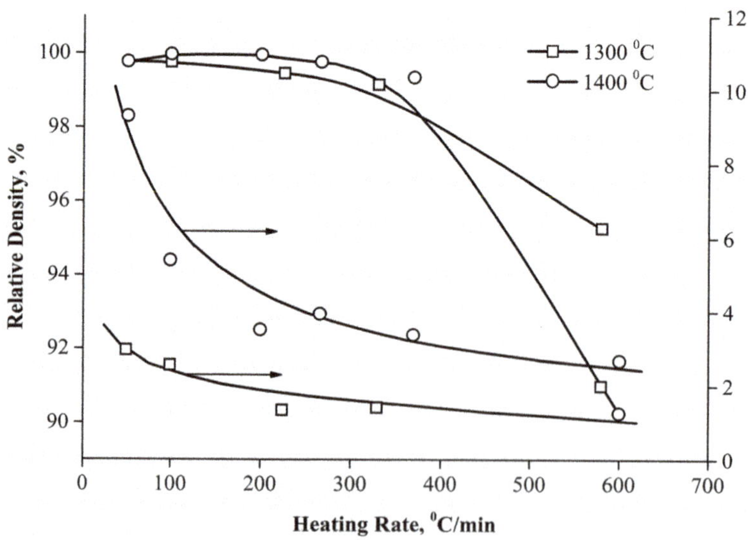

Fig. 4.43 Dependences of the relative density and grain size on the heating rate for spark plasma sintered alumina. (Reprinted from Shen et al. [56], Copyright (2002) with permission of John Wiley & Sons)

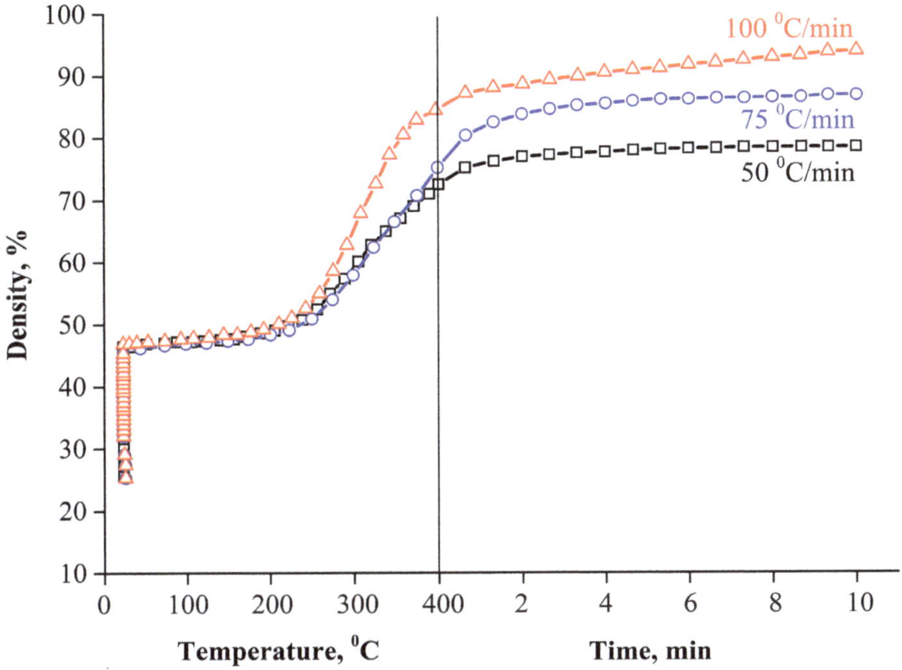

Fig. 4.44 Relative densities of ZnO compacts as a function of temperature obtained at three different heating rates; the vertical line marks the beginning of the isothermal dwell period. (Reprinted from Schwarz et al. [76], Copyright (2012) with permission of John Wiley & Sons)

responsible for the enhanced densification. Water adsorbed by the particle surfaces was presumably kept up to 400 °C in the compact when it was rapidly heated. The physical influence of water on the densification can be related to the lubrication effect facilitating particle rearrangement and sliding at the beginning of the densification process.

4.9 Modeling of the SPS Processes

Olevsky and coworkers developed a framework for modeling of the SPS processes [77–79] based on the concepts of the continuum theory of sintering [80–85]. Traditional (discrete) approaches used for sintering are based on the analysis of the local kinetics of the processes and studies of the mechanisms of the inter-particle interaction. The results of the analysis are then extended to describe the behavior of a macroscopic porous body. However, the sintering kinetics of porous bodies is influenced not only by the physical properties of the powder material and the nature of the inter-particle interaction but also by the macroscopic factors, such as kinematic limits imposed by external forces and inhomogeneity of the properties in the

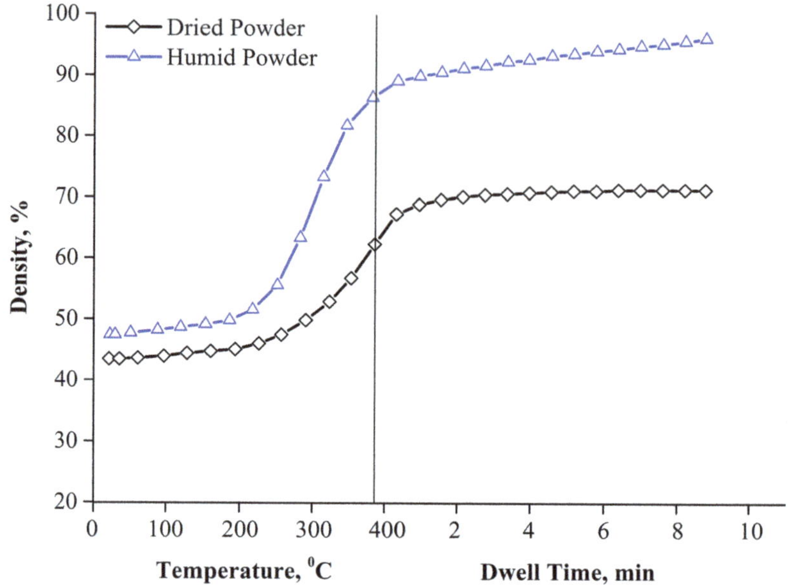

Fig. 4.45 Densification of two ZnO powder samples, one of which was kept in a humid and the other in a dry environment before sintering. (Reprinted from Schwarz et al. [76], Copyright (2012) with permission of John Wiley & Sons)

volume of the macroscopic body. The developed framework was used to study the roles of high heating rates and thermal diffusion in the SPS processes and contribution of electromigration to the electric current-accelerated diffusion.

4.9.1 Macroscopic Level of Analysis

In the continuum theory of sintering [84], the macroscopic behavior of a porous body is described by relating an external load corresponding to the stress tensor components σ_{ij} to the strain rate tensor components $\dot{\varepsilon}_{ij}$ by the nonlinear viscous constitutive relationship:

$$\sigma_{ij} = \frac{\sigma(W)}{W}\left[\varphi\dot{\varepsilon}_{ij} + \left(\psi - \frac{1}{3}\varphi\right)\dot{e}\,\delta_{ij}\right] + P_{\mathrm{L}}\delta_{ij}, \tag{4.1}$$

where W is the "equivalent strain rate" and $\sigma(W)$ is the "equivalent stress"; φ and ψ are normalized shear and bulk viscosities; δ_{ij} is the Kronecker symbol ($\delta_{ij} = 1$ if $i = j$ and $\delta_{ij} = 0$ if $i \neq j$); and $\dot{e}$ is the first invariant of the strain rate tensor, i.e., sum of tensor diagonal components: $\dot{e} = \dot{\varepsilon}_{ii} = \dot{\varepsilon}_{11} + \dot{\varepsilon}_{22} + \dot{\varepsilon}_{33}$. Physically, $\dot{e}$ represents the volume change rate of a porous body. The porosity θ is defined as the volume fraction of voids in a porous body.

Effective equivalent strain rate W depends on the invariants of the strain rate tensor

$$W = \frac{1}{\sqrt{1-\theta}}\sqrt{\varphi\dot{\gamma}^2 + \psi\dot{e}^2}, \tag{4.2}$$

where $\dot{\gamma}$ is the second invariant of the strain rate tensor deviator and represents, physically, the shape change rate of a porous body

$$\dot{\gamma} = \left[\left(\dot{\varepsilon}_{ij} - \frac{1}{3}\dot{e}\,\delta_{ij}\right)\left(\dot{\varepsilon}_{ij} - \frac{1}{3}\dot{e}\,\delta_{ij}\right)\right]^{1/2} \tag{4.3}$$

It can be expressed in terms of the principal elongation rates: $\dot{\varepsilon}_1$, $\dot{\varepsilon}_2$, $\dot{\varepsilon}_3$,

$$\dot{\gamma} = \frac{1}{\sqrt{3}}\sqrt{\left(\dot{\varepsilon}_1 - \dot{\varepsilon}_2\right)^2 + \left(\dot{\varepsilon}_2 - \dot{\varepsilon}_3\right)^2 + \left(\dot{\varepsilon}_3 - \dot{\varepsilon}_1\right)^2} \tag{4.4}$$

Parameter $\sigma(W)$ is the effective stress responsible for the constitutive behavior of a porous material. If $\sigma(W)$ is given by the linear relationship $\sigma(W) = 2\eta_0 W$, where η_0 is the shear viscosity of a fully dense material, Eq. (4.1) is transformed into

$$\sigma_{ij} = 2\eta_0\left[\varphi\dot{\varepsilon}_{ij} + \left(\psi - \frac{1}{3}\varphi\right)\dot{e}\,\delta_{ij}\right] + P_{\mathrm{L}}\delta_{ij} \tag{4.5}$$

For diffusional sintering of crystalline materials, η_0 is related to the effective diffusion coefficient, calculated based on Raj and Ashby [86] or Maximenko and Olevsky [87] approximations. In a generalized form

$$\eta_0 = \frac{ktG^2}{\Omega}\frac{1}{D_{\mathrm{eff}}}, D_{\mathrm{eff}} = D_{\mathrm{eff}}\left(D_{\mathrm{gb}}, D_{\mathrm{v}}, D_{\mathrm{s}}\right), \tag{4.6}$$

where k is the Boltzmann constant, T is the absolute temperature, G is the grain size, Ω is the atomic volume, and the effective diffusion coefficient D_{eff} depends on the coefficients D_{gb}, D_{v}, and D_{s} of the grain-boundary, volume, and surface diffusion, respectively.

Effective sintering stress P_{L} is the product of the local sintering stress $P_{\mathrm{L}_0} = 3\alpha/r_0$ (α is the surface tension; r_0 is the average particle radius) and the normalized effective sintering stress $\bar{P}_{\mathrm{L}}$,

$$P_{\mathrm{L}} = P_{\mathrm{L}_0}\bar{P}_{\mathrm{L}} = \frac{3\alpha}{r_0}\bar{P}_L \tag{4.7}$$

In the simplest case, in constitutive relationship (4.1), three main parameters P_{L}, φ, and ψ are functions of porosity (P_{L} can also incorporate an additional term responsible for the in-pore gas pressure). Skorohod's rheological model [88] postulates that

$$\varphi = (1-\theta)^2,\ \psi = \frac{2}{3}\frac{(1-\theta)^3}{\theta}, \bar{P}_{\mathrm{L}} = (1-\theta)^2 \tag{4.8}$$

In the multi-scale theory of sintering [67, 89–92], P_{L}, φ, and ψ have to be determined based on the consideration of a mesoscale pore-grain structure evolving during sintering. For the calculation of the porosity (relative density) evolution, the following continuity (mass conservation) equation is used:

$$\frac{\dot{\theta}}{1-\theta} = \dot{e} \tag{4.9}$$

Equations (4.1–4.9) describe the continuum theory of sintering framework, which does not include SPS-specific factors. This framework, however, can serve as a basis of the modeling expansion including factors of both thermal and nonthermal nature influencing mass transport under SPS conditions of processing.

4.9.2 *Microscopic Level of Analysis: Grain-Boundary Diffusion Driven by Externally Applied Load and Surface Tension*

Olevsky et al. [77–79] consider two major densification-contributing mechanisms of mass flux for SPS, grain-boundary diffusion and power-law dislocation creep, and use the previously developed micromechanic framework for modeling the grain-boundary diffusion in sintering under applied pressure [67] combined with the earlier developed model of power-law creep under hot deformation of powder bodies [81, 82, 84, 85] as the theoretical background for the consideration of SPS. A micromechanical model for a powder compact of simple-packed, rectangular grains with semiaxes a and c and elliptical pores located at the grain quadrajunctions was used (Fig. 4.46).

The maximum and minimum curvature radii r_{a} and r_{c} of the elliptical pore contour are defined as

$$r_{\mathrm{a}} = \frac{c_{\mathrm{p}}^2}{a_{\mathrm{p}}},\quad r_c = \frac{a_{\mathrm{p}}^2}{c_{\mathrm{p}}}, \tag{4.10}$$

where a_{p} and c_{p} are the semiaxes of the pores. Based on the results of Olevsky et al. [67], stresses in the x direction can be expressed as follows:

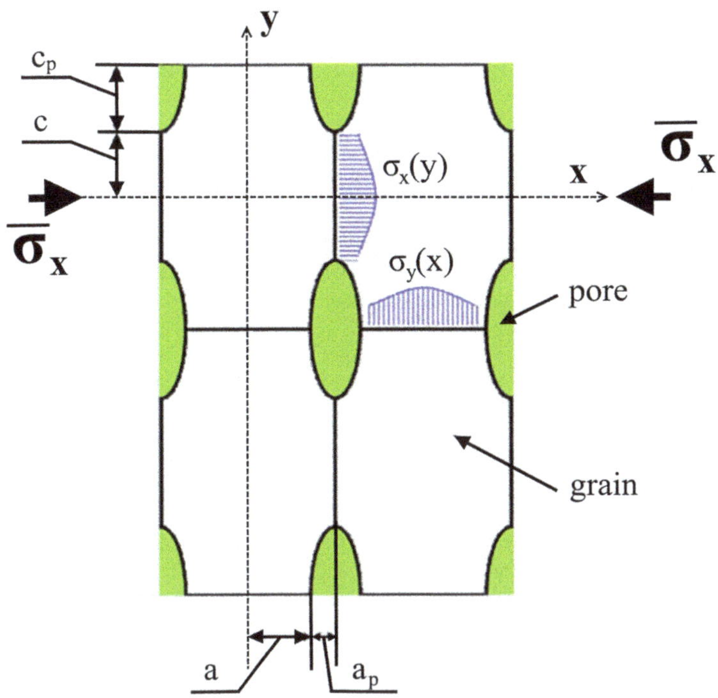

Fig. 4.46 Pore-grain structure unit cell. Rectangular grains with semiaxes a and c contain elliptical pores with semiaxes a_p and c_p located at the grain quadrajunctions. (Reprinted from Olevsky et al. [78], Copyright (2007) with permission of AIP Publishing)

$$\sigma_x = \left\{ \frac{3\alpha}{2c^2}\left[\frac{1}{r_c} + \frac{1}{c}\sin\frac{\phi}{2}\right] + \frac{3}{2c^3}\overline{\sigma}_x\left(c + c_\mathrm{p}\right)\right\} y^2 + \alpha\left[-\frac{3}{2c}\sin\frac{\phi}{2} - \frac{1}{2}\frac{1}{r_\mathrm{c}}\right] - \frac{3}{2c}\overline{\sigma}_x\left(c + c_\mathrm{p}\right), \quad (4.11)$$

where α is the surface tension, ϕ is the dihedral angle, $\overline{\sigma}_x$ is the effective (far-field) external tress in the x direction (compressive $\overline{\sigma}_x$ is negative), and y is the coordinate. Parameter $\overline{\sigma}_x\left[\frac{(c + c_\mathrm{p})}{c}\right]$ is a local stress on the grain boundary with $\frac{(c+c_\mathrm{p})}{c}$ being a stress concentration factor.

In the case of structure shown in Fig. 4.39,

$$J_y^{\mathrm{gb}} = \frac{\delta_{\mathrm{gb}} D_{\mathrm{gb}}}{kT}\left[\frac{\partial \sigma_x}{\partial y}\right], \quad (4.12)$$

where J_y^{gb} is the flux of matter in the direction of the axis y caused by grain-boundary diffusion, D_{gb} is the coefficient of the grain-boundary diffusion, δ_{gb} is the grain-boundary thickness, k is the Boltzman constant, and T is the absolute temperature.

The following relationship between the flux and the strain rate $\dot{\varepsilon}_{\text{gbx}}$ ($\dot{\varepsilon}_{\text{gbx}}$ is negative in the case of shrinkage in the x direction) in the orthogonal direction should be taken into account:

$$\dot{\varepsilon}_{\text{gbx}} = -\frac{J_y^{\text{gb}}(c)}{2(a + a_{\text{p}})} \frac{\Omega}{(c + c_{\text{p}})}, \tag{4.13}$$

where Ω is the atomic volume, to obtain the expression for the shrinkage kinetics during SPS (as SPS is carried out in a rigid die, the axial components of the shrinkage equal the overall volume shrinkage rate):

$$\dot{\varepsilon}_{\text{gbx}} = -\frac{\delta_{\text{gb}} D_{\text{gb}}}{kT} \frac{\Omega}{(a + a_{\text{p}})(c + c_{\text{p}})} \left\{ \frac{3\alpha}{c} \left[\frac{1}{r_c} - \frac{1}{2} \sin \frac{\phi}{2} \right] - \overline{\sigma}_x \frac{c + c_{\text{p}}}{c^2} \right\} \tag{4.14}$$

For equiaxial grains of size $G = a = c$,

$$\dot{\varepsilon}_{\text{gbx}} = -\frac{\delta_{\text{gb}} D_{\text{gb}}}{kT} \frac{\Omega}{\left(G + r_{\text{p}} \sqrt{\xi}\right) \left(G + \frac{r_{\text{p}}}{\sqrt{\xi}}\right)} \left\{ \frac{3\alpha}{G} \left[\frac{1}{r_c} - \frac{1}{2G} \right] - \overline{\sigma}_x \frac{G + \frac{r_{\text{p}}}{\sqrt{\xi}}}{G} \right\}, \tag{4.15}$$

Here ϕ is assumed equal to 60°, $\xi = \frac{a_{\text{p}}}{c_{\text{p}}}$ is the elliptical pore aspect ratio, and r_{p} is the equivalent pore radius:

$$r_{\text{p}} = \sqrt{a_{\text{p}} c_{\text{p}}} = c_{\text{p}} \sqrt{\xi} = \frac{a_{\text{p}}}{\sqrt{\xi}}, r_c = r_{\text{p}}\, \xi^{3/2} \tag{4.16}$$

4.9.3 Microscopic Level of Analysis: Power-Law Creep Driven by Externally Applied Load and Surface Tension

Based on the continuum theory of sintering [84],

$$\overline{\sigma}_x = A W^{m-1} \left[\varphi \dot{\varepsilon}_{\text{crx}} + \left(\psi - \frac{1}{3} \varphi \right) \left(\dot{\varepsilon}_{\text{crx}} + \dot{\varepsilon}_{\text{cry}} \right) \right] + P_{\text{L}}, \tag{4.17}$$

where $\dot{\varepsilon}_{\text{crx}}$ and $\dot{\varepsilon}_{\text{cry}}$ are components of the shrinkage rate corresponding to the mechanism of power-law creep; A and m are the power-law creep frequency factor and the power-law creep exponent, respectively; and W is the equivalent effective strain rate, which can be expressed as

$$W = \frac{1}{\sqrt{1 - \theta}} \sqrt{\frac{2}{3} \varphi \left| \dot{\varepsilon}_{\text{crx}} - \dot{\varepsilon}_{\text{cry}} \right|^2 + \psi \left(\dot{\varepsilon}_{\text{crx}} + \dot{\varepsilon}_{\text{cry}} \right)^2} \tag{4.18}$$

For the case of pressing in a rigid die, the deformation occurs only in the x direction, while $\dot{\varepsilon}_{\mathrm{cry}} = 0$. Taking into account the negative signs of the shrinkage rate and of the compressive axial stress, the following expression was obtained:

$$\dot{\varepsilon}_{\mathrm{crx}} = -\left\{ \left(\sqrt{1-\theta}\right)^{m-1} \left[P_{\mathrm{L}} - \overline{\sigma}_x\right] / A \left(\sqrt{\frac{2}{3}\varphi + \psi} \right)^{m+1} \right\}^{1/m} \tag{4.19}$$

Using $\varphi = (1-\theta)^2, \psi = \frac{2}{3}\frac{(1-\theta)^3}{\theta}, P_{\mathrm{L}} = \frac{3\alpha}{2G}(1-\theta)^2$, where $G = a = c$ is the grain size, $\dot{\varepsilon}_{\mathrm{crx}}$ can be expressed as

$$\dot{\varepsilon}_{\mathrm{crx}} = -\left\{ \left(\frac{3\theta}{2}\right)^{(m+1)/2} \left[\frac{3\alpha}{2G}(1-\theta)^2 - \overline{\sigma}_x\right] / A(1-\theta)^{(m+3)/2} \right\}^{1/m} \tag{4.20}$$

Following [93], the power-law frequency factor can be written as

$$A = A_0 \exp\left(\frac{Q_{\mathrm{cr}}}{RT}\right), \tag{4.21}$$

where Q_{cr} is the creep activation energy.

4.9.4 Theoretical Analysis of the Effect of High Heating Rates in the SPS

Olevsky et al. investigated the role of heating rates in the SPS by developing a combined surface-grain-boundary diffusion power-law creep model [78]. The model incorporates two mechanisms of mass transport contributing to densification: grain-boundary diffusion and power-law dislocation creep. Surface diffusion contributes to the evolution of the pore surface curvature, thereby influencing the intensity of the grain-boundary diffusion at the early stages of sintering. Figure 4.47 shows the assumed initial pore shape and its evolution due to surface and grain-boundary diffusion.

The curvature radius r_c increases due to the mass redistribution along the pore surface caused by surface diffusion. This increase in the curvatures of the tips slows down the grain-boundary diffusion according to Eq. (4.15). In order to verify the assumed mechanism of the sinterability enhancement, the impact of the reduction of the initial time period of the surface diffusion's domination due to higher heating rates on the increase of the grain-boundary diffusion's contribution to shrinkage has to be assessed. Regarding densification by grain-boundary diffusion, the main events happen in the areas of the pore tips located at grain junctions. The morphological changes of the pore surface in these areas due to the action of the surface diffusion were assumed to be similar to those experienced by elliptical pores with the same

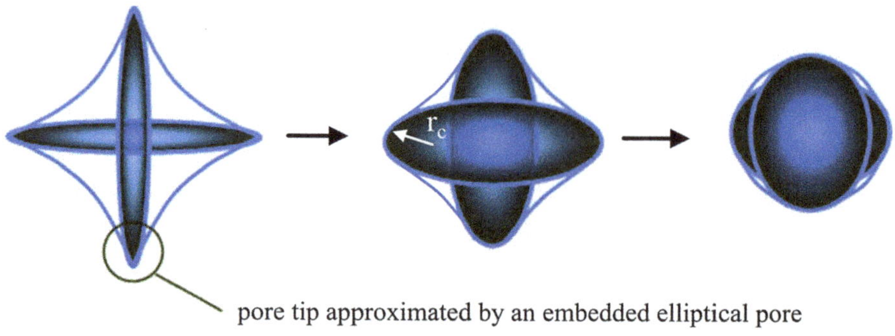

Fig. 4.47 Evolution of the pore morphology. (Reprinted from Olevsky et al. [78], Copyright (2007) with permission of AIP Publishing)

tips' configuration. The surface diffusion-based evolution of the elliptical pore with the pores' semiaxes a_p and c_p was described by Olevsky et al. [67]:

$$\dot{a}_p = -3\frac{\delta_\mathrm{s} D_\mathrm{s}\alpha\Omega}{kT}\frac{\left(c_\mathrm{p}^2 - a_\mathrm{p}^2\right)}{c_\mathrm{p}^3 a_\mathrm{p}^2}, \tag{4.22}$$

$$\dot{c}_p = 3\frac{\delta_\mathrm{s} D_\mathrm{s}\alpha\Omega}{kT}\frac{\left(a_\mathrm{p}^2 - c_\mathrm{p}^2\right)}{a_\mathrm{p}^3 c_\mathrm{p}^2} \tag{4.23}$$

where δ_s is the surface diffusive width ($\delta_\mathrm{s} = \Omega \times$ the number of atoms per unit area) and D_s is the coefficient of surface diffusion. It is possible to show that

$$\dot{\xi} = \frac{6\delta_\mathrm{s} D_\mathrm{s}\alpha\Omega}{kT}\frac{1-\xi^2}{r_\mathrm{p}^4} \tag{4.24}$$

Factor $\dot{\xi}$ corresponds to the rate of the pore spheroidization, and ξ characterizes the "sharpness" of the pore tip.

The total shrinkage during the SPS is superposition of the shrinkage rates corresponding to the grain-boundary diffusion and lower-law creep mechanisms:

$$\dot{\varepsilon}_x = \dot{\varepsilon}_{\mathrm{gbx}} + \dot{\varepsilon}_{\mathrm{crx}} \tag{4.25}$$

The equation $\dot{\xi}$ can be thus transformed into

$$\dot{\xi} = \frac{3\delta_\mathrm{s} D_\mathrm{s}\alpha\Omega}{8kTG^4\xi^2}\left(1-\xi^2\right)\left[\sqrt{(1-\xi)^2 + \frac{\pi}{\theta}\xi} - (1+\xi)\right]^4, \tag{4.26}$$

while $\dot{\theta}$ can be expressed as follows:

$$\dot{\theta} = -(1-\theta) \times \left(\frac{\delta_{gb} D_{gb} \Omega \alpha \theta}{kTG^4 \xi} \left[\sqrt{(1-\xi)^2 + \frac{\pi}{\theta}\xi} - (1+\xi) \right]^2 \right. \left\{ \frac{3}{2\xi^2} \left[\sqrt{(1-\xi)^2 + \frac{\pi}{\theta}\xi} - (\xi^2 + \xi + 1) \right] - \left(\overline{\sigma}_x \frac{G}{\alpha} \right) \left[\frac{\sqrt{(1-\xi)^2 + \frac{\pi}{\theta\xi}} - \xi + 1}{\sqrt{(1-\xi)^2 + \frac{\pi}{\theta\xi}} - \xi - 1} \right] \right\} \left. + \left\{ \left(\tfrac{3\theta}{2}\right)^{(m+1)/2} \left[\tfrac{3\alpha}{2G}(1-\theta)^2 - \overline{\sigma}_x \right] / A_0 \exp\left(\tfrac{Q_{cr}}{RT}\right)(1-\theta)^{(m+3)/2} \right\}^{1/m} \right) \quad (4.27)$$

The grain growth is an important phenomenon, which can considerably influence the kinetics of densification. Despite numerous publications on the grain structure evolution during high-temperature processing, the regularities of the grain growth kinetics in hot pressure-assisted consolidation of powders remain under discussion. There is no unified opinion on the impact of externally applied stresses on grain growth. In order to deconvolute the possible impact of the external stresses on the grain growth and, in turn, on the densification kinetics, we employ a grain growth law obtained by an asymptotic approximation of Hsueh et al. [94] and Du and Cocks results [95]. This enables an independent consideration of the primary conceptual idea – the dependence of the sinterability on the heating rate as a result of the interplay of surface diffusion, grain-boundary diffusion, and power-law dislocation creep. Aluminum was taken as an example to perform the modeling studies. The grain growth kinetics of aluminum has been studied by Beck et al. [96]. Based on [93, 95, 96], it is possible to derive the following equation for the grain growth of aluminum:

$$\dot{G} = 6.67 \times 10^{-4} \left(\frac{1}{K}\right) \varpi \ln \frac{235\mathrm{K/s}}{\varpi} G_0 \left(\frac{G_0}{G}\right)^3 \theta^{-1.34}, \text{ if } T > 533\ K$$
$$\text{and}$$
$$\dot{G} = 0, \text{ if } T \leq 533\ K \quad (4.28)$$

where ϖ is the heating rate.

The ordinary differential Eqs. (4.26), (4.27), and (4.28) were solved numerically. The calculations were carried out for an aluminum powder for the conditions of a constant heating rate starting from 25 °C until reaching the maximum temperature of 600 °C without isothermal holding periods using five different heating rates: 10, 25, 50, 100, and 200 °C/min (Fig. 4.48). The results showed that an increase of heating rate with the same maximum temperature and a corresponding time decrease of sintering time allow obtaining the same final density. This indicates the ability of the conventional sintering and hot-pressing models to describe the acceleration of the densification and deceleration of the grain growth under high heating rate conditions,

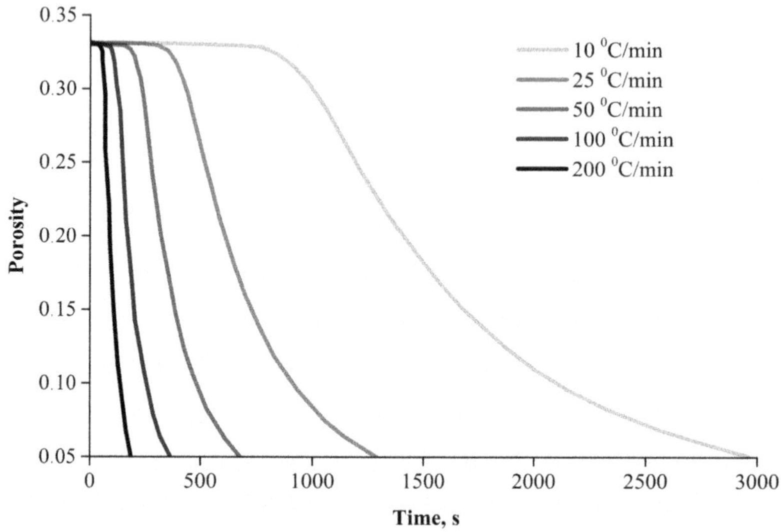

Fig. 4.48 Shrinkage kinetics of an aluminum powder heated at different rates (calculation results). (Reprinted from Olevsky et al. [78], Copyright (2007) with permission of AIP Publishing)

such as those utilized during SPS, without the explicit employment of the heating rate as an additional governing parameter.

The higher the heating rate is, the smaller the increase of the radius of curvature of the pore tip is, and, in turn, the higher the driving force for grain-boundary diffusion is obtained within the given period of time. In other words, for high rates of heating, surface diffusion does not have enough time to "smooth out" pore surfaces so that densification-contributing grain-boundary diffusion, which is activated at higher temperatures, is driven by still very substantial stresses inversely proportional to the radius of curvature r_c. At the same time, the well-known pinning effect of pores restricting grain growth disappears when pores become small due to densification. If after this critical point sintering time is long enough, substantial grain growth occurs. For high heating rates, the grain growth may be diminished due to short processing times. The time, however, must be sufficient to complete densification. Figure 4.49a, b shows the evolution of the grain size with respect to time and temperature for different heating rates. The calculation results indicate that the higher heating rate promotes grain size retention. For the fixed final porosity level (5%), however, the final grain size manifests a nonmonotonous dependence on the heating rate (Fig. 4.49a). This is the result of the competition between the grain growth rate and the available time for grain growth.

Figure 4.50 shows the evolution of the pore tip sharpness as the result of the application of different heating rates. It can be seen that the higher heating rates postpone the pore spheroidization; this, in turn, favors materials' sinterability for higher heating rates. This effect is more pronounced for larger initial grain sizes. In this case, the contribution to densification by grain-boundary diffusion increases due

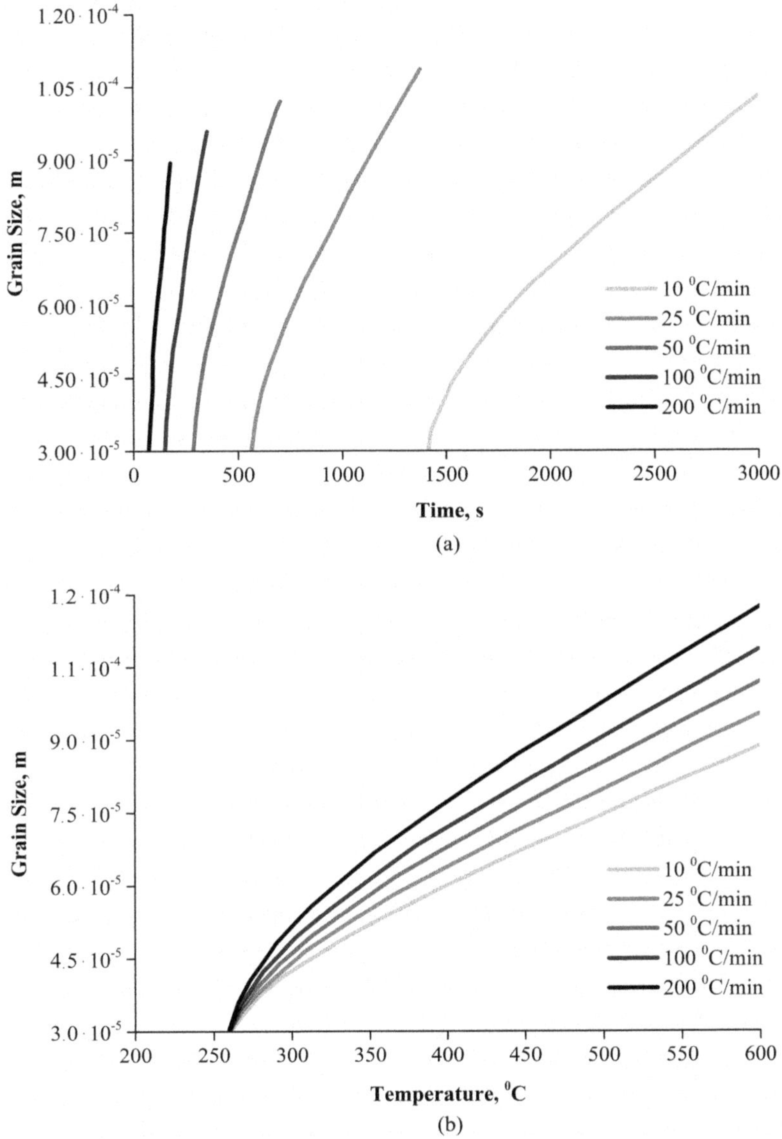

Fig. 4.49 The calculated grain growth kinetics of aluminum powder. (Reprinted from Olevsky et al. [78], Copyright (2007) with permission of AIP Publishing)

to the postponed pore spheroidization; indeed, the rate of increase of the pore aspect ratio is inversely proportional to the fourth power of the grain size (Eq. (4.24)).

Figure 4.51 shows the calculated dependence of the shrinkage rate on the temperature. The observed nonmonotonous dependence is characteristic for powder densification processes, when after the initial increase, the rate of the process decays

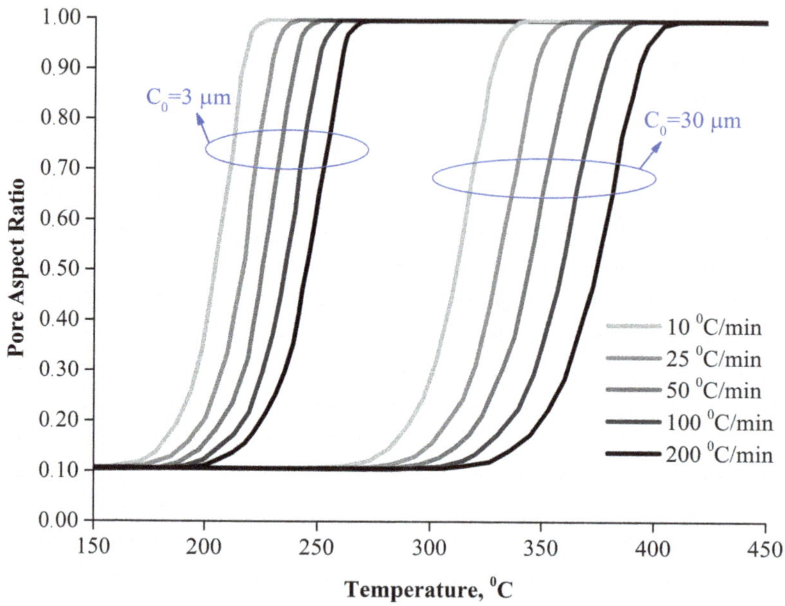

Fig. 4.50 Evolution of the pore aspect ratio (pore tip sharpness) for different average initial grain sizes and heating rates. (Reprinted from Olevsky et al. [78], Copyright (2007) with permission of AIP Publishing)

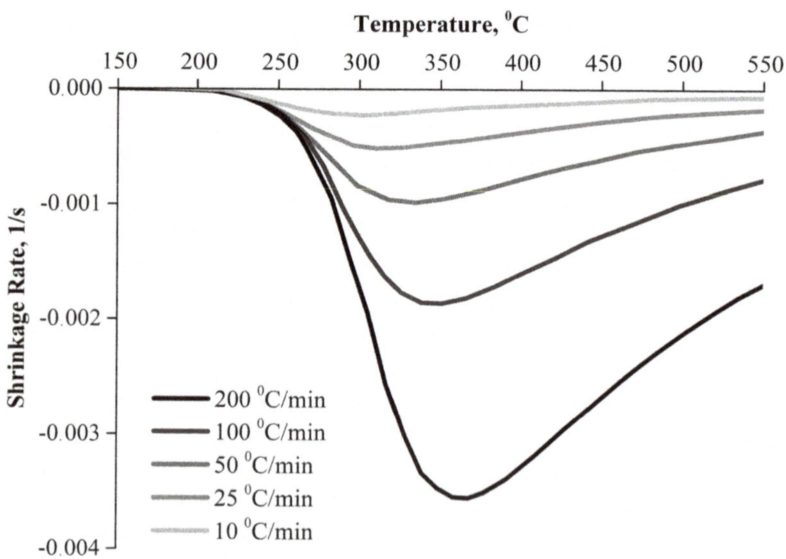

Fig. 4.51 Calculated shrinkage rates vs. temperature for an aluminum power. (Reprinted from Olevsky et al. [78], Copyright (2007) with permission of AIP Publishing)

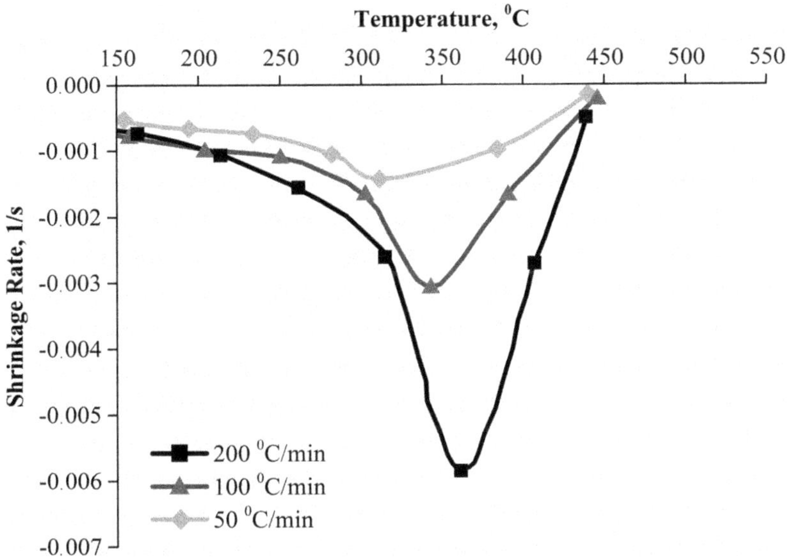

Fig. 4.52 Experimental shrinkage rates vs. temperature for an aluminum alloy powder. (Reprinted from Olevsky et al. [78], Copyright (2007) with permission of AIP Publishing)

due to the decrease of porosity and the overall material strengthening. The results of modeling qualitatively agree very well with the experimental data on the SPS of an aluminum alloy shown in Fig. 4.52.

The results of modeling suggest the strong dependence of the shrinkage rate during SPS on the two particular parameters: heating rate and the initial grain particle size. For the purpose of the verification, the experiments have been conducted to study the impact of these parameters on the densification kinetics of an aluminum powder. The results of these experiments are shown in Fig. 4.53 in the form of the dependence of the shrinkage rate on the temperature for two different heating rates (50 and 150 °C min^{-1}) and two different average particle sizes (75 and smaller than 50 μm). The results indicate a bigger difference in the shrinkage maximum rates as well as the bigger shift of the shrinkage maximum rate to higher temperatures for the larger average particle size of 75 μm. This confirms the previously noted fact, based on the modeling results, that the interaction between surface diffusion and grain-boundary mechanisms of mass transport in SPS and, in turn, the impact of heating rate on the shrinkage kinetics is more pronounced for larger particle sizes. For a certain range of grain sizes and for high heating rates, sintering can be accelerated if coarser particles are used. This phenomenon vanishes with the decrease of the heating rate and should be reversed for conventional rates of powder processing.

Experimental verification of the model was carried out to reveal that the model predictions for the shrinkage kinetics are in good qualitative agreement with the experimental data for an aluminum powder [78]. Some quantitative discrepancies between the model and experimental data may be explained by omitting a number of

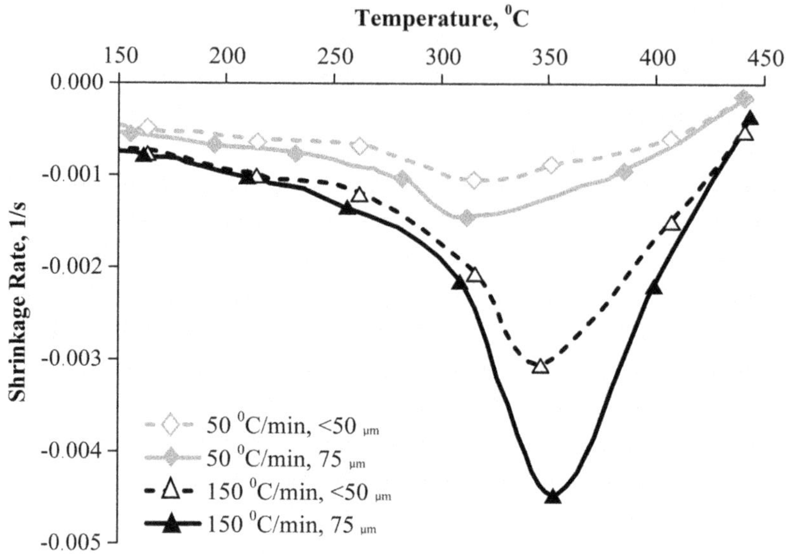

Fig. 4.53 Effect of the particle size on the densification kinetics (experimental data). (Reprinted from Olevsky et al. [78], Copyright (2007) with permission of AIP Publishing)

phenomena contributing to the mechanisms of SPS consolidation. It should be noted that sintering enhancement by high heating rates is only one of a number of possible contributing factors. In particular, no enhancement of the densification by electromagnetic field was considered in model presented in Ref. [78]. However, a closer correlation with the experimental data can be obtained if the contribution of electromigration is taken into account, which will be discussed below. The interaction of surface and grain-boundary diffusion is one of the major mechanisms enabling the impact of heating rates on the consolidation kinetics during SPS. Another heat rate-sensitive contribution to the densification is related to the shrinkage kinetics by power-law creep accelerated for rapid increases of temperature. The model predictions agree well with the experiments revealing the shrinkage rate dependence on temperature for different heating rates and average initial grain sizes of an aluminum alloy powder. Both the calculations and the SPS experiments show the increase of the maximum shrinkage rate for larger particle sizes. This mostly unusual for conventional powder sintering and hot-pressing phenomenon is explained under the conditions of the rapid heating by the delayed pore tip spheroidization and, in turn, the intensification of grain-boundary diffusion at the early stages of sintering for larger grain sizes. For lower heating rates, this effect should be inverted. The developed model takes into consideration the influences of rapid heating without distinguishing the source of heat, which makes it suitable not only for the description of SPS but applicable for the analysis of any rapid sintering techniques such as microwave sintering or self-propagating high-temperature (SHS)-enabled sintering [97]. For the description of SPS, however, the model emphasizes only one factor of thermal nature – heating rate – of many factors, which may be responsible for the enhancement of the consolidation during SPS compared to conventional powder processing techniques.

4.9.5 *Influence of Thermal Diffusion*

In contrast to heating rates, the role of temperature spatial gradients as independent governing parameters influencing the constitutive behavior during sintering may be significant. This is related to the Ludwig–Soret effect [98] of thermal diffusion, which causes concentration gradients in initially homogeneous two-component systems subjected to a temperature gradient. The thermal diffusion-based constitutive mechanism of sintering results from the additional driving force due to a spatially varying temperature gradient, which causes vacancy diffusion (in contrast to the traditional "component mixing" effect of regular diffusion, thermal diffusion leads to the separation of components). This mechanism should be an addition to the free-surface curvature-driven diffusion considered in conventional sintering theories. Based on the results of Schottky [99], the flux (along various paths) equation corresponding to thermal diffusion is given by

$$J = -D\left(\nabla C_{\mathrm{v}} + C_{\mathrm{v}}\frac{Q^*}{kT}\frac{\nabla T}{T}\right), \tag{4.29}$$

where J is the vacancy flux, D is the diffusion coefficient, C_{v} is the vacancy concentration, ∇C_{v} is the vacancy concentration gradient, Q^* is the heat of vacancy transport, and ∇T is the temperature gradient.

Equation (4.29) combines two temperature gradient-driven phenomena: the first, corresponding to the term, $-D\nabla C_{\mathrm{v}}$, describes the flux of vacancies caused by their nonuniform spatially distributed concentration due to the temperature spatial nonuniformity and the second, corresponding to the term, $DC_{\mathrm{v}}\frac{Q^*}{kT}\frac{\nabla T}{T}$, is the direct manifestation of the Ludwig–Soret effect in crystalline solids. It is noteworthy that in conventional sintering theories, the term ∇C_{v} is considered to be caused by surface tension, which makes it different from the one used in Eq. (4.29). In the analysis below, it is assumed that vacancy gradients developed from the surface curvature and from the temperature gradients act independently in contributing to sintering shrinkage.

The equilibrium vacancy concentration is determined as $C_{\mathrm{v}} = \frac{N_A\rho}{A}\exp\left(-\frac{H_{\mathrm{f}}}{kT}\right)$, where N_A is the Avogadro's number, ρ is the material's theoretical density, and A is the atomic weight.

Assuming only the temperature dependence of vacancy concentration C_{v}, it was shown [100, 101] that

$$\nabla C_{\mathrm{v}} = \frac{C_{\mathrm{v}} H_{\mathrm{f}}}{kT^2} \nabla T, \tag{4.30}$$

where H_{f} is the enthalpy of vacancy formation. The equation for the flux can be rewritten as follows:

$$J = -\frac{DC_{\mathrm{v}}}{kT} (H_{\mathrm{f}} + Q^*) \frac{\nabla T}{T}. \tag{4.31}$$

Wirtz [102] suggested that $Q^* = H_m - H_{\mathrm{f}}$,

where H_m is the enthalpy of vacancy migration. Young and McPherson [100] derived

$$J = -\frac{DC_{\mathrm{v}}}{kT} H_m \frac{\nabla T}{T} \tag{4.32}$$

Kornyushin [103, 104] introduced the spatially averaged version of Eq. (4.32) by utilizing the thermal diffusion ratio κ'_{T} ($\langle T \rangle$ is the spatial average of temperature):

$$\kappa'_T = \frac{C_{\mathrm{v}}(H_m - H_{\mathrm{f}})}{kT}, \quad J = -D\frac{\kappa'_{\mathrm{T}}}{T} \nabla T \tag{4.33}$$

Kornyushin did not include the term ∇C_{v} in Eq. (4.33), thereby omitting the input of nonuniform vacancy generation caused by temperature spatial nonuniformity. Olevsky and Froyen [79] used the formulation derived by Young and McPherson [100] and adjusted it by added spatial averaging of the temperature introduced by Kornyushin [103, 104]:

$$J = -D\frac{\kappa_{\mathrm{T}}}{T} \nabla T, \tag{4.34}$$

where the thermal diffusion ratio κ_{T} is redefined as

$$\kappa_{\mathrm{T}} = \frac{C_{\mathrm{v}} H_m}{k\langle T \rangle} \tag{4.35}$$

The driving force for the vacancy migration is given by

$$\mathrm{div}\, J = -D\frac{\kappa_{\mathrm{T}}}{\langle T \rangle} \Delta T, \tag{4.36}$$

where ΔT is the temperature Laplacian. Taking into consideration the heat transfer equation

$$\lambda_{\mathrm{T}} \Delta T = C\frac{\partial T}{\partial t} - q, \tag{4.37}$$

where λ_T is the thermal conductivity, C is the heat capacity, t is the time, and q is the heat production per unit volume of the material and per unit time, which in the case of SPS can be represented as $q = \lambda_e E^2$, where λ_e is the specific electrical conductivity and E is the electric field intensity.

Equation (4.36) can be rewritten as

$$\operatorname{div} J = -D\frac{\kappa_T}{\lambda_T\langle T\rangle}\left(C\frac{\partial T}{\partial t} - \lambda_e E^2\right) \tag{4.38}$$

In general, the heat production equation should include the impact of the Thomson thermoelectric effect [105]. However, as was estimated for aluminum at $\nabla T \approx 10^6$ K/m and $E = 400$ V/m, the contribution of the Thomson effect constitutes only 0.01 of the Joule heat generation. For higher electric fields, the Thomson effect can reach considerable values; in addition, for composite powder systems, it may become necessary to take into account the Seebeck (thermopower) and Peltier thermoelectric effects.

It follows from Eq. (4.38) that the thermal diffusion's impact on SPS is significant, if

$$\left|C\frac{\partial T}{\partial t} - \lambda_e E^2\right| \gg 0 \tag{4.39}$$

It is evident that the Ludwig–Soret effect will result in the maximum additional driving force for sintering during isothermal periods (when $\frac{\partial T}{\partial t} = 0$), which is possible if cooling from the specimen's surface compensates its heating by the electric current. For electric conductors, even for moderate values of E (e.g., 50 V/m), the term $\lambda_e E^2$ should always exceed the term $C\frac{\partial T}{\partial t}$ by several orders of magnitude due to high values of λ_e. For materials with lower electric conductivity, the contribution of the thermal diffusion mechanism depends on the balance between the amount of extracted Joule heat and the amount of energy absorbed by the powder through its heat capacity. In SPS, due to the use of conductive (mostly graphite) dies, the heat source term is always significant even for non-conductive specimens. It should be noted that Johnson [106] exploring the role of temperature gradients in microwave sintering did not include the heat source term (such as $q = \lambda_e E^2$) in the heat balance equation and concluded that along-the-grain-boundary temperature gradients in field-assisted sintering are negligible. It should also be noted that convection and radiation heat losses are omitted in Eq. (4.37), which leads to the overestimation of the temperature Laplacian. Because the driving force for the mechanism of thermal diffusion also depends on the temperature and the temperature-dependent coefficient of diffusion (Eq. (4.38)), the impact of omitting the heat losses on the intensity of the thermal diffusion contribution to the sintering shrinkage is ambiguous and should depend on the boundary conditions. It is reasonable to assume that during a short pulse in the SPS the heat conductivity of the material is not sufficiently large to lead to significant heat losses in the considered local material volume. Even for conductive metallic powders, this assumption should provide a sufficient degree of accuracy due to the presence of oxide films at the particle boundaries. With respect to temperature gradients, this assumption may, in fact, lead to their underestimation, thereby diminishing somewhat the assessed effect of thermal diffusion.

Olevsky and Froyen [79] analyzed the enhancement of the grain-boundary diffusion flux (in general, the impact of the Ludwig–Soret effect may be different for other mass transport mechanisms, such as volume diffusion). The driving force given by Eq. (4.38) should result in the flux of vacancies complementary to the curvature-driven mechanism:

$$J_{\mathrm{td}}^{\mathrm{gb}} = 2(\operatorname{div} J)G = 2\delta_{\mathrm{gb}} D_{\mathrm{gb}} \frac{\kappa_{\mathrm{T}}}{\lambda_{\mathrm{T}} \langle T \rangle} \left(C \frac{\partial T}{\partial t} - \lambda_{\mathrm{e}} E^2 \right) G \tag{4.40}$$

The strain rate component corresponding to the thermal diffusion mechanism is given by

$$\dot{\varepsilon}_{\mathrm{gbx}}^{\mathrm{td}} = -\frac{J_{\mathrm{td}}^{\mathrm{gb}} \Omega}{2(G + r_{\mathrm{p}})^2} = \frac{\delta_{\mathrm{gb}} D_{\mathrm{gb}} \kappa_{\mathrm{T}} \Omega}{\lambda_{\mathrm{T}} \langle T \rangle} \left(C \frac{\partial T}{\partial t} - \lambda_{\mathrm{e}} E^2 \right) \frac{G}{(G + r_{\mathrm{p}})^2} \tag{4.41}$$

In order to convert the mesoscopic strain rate $\dot{\varepsilon}_{\mathrm{gbx}}^{\mathrm{td}}$ into the macroscopic strain rate $\overline{\dot{\varepsilon}_{\mathrm{gbx}}^{\mathrm{td}}}$, a homogenization procedure is used, which includes the mass conservation equation

$$r_{\mathrm{p}} = \frac{2G\sqrt{\theta}}{\sqrt{\pi} - 2\sqrt{\theta}} \tag{4.42}$$

and the linear-viscous consequence of Eqs. (4.1) and (4.8):

$$\overline{\dot{\varepsilon}_{\mathrm{gbx}}^{\mathrm{td}}} = \dot{\varepsilon}_{\mathrm{gbx}}^{\mathrm{td}} \frac{\theta}{2(1 - \theta)^3} \text{ (for uniaxial loading).} \tag{4.43}$$

The following equation can be thus obtained for the macroscopic contribution of thermal diffusion to the shrinkage rate

$$\overline{\dot{\varepsilon}_{\mathrm{gbx}}^{\mathrm{td}}} = \frac{\delta_{\mathrm{gb}} D_{\mathrm{gb}} \Omega C_{\mathrm{v}} H_m}{\lambda_{\mathrm{T}} k T^2} \left(C \frac{\partial T}{\partial t} - \lambda_{\mathrm{e}} E^2 \right) \times \frac{3\left(\sqrt{\pi} - 2\sqrt{\theta}\right)^2 \theta}{2\pi G (1 - \theta)} \tag{4.44}$$

Gostomelskii and Krupnova [107] derived a non-steady-state equation for the pulsed electric heating of a volume containing a pore assuming that the current pulses are sufficiently short such that the heat conductivity cannot redistribute the heat away from the pore. This can be accepted as a ball-park approximation for the local temperature gradients in the inter-particle contact areas. The local temperature gradients after n pulses can be roughly assessed as

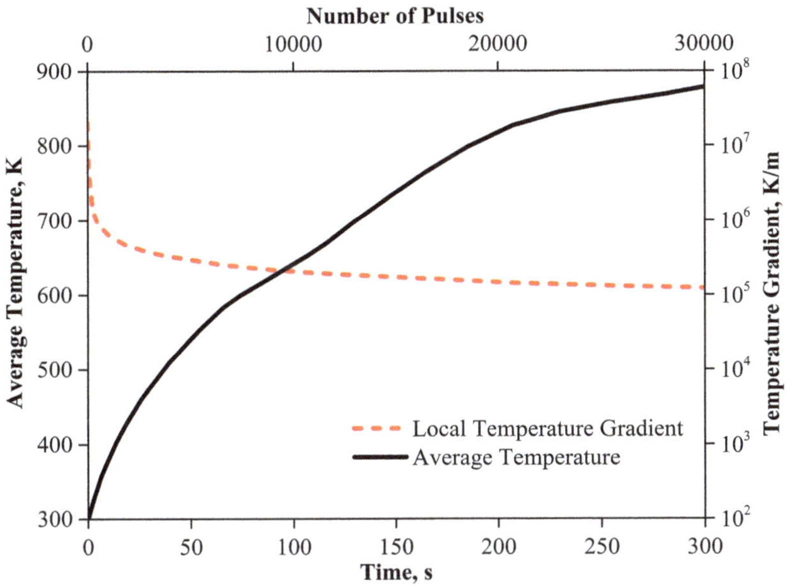

Fig. 4.54 Temperature evolution and the local temperature gradient of an alumina powder specimen subjected to SPS ($E = 300\frac{\mathrm{V}}{\mathrm{m}}$, initial temperature $T_0 = 298$ K, pulse "on" and "off" duration $\Delta t = 5$ ms, grain size $G = 50$ μm, pore size $r_p = 30$ μm). (Reprinted from Olevsky and Froyen [79], Copyright (2008) with permission of John Wiley & Sons)

$$|\nabla T|_{\mathrm{local}} = \frac{1}{(G + r_{\mathrm{p}})} \sqrt{\frac{\lambda_{\mathrm{e}} E^2}{2C} \frac{T_0}{n} \Delta t}, \tag{4.45}$$

where Δt is the duration of the electric pulse and n is the number of pulses.

The macroscopic temperature gradients in a cylindrical alumina specimen subjected to pulse Joule heating can reach 10^3 K m^{-1} [79]. The calculations of the local temperature gradients for an alumina specimen based on Eq. (4.45) are shown in Fig. 4.54. They indicate that the temperature gradients concentrated in the area of inter-particle contacts can exceed the macroscopic gradients by 3–4 orders of magnitude.

An expression for the thermal diffusion-enabled pulse duration dependent shrinkage rate can be obtained:

$$\dot{\varepsilon}^{\mathrm{td}}_{\mathrm{gbx}} = -\frac{C_{\mathrm{v}} H_m}{k \langle T \rangle^2} \frac{\delta_{\mathrm{gb}} D_{\mathrm{gb}} \Omega}{(G + r_{\mathrm{p}})^3} \sqrt{\frac{\lambda_{\mathrm{e}} E^2}{2C} \frac{T_0}{n} \Delta t}. \tag{4.46}$$

Equation (4.46) can serve as an approximation of the contribution of the thermal diffusion (caused both by the Ludwig–Soret effect and nonuniform vacancy generation due to temperature gradients) to the shrinkage rate during SPS. Equation (4.46) simplifies the impact of frequency on the thermal diffusion due to neglecting heat conductivity and heat losses. It may be useful if the direct assessment of the impact of the electric pulse duration and frequency on the shrinkage intensity during SPS is necessary. It should be noted that the contribution to SPS of the thermal diffusion mechanism related to the frequency of pulsing may be more complex than predicted by Eq. (4.46). It is logical to assume that for higher frequencies (shorter periods of pulse time), the temperature gradients developed are higher, because heat transfer has less time to homogenize the local temperature distributions. Hence, the intensity of thermal diffusion increases for higher pulse frequencies. As has been mentioned earlier, the thermal diffusion promotes the separation of atoms and vacancies. At early stages of sintering, this should lead to the growth of inter-particle necks, which corresponds to the enhancement of sintering. At the final stages of sintering, however, the pores may serve as vacancy sinks under thermal diffusion conditions, which impedes sintering. This analysis leads to an interesting conclusion that increased pulse frequencies enhance sintering at the early stages of SPS and hinder sintering at the late stages of the SPS process. This assessment may be used for the explanation of the fact that in some experimental studies the pulse frequency was found to have a limited impact on SPS results [23]. Indeed, its contributions at early and late stages of SPS could offset each other. A study by Chen et al. [108] of the effect of pulsing pattern on the reactivity between layers of Si and Mo brought in contact in the SPS without a die led to a conclusion that the growth rate of the reaction product $MoSi_2$ in the temperature range 1070–1270 °C did not depend did not depend on the pulsing pattern (Fig. 4.55), which was varied by changing the relative duration of the "on" and "off" periods.

Xie et al. did not find any significant effects of the pulse frequency during the sintering process on the density and tensile properties of consolidated aluminum compacts [109]. An interesting experiment has been conducted by Shen et al. [56], who showed that the pulse on/off sequence in the SPS influences the shrinkage rates and the temperature of the shrinkage rate maximum. In order to ensure a certain heating rate, a lower current and voltage output is needed for a higher on/off pulse ratio. A lower output will cause lower shrinkage rates at the same temperatures and a shift of the shrinkage rate maximum to higher temperatures. Worth noting here is only a slight shift of the sintering shrinkage rate maximum (approximately 50 °C) upon a change of the pulse sequence ratio from 1.1 to 18 (Fig. 4.56).

Calculations using a combined constitutive model [79] taking into account the surface curvature- and external pressure-driven grain-boundary diffusion, power-law creep, and thermal diffusion, such that the shrinkage rate is given by

$$\dot{\varepsilon}_x = \overline{\dot{\varepsilon}_{gbx}^{st}} + \dot{\varepsilon}_{crx} + \overline{\dot{\varepsilon}_{gbx}^{td}}, \tag{4.47}$$

show that the contribution of thermal diffusions to the shrinkage rate can be significant, especially for high temperatures and smaller particle sizes (Fig. 4.57). The comparative results for alumina (Fig. 4.58) indicate some shrinkage underestimation by modeling. This discrepancy between the modeling and experimental results could have been caused by a number of factors omitted in the analysis.

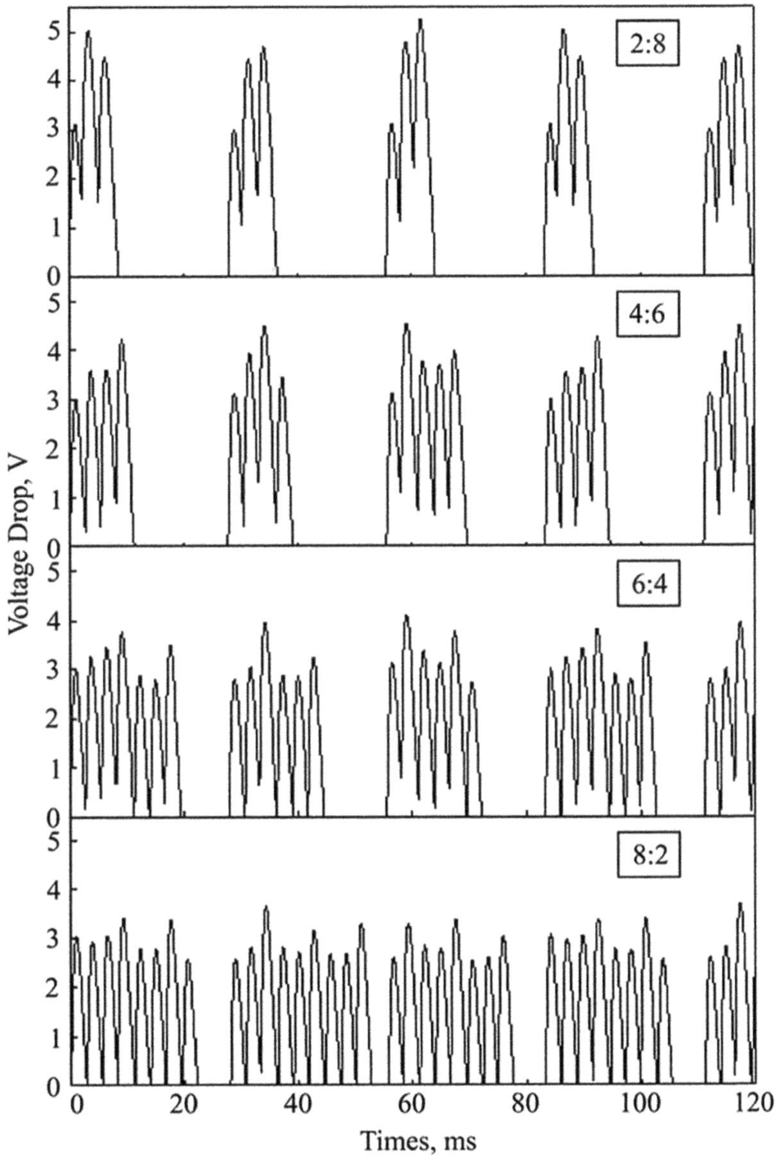

Fig. 4.55 Pulsing patterns for various combination of "on" and "off" periods. (Reprinted from Chen et al. [108], Copyright (2004) with permission from Elsevier)

4.9.6 Contribution of Electromigration

The first attempt to take into consideration a direct contribution of electric field to the densification during SPS has been undertaken by Olevsky and Froyen [77], who

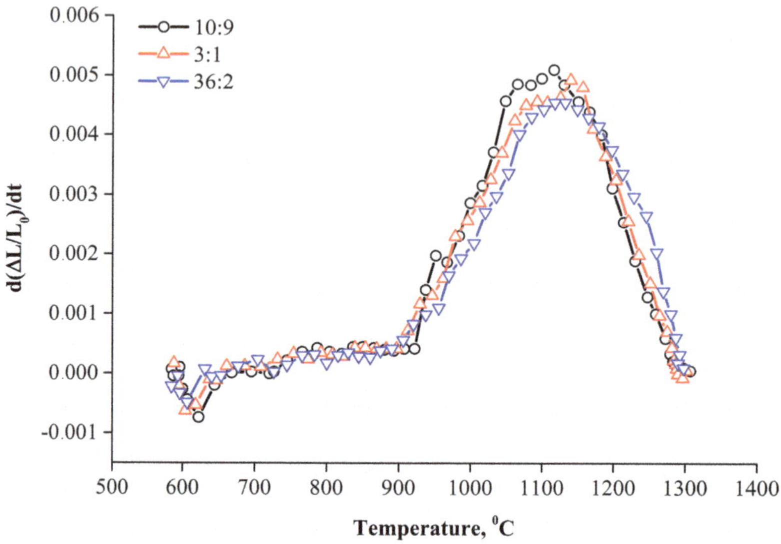

Fig. 4.56 Shrinkage rates of spark plasma sintered alumina observed at different pulse on/off sequences. (Reprinted from Shen et al. [56], Copyright (2002) with permission of John Wiley & Sons)

developed a constitutive model for SPS taking into account the contributions of sintering stress (surface tension), external load, and electromigration to sintering shrinkage. It has been shown that electromigration-related material flux can be a significant component of the electric current-accelerated diffusion. The flux of the matter $\overrightarrow{\mathbf{J}}$ caused by the grain-boundary diffusion is determined by Nernst–Einstein equation [110]:

$$\overrightarrow{\mathbf{J}} = C_E \overrightarrow{\mathbf{E}} + C_\sigma \overrightarrow{\nabla} \sigma, \tag{4.48}$$

where $\overrightarrow{\mathbf{E}}$ is the component of the electric field in the tangent plane of the grain-boundary and $\overrightarrow{\nabla} \sigma$ is the gradient of stresses normal to the grain boundary, $C_\sigma = \frac{\delta_{gb} D_{gb}}{kT}$, where D_{gb} is the coefficient of the grain-boundary diffusion and δ_{gb} is the grain-boundary thickness. C_E is determined by the following formula [111]:

$$C_E = \frac{\delta_{gb} D_{gb}}{\Omega kT} Z^* e_q, \tag{4.49}$$

where Z^* is the valence of a migrating ion and e_q is the electron charge. In the case shown in Fig. 4.59 (see also Fig. 4.46)

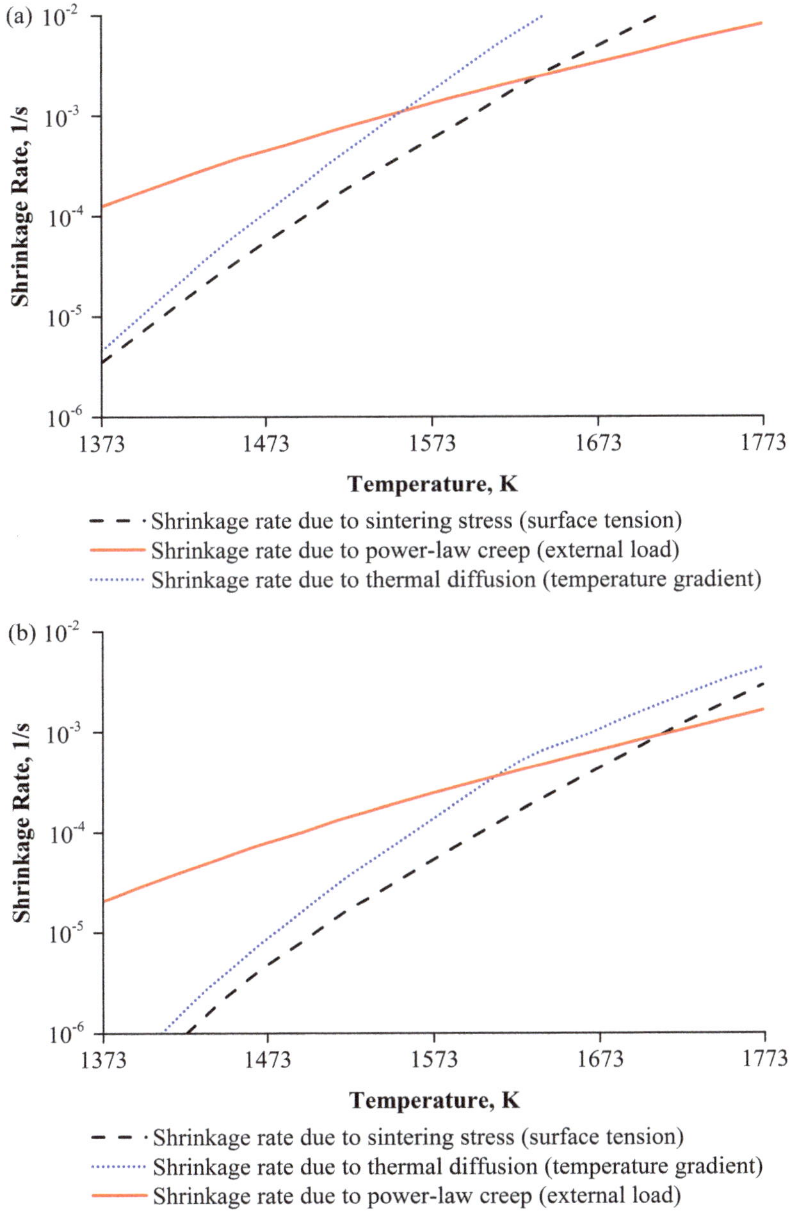

Fig. 4.57 Contribution to shrinkage rate from different mechanisms of mass transport for an alumina powder, $E = 300\frac{\mathrm{V}}{\mathrm{m}}$, $\Delta t = 5$ ms, applied stress $\overline{\sigma}_x = 30$ MPa, porosity $\theta = 0.3$, heating rate, grain size $G = 0.5$ μm (**a**) and $G = 10$ μm (**b**). (Reprinted from Olevsky and Froyen [79], Copyright (2008) with permission of John Wiley & Sons)

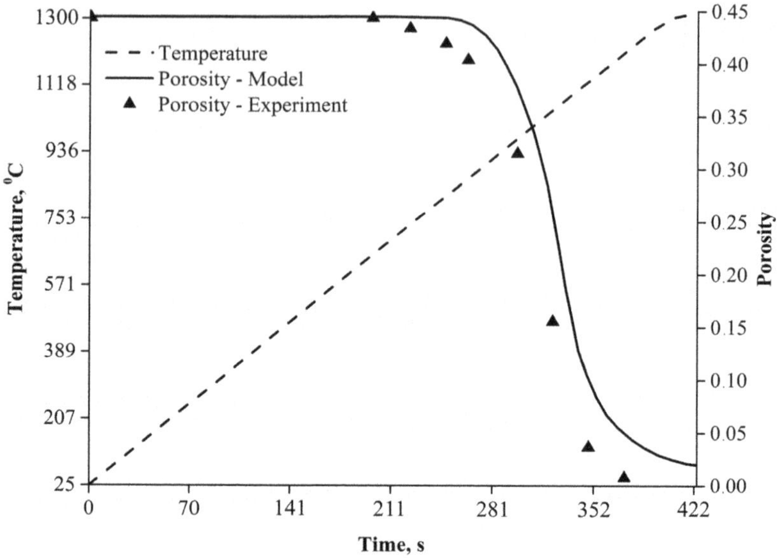

Fig. 4.58 Porosity kinetics during SPS of alumina powder. Comparison of the developed model taking into account the impact of thermal diffusion with experimental data of Shen et al. [56]. (Reprinted from Olevsky and Froyen [79], Copyright (2008) with permission of John Wiley & Sons)

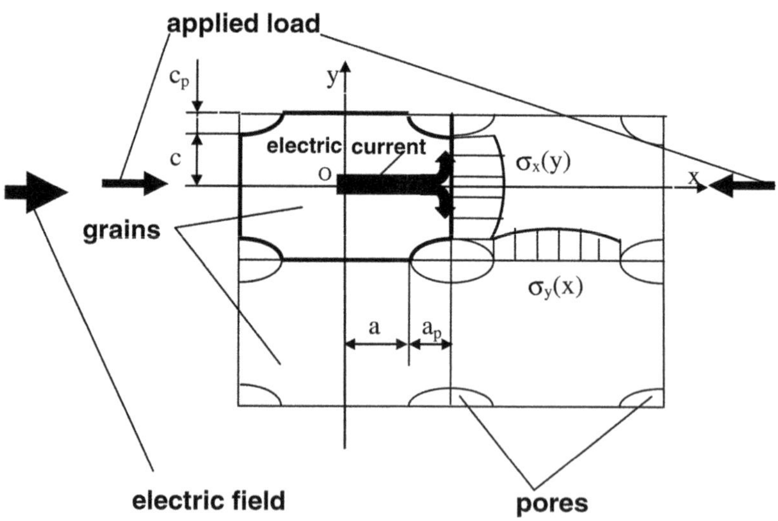

Fig. 4.59 Representative unit cell. The pore-grain structure is subjected to a simultaneous action of external load, surface tension, and electromigration. The electric field is macroscopically unidirectional and can branch locally. (Reprinted from Olevsky and Froyen [77], Copyright (2006) with permission from Elsevier)

$$J_y^{gb} = \frac{\delta_{gb}D_{gb}}{kT}\left[\frac{Z^* e_q}{\Omega}\frac{U}{l} + \frac{\partial \sigma_x}{\partial y}\right], \quad (4.50)$$

where U is the electric potential and l is the characteristic length along the electric field. Then $\dot{\varepsilon}_{gbx}$ can be expressed as follows:

$$\dot{\varepsilon}_{gbx} = -\frac{\delta_{gb}D_{gb}}{kT}\frac{\Omega}{(G+r_p)^2}\left\{\frac{Z^* e_q}{\Omega}\frac{U}{l} + \frac{3\alpha}{G}\left[\frac{1}{r_c} - \frac{1}{2G}\right] - \overline{\sigma}_x\frac{G+r_p}{G^2}\right\}, \quad (4.51)$$

where $r_p = a_p = c_p$ is the pore radius (Fig. 4.52). This equation incorporates three contributions to densification during the SPS due to electromigration $\dot{\varepsilon}_{gbx}^{em}$, surface tension $\dot{\varepsilon}_{gbx}^{st}$, and external load $\dot{\varepsilon}_{gbx}^{dl}$:

$$\dot{\varepsilon}_{gbx}^{em} = -\frac{\delta_{gb}D_{gb}}{kT}\frac{Z^* e_q}{(G+r_p)^2}\frac{U}{l}; \quad (4.52)$$

$$\dot{\varepsilon}_{gbx}^{st} = -\frac{\delta_{gb}D_{gb}}{kT}\frac{\Omega}{(G+r_p)^2}\frac{\alpha}{G}\left[\frac{1}{r_c} - \frac{1}{2G}\right], \quad (4.53)$$

$$\dot{\varepsilon}_{gbx}^{dl} = \frac{\delta_{gb}D_{gb}}{kT}\frac{\Omega}{(G+r_p)}\frac{\overline{\sigma}_x}{G^2} \quad (4.54)$$

The derived expressions for the component of the axial strain rate are valid for the SPS pressing schematics due to the proximity to zero of the equivalent Poisson's ratio's viscous analogy [67].

The total shrinkage rate during the SPS is equal to the superposition of the shrinkage rate corresponding to the grain-boundary diffusion and power-law creep mechanisms:

$$\dot{\varepsilon}_x = \dot{\varepsilon}_{gbx} + \dot{\varepsilon}_{crx} == -\frac{\delta_{gb}D_{gb}}{kT}\frac{\Omega}{(G+r_p)^2}\left\{\frac{Z^* e_q}{\Omega}\frac{U}{l} + \frac{3\alpha}{G}\left[\frac{1}{r_c} - \frac{1}{2G}\right] - \overline{\sigma}_x\frac{G+r_p}{G^2}\right\}$$
$$-\left\{\left(\tfrac{3\theta}{2}\right)^{(m+1)/2}\left[\tfrac{3\alpha}{2G}(1-\theta)^2 - \overline{\sigma}_x\right]/A(1-\theta)^{(m-3)/2}\right\}^{1/m} \quad (4.55)$$

In Eq. (4.55) the power-law creep component is related to the macroscopic powder volume subjected to SPS (this component depends on porosity). The macroscopic homogenization of the component corresponding to the grain-boundary diffusion in a simplified way can be achieved by introducing the following the dependence of the pore radius on the grain size and porosity [74]:

$$r_p = G\sqrt[3]{\theta/[6(1-\theta]} \quad (4.56)$$

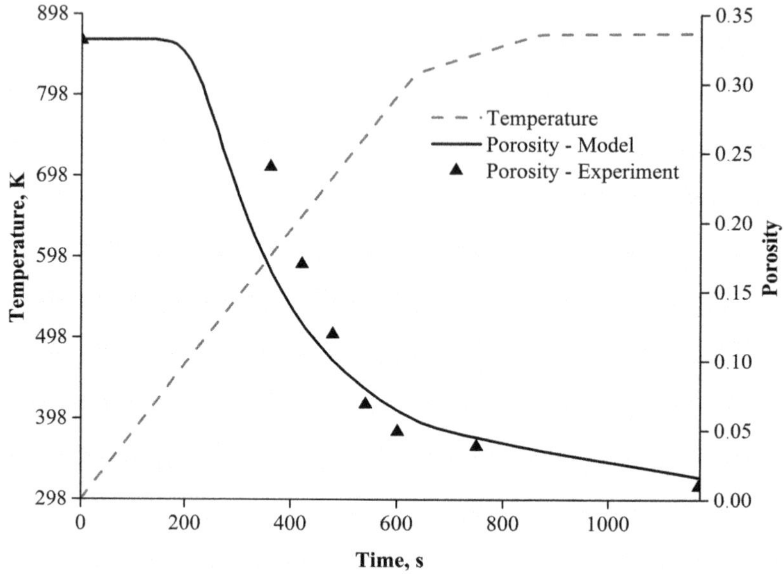

Fig. 4.60 Evolution of porosity during the SPS of Al powder calculated according to the model developed by Olevsky and Froyen [77] and measured experimentally by Xie et al. [112]. (Reprinted from Olevsky and Froyen [77], Copyright (2006) with permission from Elsevier)

Equation (4.55) allows the direct comparison of the contributions of different mechanisms to the SPS shrinkage kinetics. The analysis indicates how the magnitude of the contributions of three different factors changes with changing the grain size. For conventional, micron-sized powders ($G = 40$ μm), power-law is the dominant mechanism of material transport. For smaller porosities and ultrafine powders ($G < 1$ μm), the electromigration plays the dominant role, while for smaller porosity in the nanopowder range, sintering stress dominates. The abovementioned interplay is possible when porosity is lower than 30% (for larger void volume fractions, power-law creep is always the main mechanism of the material flow). For porosities lower than 30%, both surface tension (sintering stress) and electromigration can become the main contributors to shrinkage depending on the average grain size. Here we should note a possible limitation of this preliminary model assessment. Since both locally and macroscopically the generation of Joule heat (and, hence, temperature distribution) cannot be decoupled from the electric current, parameter $\frac{U}{I}$ should evolve with densification. Therefore, for very small porosities in the electromigration-dominating zone, the ultimate collapse of voids may require externally applied load as the primary factor.

The densification kinetics is described by the following equation [77]:

$$\dot{\theta} = (1 - \theta)\left(\dot{\varepsilon}_x + \dot{\varepsilon}_y\right) = (1 - \theta)\left(\dot{\varepsilon}_x\right) \tag{4.57}$$

The numerical solution of the differential equation is shown in Fig. 4.60 together with the experimental data obtained by Xie et al. [112]. The temperature and

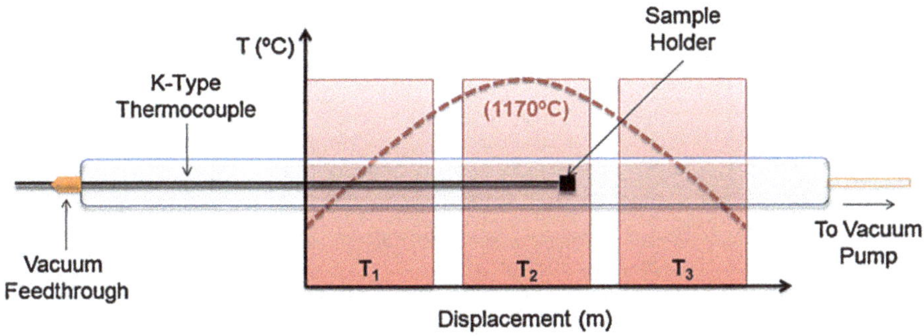

Fig. 4.61 Schematic of the tube furnace used for free sintering at high heating rates in the absence of field. (Reprinted from Holland et al. [113], Copyright (2012) with permission from Elsevier)

pressure cyclograms used in the calculations replicate the conditions of the experiments. The comparison of the modeling and experimental results indicates their satisfactory agreement.

Holland et al. [113] conducted experiments on sintering of hydroxyapatite and zinc oxide nanopowders at high heating rates with the application of a field and in the absence of field without externally applied pressure. Pressureless sintering in the absence of a field was performed in a three-zone tube furnace (Fig. 4.61). The sample pellet was first placed in a graphite crucible that was connected to a long thermocouple enabling the temperature measurements and providing a tool, by which the graphite crucible containing the powder compact was inserted in and removed from the tube furnace. The heating rate was controlled by varying the rate of insertion of the crucible into the furnace. Using this arrangement, it is necessary to closely monitor the heating rate and adjust the crucible insertion depth frequently during heating. A vacuum level of approximately 20 Pa, which is close to that used in a typical SPS experiment, was maintained by inserting the thermocouple body through a sliding O-ring seal and connecting the opposite end of the tube to a vacuum pump. For comparison, pressureless sintering with an applied, non-contacting, field was performed using a SPS facility.

While the density of the hydroxyapatite specimens was nearly equal when heated to 950 °C at 50 °C min^{-1} or 400 °C min^{-1} (67–68%), heating at 800 °C min^{-1} reproducibly yielded a density of ~88% with no hold time in field-assisted tests (Fig. 4.62). In the field-free test, the application of a heating rate of 800 °C min^{-1} resulted in the rapid formation of a sintered compact with a relative density about 10% lower compared with that in the field-assisted test. With an applied field, the 400 °C min^{-1} tests achieved closed porosity in less than 10 min, whereas the 50 °C min^{-1} tests did not within up to 20 min of holding at the target temperature. For zinc oxide nanopowders, the slowest and fastest heating rates achieved the same densities in zinc oxide when the target temperature of 850 °C was reached (Fig. 4.63). Several possible explanations were suggested to explain the observed heating rate-dependent behavior in the presence of an applied field: pore spheroidization and grain growth in the nanometric powders limited due to kinetic

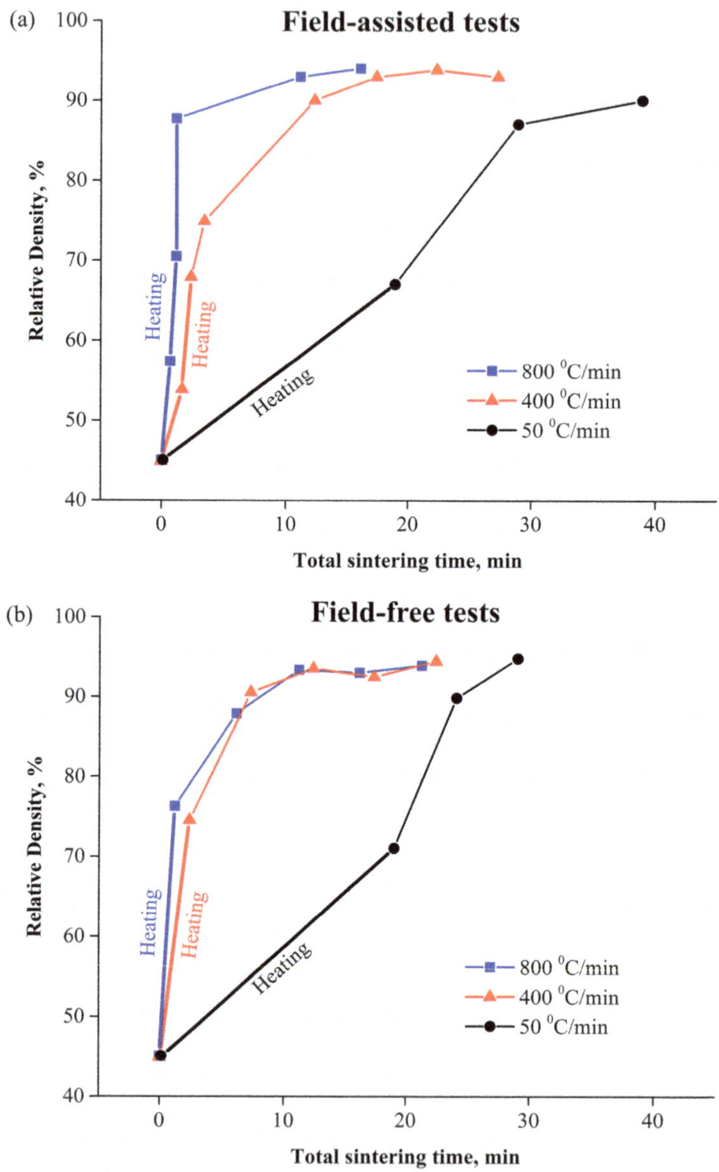

Fig. 4.62 Relative density of hydroxyapatite sintered in a SPS facility (**a**) and in a tube furnace in the absence of field (**b**) without external pressure at different heating rates. (Reprinted from Holland et al. [113], Copyright (2012) with permission from Elsevier)

influences, a time- and field strength-dependent dielectric breakdown behavior, the rate of space charge layer development as it relates to heating rate, local field strengths causing current flows to produce local heating, as was suggested by

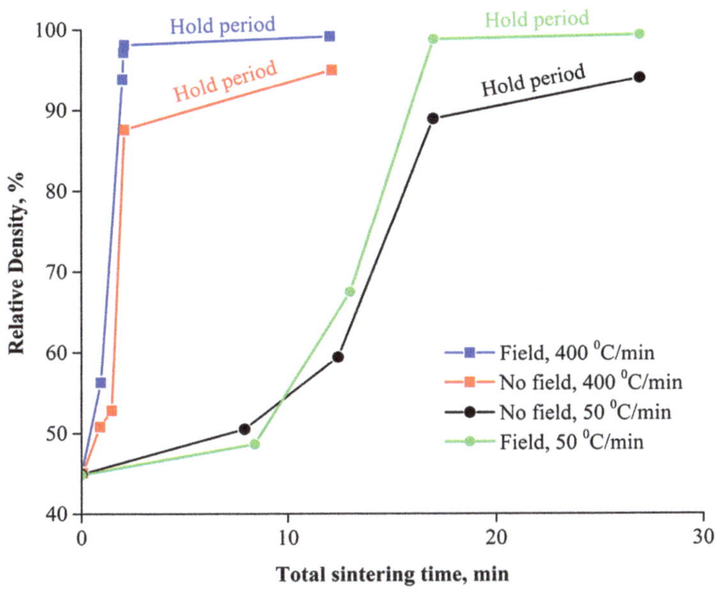

Fig. 4.63 Relative density of zinc oxide sintered in a tube furnace in the absence of field and in a SPS facility without external pressure at different heating rates. (Reprinted from Holland et al. [113], Copyright (2012) with permission from Elsevier)

Cologna et al. [114] regarding "flash sintering," and a significant increase in the diffusivity of the rate-controlling ion in the presence of the electric field. Holland et al. suggested that a synergistic relationship between field and heating rates exists during the initial stage of sintering at moderate field strengths, consistent with those encountered during typical FAST/SPS experiments.

Holland et al. [113] came to a conclusion that the heating rate alone does not provide much specific benefit to densification rates. The athermal benefits to densification were found to be considerable and heating rate-dependent.

4.9.7 Constitutive Equation of SPS Taking into Account the Enhanced Dislocation Motion by Local Resistive Heating

Lee et al. [38] modified the constitutive equation of the continuum theory of sintering to include the electric current effect term based on the recent results of the electroplasticity theory [115, 116]. It was assumed that the reduction of the particle material yield stress by an electric current is a major mechanism of the rapid densification of conductive powders compacted via an electric current-assisted consolidation method. When the electric current flows through a powder during deformation, the local "hot spots" caused by the greater electron scattering at defects, such as dislocations, voids, impurities, and grain boundaries, largely enhance the lattice vibration energy in the defect vicinity. As a result, dislocations can move more easily along the slip plane bypassing lattice obstacles with smaller resistance. Therefore, the stress required to move dislocations is reduced. This heating occurs at the atomic level within the lattice contributing to the total heating of the sample and expanding the lattice locally [117, 118]. This effect can lead to faster defect motion under SPS compared with hot-pressing conditions. While HP needs external energy input that have to heat the entire material to allow for the dislocations to diffuse to a sink such as grain boundaries, SPS provides a greater amount of energy in a faster mode and directly to the dislocations.

In Refs. [119, 120], it was shown that the electron wind effect is not sufficient to affect the dislocation activities. Therefore, it was assumed that the main source of the reduction of the flow stress by the electric current is from the indirect local vibration energy from electron scattering, not from the direct momentum transfer to the dislocation.

Based on the following considerations, a constitutive equation of SPS taking into account the effect of the electric current effect was developed taking into consideration the enhanced dislocation motion by local resistive heating. The stress for high-temperature deformation of crystalline materials can be described as a function of strain rate ($\dot{\varepsilon}$) and absolute temperature (T) as expressed by Mukherjee–Bird–Dorn equation [121]:

$$\left(\frac{\sigma}{G}\right)^n = \frac{\dot{\varepsilon}\,kT}{A_{\mathrm{Cr}}DGb}\left(\frac{d}{b}\right)^p \tag{4.58}$$

where $\dot{\varepsilon}$ is the steady-state creep rate (*1/s*), k is the Boltzmann constant (J/K), D is the diffusion coefficient (cm^2/s), G is the shear modulus (MPa), T is the absolute temperature (K), b is the Burgers vector (m), σ is the applied stress (MPa), A_{cr} is the material creep constant (unitless), d is the grain size (m), p is the grain size exponent, and n is the stress exponent (n is equal to *1/m*, where m is the strain rate sensitivity).

The dependence of the diffusion coefficient D on temperature T can be expressed through

$$D = D_0 \exp\left(\frac{-Q}{RT}\right) \tag{4.59}$$

where D_0 is the diffusion constant (cm^2/s), Q is the creep activation energy (KJ/mol), and R is the gas constant.

For a porous material [84], Mukherjee–Bird–Dorn equation can be converted to the power-law relationship in terms of the strain rate as follows:

$$\sigma(W) = G\left\{\frac{A_0 T}{G}\exp\left(\frac{Q}{RT}\right)\right\}^m \left(\frac{d}{b}\right)^{pm} W^m \tag{4.60}$$

Here, A_0 (Pa·s/K) is given by

$$A_0 = \frac{k}{A_{Cr} b D_0} \tag{4.61}$$

where $\sigma(W)$ is the effective stress and W is the effective equivalent strain rate of the porous material.

On the basis of the continuum theory of sintering, the following equation can be derived to describe hot pressing of a nonlinear viscous porous material in a rigid die that can be written as [38]

$$\dot{\theta} = -\left[\frac{G}{A_0 T}\left(\frac{b}{d}\right)^p \exp\left(\frac{-Q}{RT}\right)\right]\left(\frac{3\theta}{2}\right)^{\frac{m+1}{2m}}(1-\theta)^{\frac{m-3}{2m}}\left(\frac{\sigma_z}{G}\right)^{\frac{1}{m}} \tag{4.62}$$

where $\dot{\theta}$ is the densification rate (*1/s*), G is the shear modulus (MPa), d is the grain size (m), p is the grain size exponent, Q is the creep activation energy (KJ/mol), and R is the gas constant (J/molK), σ_z is the Z-axis applied stress (MPa), and A_0 is the material constant (Pa·s/K).

In Eq. (4.62), the function in the square brackets indicating the thermal deformability of powders during sintering can be defined as follows:

$$A_{TD} = \frac{G}{A_0 T}\left(\frac{b}{d}\right)^p \exp\left(\frac{-Q}{RT}\right) \tag{4.63}$$

Total equivalent effective strain rate (W) should be the sum of the strain rate from thermal and electric current-assisted effect during the SPS process:

$$W = W_{\text{Thermal}} + W_{\text{Electric current assisted}} \tag{4.64}$$

Equation (4.64) can be converted with the introduction of A_{ECAD} (1/s) which indicates the electric current-assisted deformability of the powders as follows:

$$W = (A_{TD} + A_{ECAD})\left(\frac{\sigma_z}{G}\right)^{\frac{1}{m}} \tag{4.65}$$

Taking into account that $\dot{\varepsilon}_r = 0$ (uniaxial die compaction boundary condition), the constitutive equation for SPS can be written as

$$\dot{\theta} = -[A_{TD} + A_{ECAD}]\left(\frac{3\theta}{2}\right)^{\frac{m+1}{2m}}(1-\theta)^{\frac{m-3}{2m}}\left(\frac{\sigma_z}{G}\right)^{\frac{1}{m}} \tag{4.66}$$

The following analysis was used to find A_{ECAD} [38]. Local Joule heating can provide the vibration energy to the dislocation movement [117]. High temperature of the locally overheated region may be sufficient to activate additional intrinsic defects [122]. These generated defects can enhance the mobility of dislocations by interacting with elastic and electronic fields of the dislocations [123–125]. Moreover, they can be also the source of the local Joule heating, making the local temperature to increase again and thereby starting the cascade effects. However, due to the relatively low electric current density under regular SPS conditions and being coupled with vacancy annihilation by dislocation movement, these cascade effects are limited compared to flash sintering. Therefore, it was assumed that the energies of the local Joule heating are the original source for the electric current effects. Joule heating effects like the defect generation should be accumulated with sintering time during SPS due to the cascade characteristics. The localized energy provided to the dislocation core due to the Joule heating effect (per dislocation core volume) is given by [117]

$$E_{\mathrm{J}} = J^2 \lambda_{\mathrm{d}} \Delta t \tag{4.67}$$

where J is electric current density, λ_{d} is electrical resistivity of the dislocation core, and Δt is the time duration when the electric current is applied. Parameter λ_{d} was determined to be around 6–8 times the electric resistance of the defect-free lattice (λ) [126]. Also, the dislocation movement by Joule heating can be hindered by Peierls–Nabarro stress (τ_{PN}) of each material.

Including the terms that affect the dislocation movement considered above, A_{ECAD} can be written as

$$A_{\mathrm{ECAD}} = \left[\int_{t_0}^{t_f} \frac{J^2 \lambda_{\mathrm{d}}}{\tau_{\mathrm{PN}}} dt \right]^{\omega} \tag{4.68}$$

where t_0 and t_{f} are the starting and final time for SPS and ω is the electric current sensitivity exponent (material constant).

Current density in a powder sample is affected by the neck area between particles during the SPS. The neck area changes with porosity evolution. Current density can be large at the initial stage of sintering (small neck) and reduces with decreasing porosity (large neck) during sintering. For an irregular packing of the spherical particles, the neck area evolution with porosity can be estimated as shown by Helle et al. [127]:

$$\frac{AZ}{4\pi R^2} = \frac{(\theta_0 - \theta)(1 - \theta)}{\theta_0} \tag{4.69}$$

where A is average area of contacts between two particles, Z *is* coordination number, R is particle radius, θ_0 is initial porosity, and θ is porosity. When R is constant, $AZ/4\pi R^2$ value is increased from 0 to 1 when porosity is changed from θ_0 to 0. It can be assumed that the neck area evolution of one particle is proportional to the total contact area change of the total powder specimen's volume. Therefore, we consider

$AZ/4\pi R^2$ to be the ratio of the total neck area to the total powder surface area. Overall local electric current density (OLECD) is the summation of the electric current density of all necks in the powder volume. Using the previous consideration, J_{OL} (OLECD) of the sample can be calculated as follows:

$$J_{\mathrm{OL}} = \frac{I_s}{A_{\mathrm{cross\text{-}section}}} \times \frac{4\pi R^2}{AZ} \tag{4.70}$$

where $A_{\mathrm{cross\text{-}section}}$ is the cross-sectional area of the full density sample.

By including the electric current effect coefficient (β), which involves the relationships between J_{OL} and actual J in the sample, λ_{d} and actual λ, and τ_{PN} and G, Eq. (4.68) can be written as

$$A_{\mathrm{ECAD}} = \beta^{\omega}\left[\int_{t_0}^{t_f} \frac{{J_{\mathrm{OL}}}^2 \lambda}{G} dt\right]^{\omega} \tag{4.71}$$

Therefore, the constitutive equation for SPS, Eq. (4.66), can be expressed as

$$\dot{\theta} = -\left[\frac{G}{A_0 T}\left(\frac{b}{d}\right)^p \exp\left(\frac{-Q}{RT}\right) + \beta^{\omega}\left[\int_{t_0}^{t_f} \frac{{J_{\mathrm{OL}}}^2 \lambda}{G} dt\right]^{\omega}\right]\left(\frac{3\theta}{2}\right)^{\frac{m+1}{2m}} (1-\theta)^{\frac{m-3}{2m}} \left(\frac{\sigma_z}{G}\right)^{\frac{1}{m}} \tag{4.72}$$

Thus, a constitutive equation of SPS taking into account the effect of electric current was developed taking into consideration the enhanced dislocations motion by local resistive heating. The SPS and hot-pressing experiments with a ZrN powder showed that electric current passing through the sample affects the densification behavior of this powder [38]. Figure 4.64 shows porosity evolution during the whole processing cycle for the sintered specimens obtained from the two experiments: SPS at 1100 °C and HP at 1200 °C. Experimental conditions, including heating rate,

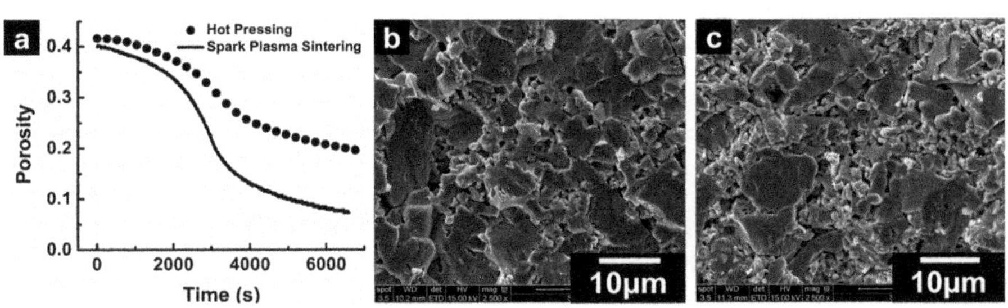

Fig. 4.64 (**a**) Porosity variation of ZrN powder densified (60 MPa and 10 °C/min) by SPS (1100 °C) and hot pressing (1200 °C); images of the fractured surface of samples densified by (**b**) SPS at 1100 °C and (**c**) hot pressing at 1200 °C. (Reprinted from Lee et al. [38], Copyright (2017) with permission from Elsevier)

applied uniaxial pressure, and atmosphere, as well as the amounts of powder utilized per specimen, were maintained the same. For the SPS, the actual ZrN temperature inside of the die at the holding stage was 1225 °C, which was 25 °C higher compared with the hot-pressing experiment (see Sect. 4.4). SPS showed faster densification (Fig. 4.64a) and a more consolidated structure compared with hot pressing (Fig. 4.64b–c). It is clear that the 25 °C temperature difference is not sufficient to explain the final density difference of 12.30% between compacts obtained by SPS and hot pressing. Therefore, the electric current flowing into the ZrN powder during SPS can be a possible factor contributing to the porosity evolution. The observed differences between the behavior of samples obtained by SPS and hot pressing can be related to the effect of electric current on the dislocation motion (electroplasticity phenomena).

When fitting Eq. (4.62) into the experimental porosity evolution data from hot pressing and SPS, if the m value for SPS is not the same as for hot pressing, then the densification mechanism is affected by electric current. HP results showed $m = 0.23$–0.24 and $Q = 114.25$–220.73 kJ/mol, and SPS rendered $m = 0.33$–0.36 and $Q = 109.2$–207.04 kJ/mol. The obtained m and Q in the hot-pressing experiments indicate that ZrN consolidation is based on the dislocation climb-controlled creep mechanism. The values of m and Q for SPS show that the densification mechanism is close to the glide-controlled creep ($m = 0.33$). The electric current effect, such as localized heating at defects, and defect generation may facilitate the diffusion of barriers encountered by a mobile dislocation. Also, a mobile dislocation that obtains energy from local Joule heating or an elastic field effect from vacancies generated by an electric current can overcome obstacles more easily. Vacancy formation by local Joule heating can promote the dislocation climb, and the overall rate-limiting step of ZrN powder deformation during SPS can be changed from the dislocation climb-controlled to the glide-controlled mechanism.

Alternatively, it can be assumed that the creep parameters are the same for both hot pressing and SPS. It was found that the electric current term in Eq. (4.72) can fit the experimentally observed porosity evolution corresponding to the SPS results. So, the constitutive equation for SPS equation showed a possibility of the facilitation of the dislocation movement by an intrinsic current effect other than a change of the densification mechanism. This equation for SPS can be applied to study densification mechanisms for electrically conductive materials in field-assisted sintering processes.

4.10 Selected Examples of Processes and Materials Developed Using SPS

In this section, processing methods developed using SPS equipment (other than regular consolidation of powders into shapes corresponding to the shapes of the die cavities) will be discussed to emphasize the scientific and technological potential of combinations of SPS with other techniques and approaches. Examples of successful fabrication of dense materials with improved properties by SPS will be provided. Approaches used to obtain porous materials using SPS will also be presented, as this area is rapidly growing. The list of examples given in this section is not claimed to fully cover the activities in the corresponding areas of functional materials. However, they were selected to highlight new directions and promising ideas in the current SPS research from the viewpoint of obtaining materials with unusual microstructures, architectures, and phase compositions. For broad, systematic, and encyclopedic overview of materials processed by SPS, the reader is directed to an excellent review by Orrù et al. [26]. Metallic materials and metal matrix composites obtained by SPS have been reviewed by Saheb et al. [128].

4.10.1 Processing and Testing Methods Developed Using SPS Equipment

Hulbert et al. [129] and Jiang et al. [130] used SPS for simultaneous sintering and forming of ceramics. Two strategies were employed to make the shaped parts. The first one was a traditional approach to shape a pre-sintered material (bulk forming) [129]. According to the displacement curve recorded during SPS, forming of the pre-sintered Al_2O_3–ZrO_2–$MgAl_2O_4$ blank occurred at a deformation rate of approximately $10^{-2}\ s^{-1}$. The forming temperature was 1150 °C; the sample was held at this temperature for 3 min. Such forming conditions are commercially attractive and incur low electrical energy costs. The aligned structure in the shear regions of the formed part indicated that boundary sliding occurred in groups of grains. The second approach started from a ceramic material in the powder form and combined sintering and forming in a single step in a SPS run with the same maximum temperatures and dwell time at the maximum temperature [130]. Figure 4.65 shows general views and a cross-section of the Al_2O_3–ZrO_2–$MgAl_2O_4$ composite ceramics after powder forming. While significant grain growth was observed in the bulk formed Al_2O_3–

(a) (b) (c)

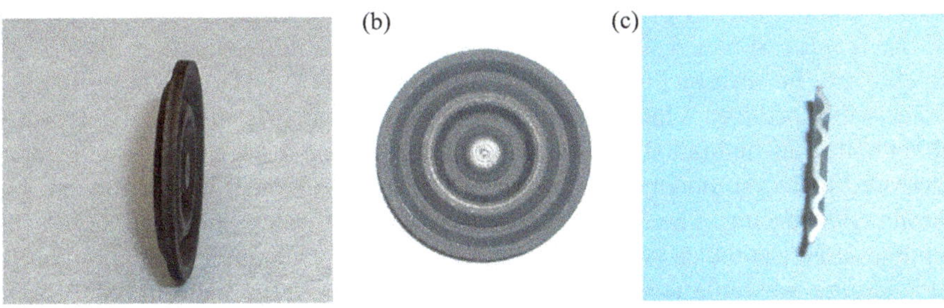

Fig. 4.65 Al_2O_3–ZrO_2–$MgAl_2O_4$ composite ceramics after powder forming: (**a**) side view, (**b**) top view, (**c**) cross-sectional view. (Reprinted from Jiang et al. [130], Copyright (2006) with permission from Elsevier)

Fig. 4.66 Microstructure of Al_2O_3–ZrO_2–$MgAl_2O_4$ composite ceramics produced by powder forming (**a**) and bulk forming (**b**), A–Al_2O_3 grains, Z–ZrO_2 grains, S-spinel grains. (Reprinted from Jiang et al. [130], Copyright (2006) with permission from Elsevier)

ZrO_2–$MgAl_2O_4$ composite relative to the pre-sintered state, the powder formed composite remained fine-grained (Fig. 4.66). This difference in the grain size can be explained by a shorter exposure to high temperatures of the powder formed sample relative to the bulk formed sample, as the latter was first sintered and then shaped in two separate processes.

An interesting direction in the utilization of SPS facilities is conducting electric field-assisted sinter-forging [131–133]. Zhan et al. [131] conducted sinter-forging in a SPS apparatus to produce nanostructured 50vol.%Al_2O_3–50vol.%$MgAl_2O_4$ composite with a high infrared transmittance. A piece of the porous compact (1/4) was placed in the same die that was used for its pre-consolidation, and the sinter-forging experiments were carried out. It was found that starting with a porous (pre-consolidated) specimen and conducting sinter-forging in the SPS chamber, materials with a finer grain size could be obtained compared with materials produced at the same temperature without deformation (in a classic SPS assembly). Wei et al. [133] conducted both experimental and theoretical studies of the densification kinetics during SPS-forging. The theoretical studies were based on the continuum theory of sintering. Using pre-consolidated porous zirconium carbide (ZrC) as an example, the authors found that the loading mode plays an important role in the densification efficiency of porous ZrC during SPS. The differences in the geometry of the regular SPS and SPS-forging assemblies and corresponding densification kinetics can be seen in Figs. 4.67 and 4.68, respectively. The SPS-forging design isolated the specimen's lateral surface from the die wall and allowed displacements along both axial and radial directions. The axial load on the specimen was gradually increased to accommodate the change in the cross-sectional area of the specimen during SPS-forging. The SPS-forging-processed specimens showed a higher final relative density at the end of the process. Consequently, compared with regular SPS, SPS-forging was able to enhance densification during later sintering stages more significantly. The developed analytical models were able to distinguish the effects of loading modes on the densification behavior of ZrC. The porosity difference between the numerical solutions was found to be smaller than that between the

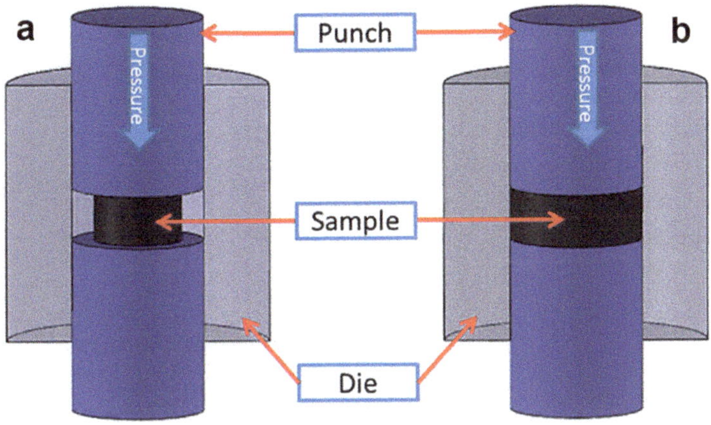

Fig. 4.67 Schematic of SPS-forging (**a**) and regular SPS (**b**) setups. (Reprinted from Wei et al. [133], Copyright (2017) by permission of Taylor & Francis Ltd., www.tandfonline.com)

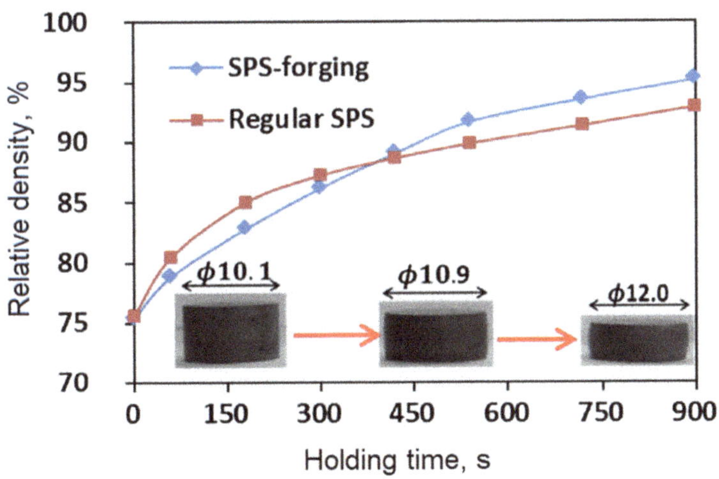

Fig. 4.68 Densification kinetics of ZrC during SPS-forging and regular SPS at 1700 °C and 60 MPa. (Reprinted from Wei et al. [133], Copyright (2017) by permission of Taylor & Francis Ltd., www.tandfonline.com)

experimental sets, which can be attributed to the fact that SPS-forging experiments were influenced by friction at the contact surfaces, while the analytical equations were derived assuming ideal contact sliding conditions. Using results of this study, SPS-forging and regular SPS modes were successfully combined to form a new SPS hybrid loading scheme, in which SPS-forging was followed by regular SPS in order to reduce shape irregularity in the specimens forged during the SPS. In the hybrid SPS experiment, the pre-sintered cylinder was first to reach the die wall through the forging process so that the specimen could follow the die wall in the next step. In the

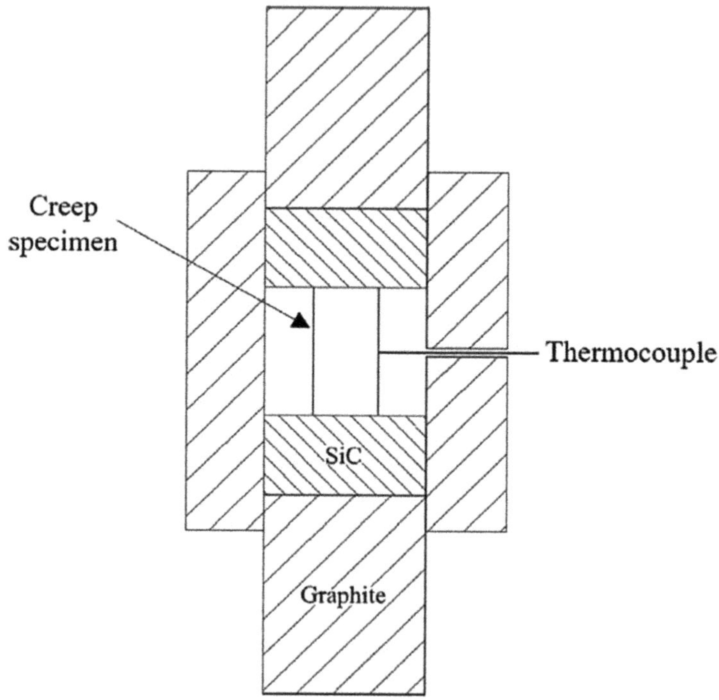

Fig. 4.69 Tooling configuration for creep tests to be conducted with the help of SPS facilities. (Reprinted from Ratzker et al. [135], Copyright (2017) with permission from Elsevier)

hybrid loading mode, it became possible to manage the structure and densification intensity of the specimens.

Ratzker et al. [134, 135] suggested using SPS facilities as tools in compressive creep studies. Figure 4.69 shows a tooling configuration designed for creep tests. When the behavior of fine-grained alumina was studied, the measured creep rates, stress exponent, and apparent activation energy were in a good agreement with the data previously reported for this material. The creep parameters together with microstructural investigations suggested that SPS apparatus can be employed as accurate high-temperature creep testing systems. In order to correctly use SPS for creep studies of other materials, it is important to know whether the applied field strength and the composition of the atmosphere of the SPS chamber influence the creep behavior of the tested material.

The challenge of sintering complex shapes by the SPS technique was tackled by Manière et al. [136]. It was shown that parts with complex shapes and large thickness differences between the elements can be sintered to full density using a sacrificial material approach. The sintered material had a homogeneous microstructure. The presence of a sacrificial part homogenizes the shrinkage distance in all areas of the part. The sacrificial element can be made from the same material as the main part or from other materials. This approach is restrained to shapes with a limited number of

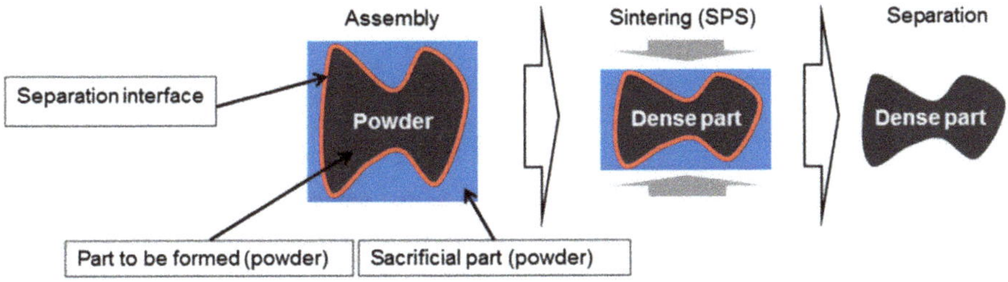

Fig. 4.70 SPS of parts with complex shapes using the deformed interfaces approach. (Reprinted from Manière et al. [137], Copyright (2017) with permission from Elsevier)

thickness differences. For shapes with a continuous variation or a large number of different thicknesses, the sacrificial material should be highly segmented, and the described method becomes difficult to perform.

Another approach suggested by Manière et al. [137] for the fabrication of parts of complex shapes by SPS is based on the assemblies that include interfaces that change their shape. The method of deformed interfaces consists of assembling a minimum of two porous areas separated by interfaces. The role of the interface is to maintain a physical and chemical separation between the two porous bodies during their densification and to allow the post-sintering separation of the parts (Fig. 4.70). The separation interface was made by either the imprint method or using graphite foil. In the imprint method, a boron nitride (BN) coating was used to enable the post-sintering separation process. The graphite foil approach is close to the imprint approach. The difference between these approaches is in the fact that the powder is filled on both sides of the graphite foil interface. The graphite foil is folded to acquire a stretched shape of the sample before the powder-filling step. The graphite foil can be rendered into a desired shape by cold folding or forming processes.

A new route to materials with a designed architecture was proposed by Martin et al. [138], who used a combination of Electron Beam Melting (EBM) and SPS to generate an architectured Ti–Ti composite (Fig. 4.71). A lattice structure was first fabricated by EBM from a titanium powder. Then the space between the struts was filled with a Ti powder, and the assembly was subjected to SPS. The consolidated material consisted of lattice elements, whose microstructure was predetermined by the EBM conditions, and the matrix, which formed during the SPS step. The choice of SPS was dictated by a possibility of preserving the microstructure of the material obtained by EBM (Fig. 4.72). In view of rapid development of 3D manufacturing technologies, this study introduces an interesting research direction, which may further widen the application areas of materials containing 3D elements. In the combination of a 3D manufacturing technology with SPS, new possibilities can be found of producing materials with microstructure and properties tailored at different length scales. While the 3D manufacturing technology is responsible for the special arrangement of the struts and the formation of their microstructure, SPS makes it possible to embed such a 3D structure into a matrix and densifies the matrix while preserving the microstructure of the struts. Also, the struts and the matrix can be

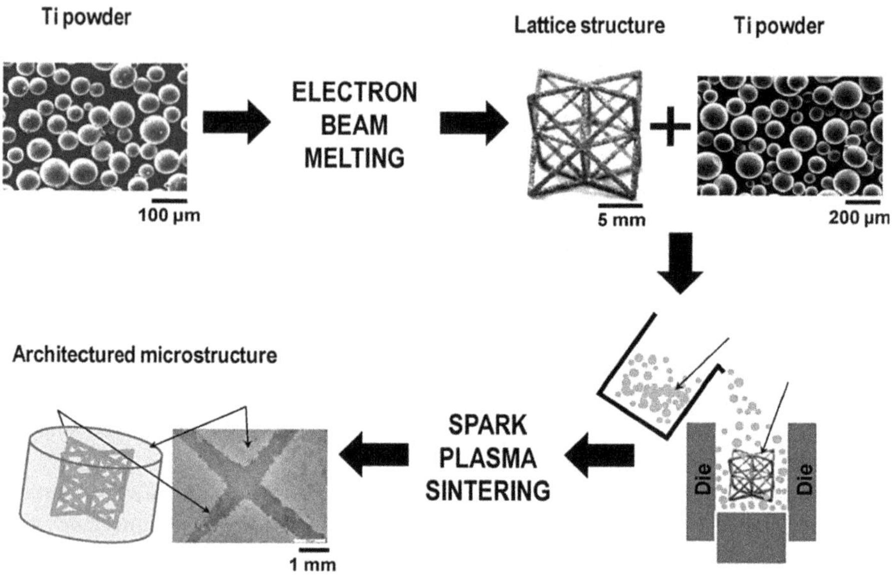

Fig. 4.71 Processing route of Ti–Ti materials with an architecture microstructure. (Reprinted from Martin et al. [138], Copyright (2016) with permission from Elsevier)

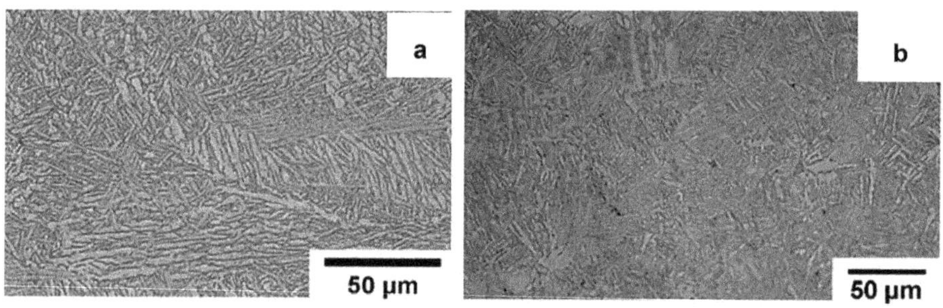

Fig. 4.72 Microstructure of Ti alloy struts in the as-obtained 3D lattice (**a**) and lattice as part of the architectured composite sintered by SPS (**b**). (Reprinted from Martin et al. [138], Copyright (2016) with permission from Elsevier)

selected to have different chemical and phase compositions. In such cases, the differences in the properties of the two materials should be thoroughly analyzed to avoid situations, in which the presence of the lattice will hinder the densification of the matrix.

4.10.2 *Joining of Materials by SPS*

The uniaxial pressure and pulsed DC supplied by the SPS facilities can be used for joining of macro-objects [139–142]. Intermediate layers (fillers) in the form of a powder or a foil were suggested. A steel powder layer placed between steel cylinders was used to join them together by passing an electric current through the assembly [139]. The powder layer was a region of a higher resistance in the assembly and, therefore, was heated more intensely than the cylinders, acting as an effective heating source (Fig. 4.73). The amount of powder was such that no die was necessary for the intermediate layer confinement. By adjusting the electric current and the heating time, joints with a high mechanical strength were obtained. An example of the joint made by the SPS is shown in Fig. 4.74. As the resistance of the powder layer depends on the particle size, particles with a smaller size produced joints of a higher strength. It was possible to reduce the size of the heat-affected zone by conducting joining within several seconds.

The composition of the filler is of great importance for the outcome of the joining process. Dong et al. [142] compared the behavior of intermediate layers of different compositions in the process of joining of SiC parts by SPS. The maximum room temperature flexural strength of the samples was 66 MPa and 133 MPa for the Ti_3SiC_2 filler and 3Ti–1.2Si–2C–0.2Al powder mixture filler, respectively. The good sinterability of the 3Ti–1.2Si–2C–0.2Al powder mixture and chemical reactions between the filler components and SiC resulted in the formation of a joint with a higher flexural strength as compared with that of the joint formed with Ti_3SiC_2 as a filler.

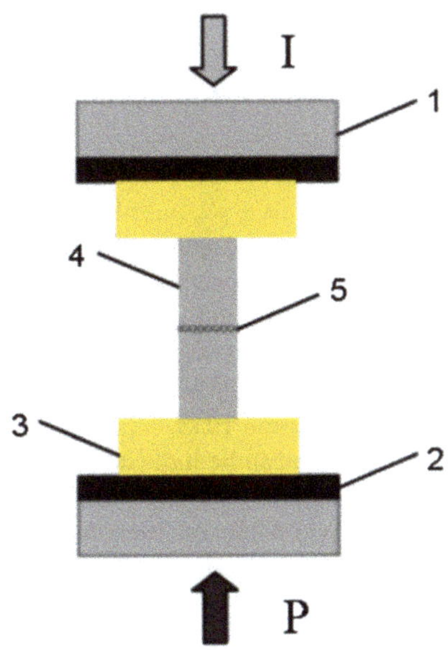

Fig. 4.73 Schematic of the SPS setup used for the joining of rods: 1, electrode; 2, graphite spacer; 3, copper spacer; 4, 316 L stainless steel rod; 5, 316 L stainless steel powder. (Reprinted from Yang et al. [139], Copyright (2013) with permission from Elsevier)

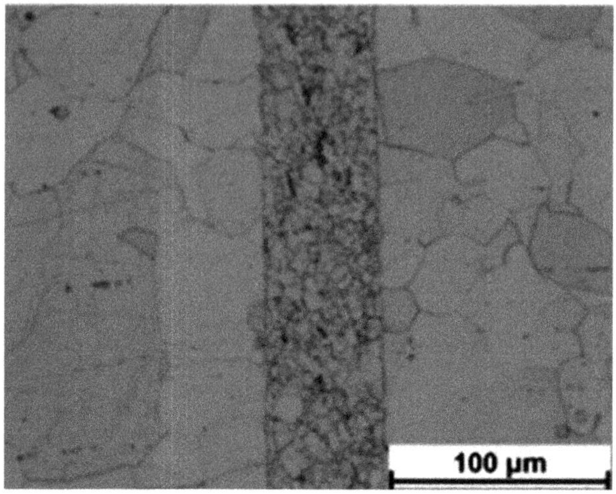

Fig. 4.74 Microstructure of the joint between the steel rods obtained by SPS using an intermediate powder layer. (Reprinted from Yang et al. [139], Copyright (2013) with permission from Elsevier)

SPS facilities can also be used for annealing of powder and compact samples without the application of pressure [143–145]. It was shown that for short times of annealing, the phase and microstructure changes of the sample depend on the geometric features of its position in the graphite die [145]. Structures with gradients in the microstructure and phase composition (in the case of reactive systems) can be obtained when the sample is heated from the bottom as a result of contacting the lower punch and not contacting any punch from above. In such cases, the gradients will disappear with the time of annealing, so the annealing or synthesis time can be a process control parameter.

4.10.3 Surface Engineering by SPS

Consolidation of a powder layer by EDS with a simultaneous joining of the layer to a monolithic piece was used as early as the 1970s by Rymorov et al. [146]. This method was suggested as efficient surface strengthening, as exemplified in Fig. 4.75 showing the engineered surfaces on toothed rack of a rotary excavator.

The same method was used to produce porous surface layers on heat-dissipating elements [4]. Sintering-based welding was shown to be a promising method to butt-weld steel sheets [147] and to join porous parts by means of a porous connecting layer [4]. During EDS, infiltration of the molten metal into the porous materials of the electrode can be induced [148]. This happens when Joule heating is sufficient to partially melt the metal powder. The pressure transferred to the powder compact by the electrode itself squeezes the melt out of the compact so that it penetrates into the pores of the electrode open to the surface. This effect offers an interesting possibility to join materials together. Such joints can be expected to be strong establishing high

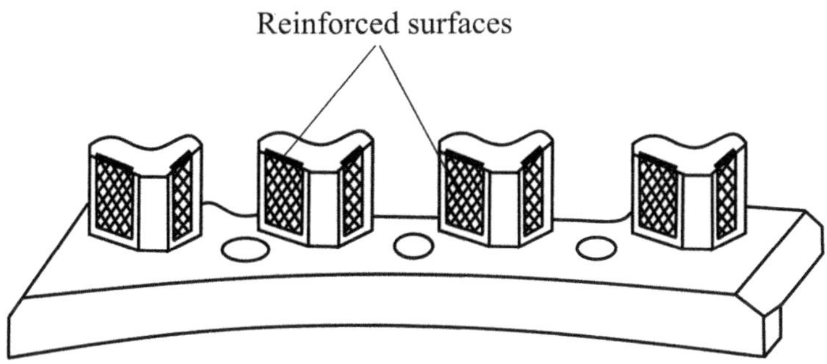

Fig. 4.75 Reinforced surfaces on toothed rack of a rotary excavator. (Reprinted from Rymorov and Dyn'kin [146], Copyright (1974) with permission of Springer)

adhesion of the compact to the electrode, which may exceed the cohesion strength of the electrode material.

More deposition flexibility is anticipated from the use of SPS facilities for the coating deposition purposes. Describing the advantages of the SPS method as a coating deposition process, Mulukutla et al. [149] outlined the limitations of the previously known deposition techniques for coatings of metastable structure, including amorphous coatings, due to undesirable crystallization, phase transformations, chemical reactions, and grain growth induced at high deposition temperatures. Three approaches to multiscale surface engineering by SPS were outlined in Ref. [149] (Fig. 4.76). In the first approach, the powder, from which the coating is formed, is loaded above the substrate disk. In the second approach, the coating layer and the substrate disk are sintered in the SPS separately and then joined together in another sintering process. The third approach uses powders as initial materials for both the substrate and the coating, which both form in a single SPS run, and bonding between them is also established. A possibility of preserving an amorphous phase in the coating manufactured by SPS has been demonstrated; however, as could have been expected, such process is very sensitive to temperature. As is seen in Fig. 4.77, a Fe-based amorphous phase is present in the coating processed at 575 °C, but at 590 °C, it crystallizes. An increased intensity of the Al peaks in the XRD patterns of the samples processed at a higher temperature indicates penetration of the substrate material into the coating, which has also been confirmed by scanning electron microscopy (SEM) studies of the substrate/coating interface.

4.10.4 Dense Materials with Improved Properties Obtained by SPS

To date, a wide variety of structural and functional materials have been processed by SPS. Several examples will be briefly discussed below. Successes of SPS include, but are not limited to, the production of bulk amorphous materials from powders

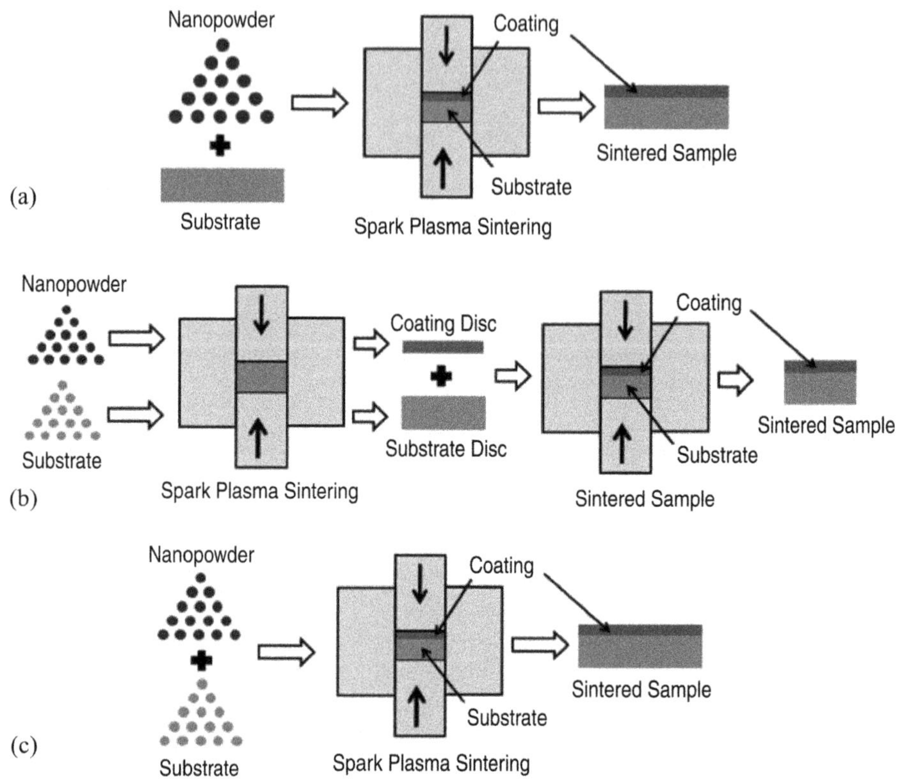

Fig. 4.76 Three possible approaches of multi-scale surface engineering by SPS: (**a**) the powder forming the coating is loaded above the substrate disk, (**b**) the coating layer and the substrate disk are sintered in the SPS separately and then joined together in another sintering process, (**c**) the substrate and the coating form simultaneously in a single SPS run. (Reprinted from Mulukutla et al. [149], Copyright (2010) with permission of Springer)

[150], metallic glass-reinforced metal matrix composites [151], fracture tough ceramics [152–155], ultra-high-temperature ceramic materials [156, 157], transparent ceramics [158–161], thermoelectric materials [162, 163], and diamond-based composites [164].

SPS was shown to be capable of sintering tantalum carbide TaC up to a relative density of 97% within 10 min of holding at 2400 °C [156]. Reaching full density was accompanied by a certain degree of grain growth: while the grain size of the 68%-dense compact sintered at 1900 °C and 30 MPa was 330 nm, that of the 97%-dense compact sintered at 2400 °C and 75 MPa was 9 μm (Fig. 4.78). Considering a very high melting point of TaC (3880 °C), the results of SPS are indicative of the suitability of this technique for the preparation of nearly fully dense ultra-high-temperature ceramics with a relatively small grain size.

The SPS-forging approach was shown to be promising not only for enhancing densification but also for tailoring the functional properties of materials. Noudem

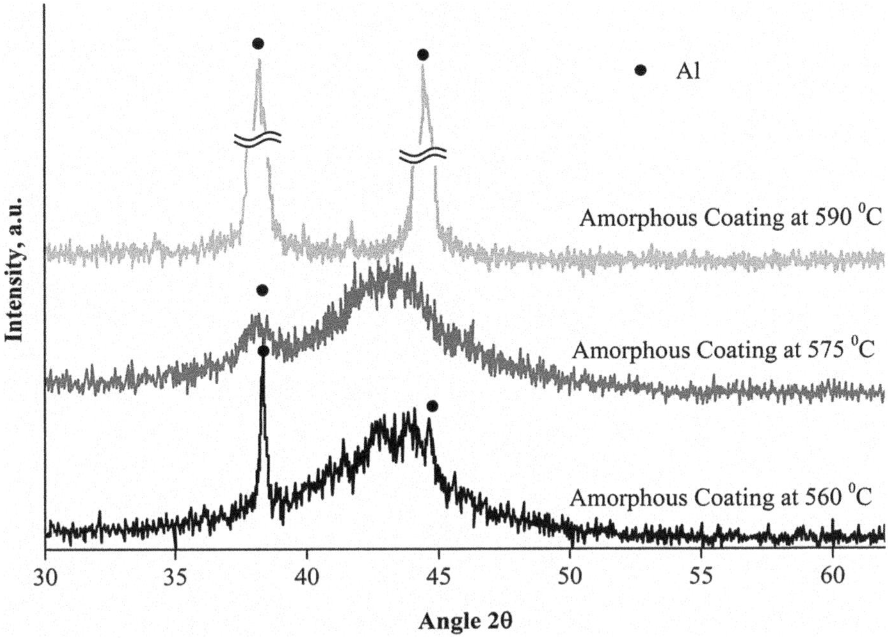

Fig. 4.77 XRD patterns of the coatings produced by SPS from an amorphous Fe-based alloy on an Al substrate. (Reprinted from Mulukutla et al. [149], Copyright (2010) with permission of Springer)

Fig. 4.78 Fracture surface of 97%-dense TaC obtained by SPS at 2400 °C. (Reprinted from Khaleghi et al. [156], Copyright (2010) with permission of Elsevier)

et al. [162] obtained an anisotropic $Ca_3Co_4O_9$ thermoelectric material in this manner and called the process edge-free spark plasma texturing (SPT), as the pre-consolidated specimen had a diameter smaller than that of the die and the processing resulted in the formation of a strong texture. Microstructures of the $Ca_3Co_4O_9$ materials consolidated by conventional sintering, SPS in a regular assembly, and SPT show significant differences (Fig. 4.79a–c). The temperature dependences of the electrical resistivity and the power factor of the $Ca_3Co_4O_9$ materials consolidated by conventional sintering, SPS in a regular assembly, and SPT are

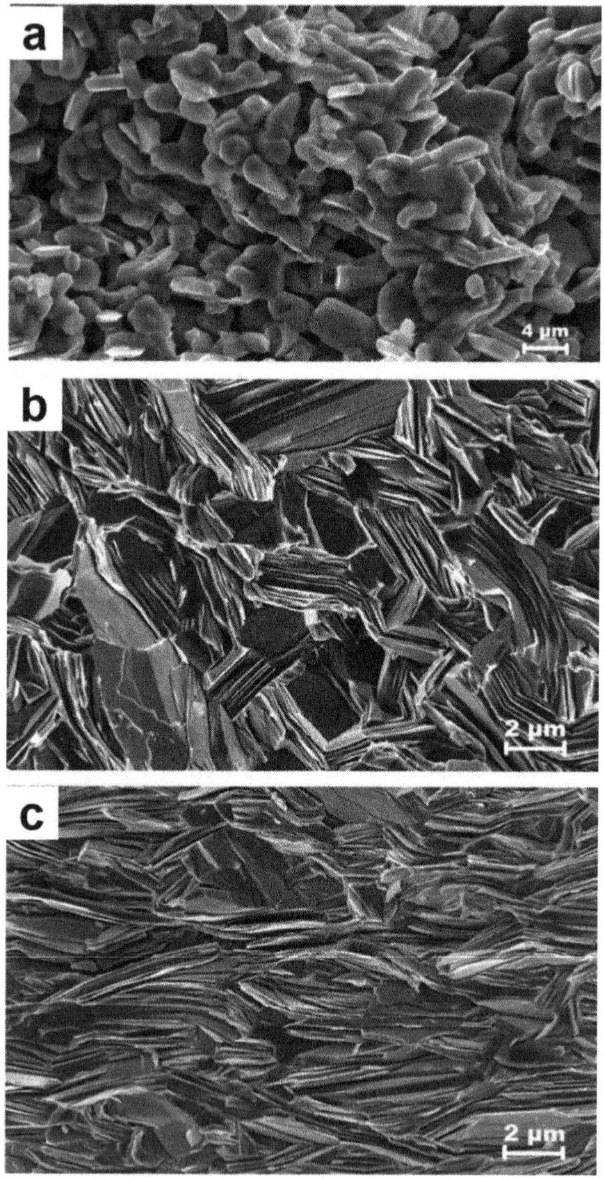

Fig. 4.79 Microstructure of the $Ca_3Co_4O_9$ materials consolidated by conventional sintering (**a**), SPS in a regular assembly (**b**) and SPT (**c**). (Reprinted from Noudem et al. [162], Copyright (2011) with permission from Elsevier)

given in Fig. 4.80. The measurements were conducted on the bars cut from the sintered specimens in the direction normal to the pressing axis. The lowest resistivity values were observed for the SPT material, which correlates with its microstructural features (Fig. 4.79c): platelike morphology of the grains, a reduction in the grain-boundary densities in the plane normal to the pressing direction, and a stronger texture. At 550 °C, the maximum power factor reached by the SPT material is 30%

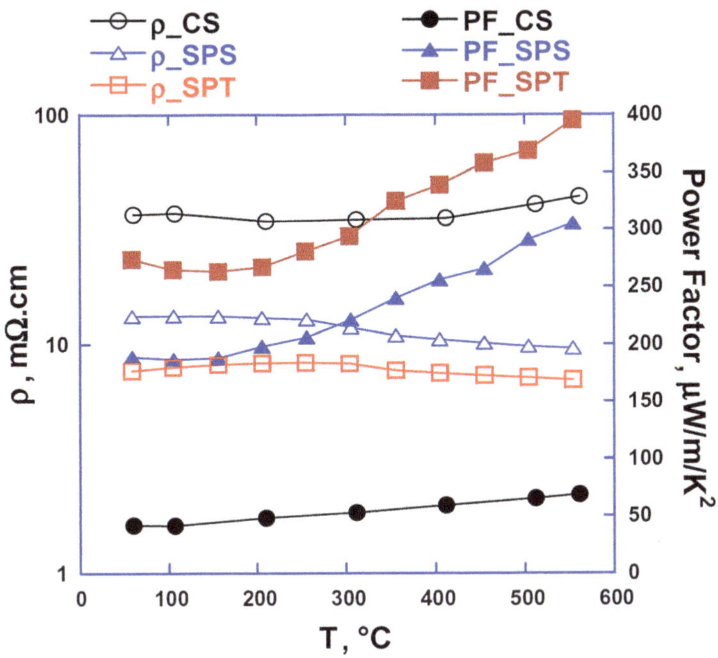

Fig. 4.80 Electrical resistivity and power factor of the $Ca_3Co_4O_9$ materials consolidated by conventional sintering, SPS in a regular assembly, and SPT vs. temperature. (Reprinted from Noudem et al. [162], Copyright (2011) with permission from Elsevier)

higher than the corresponding value for the SPS-processed material and eight times greater than that for the $Ca_3Co_4O_9$ material processed by conventional sintering (Fig. 4.80).

SPS was successfully used for the production of metallic glass-reinforced metal matrix composites – an emerging class of composites [165]. As is demonstrated in Chap. 8, consolidation of such composites requires moderate temperatures. Wang et al. [151] consolidated 15vol.%$Zr_{65}Cu_{18}Ni_7Al_{10}$–Al7075 composites by SPS. The $Zr_{65}Cu_{18}Ni_7Al_{10}$ particles and fibers of amorphous structure were distributed in a matrix of the Al7075 alloy using mechanical milling. SPS of this powder mixture at 300 °C for 10 min resulted in the formation of fully dense compacts (Fig. 4.81). In order to achieve densification at such a low temperature, a pressure as high as 600 MPa was used during the whole sintering cycle. Due to a moderate temperature and a short sintering time, the reinforcements retained their amorphous structure, as was confirmed by the XRD (Fig. 4.82) and TEM (Fig. 4.83) analyses. The metallic glass fibers and particles were distributed homogeneously in the matrix and had good bonding with the Al7075 alloy matrix. The 15vol.%$Zr_{65}Cu_{18}Ni_7Al_{10}$–Al7075 composites exhibited a significant increase in the yield strength: from 168 MPa for Al7075 alloy to 366 MPa for the composite (Fig. 4.84). At the same time, the composites showed about 25% of plastic strain before fracture. Overall, the

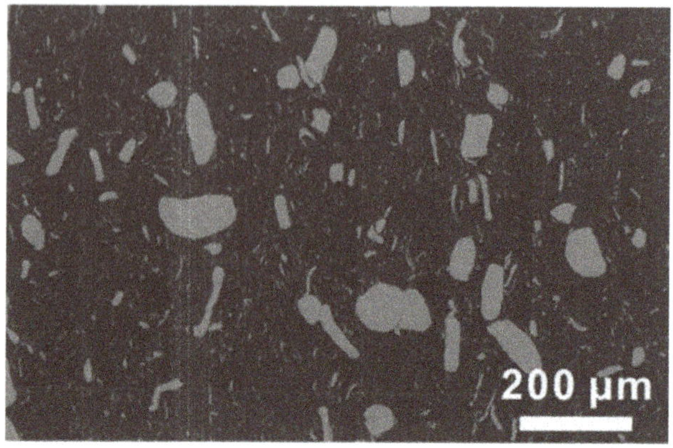

Fig. 4.81 Microstructure of the spark plasma sintered 15vol.%$Zr_{65}Cu_{18}Ni_7Al_{10}$–Al7075 composite. (Reprinted from Wang et al. [151], Copyright (2016), Rights managed by Nature Publishing Group. This work is licensed under a Creative Commons Attribution License, http://creativecommons.org/licenses/by/4.0/)

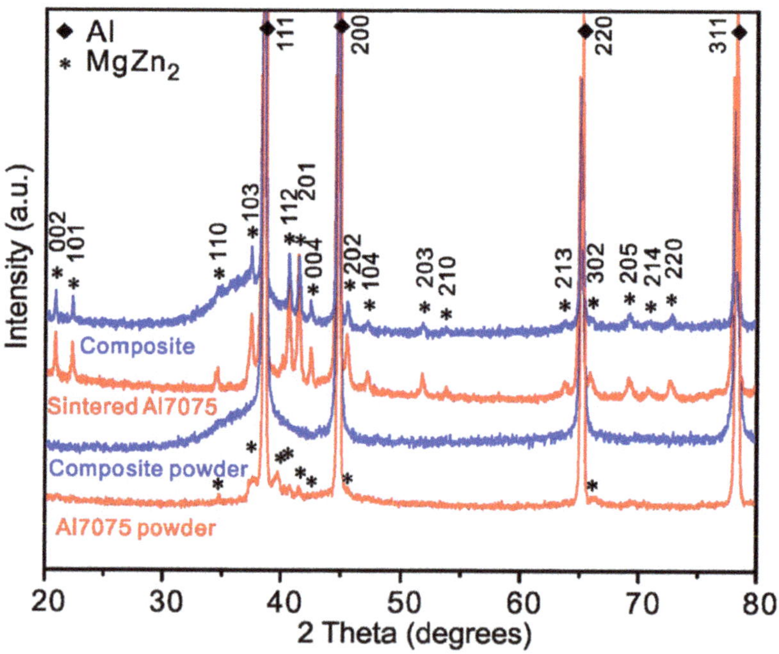

Fig. 4.82 X-ray diffraction patterns of the Al7075 alloy powder, mechanically milled powder 15vol.%$Zr_{65}Cu_{18}Ni_7Al_{10}$–Al7075, the spark plasma sintered 15vol.%$Zr_{65}Cu_{18}Ni_7Al_{10}$–Al7075 composite, and sintered Al7075 alloy without glassy reinforcements. (Reprinted from Wang et al. [151], Copyright (2016), Rights managed by Nature Publishing Group. This work is licensed under a Creative Commons Attribution License, http://creativecommons.org/licenses/by/4.0/)

introduction of metallic glass fiber reinforcements in Al-based matrices by means of ball milling and SPS significantly increase the mechanical strength of the composites through several strengthening mechanisms.

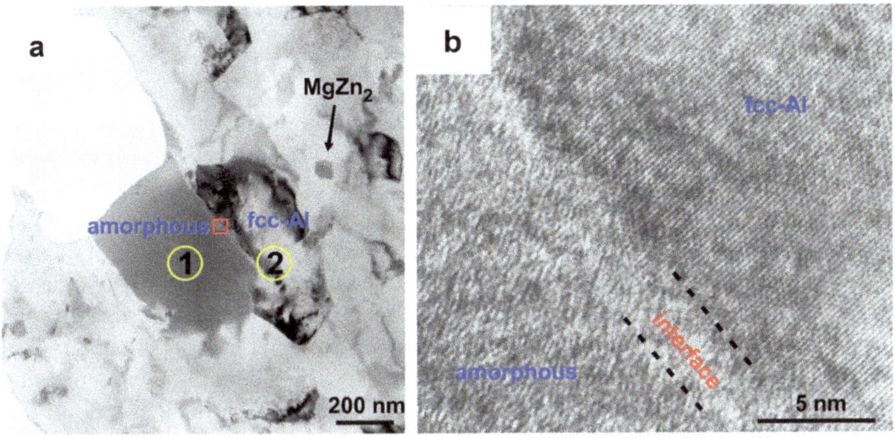

Fig. 4.83 Bright-field TEM image of the 15vol.%$Zr_{65}Cu_{18}Ni_7Al_{10}$–Al7075 sintered composite (**a**), high-resolution TEM image of the interface between the metallic glass and the fcc-Al matrix (**b**). (Reprinted from Wang et al. [151], Copyright (2016), Rights managed by Nature Publishing Group. This work is licensed under a Creative Commons Attribution License, http://creativecommons.org/licenses/by/4.0/)

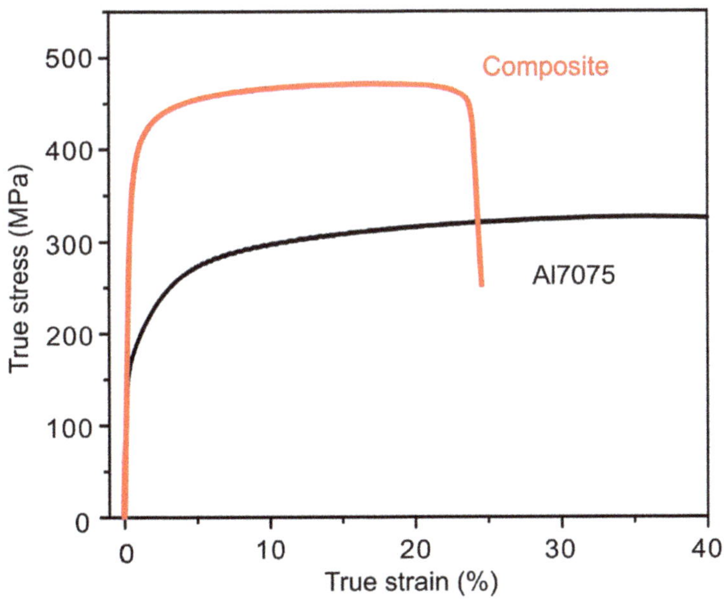

Fig. 4.84 Mechanical properties of the spark plasma sintered 15vol.%$Zr_{65}Cu_{18}Ni_7Al_{10}$–Al7075 composite and the Al7075 alloy without glassy reinforcements. (Reprinted from Wang et al. [151], Copyright (2016), Rights managed by Nature Publishing Group. This work is licensed under a Creative Commons Attribution License, http://creativecommons.org/licenses/by/4.0/)

4.10.5 Porous Materials by SPS

In powder metallurgy-based manufacturing of porous metals, the main challenge of making a consolidated porous body is establishing bonding between the powder particles while maintaining high levels of porosity. The number of studies on the preparation of porous metals by SPS is much smaller than the number of studies aimed at dense materials. As SPS is capable of producing dense multiphase materials, then a route to a porous material will be a selective removal of one of them to create pores and form a porous structure out of the other. The main porosity-creating processes in the initially dense materials are dealloying and phase separation; the former is the removal of an element from a phase, while the latter is the removal of grains or particles of a phase, leaving behind the other one, both initially comprising a composite. The removal of a template does not bring about shrinkage, and, for this reason, the particles and compacts are not prone to the crack formation rather typical of dealloying, in which shrinkage often accompanies the reorganization of the crystalline lattice. The in situ porosity forms when pressureless SPS is used, the architecture of the porous skeleton depending strongly on the sintering temperature. The ex situ porosity is generated by post-sintering selective dissolution treatment of a dense two-phase nanocomposite. In both cases, consolidation by SPS plays a key role, ensuring the connectivity of the particles in the skeleton during the in situ generation of the porosity and densification of the nanocomposite itself without losing its fine-grained structure, which is crucial for the generation of nanoporosity ex situ through selective dissolution.

Hakamada et al. [166] proposed making porous aluminum using the spacer method with particles of NaCl added to the Al powder and removed after sintering. However, their approach was utilized for making large pores (850–1000 μm), the size of which was controlled by the size of the spacer particles. Another kind of porosity formed in the walls of the porous structure between sintered only to a certain extent Al particles, but was not the target kind of porosity and was purposefully removed.

The feasibility of using a two-phase composite with an interpenetrating structure sintered by SPS as a precursor for porous bodies was demonstrated on TiB_2–Cu metal–ceramic composites, which were electrochemically etched to selectively remove copper and form a nanograined titanium diboride skeleton (Fig. 4.85) [167–170]. The processing excluded infiltration operations and thus eliminated the problem of closed pores and low wettability between the ceramic and metal phases in the composite.

Bokhonov and Dudina [171] demonstrated a possibility of forming porous metals with pores 50–200 nm in size by selective dissolution treatment of nanostructured compacts consolidated from mechanically milled metallic powders by SPS. In this way, a preparation method of nanoporous silver by phase separation in nanocomposite mixtures of Ag–Fe and Ag–Ni metals, which are immiscible in both solid and liquid states, was suggested. Consolidation and sintering of agglomerated powders produced by mechanical milling into dense compacts present serious

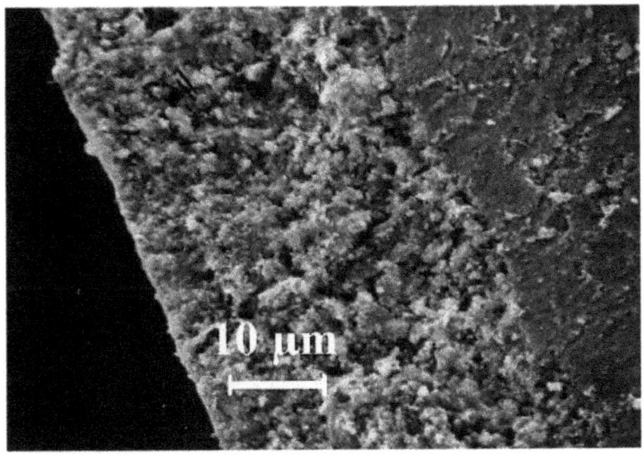

Fig. 4.85 A porous titanium diboride-rich layer formed after removal of copper from the spark plasma sintered TiB_2–Cu composite through electrochemical etching. (Reprinted from Kim et al. [169], Copyright (2006) with permission from Elsevier)

challenges. Particles forming agglomerates do not have as much flexibility of rearrangement during compaction as non-agglomerated particles. When agglomerates are nevertheless to be sintered, special measures are required to eliminate inter- and intra-agglomerate porosity. The agglomerates of mechanically milled Ag–Fe and Ag–Ni alloys were dense, so it was only the inter-agglomerate pores that needed to be removed. Few microstructure changes in the interior regions of the Ag–Fe and Ag–Ni agglomerates could be expected as the duration of the temperature exposure during the chosen SPS cycle was very short. However, a feature typical of pulsed current sintering and of SPS in particular is a local temperature rise at the inter-particle contacts during the passage of current in the beginning of sintering of large particles. In the case of sintering of powder agglomerates, this SPS feature can solve the problem of inter-agglomerate porosity and ensure more efficient bonding between the particles. As can be concluded from a characteristic microstructure of the Ag–Fe compact (Fig. 4.86a), higher local temperatures at the agglomerate contacts caused melting of silver, and a slight redistribution of the phases was indicated by the presence of silver streaks, the distances between which were comparable to the size of the powder agglomerates. The presence of these silver layers exerts only a minor influence on the bulk nanoporous structure produced from the sintered compact by the phase separation, as the volume fraction of silver experiencing melting and solidification is limited to a few percent as was estimated from the SEM images. Selective dissolution of iron from the compact resulted in the formation of nanoporous silver with pore and ligament sizes ranging from 100 to 200 nm (Fig. 4.86b). The importance of using SPS rather than conventional sintering is clearly demonstrated by the microstructural features of the Ag–Fe composites annealed in hydrogen at 600 °C (Fig. 4.87). The redistribution of phases and silver coalescence and growth into crystals several microns in size are unavoidable in conventional heat treatment processes. When single-phase Ag nanopowders are sintered by SPS, it is very challenging to keep their grain size unaltered [172]: SPS at 300 °C for 5 min gives a 98%-dense compact, but densification is accompanied by grain growth from 70 nm in the powder up to 300 nm in the bulk. After the

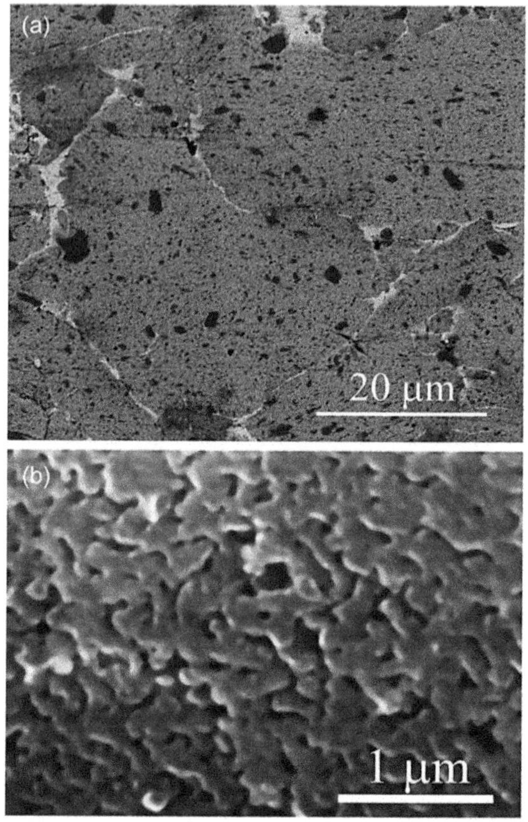

Fig. 4.86 Cross-sectional view of the spark plasma sintered Ag–Fe compact (**a**) and nanoporous silver obtained through phase separation by selective dissolution of iron in HCl solution (**b**). (Reprinted from Bokhonov and Dudina [171], Copyright (2013))

Fig. 4.87 Microstructure of the Ag–Fe composites mechanically milled for 60 min and annealed in hydrogen at 600 °C for 1 h. Microstructure evolution of the nanocomposites upon annealing leads to phase redistribution and extensive coalescence and growth of silver. (Reprinted from Bokhonov and Dudina [171], Copyright (2013)

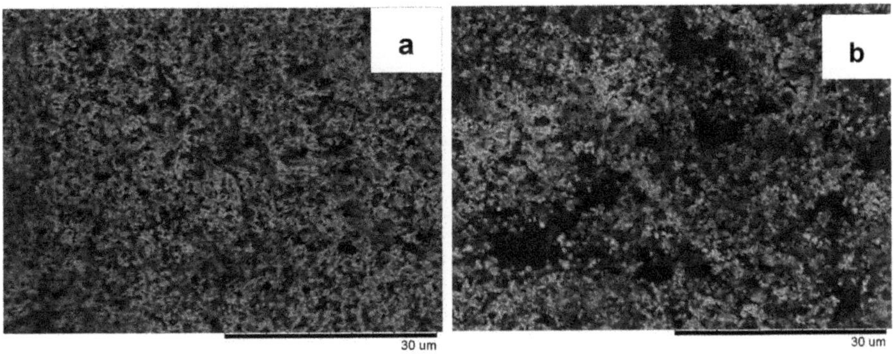

Fig. 4.88 Fracture surface of the compacts obtained by SPS of the mechanically milled (**a**) and mixed (**b**) Ni–C_{am} powders at 1000 °C. (Reprinted from Bokhonov et al. [174], Copyright (2014) with permission from Elsevier)

SPS at 600 °C, even larger grains are observed (up to 700 nm). In Ref. [171], Ag–Fe and Ag–Ni alloys were consolidated by SPS into compacts retaining silver nanograins and showing only limited phase redistribution caused by melting. Therefore, selective dissolution performed on these compacts can be used to produce porous silver bodies of selected geometry.

Similarly, porous graphitic materials can be obtained by combining SPS and selective dissolution [173–176]. SPS of powder mixtures containing amorphous carbon and metals – graphitization catalysts (nickel or iron) – was conducted to obtain the precursors of porous carbon materials. Sintered compacts were further immersed in HCl solution for dissolving the metals. In the case of nickel catalyst, no carbide phase was found in the compacts, and metallic nickel could be fully dissolved so that porous graphitic materials could be obtained. In the case of iron catalyst, the formation of the Fe_3C phase made the graphitization process more complicated [176]. In addition, Fe_3C remained in the compacts after HCl treatment. After the SPS, carbon remained poorly graphitized. Graphitization was enhanced when Fe-containing compacts were annealed after SPS.

Using nickel–carbon (Ni–C_{am}) mixtures, it was shown that it is possible to control the structure of porous graphitic materials by altering the structure of the initial powder mixture (and, hence, that of the sintered nickel–carbon composite (Fig. 4.88)) [174]. The porous graphitic material obtained from the mechanically milled nickel–carbon mixture had a more uniform structure than the porous material obtained from powders prepared by simple mixing (Fig. 4.89). The graphitization degree was higher in the compact sintered from the mechanically milled mixture as compared with that sintered from a mixture prepared in a mortar (Fig. 4.90).

Carbonyl iron, a powder of spherical morphology, is a convenient system to trace the morphological feature of the formation of porous compacts during SPS. A study of the formation porous iron during pressureless SPS showed that by varying the sintering temperature, different architectures of the iron skeleton can be obtained (Fig. 4.91) [177].

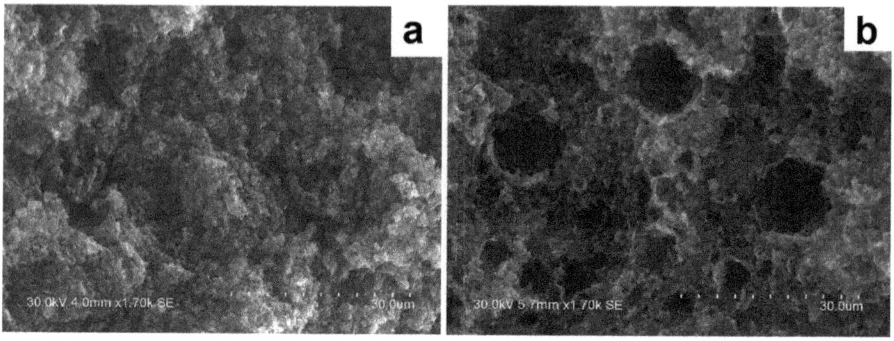

Fig. 4.89 Fracture surface of porous graphitic material obtained by selective dissolution of nickel from the compacts sintered by SPS from the mechanically milled (**a**) and mixed (**b**) Ni–C_{am} powders. (Reprinted from Bokhonov et al. [174], Copyright (2014) with permission from Elsevier)

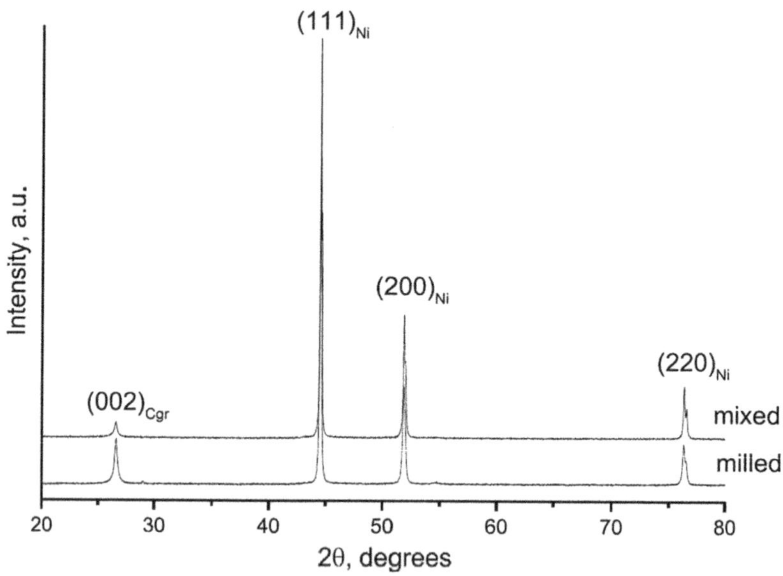

Fig. 4.90 XRD patterns of the compacts obtained by SPS at 1000 °C from mixed and mechanically milled Ni–C_{am} powders. (Reprinted from Bokhonov et al. [174], Copyright (2014) with permission from Elsevier)

Consolidation of nanodiamonds by SPS can be used for developing porous electrically conductive materials with tailored specific surface area [178]. The compact formation was accompanied by graphitization, as nanodiamond particles gradually transform into carbon onion structures upon annealing (Fig. 4.92). Robust defect-free compacts with different degrees of graphitization were obtained by SPS (Fig. 4.93). Despite consolidation that occurred in parallel to graphitization, an increase in the specific surface area (up to 480 $m^2 \cdot g^{-1}$) relative to the nanodiamond

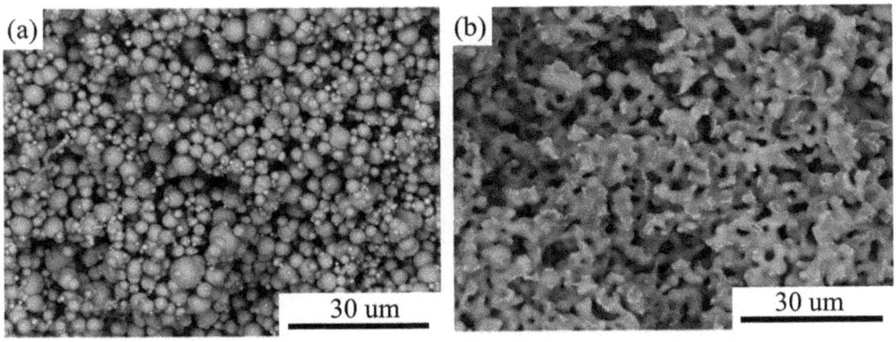

Fig. 4.91 Porous iron obtained by pressureless SPS from a carbonyl iron powder (fracture surface of the sintered compacts): (**a**) SPS temperature 600 °C, (**b**) SPS temperature 1000 °C

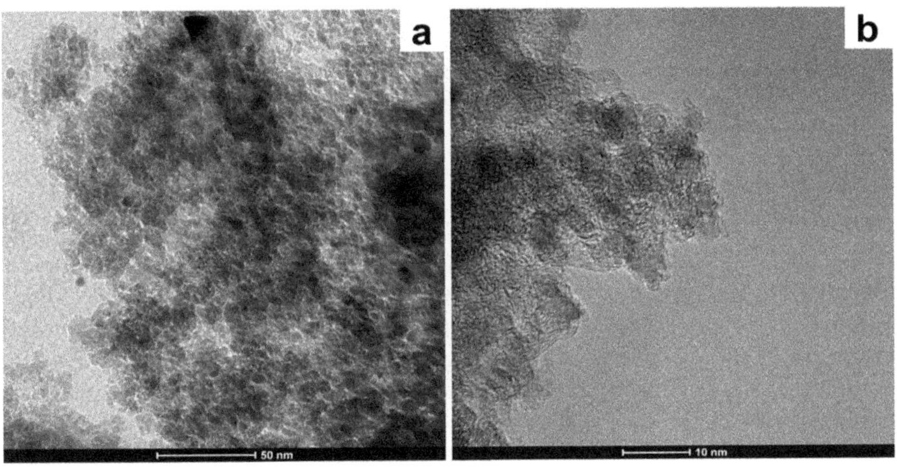

Fig. 4.92 TEM images of the nanodiamond powder (**a**) and a piece of the compact obtained by SPS at 1200 °C (**b**). (Reprinted from Ukhina et al. [178], Copyright (2015) with permission from Elsevier)

powder (360 $m^2 \cdot g^{-1}$) was observed in compacts sintered at 1200 °C. The electrical conductivity of these compacts was an order of magnitude higher than that of the samples obtained by filling the annealed powder into a glass tube [179]. As graphite shells on the surface of nanodiamond particles grew upon heating, a conductive phase was created in situ during the process. Sintering between particles of partially graphitized nanodiamonds presents an opportunity to increase the electrical conductivity of the material by establishing and improving the inter-particle contacts.

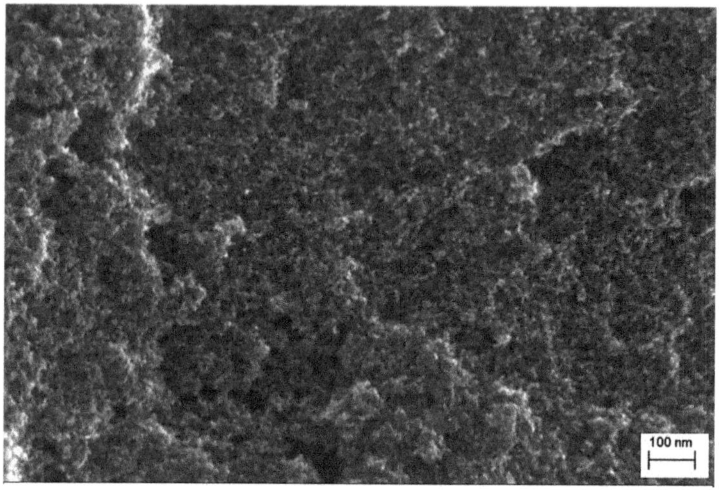

Fig. 4.93 Fracture surface of the compact obtained by SPS of the nanodiamond powder at 1200 °C. (Reprinted from Ukhina et al. [178], Copyright (2015) with permission from Elsevier)

4.11 Summary

Low-voltage electric pulse sintering is usually associated with the most frequently used mode of this type of processing known as spark plasma sintering (SPS). SPS is an emerging powder consolidation technique, which provides significant advantages to the processing of materials into configurations unattainable under conventional powder pressing or sintering. SPS consists essentially of conjoint application of fast heating rates, high axial pressure, and field (electric current-based)-assisted sintering. SPS gains particular prominence in connection with its exceptional potential of rapid and ultra-rapid processing of very hard-to-deform materials, which would typically require lengthy consolidation times at significantly elevated temperatures under conditions of conventional powder pressing or sintering.

In terms of the fundamental analysis, a number of important questions are still open and are being actively researched. Those include the contributions of thermal and nonthermal phenomena to mass transport under SPS processing conditions. Despite some progress achieved in recent years in this area, still the role of SPS-specific factors in the processing outcomes distinguishing SPS from conventional techniques of powder materials processing, such as hot pressing and traditional sintering, is not completely clear. For SPS, the influence of multiple phenomena of transient nature is significant. Therefore, SPS, similarly to other field-assisted sintering techniques, heavily relies on the *efficient control of non-equilibrium* processes.

In terms of practical applications, net-shape capability and scalability of SPS are the areas for intensive research and improvement. The important current studies include SPS-based joining of materials, SPS surface engineering, and fabrication of porous materials using SPS.

References

1. Inoue K (1962) Electric Discharge sintering. US Patent 3241965
2. Inoue K (1966) Apparatus for electrically sintering discrete bodies. US Patent 3250892
3. Raichenko AI, Burenkov GL, Khrienko AF, Litvinenko VP (1976) Electric discharge sintering of binary powder mixtures. Sov Powder Metall Met Ceram 15:602–606
4. Raichenko AI (1987) Basics of electric current-assisted sintering. Moscow, Metallurgiya, 128 p (in Russian)
5. Slosman AI, Matrenin SV (1994) Electric discharge sintering of ceramics based on zirconium dioxide. Refractories 35:296–297 (in Russian)
6. Matrenin SV, Slosman AI, Myachin YV (2005) Electric discharge sintering of a Fe-Ti antifriction alloy. News Tomsk State Polytech Univ 308:74–77 (in Russian)
7. Zamula MV, Derevyanko AV, Kolesnichenko VG, Samelyuk AV, Zgalat-Lozinskii OB, Ragulya AV (2007) Electric discharge sintering of TiN-AlN nanocomposites. Powder Metall Met Ceram 46:325–331
8. Petukhov AS, Khobta IV, Ragulya AV, Derevyanko AV, Raichenko AI, Isaeva LP, Koval'chenko AM (2007) Reactive electric discharge sintering of $TiN\text{-}TiB_2$. Powder Metall Met Ceram 46:525–532
9. Popov VP, Raichenko AI (2000) Kinetics of electric discharge sintering of high-speed steel powders. Powder Metall Met Ceram 39:545–548
10. Petukhov AS (2009) The influence of various technological conditions on the processes of reactive electric discharge sintering composition $TiN\text{-}TiB_2$. Powder Metall Met Ceram 11–12:13–22 (in Russian)
11. Kolesnichenko VG, Popov VP, Zgalat-Lozinskii OB, Klochkov LA, Lobunets TF, Raichenko AI (2011) Field assisted sintering of nanocrystalline titanium nitride powder. Powder Metall Met Ceram 50:3–4
12. Olevsky EA, Aleksandrova EV, Ilyina AM, Dudina DV, Novoselov AN, Pelve KY, Grigoryev EG (2013) Outside mainstream electronic databases: review of studies conducted in the USSR and post-soviet countries on electric current-assisted consolidation of powder materials. Materials 6:4375–4440
13. Crivelli IV, Esposito E, Mele G, Siniscalchi A (1973) Formatura per Spark Sintering. Metallurgia Italiana 65(11):611–618
14. Olevsky E, Bogachev I, Maximenko A (2013) Spark-plasma sintering efficiency control by inter-particle contact area growth: a viewpoint. Scr Mater 69(2):112–116
15. Sukhov OV, Baidenko AA, Istomina TI, Raichenko AI, Popov VP, Svechkov AV, Goldberg MS (1987) Densification kinetics of a copper-tin powder composite during electric discharge sintering. Sov Powder Metall Met Ceram 26:530–532
16. Raichenko AI, Kol'chinskii MZ, Levina DA (1976) Electric discharge sintering of oxidized metal powders. Sov Powder Metall Met Ceram 15:754–759
17. Raichenko AI, Istomina TI, Troyan IA (2000) Powder sintering with application of an electric current and periodic mechanical pulses. Powder Metall Met Ceram 39(3–4):198–201
18. Tokita M (1993) Trends in advanced SPS spark plasma sintering systems and technology. J Soc Powder Technol Japan 30(11):790–804

19. Tokita M (2013) Spark plasma sintering (SPS) method, systems and applications. In: Editor-in-Chief Somiya S (ed) Handbook of advanced ceramics: materials, applications, processing and properties. 2nd edn. Academic Press, USA, pp 1149–1178 (in Russian)
20. Omori M (2000) Basic research and industrial production using the spark plasma system (SPS). Mater Sci Eng A 287(2):183–188
21. Groza JR, Zavaliangos A (2000) Sintering activation by external electrical field. Mater Sci Eng A 287(2):171–177
22. Groza JR, Zavaliangos A (2003) Nanostructured bulk solids by field activated sintering. Rev Adv Mater Sci 5:24–33
23. Munir ZA, Anselmi-Tamburini U, Ohyanagi M (2006) The effect of electric field and pressure on the synthesis and consolidation of materials: a review of the spark plasma sintering method. J Mater Sci 41(3):763–777
24. Munir ZA, Quach D, Ohyanagi M (2011) Electric current activation of sintering: a review of the pulsed electric current sintering process. J Am Ceram Soc 94(1):1–19
25. Garay JE (2010) Current-activated, pressure-assisted densification of materials. Annu Rev Mater Res 40:445–468
26. Orrù R, Licheri R, Locci AM, Cincotti A, Cao G (2009) Consolidation/synthesis of materials by electric current activated/assisted sintering. Mater Sci Eng R 63(4–6):127–287
27. Dudina DV, Mukherjee AK (2013) Reactive spark plasma sintering: successes and challenges of nanomaterial synthesis. J Nanomater 625218, 12
28. Dudina DV, Mukherjee AK (2013) Reactive spark plasma sintering for the production of nanostructured materials. In: Sinha S, Navani NK (eds) Nanotechnology series, Nanomaterials and nanostructures, vol 4. Studium Press LLC, pp 237–264
29. Locci AM, Orrù R, Cao G, Munir ZA (2003) Field-activated pressure-assisted synthesis of NiTi. Intermetallics 11:555–571
30. Gauthier V, Bernard F, Gaffet E, Munir ZA, Larpin JP (2001) Synthesis of nanocrystalline $NbAl_3$ by mechanical and field activation. Intermetallics 9:571–580
31. Meng QS, Fan WH, Chen RX, Munir ZA (2011) Thermoelectric properties of Sc- and Y-doped Mg_2Si prepared by field-activated and pressure-assisted reactive sintering. J Alloys Compd 509:7922–7926
32. Anselmi-Tamburini U, Gennari S, Garay JE, Munir ZA (2005) Fundamental investigations on the spark plasma sintering/synthesis process II. Modeling of current and temperature distributions. Mater Sci Eng A 394(1–2):139–148
33. Vanmeensel K, Laptev A, Hennicke J, Vleugels J, Van der Biest O (2005) Modelling of the temperature distribution during field assisted sintering. Acta Mater 53:4379–4388
34. Langer J, Quach DV, Groza JR, Guillon O (2011) A comparison between FAST and SPS apparatuses based on the sintering of oxide ceramics. Int J Appl Ceram Technol 8 (6):1459–1467
35. Langer J, Hoffmann MJ, Guillon O (2009) Direct comparison between hot pressing and electric field-assisted sintering of submicron alumina. Acta Mater 57(18):5454–5465
36. Langer J, Hoffmann MJ, Guillon O (2011) Electric field assisted sintering in comparison to hot pressing of yttria stabilized zirconia. J Am Ceram Soc 94(1):131–138
37. Langer J, Hoffmann MJ, Guillon O (2011) Electric field-assisted sintering and hot pressing of semiconductive zinc oxide: a comparative study. J Am Ceram Soc 94(8):2344–2353
38. Lee G, Olevsky EA, Manière C, Maximenko A, Izhvanov O, Back C, McKittrick J (2018) Effect of electric current on densification behavior of conductive ceramic powders consolidated by spark plasma sintering. Acta Mater 144:524–533
39. Wei X, Giuntini D, Maximenko AL, Haines CD, Olevsky EA (2015) Experimental investigation of electric contact resistance in spark plasma sintering tooling setup. J Am Ceram Soc 98(11):3553–3560
40. Manière C, Durand L, Brisson E, Desplats H, Carré P, Rogeon P, Estournès C (2016) Contact resistances in spark plasma sintering: from in-situ and ex-situ determinations to an extended model for the scale up of the process. J Eur Ceram Soc 37(4):1593–1605

41. Zhang F, Reich M, Kessler O, Burkel E (2013) The potential of rapid cooling spark plasma sintering for metallic materials. Mater Today 16(5):192–197
42. Manière C, Lee G, Olevsky EA (2017) Proportional integral derivative, modeling and ways of stabilization for the spark plasma sintering process. Results Physics 7:1494–1497
43. Hulbert DM, Anders A, Dudina DV, Andersson J, Jiang D, Unuvar C, Anselmi-Tamburini U, Lavernia EJ, Mukherjee AK (2008) The absence of plasma in "spark plasma sintering". J Appl Phys 104:033305 7 p
44. Saunders T, Grasso S, Reece MJ (2015) Plasma formation during electric discharge (50 V) through conductive powder compacts. J Eur Ceram Soc 35:871–877
45. Song X, Liu X, Zhang J (2006) Neck formation and self-adjusting mechanism of neck growth of conducting powders in spark plasma sintering. J Am Ceram Soc 89(2):494–500
46. Aman Y, Garnier V, Djurado E (2012) Pressure-less spark plasma sintering effect on non-conventional necking process during the initial stage of sintering of copper and alumina. J Mater Sci 47:5766–5773
47. Yanagisawa O, Kuramoto H, Matsugi K, Komatsu M (2003) Observation of particle behavior in copper powder compact during pulsed electric discharge. Mater Sci Eng A 350:184–189
48. Vasiliev P, Akhtar F, Grins J, Mouzon J, Andersson C, Hedlund J, Bergström L (2010) Strong hierarchically porous monoliths by pulsed current processing of zeolite powder assemblies. Appl Mater Interfaces 2(3):732–737
49. Dudina DV, Mali VI, Anisimov AG, Bulina NV, Korchagin MA, Lomovsky OI, Bataev IA, Bataev VA (2013) Ti_3SiC_2-Cu composites by mechanical milling and spark plasma sintering: possible microstructure formation scenarios. Met Mater Int 19(6):1235–1241
50. Zhao S, Song X, Zhang J, Liu X (2008) Effects of scale combination and contact condition of raw powders on SPS sintered near-nanocrystalline WC-Co. Mater Sci Eng A 473:323–329
51. Holland TB, Thron AM, Bonifacio CS, Mukherjee AK, van Benthem K (2010) Field assisted sintering of nickel nanoparticles during in situ transmission electron microscopy. Appl Phys Lett 96:243106 3 p
52. Matsuno M, Bonifacio CS, Rufner JF, Thron AM, Holland TB, Mukherjee AK, van Benthem K (2012) In situ transmission electron microscopic investigations of reduction-oxidation reactions during densification of nickel nanoparticles. J Mater Res 18:2431–2440
53. Bonifacio CS, Rufner JF, Holland TB, van Benthem K (2012) In situ transmission electron microscopy study of dielectric breakdown of surface oxides during electric field-assisted sintering of nickel nanoparticles. Appl Phys Lett 101:093107 5 p
54. Bonifacio CS, Holland TB, van Benthem K (2013) Evidence of surface cleaning during electric field assisted sintering. Scr Mater 69:769–772
55. Zhou Y, Hirao K, Yamauchi Y, Kanzaki S (2003) Effects of heating rate and particle size on pulse electric current sintering of alumina. Scr Mater 48:1631–1636
56. Shen Z, Johnsson M, Zhao Z, Nygren M (2002) Spark plasma sintering of alumina. J Am Ceram Soc 85(8):1921–1927
57. Chu MY, Rahaman MN, Dejonghe LC, Brook RJ (1991) Effect of heating rate on sintering and coarsening. J Am Ceram Soc 74:1217–1225
58. Stanciu LA, Kodash VY, Groza JR (2001) Effects of heating rate on densification and grain growth during field-assisted sintering of α-Al_2O_3 and $MoSi_2$ powders. Metall Mater Trans A 32:2633–2638
59. Ivensen VA (1973) Densification of metal powders during sintering. Consultants Bureau, New York
60. Dabhade VV, Mohan TRR, Ramakrishnan P (2007) Dilatometric sintering study of titanium–titanium nitride nano/nanocomposite powders. Powder Metall 50:33–39
61. Bhattacharjee PP, Sinha SK, Upadhyaya A (2007) Effect of sintering temperature on grain boundary character distribution in pure nickel. Scr Mater 56:13–16
62. Panda PC, Mobley WM, Raj R (1989) Effect of the heating rate on the relative rates of sintering and crystallization in glass. J Am Ceram Soc 72:2361–2364

63. Panda PC, Raj R (1989) Sintering and crystallization of glass at constant heating rates. J Am Ceram Soc 72:1564–1566
64. Gillia O, Bouvard D (2000) Phenomenological analysis of densification kinetics during sintering: application to WC–co mixture. Mater Sci Eng A 279:185–191
65. Johnson DL (1990) Temperature-gradient-driven diffusion in rapid-rate sintering—comment. J Am Ceram Soc 73(8):2576–2578
66. Roura P, Costa J, Farjas J (2002) Is sintering enhanced under non-isothermal conditions? Mater Sci Eng A 337(1–2):248–253
67. Olevsky EA, Kushnarev B, Maximenko A, Tikare V, Braginsky M (2005) Modelling of anisotropic sintering in crystalline ceramics. Philos Mag 85(19):2123–2146
68. Bross P, Exner HE (1979) Computer simulation of sintering processes. Acta Metall 27:1013–1020
69. Exner HE (1987) Neck shape and limiting GBD/SD ratios in solid state sintering. Acta Metall 35:587–591
70. Takahashi Y, Ueno F, Nishiguchi K (1988) A numerical analysis of the void-shrinkage process controlled by surface-diffusion. Acta Metall 36:3007–3018
71. Bouvard D, McMeeking RM (1996) Deformation of interparticle necks by diffusion-controlled creep. J Am Ceram Soc 79:666–672
72. Zhang W, Schneibel JH (1995) The sintering of two particles by surface and grain boundary diffusion—a two-dimensional numerical study. Acta Metall Mater 43:4377–4386
73. Swinkels FB, Ashby MF (1980) Role of surface redistribution in sintering by grain boundary transport. Powder Metall 23(1):1–6
74. Swinkels FB, Ashby MF (1981) A second report on sintering diagrams. Acta Metall 29:259–281
75. Svoboda J, Riedel H (1995) Quasi-equilibrium sintering for coupled grain-boundary and surface diffusion. Acta Metall Mater 43(2):499–506
76. Schwarz S, Thron AM, Rufner J, van Benthem K, Guillon O (2012) Low temperature sintering of nanocrystalline zinc oxide: effect of heating rate achieved by field assisted sintering/spark plasma sintering. J Am Ceram Soc 95(8):2451–2457
77. Olevsky EA, Froyen L (2006) Constitutive modeling of spark-plasma sintering of conductive materials. Scr Mater 55(12):1175–1178
78. Olevsky EA, Kandukuri S, Froyen L (2007) Consolidation enhancement in spark-plasma sintering: impact of high heating rates. J Appl Phys 102:114913 *12 p*
79. Olevsky EA, Froyen L (2009) Impact of thermal diffusion on densification during SPS. J Am Ceram Soc 92S:122–132
80. Skorokhod VV, Olevskii EA, Shtern MB (1993) Continuum theory of sintering. I. Phenomenological model. Analysis of the effect of external forces on the kinetics of sintering. Powder Metall Met Ceram 32(1):21–26
81. Olevsky E, Dudek HJ, Kaysser WA (1996) Hiping conditions for processing of metal matrix composites using the continuum theory for sintering—I. Theoretical analysis. Acta Mater 44:707–713
82. Olevsky E, Dudek HJ, Kaysser WA (1996) Hiping conditions for processing of metal matrix composites using continuum theory for sintering—II. Application to fibre reinforced titanium alloys. Acta Mater 44:715–724
83. Olevsky E, Skorohod V, Petzow G (1997) Densification by sintering incorporating phase transformations. Scr Mater 37(5):635–643
84. Olevsky EA (1998) Theory of sintering: from discrete to continuum. Mater Sci Eng R 23 (2):41–100
85. Olevsky EA, Ma J, LaSalvia JC, Meyers MA (2007) Densification of porous bodies in a granular pressure-transmitting medium. Acta Mater 55:1351–1366
86. Raj R, Ashby MF (1971) On grain boundary sliding and diffusional creep. Metall Trans 2:1113–1127

87. Maximenko AL, Olevsky EA (2004) Effective diffusion coefficients in solid-state sintering. Acta Mater 52:2953–2963
88. Skorokhod VV (1972) Reologicheskie Osnovy Teorii Spekaniya (Rheological Basis of Theory of Sintering) Naukova Dumka, Kiev (in Russian)
89. Olevsky EA, Tikare V, Garino T (2006) Multi-scale study of sintering: a review. J Am Ceram Soc 89:1914–1922
90. Tikare V, Braginsky M, Olevsky E, Johnson DL (2005) Numerical simulation of anisotropic shrinkage in a 2D compact of elongated particles. J Am Ceram Soc 88(1):59–65
91. Braginsky M, Tikare V, Olevsky E (2005) Numerical simulation of solid state sintering. Int J Solids Struct 42:621–636
92. Tikare V, Braginsky M, Olevsky EA (2003) Numerical simulation of solid-state sintering: I. Sintering of three particles. J Am Ceram Soc 86:49–53
93. Frost HJ, Ashby MF (1982) Deformation-mechanism maps. Pergamon Press, New York
94. Hsueh CH, Evans AG, Cannon RM, Brook RJ (1986) Viscoelastic stresses and sintering damage in heterogeneous powder compacts. Acta Metall 34(5):927–936
95. Du ZZ, Cocks ACF (1992) Constitutive models for the sintering of ceramic components—I. Material models. Acta Metall Mater 40(8):1969–1979
96. Beck PA, Kremer JC, Demer L (1947) Grain growth in high purity aluminum. Phys Rev 71:555
97. Olevsky EA, Strutt ER, Meyers MA (2002) Combustion synthesis and quasi-isostatic densification of powder cermets. J Mater Process Technol 121(1):157–166
98. Chipman J (1926) The Soret effect. J Am Chem Soc 48:2577–2589
99. Schottky G (1965) A theory of thermal diffusion based on lattice dynamics of a linear chain. Phys Status Solidi 8(1):357–368
100. Young RM, McPherson R (1989) Temperature-gradient-driven diffusion in rapid-rate sintering. J Am Ceram Soc 72(6):1080–1084
101. Matlock JH, Stark JP (1971) Thermal diffusion of vacancies in aluminum. Acta Metall 19 (9):923–931
102. Wirtz K (1943) The kinetic theory of thermodiffusion in crystal lattices. Phys Z 44:221–231
103. Kornyushin YV (1980) Influence of external magnetic and electric fields on sintering, structure and properties. J Mater Sci 15(3):799–801
104. Kornyushin YV (2004) Phenomenological theory of sintering and its application to swelling. Sci Sinter 36:143–154
105. Rowe DM (2006) Thermoelectrics handbook: macro to Nano. CRC/Taylor & Francis, Boca Raton
106. Johnson DL (1991) Microwave heating of grain boundaries in ceramics. J Am Ceram Soc 74 (4):849–850
107. Gostomelskiy VS, Krupnova LV (1985) Growth and healing of pores in metals under the action of current pulses. Phys Chem Mater Treat (Fizika I Khimiya obrabotki Materialov) 4:82–87 (in Russian)
108. Chen W, Anselmi-Tamburini U, Garay JE, Groza JR, Munir ZA (2005) Fundamental investigations on the spark-plasma sintering/synthesis process I. Effect of dc pulsing on reactivity. Mater Sci Eng A 394(1–2):132–138
109. Xie G, Ohashi O, Chiba K, Yamaguchi N, Song M, Furuya K, Noda T (2003) Frequency effect on pulse electric current sintering process of pure aluminum powder. Mater Sci Eng A 359 (1–2):384–390
110. Scherge M, Bauer CL, Mullins WW (1995) Stress distribution and mass transport along grain boundaries during steady-state electromigration. Acta Metall Mater 43(9):3525–3538
111. Blech IA, Herring C (1976) Stress generation by electromigration. Appl Phys Lett 29 (3):131–133
112. Xie G, Ohashi O, Yamaguchi N (2002) Sintering behavior of aluminum powder by Spark Plasma Sintering. Trans Mater Res Soc Jpn 27:743–746

113. Holland TB, Tran TB, Quach DV, Anselmi-Tamburini U, Groza JR, Mukherjee AK (2012) Athermal and thermal mechanisms of sintering at high heating rates in the presence and absence of an externally applied field. J Eur Ceram Soc 32(14):3675–3683
114. Cologna M, Rashkova B, Raj R (2010) Flash sintering of nanograin zirconia in < 5 s at 850 degrees C. J Am Ceram Soc 93:3556–3559
115. Conrad H (2002) Thermally activated plastic flow of metals and ceramics with an electric field or current. Mater Sci Eng A 322(1–2):100–107
116. Ruszkiewicz B, Grimm T, Ragai I, Mears L, Roth JT (2017) A review of electrically-assisted manufacturing with emphasis on modeling and understanding of the electroplastic effect. J Manuf Sci Eng 139(11):110801
117. Salandro WA, Jones JJ, Bunget C, Mears L, Roth JT (2015) The effect of electric current on metals, electrically assisted forming: modeling and control. Springer International Publishing, Cham, pp 37–54
118. Roth JT, Loker I, Mauck D, Warner M, Golovashchenko SF, Krause A (2008) Enhanced formability of 5754 aluminum sheet metal using electric pulsing. Trans North Am Manuf Res Inst SME 36:405–412
119. Kang W, Beniam I, Qidwai SM (2016) In situ electron microscopy studies of electromechanical behavior in metals at the nanoscale using a novel microdevice-based system. Rev Sci Instrum 87(9):095001
120. Kim SJ, Kim SD, Yoo D, Lee J, Rhyim Y, Kim D (2016) Evaluation of the athermal effect of electric pulsing on the recovery behavior of magnesium alloy. Met Mater Trans A 47 (12):6368–6373
121. Mukherjee AK (2002) An examination of the constitutive equation for elevated temperature plasticity. Mater Sci Eng A 322:1–22
122. Zlokazov VO, Potemkin VV, Stepanov AV, Chikin DG (1992) Influence of structural defects on 1/f in thin niobium films. Radiophys Quantum Electron 35(11):610–613
123. Narayan J (2013) A new mechanism for field-assisted processing and flash sintering of materials. Scr Mater 69(2):107–111
124. Kumar M, Umezawa N, Ishii S, Nagao T (2016) Examining the performance of refractory conductive ceramics as plasmonic materials: a theoretical approach. ACS Photonics 3 (1):43–50
125. Garay JE, Glade SC, Anselmi-Tamburini U, Asoka-Kumar P, Munir ZA (2004) Electric current enhanced defect mobility in Ni_3Ti intermetallics. Appl Phys Lett 85(4):573–575
126. Kino T, Endo T, Kawata S (1974) Deviations from Matthiessen's rule of the electrical resistivity of dislocations in aluminum. J Phys Soc Jpn 36(3):698–705
127. Helle AS, Easterling KE, Ashby MF (1985) Hot-isostatic pressing diagrams: new developments. Acta Metall 33(12):2163–2174
128. Saheb N, Iqbal Z, Khalil A, Hakeem AS, Al Aqeeli N, Laoui T, Al-Qutub A, Kirchner R (2012) Spark plasma sintering of metals and metal matrix nanocomposites: a review. J Nanomater 2012(2012):983470 13 p
129. Hulbert DM, Jiang D, Kuntz JD, Kodera Y, Mukherjee AK (2007) A low-temperature high-strain-rate formable nanocrystalline superplastic ceramic. Scr Mater 56(12):1103–1106
130. Jiang D, Hulbert DM, Kuntz JD, Anselmi-Tamburini U, Mukherjee AK (2007) Spark plasma sintering: a high strain rate low-temperature forming tool for ceramics. Mater Sci Eng A 463:89–93
131. Zhan GD, Garay JE, Mukherjee AK (2005) Ultralow-temperature superplasticity in nanoceramic composites. Nano Lett 5(12):2593–2597
132. Aleksandrova EV, Ilyina AM, Grigoryev EG, Olevsky EA (2015) Contribution of electric current into densification kinetics during spark plasma sintering of conductive powder. J Am Ceram Soc 98(11):3509–3517
133. Wei X, Maximenko AL, Back C, Izhvanov O, Olevsky EA (2017) Effects of loading modes on densification efficiency of spark plasma sintering: sample study of zirconium carbide consolidation. Philos Mag Lett 97:265–272

134. Ratzker B, Sokol M, Kalabukhov S, Frage N (2016) Creep of polycrystalline magnesium aluminate spinel studied by an SPS apparatus. Materials 9(6):493
135. Ratzker B, Sokol M, Kalabukhov S, Frage N (2017) Using a spark plasma sintering apparatus as a tool in a compressive creep study of fine-grained alumina. Ceram Int 43:9369–9376
136. Manière C, Durand L, Weibel A, Chevallier G, Estournès C (2016) A sacrificial material approach for spark plasma sintering of complex shapes. Scr Mater 124:126–128
137. Manière C, Nigito E, Durand L, Weibel A, Beynet Y, Estournès C (2017) Spark plasma sintering and complex shapes: the deformed interfaces approach. Powder Technol 320:340–345
138. Martin G, Fabrègue D, Mercier F, Chafino-Aixa JA, Dendievel R, Blandin JJ (2016) Coupling electron beam melting and spark plasma sintering: a new processing route for achieving titanium architectured microstructures. Scr Mater 122:5–9
139. Yang J, Trapp J, Guo Q, Kieback B (2013) Joining of 316L stainless steel by using spark plasma sintering method. Mater Des 52:179–189
140. Rizzo S, Grasso S, Salvo M, Casalegno V, Reece MJ, Ferraris M (2014) Joining of C/SiC composites by spark plasma sintering technique. J Eur Ceram Soc 3:903–913
141. Grasso S, Tatarko P, Rizzo S, Porwal H, Hu C, Katoh Y, Salvo M, Reece MJ, Ferraris M (2014) Joining of beta-SiC by spark plasma sintering. J Eur Ceram Soc 34:1681–1686
142. Dong H, Yu Y, Jin X, Tian X, He W, Ma W (2016) Microstructure and mechanical properties of SiC-SiC joints joined by spark plasma sintering. Ceram Int 42:14463–14468
143. Cui G, Wei X, Olevsky EA, German RM, Chen J (2016) The manufacturing of high porosity iron with an ultra-fine microstructure via free pressureless spark plasma sintering. Materials 9:495
144. Giuntini D, Wei X, Maximenko AL, Wei L, Ilyina AM, Olevsky EA (2013) Initial stage of free pressureless spark-plasma sintering of vanadium carbide: determination of surface diffusion parameters. Int J Refract Met Hard Mater 41:501–506
145. Dudina DV (2017) Application of a spark plasma sintering facility for the heat treatment of compact and powder materials. Inorg Mater 53(6):658–663
146. Rymorov EV, Dyn'kin LK (1973) Wear resistance of parts clad with sintered layers welded on by the electric contact technique. Sov Powder Metall Met Ceram 12(7):569–572
147. Mel'nikov LA (1971) Investigation of the roller butt welding of steel sheet 1–3 mm thick using iron powders. Synopsis of thesis, Bryansk, 1971 (in Russian)
148. Ryabinina ON, Raichenko AI, Pushkarev VV (1982) Infiltration of graphite by aluminum during electric-discharge sintering. Sov Powder Metall Met Ceram 21(3):179–181
149. Mulukutla M, Singh A, Harimkar S (2010) Spark plasma sintering for multi-scale surface engineering of materials. JOM 62(6):65–71
150. Xie G, Louzguine-Luzgin DV, Kimura H, Inoue A (2007) Nearly full density $Ni_{52.5}Nb_{10}Zr_{15}Ti_{15}Pt_{7.5}Ni_{52.5}Nb_{10}Zr_{15}Ti_{15}Pt_{7.5}$ bulk metallic glass obtained by spark plasma sintering of gas atomized powders. Appl Phys Lett 90:241902
151. Wang Z, Georgarakis K, Nakayama K, Li Y, Tsarkov A, Xie G, Dudina D, Louzguine D, Yavari AR (2016) Microstructure and mechanical behavior of metallic glass fiber-reinforced Al alloy matrix composites. Sci Rep 6:24384
152. Walker LS, Marotto VR, Rafiee MA, Koratkar N, Corral EL (2011) Toughening in graphene ceramic composites. ACS Nano 5(4):3182–3190
153. Azarniya A, Sovizi S, Azarniya A, Reza Rahmani Taji Boyuk M, Varol T, Nithyadharseni P, Reza Madaah Hosseini H, Ramakrishna S, Reddy MV (2017) Physicomechanical properties of spark plasma sintered carbon nanotube-containing ceramic matrix nanocomposites. Nanoscale 9:12779–12820
154. Demirskyi D, Sakka Y, Vasylkiv O (2015) Consolidation of B_4C-TaB_2 eutectic composites by spark plasma sintering. J Asian Ceramic Soc 3(4):369–372
155. Demirskyi D, Sakka Y (2014) In situ fabrication of B_4C-NbB_2 eutectic composites by spark plasma sintering. J Am Ceram Soc 97:2376–2378

156. Khaleghi E, Lin YS, Meyers MA, Olevsky EA (2010) Spark plasma sintering of tantalum carbide. Scr Mater 63:577–580
157. Rudolf CC, Agarwal A, Boesl B (2016) TaC–NbC formed by spark plasma sintering with the addition of sintering additives. J Ceram Soc Japan 124(4):381–387
158. Jiang D, Hulbert DM, Anselmi-Tamburini U, Ng T, Land D, Mukherjee AK (2008) Optically transparent polycrystalline Al_2O_3 produced by spark plasma sintering. J Am Ceram Soc 91 (1):151–154
159. Meir S, Kalabukhov S, Froumin N, Dariel MP, Frage N (2009) Synthesis and densification of transparent magnesium aluminate spinel by SPS processing. J Am Ceram Soc 92(2):358–364
160. Sokol M, Halabi M, Kalabukhov S, Frage N (2017) Nano-structured $MgAl_2O_4$ spinel consolidated by high pressure spark plasma sintering (HPSPS). J Eur Ceram Soc 37:755–762
161. Khasanov OL, Dvilis ES, Bikbaeva ZG, Paygin VD, Khasanov AO (2017) Relationship of optical properties and elastoplastic characteristics of transparent spark-plasma-sintered YSZ ceramics. J Ceram Sci Technol 8(1):161–168
162. Noudem JG, Kenfaui D, Chateigner D, Gomina M (2012) Toward the enhancement of thermoelectric properties of lamellar $Ca_3Co_4O_9$ by edge-free spark plasma texturing. Scr Mater 66:258–260
163. Bhame SD, Pravarthana D, Prellier W, Noudem JG (2013) Enhanced thermoelectric performance in spark plasma textured bulk n-type $BiTe_{2.7}Se_{0.3}$ and p-type $Bi_{0.5}Sb_{1.5}Te_3$. Appl Phys Lett 102:211901
164. He Z, Katsui H, Goto T (2016) High-hardness diamond composite consolidated by spark plasma sintering. J Amer Ceram Soc 99(6):1862–1865
165. Jayalakshmi S, Gupta M (2015) Metallic amorphous alloy reinforcements in light metal matrices, Springer, 2015 - Technology & Engineering, 112 p
166. Hakamada M, Yamada Y, Nomura T, Kusuda H, Chen Y, Mabuchi M (2005) Effect of sintering temperature on compressive properties of porous aluminum produced by spark plasma sintering. Mater Trans 46(2):186–188
167. Dudina DV, Lomovsky OI, Korchagin MA, Mali VI (2004) Reactions in a metal matrix: synthesis and properties of TiB_2-cu nanocomposites. Chem Sustain Dev 12:319–325
168. Kwon YS, Dudina DV, Korchagin MA, Lomovsky OI (2004) Microstructure changes in TiB_2-cu nanocomposite under sintering. J Mater Sci 39(16–17):5325–5331
169. Kim JS, Kwon YS, Lomovsky OI, Korchagin MA, Mali VI, Dudina DV (2006) A synthetic route for metal-ceramic interpenetrating phase composites. Mater Lett 60(29–30):3723–3726
170. Korchagin MA, Dudina DV (2007) Application of self-propagating high-temperature synthesis and mechanical activation for obtaining nanocomposites. Comb Expl Shock Waves 43 (2):176–187
171. Bokhonov BB, Dudina DV (2013) Recrystallisation-accompanied phase separation in ag–Fe and ag–Ni nanocomposites: a route to structure tailoring of nanoporous silver. RSC Adv 3 (31):12655–12661
172. Fu YQ, Shearwood C, Xu B, Yu LG, Khor KA (2010) Characterization of spark plasma sintered ag nanopowders. Nanotechnology 21(11):115707
173. Dudina DV, Ukhina AV, Bokhonov BB, Mali VI, Anisimov AG, Bulina NV, Skovorodin IN (2015) Nickel-graphite composites of variable architecture by graphitization-accompanied spark plasma sintering and hot pressing and their response to phase separation. Sci Sinter 47:237–248
174. Bokhonov BB, Dudina DV, Ukhina AV, Korchagin MA, Bulina NV, Mali VI, Anisimov AG (2015) Formation of self-supporting porous graphite structures by spark plasma sintering of nickel-amorphous carbon mixtures. J Phys Chem Solids 76:192–202
175. Dudina DV, Bokhonov BB, Mukherjee AK (2016) Network distribution of reinforcements in composites produced by sintering: microstructure formation and influence on consolidation behavior and properties. J Ceram Soc Jpn 124:289–295

176. Dudina DV, Ukhina AV, Bokhonov BB, Korchagin MA, Bulina NV, Kato H (2017) The influence of the formation of Fe_3C on graphitization in a carbon-rich iron-amorphous carbon mixture processed by spark plasma sintering and annealing. Ceram Int 43:11902–11906
177. Mali VI, Anisimov AG, Kurguzov VD, Dudina DV, Bokhonov BB (2013) Spark Plasma Sintering for the production of micron- and nanoscale materials. Proceedings of Taiwan-Russia Bilateral Symposium on Materials Processing at micro and nano level, Novosibirsk, p 116–119
178. Ukhina AV, Dudina DV, Anisimov AG, Mali VI, Bulina NV, Bataev IA, Skovorodin IN, Bokhonov BB (2015) Porous electrically conductive materials produced by spark plasma sintering and hot pressing of nanodiamonds. Ceram Int 41:12459–12463
179. Kuznetsov VL, Butenko YV, Chuvilin AL, Romanenko AI, Okotrub AV (2001) Electrical resistivity of graphitized ultra-disperse diamond and onion-like carbon. Chem Phys Lett 336:397–404

Chapter 5
Flash Sintering

5.1 Principle of Flash Sintering

In the past few years, flash sintering has attracted significant attention as a field-assisted phenomenon and a promising technological process [1]. The fundamental aspect of flash sintering that makes it interesting for the sintering science is a unique mechanism of the influence of electric field on densification. The important applied aspects of flash sintering include energy saving and dramatic shortening of the processing times in the technological processes of sintering. In the originally proposed format, flash sintering is based on the application of an electrical potential to the powder compact heated in a furnace [2–4]. Flash sintering has gained its name due to an extremely short densification process (a few seconds). Cologna et al. [2] reported sintering of yttria-stabilized zirconia at 850 °C within less than 5 s. This term was previously used to refer to sintering processes conducted within several minutes [5]. A DC or AC potential is applied when the sample reaches a certain temperature through furnace heating. The characteristic field strength and power dissipation values in flash sintering are 100–100 V·cm^{-1} and 10–1000 W·cm^{-3}, respectively. Flash sintering has been demonstrated for oxide ceramic materials with ionic [6–9] and electronic conductivities [10, 11] as well as for non-oxide ceramics [12–17]. An Al alloy was reported to show a flash sintering behavior when a porous compact obtained by cold isostatic pressing was heated in a furnace, while copper electrodes carried a DC current to the sample [18]. In this case, the presence of oxide layers on the powder particles played a crucial role.

A schematic of the flash sintering setup is shown in Fig. 5.1. As was originally designed, flash sintering is conducted in a vertical furnace, in which a dog-bone specimen is placed in contact with platinum electrodes.

Flash sintering is accompanied by a sudden increase in the conductivity of the sintered material. It is currently agreed that the temperature (overheating) instability plays a crucial role in the development of the flash sintering. The sintering rate depends on the applied electric field; after certain field strength, the sintering process

E. A. Olevsky, D. V. Dudina, *Field-Assisted Sintering*,
https://doi.org/10.1007/978-3-319-76032-2_5

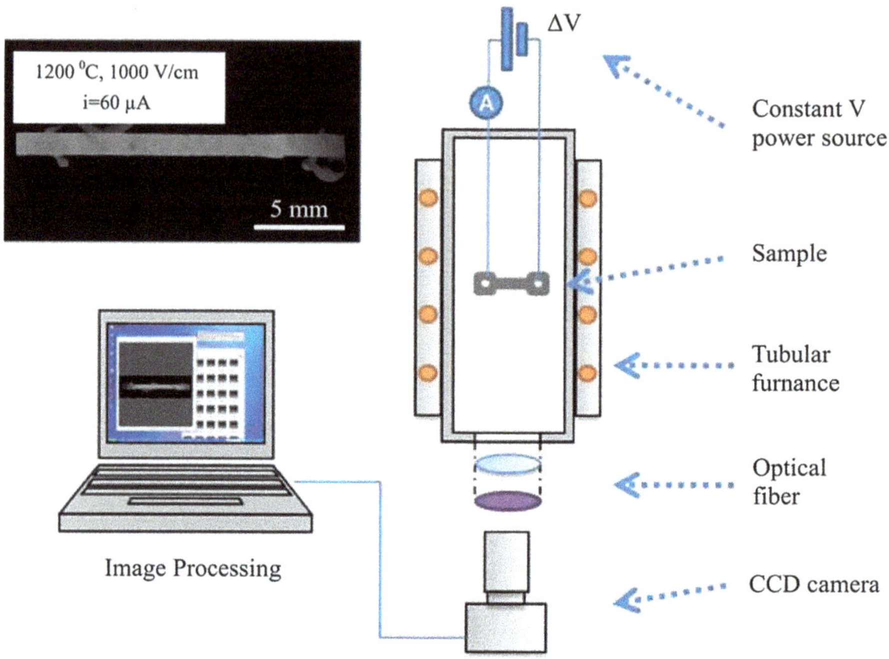

Fig. 5.1 Flash sintering setup. (Reprinted from Cologna et al. [11], Copyright (2011) with permission from Elsevier)

becomes unstable [2]. As the composition of the atmosphere can influence the conductivity of oxide materials, it can also influence the flash sintering behavior of these materials. Indeed, Zhang & Luo [19] observed a strong dependence of the onset flash sintering temperature of ZnO on the sintering atmosphere. At 300 V·cm^{-1}, the onset flash sintering temperature of ZnO was 631 °C when sintering was conducted in oxygen, 599 °C in air, 237 °C in argon, and 186 °C in Ar + 5 mol.%H_2 (Fig.5.2a). Different onset flash sintering temperatures were explained by the dependence of the electrical conductivity of ZnO on the composition of the gaseous atmosphere. The measured conductivity of ZnO compacts versus the reciprocal of the estimated specimen temperature is shown in Fig. 5.2b for flash sintering in the four atmospheres. At 1000 V·cm^{-1} in a reducing atmosphere of Ar + 5 mol.%H_2, ZnO could be sintered to relative densities greater than 97% in 30 s at furnace temperatures below 120 °C. The authors emphasize that although significant lowering of the onset flash sintering temperature can be achieved for ZnO in reducing atmospheres, the temperature must still be greater than 110 °C, as at lower temperatures the interaction between hydrogen and ZnO is limited.

Hewitt et al. [20] identified incipient thermal runaway as a necessary condition for the flash event. Dong & Chen [21] showed that the onset conditions of the flash process are the same in mold-free flash sintering and mold-assisted flash sintering. Thermal runaway follows the sample's heating by means of DC or AC. The serial

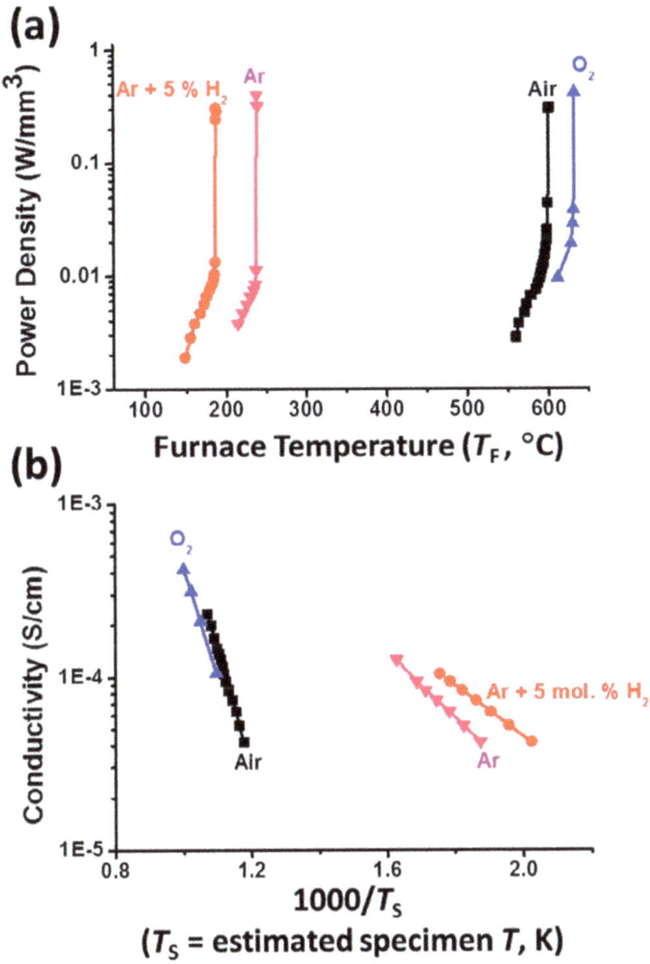

Fig. 5.2 (**a**) Measured electric power dissipation versus furnace temperature for the flash sintering of ZnO in different atmospheres at 300 V·cm^{-1}, (**b**) measured conductivity versus the reciprocal of the estimated specimen temperature. (Reprinted from Zhang and Luo [19], Copyright (2015) with permission from Elsevier)

and parallel components influence the Joule heating and conduction heat losses. In mold-assisted flash sintering, the temperature of the sample reaches a high value, but this value has a limit caused by the presence of the punches and layers of thermal insulation (Fig. 5.3). Soon after the flash, there is a drop in the voltage along the sample, as its resistance falls below those of the punches and insulations. As these conditions develop, the punches and the thermal insulation act as current limiters. So, in mold-assisted flash sintering, the maximum reachable sample temperature is limited. In both mold-free and mold-assisted flash sintering, the criterion for thermal runaway is the same: the time and temperature of thermal runaway correspond to the

Fig. 5.3 A schematic of the assembly for mold-assisted flash sintering. (Reprinted from Dong and Chen [21], Copyright (2016) with permission of John Wiley & Sons)

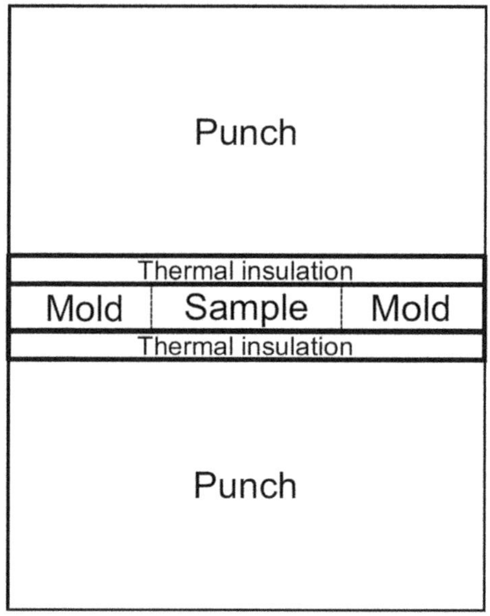

moment when the Joule heating in the sample catches up with the radiation heating from the environment. When this is set as the criterion for thermal runaway, the onset temperatures of thermal runaway can be accurately predicted. For the case of mold-assisted sintering, the criterion for thermal runaway can be expressed as

$$\frac{A_s V_a^2}{H_a \rho_0 \exp\left(\frac{E_a}{k_B T_s}\right)} > \varepsilon \sigma S_{d-s} T_d^4, \tag{5.1}$$

where V_a is the voltage along the sample/mold assembly, E_a is the activation energy, ε is the emissivity, σ is the Stefan-Boltzmann constant, S_{d-s} is the surface area of the sample placed in the mold (Fig. 5.3), T_d is the mold temperature, k_B is the Boltzmann constant, T_s is the sample temperature, ρ_0 is the room temperature resistivity of the sample, H_a is the height of the sample/mold assembly, and A_s is the cross section of the sample [21].

According to Dong and Chen [21], before the onset criterion is met, the sample is heated either by radiation from the mold or from the furnace. Once Eq. (5.1) is satisfied at a certain temperature, the Joule heating will take over, and the sample will be heated further on in an accelerated manner. Thermal runaway is suppressed when at least one of the following is observed: (1) the applied voltage is below the critical voltage, (2) the sample experiences excessive thermal conduction and loses heat to the environment, (3) the conductivity of the mold is too high, or (4) the conductivity of the mold is too low. As the spark plasma sintering (SPS) machines utilize low voltages, in order to use a SPS setup as equipment for mold-assisted flash sintering, the samples should be of low height and of large diameter. An advantage of the SPS chambers for conducting flash sintering may be the vacuum conditions and a reducing character of the sample's environment [21].

Dong & Chen suggested a simple relationship between the onset flash sintering temperature T_{on} and the applied field strength E [22]: $\frac{1}{T_{on}} \sim \log E$. A more detailed analysis allowed suggesting the following relationship [23]:

$$\ln\left(\frac{E^2}{T_{on}^4}\right) = \frac{E_a}{k_B T_{on}} + \ln\left(\frac{\varepsilon\sigma S d^2 R_0}{\beta}\right), \tag{5.2}$$

where E_a is the activation energy, k_B is the Boltzmann constant, ε is the emissivity, σ is the Stefan-Boltzmann constant, d is the sample's length, S is the sample's surface area, β is a numerical constant, and R_0 is the resistance pre-exponential factor ($R = R_0 \exp\left(\frac{E_a}{k_B T}\right)$).

Downs & Sglavo [24] have shown that cubic yttria-stabilized zirconia can be sintered in a DC electric field of 2250 V·cm^{-1} at 390 °C, which is more than 1000 °C below the sintering temperatures normally used for sintering of this ceramics. The following relationship between T_{on} and E was determined as the best fit of the experimental data:

$$T_{on}(K) = 2440E^{-1/5.85}\left(\mathrm{V}\cdot\mathrm{cm}^{-1}\right) \tag{5.3}$$

Using this equation, the critical electric field, at which the sample enters the flash sintering regime, was calculated to be 24.5 V·cm^{-1}. This value corresponds to T_{on} equal to 1408 K. At this temperature, sintering of the chosen cubic yttria-stabilized zirconia powder starts in the absence of electric field, as was determined by dilatometry measurements. A graphical determination of the critical field is demonstrated in Fig. 5.4. The critical electric field determines the boundary between the FAST (type A) and flash (type B) sintering regimes. In type A (at low E), sintering starts at the same temperature as without the field, and the presence of the electric field only facilitates sintering by accelerating it at temperatures above 1408 K. At E, exceeding the critical value, flash sintering occurs (type B).

Analyzing the shrinkage of the samples sintered at different applied voltages, Downs & Sglavo [24] found that the compacts sintered at higher electric fields show lower linear shrinkage. Two reasons were suggested to explain the observed trend: 1) the formation of short circuit paths at high voltages and 2) the value of the current density not reaching the threshold. Based on these experiments, the authors concluded that the values of the maximum power dissipation are important for the initiation of flash sintering, but they do not determine the sample shrinkage. It is possible that a certain value of the current density needs to be reached for the uniform and efficient densification.

In their proof-of-concept paper, Saunders et al. [13] proposed a contactless version of flash sintering, in which the electric current was supplied to the sample by the arc plasma. A pre-compacted sample was placed in the gap between two pairs of plasma torches (Fig. 5.5). The sample was preheated by the plasma, and the flash

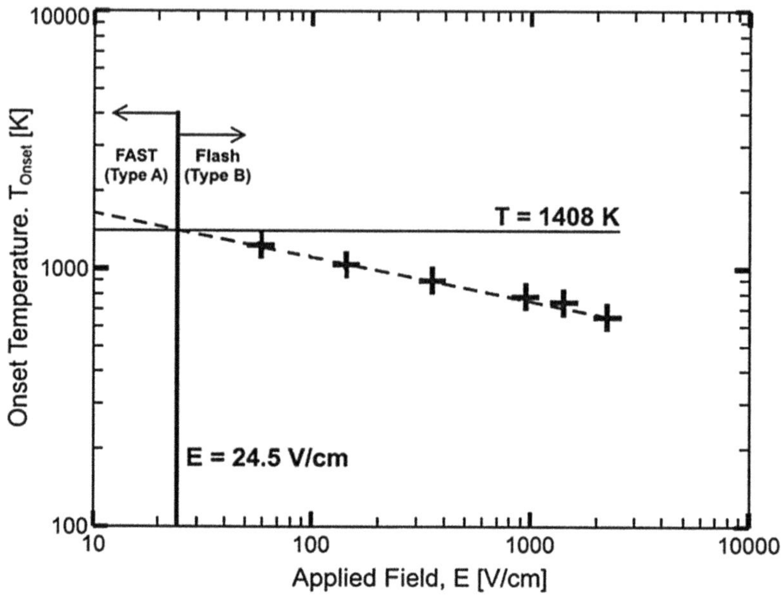

Fig. 5.4 The onset temperature of flash sintering of cubic yttria-stabilized zirconia as a log-log diagram. The dotted line corresponds to $T_{on}(K) = 2440E^{-1/5.85}$. The diagram shows a graphical method of determining the critical electric field and the boundary between the field-assisted sintering technique (FAST) and flash regimes. (Reprinted from Downs and Sglavo [24], Copyright (2013) with permission of John Wiley & Sons)

sintering itself occurred upon imposition of a flash power (Fig. 5.6). Figure 5.7 shows a general view of a B_4C–SiC sample sintered by contactless flash sintering. The compact has a dense central region after 2 s of flash sintering. In this technique, a need of direct contact of the surface of the sample with a current-carrying electrode is eliminated, and the sample can be freely moved in the gap between the electrodes. The authors emphasize that, as no electrodes are to be attached to the surface of the sample, this operation is eliminated from the sintering cycle, which becomes a great advantage of this technique over "classic" flash sintering with platinum electrodes. A transformer welder apparatus having an AC power supply was used for creating the plasma. The AC current rather than DC current was chosen to avoid a polarity-related nonuniformity of heating. Using this technique, consolidation of silicon carbide (SiC), boron carbide (B_4C), and a B_4C-SiC composite was attempted. The samples that were subjected to flash sintering were plates cut from porous compacts consolidated by (SPS). In the designed setup, conditions of flash sintering that ensure full densification of the B_4C-SiC composite and nearly full densification of single-phase B_4C were found. The geometry of the setup imposed a limitation on the sample thickness: thick samples could not be heated through the entire thickness.

A schematic of the B_4C–SiC sample subjected to contactless flash sintering in the gap between two pairs of electrodes is shown in Fig. 5.8. It was found that the

Fig. 5.5 The principle of contactless flash sintering. (Reprinted from Saunders et al. [13], Copyright (2016), Rights managed by Nature Publishing Group. This work is licensed under a Creative Commons Attribution License, http://creativecommons.org/licenses/by/4.0/)

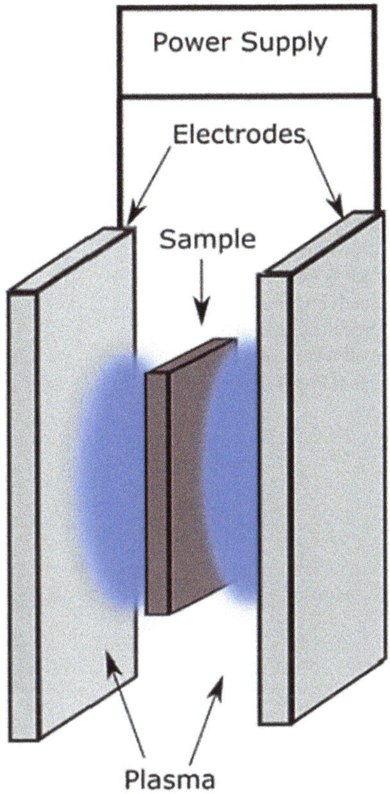

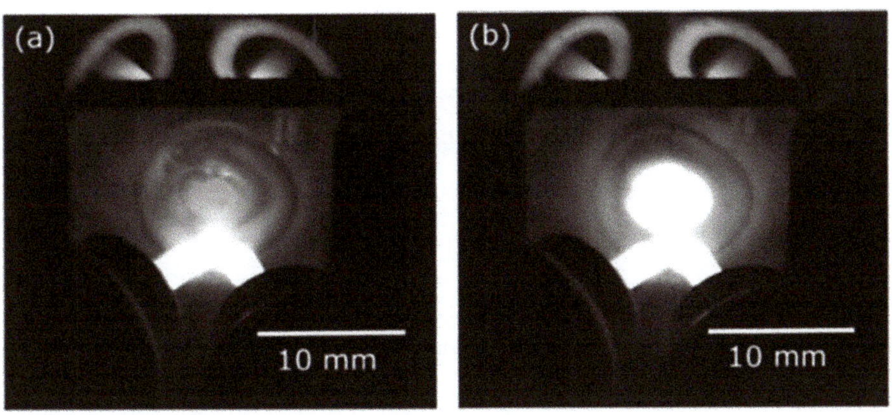

Fig. 5.6 A sample to be sintered placed in the gap between the electrodes before flash sintering (**a**) and 0.5 s after the application of the flash power (**b**). (Reprinted from Saunders et al. [13], Copyright (2016), Rights managed by Nature Publishing Group. This work is licensed under a Creative Commons Attribution License, http://creativecommons.org/licenses/by/4.0/)

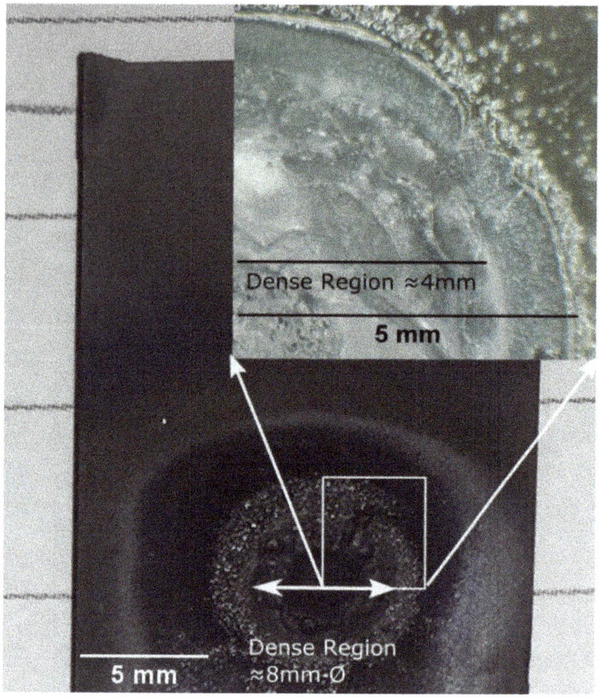

Fig. 5.7 A general view of a B_4C–SiC sample sintered by contactless flash sintering showing a dense central region; the flash sintering time was 2 s. (Reprinted from Saunders et al. [13], Copyright (2016), Rights managed by Nature Publishing Group. This work is licensed under a Creative Commons Attribution License, http://creativecommons.org/licenses/by/4.0/)

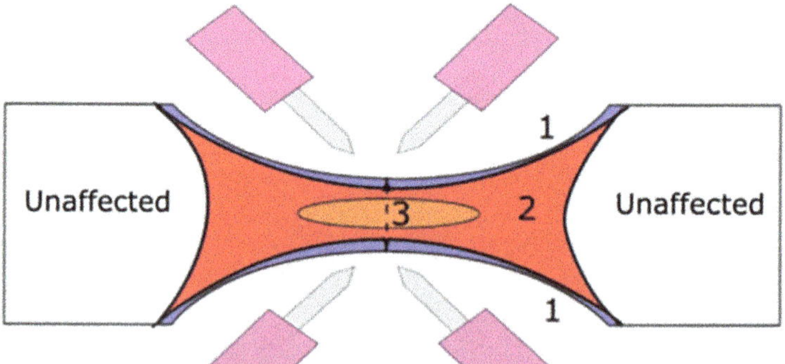

Fig. 5.8 A schematic of the B_4C–SiC sample subjected to contactless flash sintering in the gap between two pairs of electrodes (zone 1 is the zone with the highest temperature, zone 2 was densified earlier in the flash process than zone 3; as the flash sintering progressed, the boundary between zones 2 and 3 disappeared). (Reprinted from Saunders et al. [13], Copyright (2016), Rights managed by Nature Publishing Group. This work is licensed under a Creative Commons Attribution License, http://creativecommons.org/licenses/by/4.0/)

surface temperature of the sample was higher than the temperature of the interior of the sample. Zone 1 had the highest temperature and was always remelted. After 2 s from the flash sintering start, zones 2 and 3 could be distinguished by a difference between the relative densities: zone 3 has a higher porosity than zone 2. After 3 s, no difference between zones 2 and 3 could be observed, as the material was fully sintered. As the time of the flash increased, bubbles formed in zones 2–3 due to sublimation of SiC causing expansion of the material.

Flash sintering conducted using a conventional scheme is suitable for small specimens and requires platinum electrodes. Grasso et al. [14] discuss the advantages of realizing flash sintering in the SPS facilities to increase the size of the specimen, increase the productivity, and eliminate the need to use metallic electrodes. In their work, a SPS facility was successfully used for flash sintering of zirconium diboride (ZrB_2). As ZrB_2 has a higher room temperature conductivity than materials, for which flash sintering was initially developed (e.g., zirconium dioxide (ZrO_2)), it can be processed in the flash sintering mode using a low voltage. For this reason, voltages usually used in the SPS facilities are sufficient for carrying out the process of flash sintering. A pre-sintering operation was used to make a porous compact, which was placed between the punches of the assembly while no die was used. Figure 5.9 shows the displacement rate, the heating power, and the temperature developed during flash sintering of ZrB_2 with a total discharge time of 35 s.

An interesting observation was made regarding the composition of the sintered material. The initial powder of ZrB_2 contained 0.2 wt.% of carbon and 0.8 wt.% of oxygen, as the boride phase was synthesized by carbothermal reduction of ZrO_2 and the reaction was seemingly not complete. In the compact flash sintered within 35 s, substantial quantities of carbon were detected by Raman spectroscopy (Fig. 5.10a), while only negligible amounts were found in the compact produced by conventional SPS (Fig. 5.10b). In the latter, the reaction between the remaining ZrO_2 and carbon was brought to completion during sintering, and the carbon was removed in the form of carbon monoxide (CO). This example shows that full densification by flash sintering may not coincide with the synthesis completion.

Zapata-Solvas et al. [12] investigated the possibility of flash sintering of a covalent ceramic – silicon carbide (SiC). A schematic of their setup is shown in Fig. 5.11. It was possible to achieve substantial densification in a flash sintering process of SiC only when Al_2O_3 + Y_2O_3 additives were used. Specimen temperatures were concluded to be much higher than those in the rest of the furnace owing to direct Joule heating of the specimen; as a result, the densification near the specimen surfaces was limited by the heat losses. In pure SiC or SiC with an Al + B_4C + C sintering aid, high densities of the compact were unachievable under the flash sintering conditions. Zapata-Solvas et al. [12] pointed out that the evolution of the voltage and current is affected by the temperature dependence of the resistivity of the material. The flash phenomenon was observed in all compositions — pure SiC, SiC +(Al_2O_3 + Y_2O_3), and SiC+(Al + B_4C + C) (Fig. 5.12). However, for each composition, there was its own range of parameters, at which flash sintering occurred. The maximum mean density of 88% was obtained for the SiC+(Al_2O_3 + Y_2O_3) at an applied field of 100 $V \cdot cm^{-1}$. In pure SiC and SiC+(Al + B_4C + C), densification was

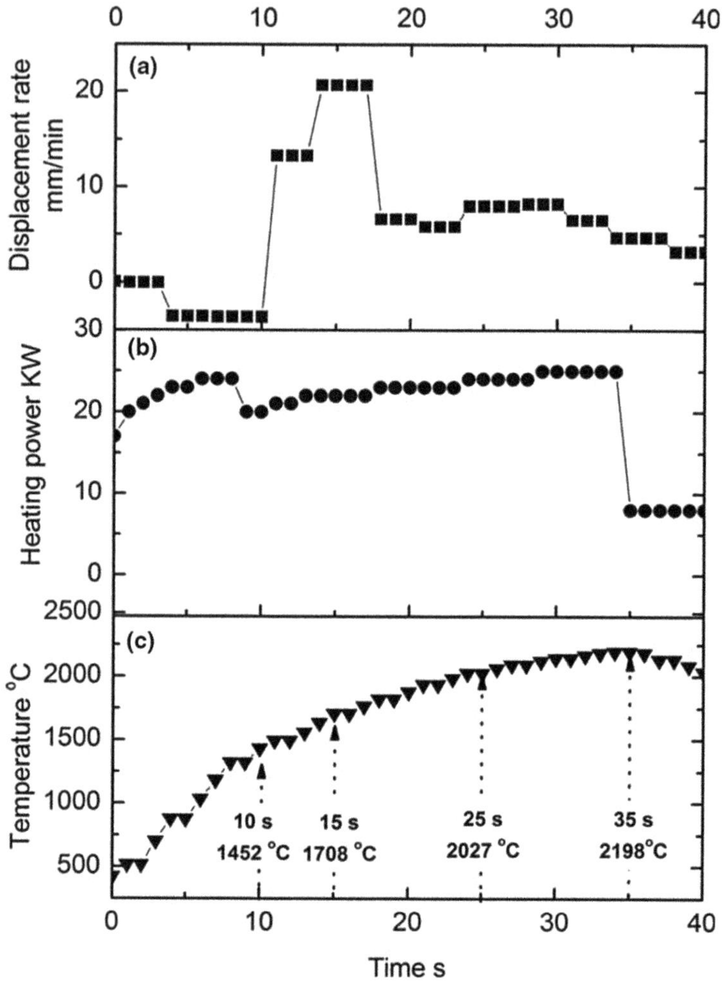

Fig. 5.9 Displacement rate (**a**), heating power (**b**), and temperature (**c**) during flash sintering of ZrB_2 (discharge time 35 s). (Reprinted from Grasso et al. [14], Copyright (2014) with permission of John Wiley & Sons)

less significant. A characteristic feature of the sintered specimens was a nonuniform density across the diameter of the pellet with the outer regions remaining porous (due to heat losses) and the core reaching nearly full density. An altered ratio of SiC observed by the authors is an evidence of thermal decomposition of silicon carbide (Fig. 5.13). From the known decomposition temperatures of SiC, it was possible to conclude that the sintered specimen experienced temperatures much higher than that of the furnace (1200 °C). The fact that the compacts heat treated before the flash sintering experiments showed a lower flash sintering temperature suggests that the initial formation of inter-particle contacts to form a continuous current pathway is

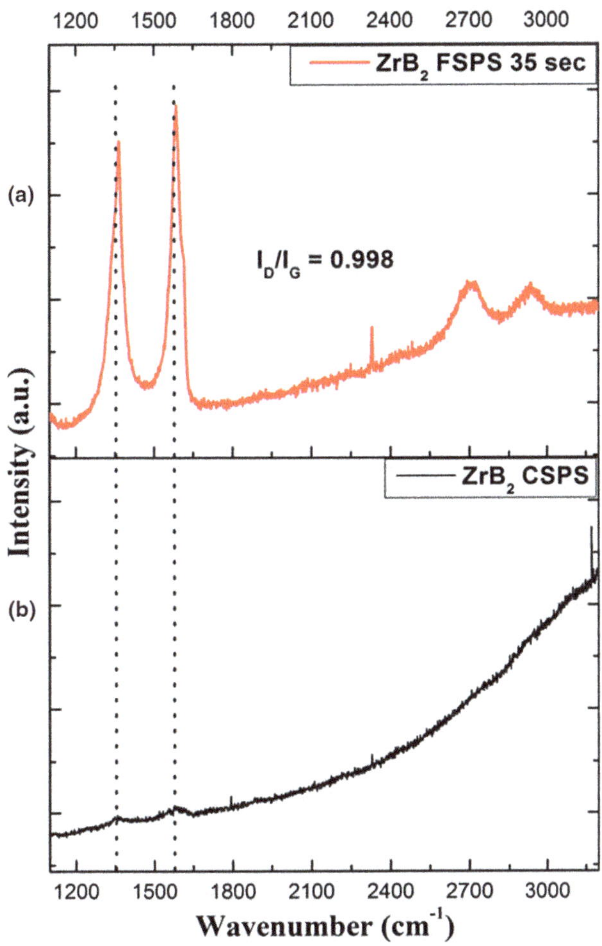

Fig. 5.10 Raman spectra of the ZrB_2-based compacts sintered from a powder containing 0.2 wt.% of carbon and 0.8 wt.% of oxygen: (**a**) flash sintering in a SPS chamber completed in 35 s, (**b**) conventional SPS. In the compact processed by SPS, the reaction between the remaining ZrO_2 and carbon is complete, and the carbon is removed in the form of CO. (Reprinted from Grasso et al. [14], Copyright (2014) with permission of John Wiley & Sons)

critical in determining the flash sintering temperature. According to the authors, the role of the sintering aids was also to establish better inter-particle contacts at low temperatures prior to flash sintering, while the main phenomenon responsible for the occurrence of flash sintering is Joule heating. It is noted that their flash sintering facility is similar to the SPS (induction heating of a graphite susceptor is used as a "furnace"; however, the sample is not in contact with it but insulated by alumina tubes) but differs from it by the absence of the die and by forcing the electric current through the sample, by means of which higher heating rates than in the SPS facilities can be achieved.

Olevsky et al. [15] proposed a modified flash sintering technique – flash hot pressing (or "flash spark plasma sintering") – in which pressure can be applied to the sample to stabilize the flash sintering process. Flash hot pressing can be conducted using conventional SPS facilities using a specially designed tooling. Alternatively, SPS facilities with hybrid heating can be used. The modified tooling included a

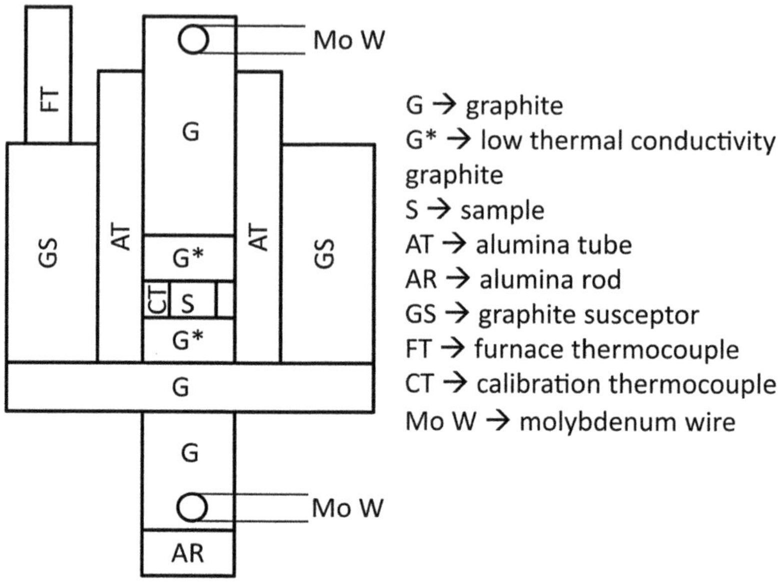

Fig. 5.11 Schematic of the flash sintering setup used in Ref. [12]. (Reprinted from Zapata-Solvas et al. [12], Copyright (2013) with permission from Elsevier)

copper collar tube, which acted as a heating element at temperatures below the melting point of copper (Fig. 5.14). At the initial stage of the process, the pre-sintered specimen is isolated from the electric current-carrying elements of the tooling and is, therefore, heated by radiation only. Once the melting point is reached, the tube collapses, so that the electrical contact between the punch and the tube is disrupted and the contact between the punch and the sample is established. The choice of copper as a material of the sacrificial collar was dictated by the fact that, at temperatures close to the melting point of copper, the major contribution to the overall electrical conductivity of SiC starts to come from its intrinsic conductivity, which increases exponentially with temperature above 1500 K. At temperatures below the melting point of copper, the conductivity of SiC has extrinsic nature and is determined by the presence of impurities. Flash hot pressing (flash spark plasma sintering) of a SiC powder allowed obtaining SiC compacts with a relative density of 99% without any noticeable grain growth (Fig. 5.15). The top edge of the sacrificial copper collar was exposed to the radiation thermometer, which was used by the SPS to regulate its temperature. Figure 5.16 shows the flash character of densification of the SiC pre-consolidated specimen: a spike in the temperature coincides with the spike in the relative density, while an increase in conductivity makes the SPS apparatus reduce the electric current.

The assembly designed for flash hot pressing has advantages of adding variable parameters to the flash sintering-based processing, which can be used to control the sintering efficiency. These parameters are the height of the copper collar and the applied pressure. Figure 5.17 shows the fracture surface of the compacts obtained using short and tall collars. In the case of a short collar, a relative density of 77% was

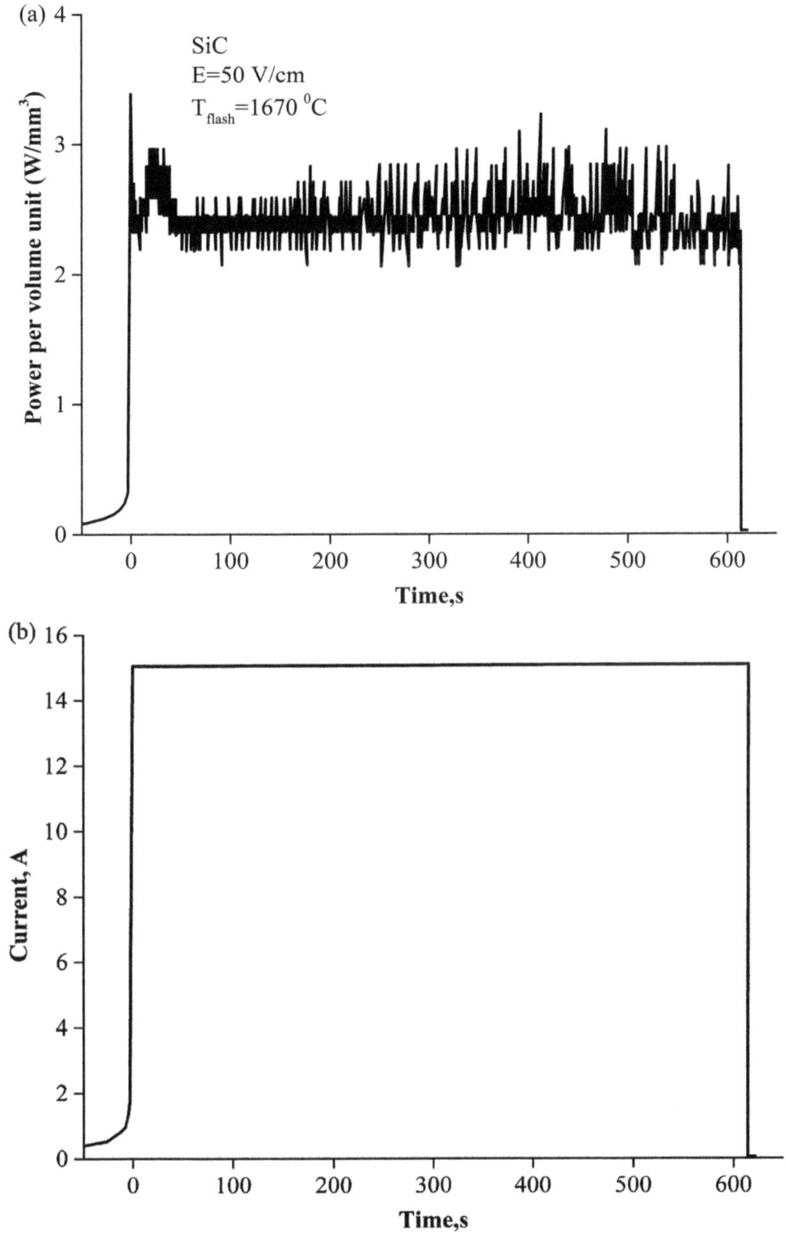

Fig. 5.12 Power per unit volume during flash sintering of SiC (**a**), SiC+(Al_2O_3 + Y_2O_3) (**c**), and SiC+ (Al + B_4C + C) (**d**); a sudden increase in current during sintering of SiC is shown in (**b**). (Reprinted from Zapata-Solvas et al. [12], Copyright (2013) with permission from Elsevier)

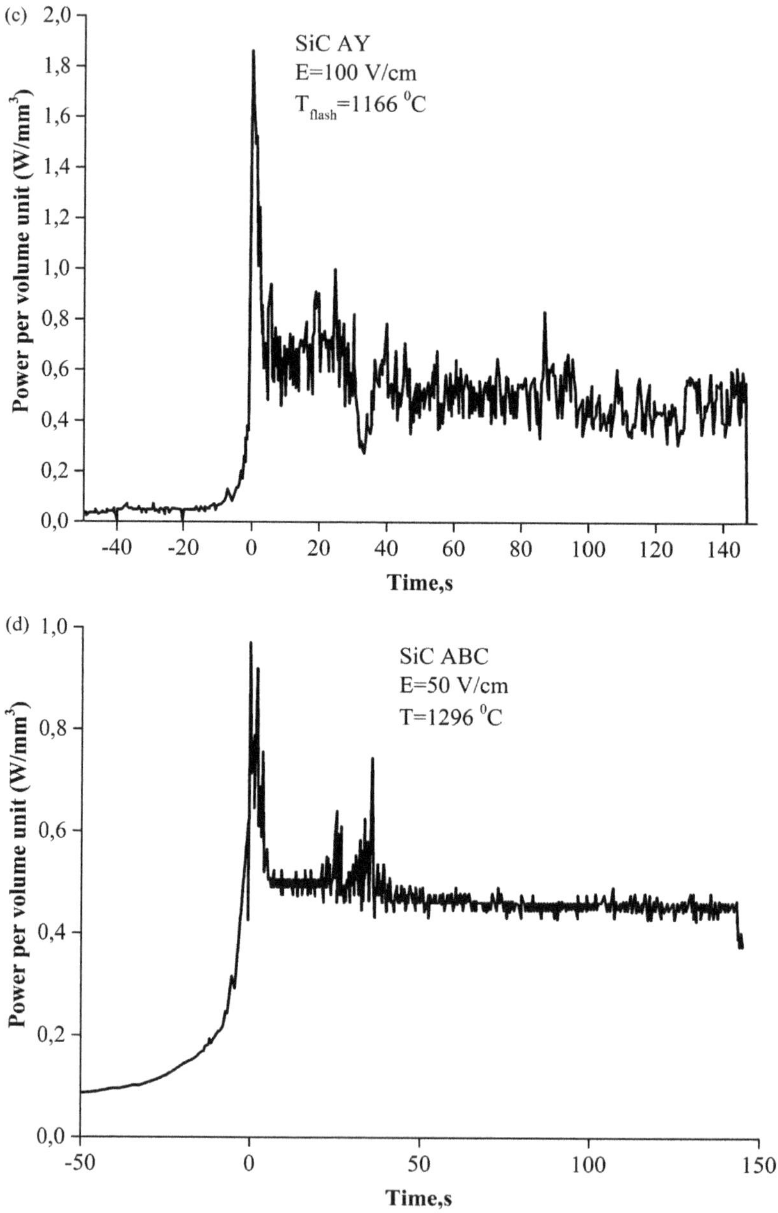

Fig. 5.12 (continued)

achieved, while the use of a long collar led to a higher relative density of 86%. Experiments with a punch with an oblique surface showed that by changing the applied pressure during flash hot pressing, the density of the sintered material can be varied. Figure 5.18 shows a SiC compact, in which zones with different values of shrinkage were formed corresponding to relative densities of 78% and 91%.

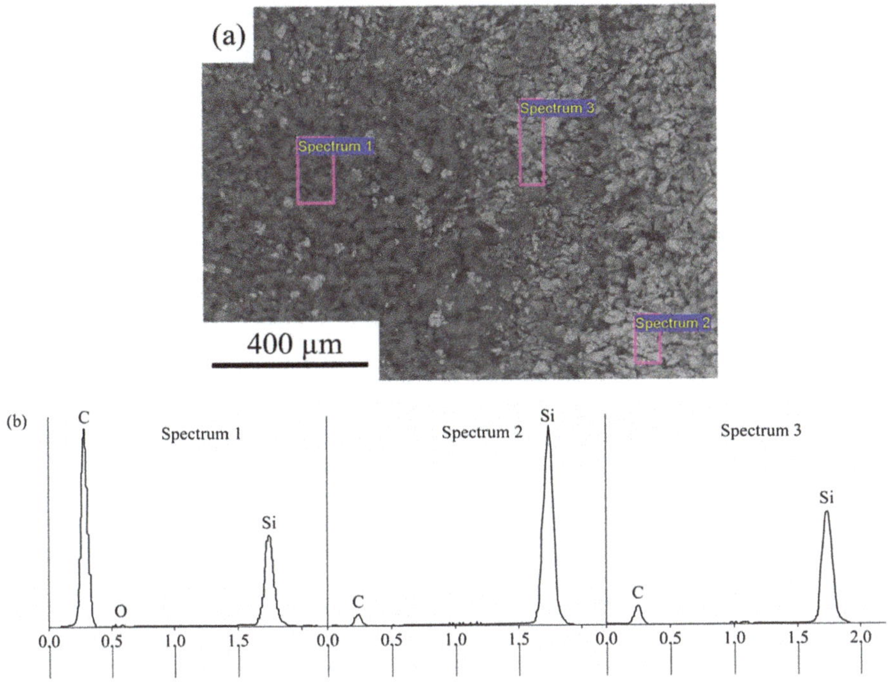

Fig. 5.13 SEM and EDX analyses of SiC flash sintered at 15 A and 200 $V \cdot cm^{-1}$. (Reprinted from Zapata-Solvas et al. [12], Copyright (2013) with permission from Elsevier)

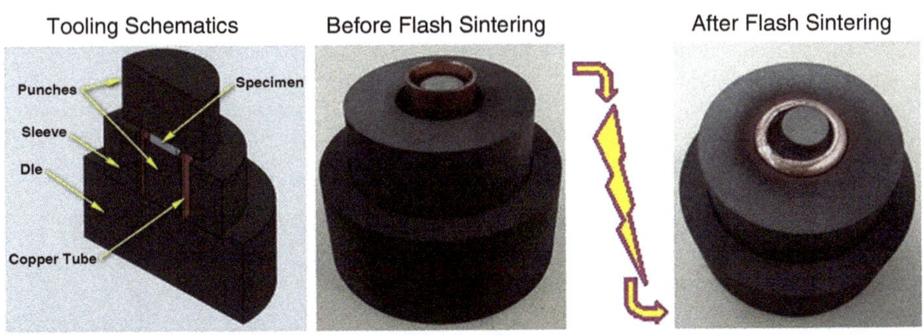

Fig. 5.14 Die design for flash hot pressing: a sacrificial copper collar is used as a supporting and a heating element for the pre-consolidated specimen at temperatures below the melting point of copper. (Reprinted from Olevsky et al. [15], Copyright (2015), Rights managed by Nature Publishing Group. This work is licensed under a Creative Commons Attribution License, http://creativecommons.org/licenses/by/4.0/)

The approach elaborated in Ref. [15] is different from die-free flash sintering of a pre-sintered pellet of an initially conductive material in a SPS facility between graphite punches by maximizing the power. In the die-free flash SPS [14], the ZrB_2 sample was preheated from the beginning of the process by the electric current

Fig. 5.15 The SiC powder (**a**) and the fracture surface of the compact (**b**) processed from this powder to a relative density of 99% by flash hot pressing (flash spark plasma sintering). (Reprinted from Olevsky et al. [15], Copyright (2015), Rights managed by Nature Publishing Group. This work is licensed under a Creative Commons Attribution License, http://creativecommons.org/licenses/by/4.0/)

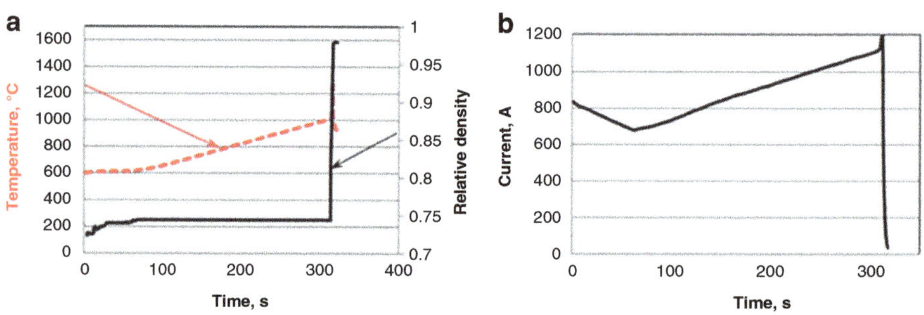

Fig. 5.16 (**a**) Temperature (dashed red line) and relative density (solid black line) of the SiC compact subjected to flash hot pressing, (**b**) evolution of electric current during flash hot pressing. (Reprinted from Olevsky et al. [15], Copyright (2015), Rights managed by Nature Publishing Group. This work is licensed under a Creative Commons Attribution License, http://creativecommons.org/licenses/by/4.0/)

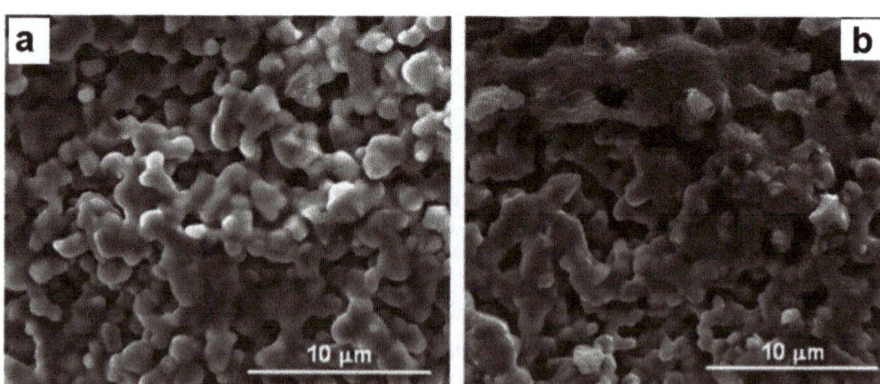

Fig. 5.17 Fracture surface of the SiC compacts consolidated by flash hot pressing (flash spark plasma sintering): (**a**) short copper collar (tube), (**b**) tall copper collar. (Reprinted from Olevsky et al. [15], Copyright (2015), Rights managed by Nature Publishing Group. This work is licensed under a Creative Commons Attribution License, http://creativecommons.org/licenses/by/4.0/)

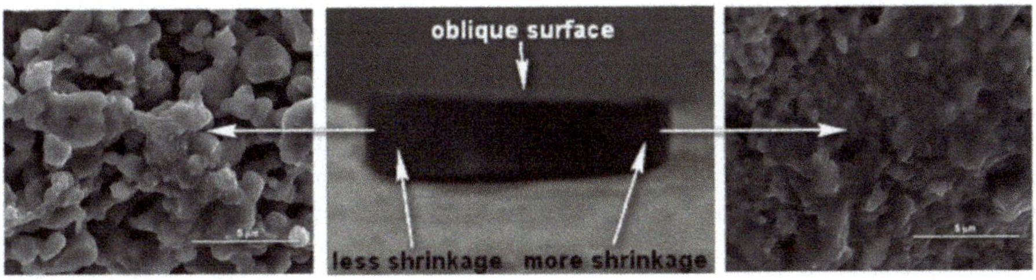

Fig. 5.18 Fracture surface and general view of the consolidated SiC compact processed by flash hot pressing using a punch with an oblique surface. (Reprinted from Olevsky et al. [15], Copyright (2015), Rights managed by Nature Publishing Group. This work is licensed under a Creative Commons Attribution License, http://creativecommons.org/licenses/by/4.0/)

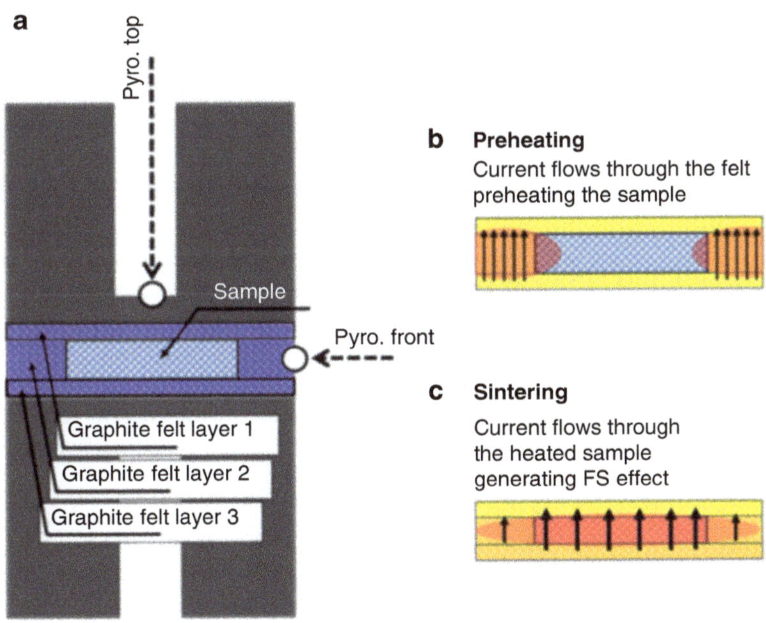

Fig. 5.19 Experimental setup for conducting flash SPS (**a**) and schemes showing the distribution of electric current at the preheating stage (**b**) and flash sintering stage (**c**). (Reprinted from Grasso et al. [16], Copyright (2016) Grasso et al., This is an open access article under the terms of the Creative Commons Attribution License, https://creativecommons.org/licenses/by/4.0/)

passing through it. Also, the use of a sacrificial copper collar makes the process presented in Ref. [15] different from the process, in which a graphite felt around the sample was used to establish conditions to preheat the pre-compacted SiC before flash sintering in a SPS facility [16, 17]. A schematic demonstrating the use of graphite felt as a conducting element at low temperatures, at which SiC is still too resistive, is shown in Fig. 5.19. At higher temperatures, the SiC compact is carrying the major fraction of the electric current.

5.2 Mechanisms of Flash Sintering

As flash sintering is usually conducted without the application of pressure, it presents a field-assisted sintering method, which allows studying the effect of electric current per se on densification and grain growth without having to overcome the difficulties of separating the sintering enhancement effects induced by pressure from those related to the passing current [18]. Raj [4] suggests that Joule heating, though involved in the process, cannot explain the phenomenon, as the increased temperature of the sintered material is still several hundred degrees lower than that required for such a fast sintering according to the calculations based on the Arrhenius equation. The conductivity of the oxide materials taken for the studies is electronic, and they behave as semiconductors upon heating. Initially, a constant voltage is applied to the compact (Fig. 5.20). A rapid increase in conductivity causes a surge in the powder dissipation. The power supply is then switched to the current control (Fig. 5.21) [4, 25]. The difficulty in explaining the flash sintering effect is, therefore, due to the fact that the conductivity and densification are dependent on different moving species — electrons and ions, respectively. Raj [4] believes that more attention should be directed to the mechanism of the field influence on densification during flash sintering, as the electric field can produce an avalanche of Frenkel pairs

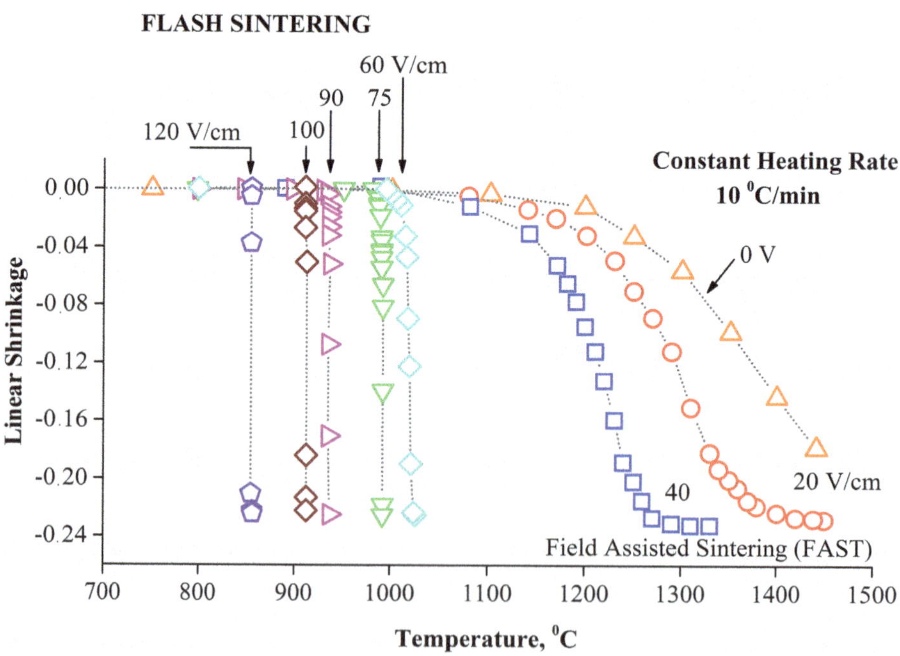

Fig. 5.20 Shrinkage rates of yttria-stabilized zirconia (3YSZ) under an applied field. The instability observed at a field strength of 40 $V{\cdot}m^{-1}$ leads to sintering in a few seconds. (Reprinted from Cologna et al. [2], Copyright (2010) with permission John Wiley & Sons)

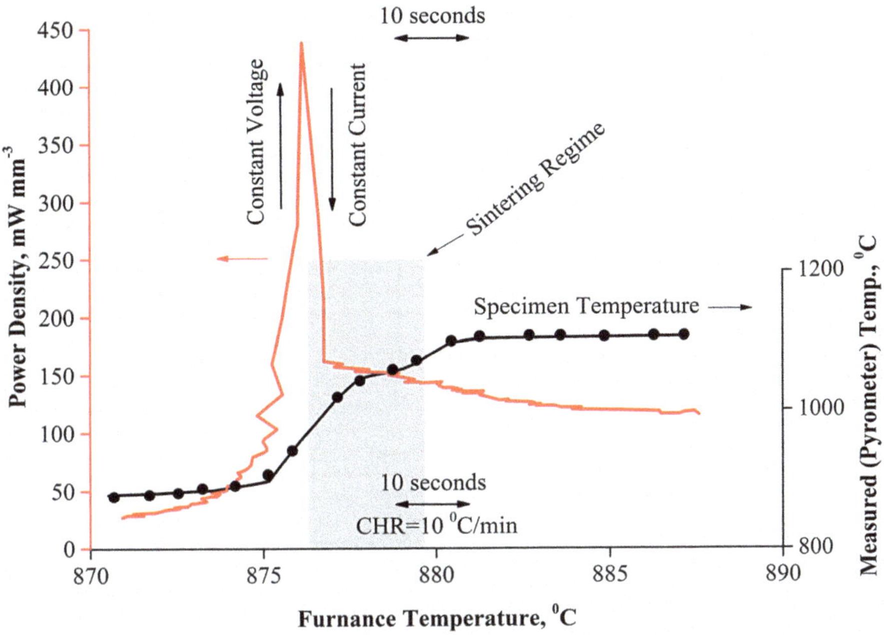

Fig. 5.21 The power density and the specimen temperature during a flash sinter-forging experiment. (Reprinted from Raj [4], Copyright (2012) with permission from Elsevier)

and induce their ionization [26]. An increased concentration of Frenkel pairs would facilitate mass transport in the sintered material.

For initially non-conducting $SrTiO_3$, higher compact densities were achieved at lower applied fields, as was reported by Karakuscu et al. [27], similar to the results presented in Ref. [24]. At an applied field of 1000 $V{\cdot}m^{-1}$, flash sintering started at 740 °C, but the final sintered density was below 70%. At 500 $V{\cdot}m^{-1}$, flash sintering occurred at 900 °C yielding a compact 76% dense. A weaker field of 150 $V{\cdot}m^{-1}$ induced flash sintering at 1200 °C, which was closer to a field-assisted process as the densification was not instantaneous (Fig. 5.22). The microstructure of the sintered compact shown in Fig. 5.23 indicates significant grain growth in the sample sintered at 1200 °C relative to that sintered at 900 °C. The ability of the $SrTiO_3$ phase to be flash sintered was attributed to an increase in its conductivity during flash sintering through the introduction of defects associated with nonstoichiometric phases. This conclusion was supported by the characterization results of the flash-sintered material by transmission electron microscopy, X-ray diffraction, and photoinduced reflectivity dynamics, which all indicated higher content of defects in the sintered material relative to the powder. Future investigations of the defect structures in the materials will shed light on the processes occurring during flash sintering and contribute to a deeper understanding of its mechanisms.

Grasso et al. [28] analyzed the temperature distribution in an yttria-stabilized zirconia dog bone-shaped compact (Fig. 5.24) under conditions of flash sintering by

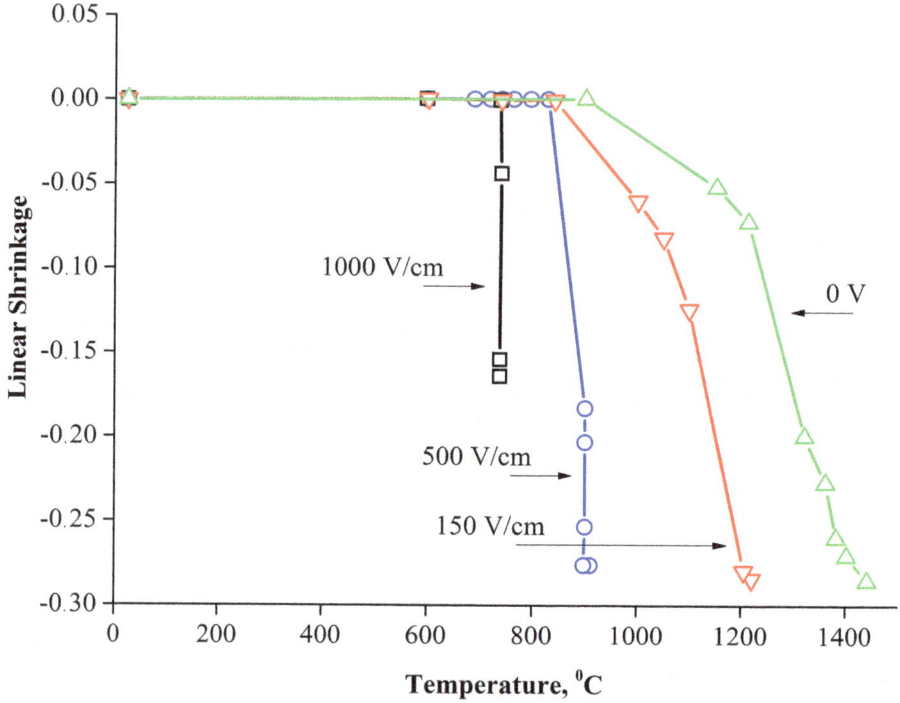

Fig. 5.22 Linear shrinkage of $SrTiO_3$ at different applied electrical fields versus furnace temperature (heating rate 10 °C·min^{-1}). (Reprinted from Karakuscu et al. [27], Copyright (2012) with permission of John Wiley & Sons)

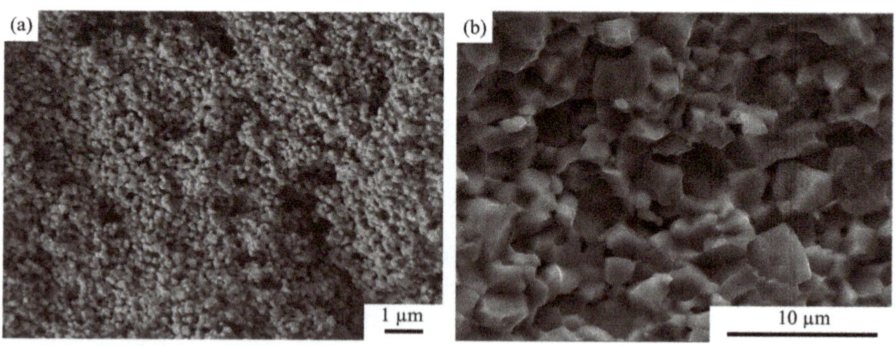

Fig. 5.23 SEM images of the flash-sintered $SrTiO_3$: (**a**) 500 V·cm^{-1}, 60 mA, flash sintering occurred at 900 °C; (**b**) 150 V·cm^{-1}, 500 mA, flash sintering occurred at 1200 °C. (Reprinted from Karakuscu et al. [27], Copyright (2012) with permission of John Wiley & Sons)

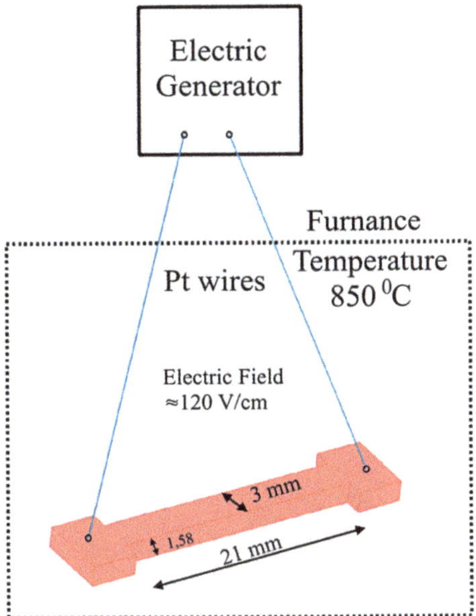

Fig. 5.24 A dog bone-shaped specimen used in flash sintering. (Reprinted from Grasso et al. [28], Copyright (2011) with permission from the Ceramic Society of Japan)

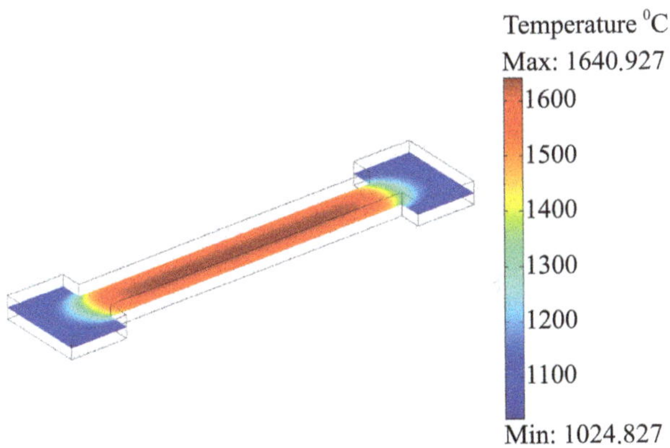

Fig. 5.25 Finite element model simulation of the temperature distribution in the yttria-stabilized zirconia dog-bone shaped compact under a power of 70 W applied for 3 s. A field intensity of 120 V·cm^{-1} was applied when a furnace temperature reached 850 °C. (Reprinted from Grasso et al. [28], Copyright (2011) with permission from the Ceramic Society of Japan)

means of finite element model simulation and conducted flash sintering experiments with doped and undoped zirconia. In the simulation, the electrical and thermal conductivities of the material as well as its heat capacity were assumed to depend on the temperature. The simulation results showed that the flash sintering phenomena should be ascribed to a sudden temperature increase inside the sample. As is seen from Fig. 5.25, the temperature rises very rapidly – from 850 to 1600 °C.

By comparing the microstructure of the tetragonal yttria-stabilized zirconia and undoped zirconia compacts after the flash sintering experiments, it was found that the former was well sintered, while the latter was still porous, which indicated the primary significance of the electrical conductivity of the material in the success of

Fig. 5.26 Fracture surface of tetragonal yttria-stabilized zirconia (**a**) and undoped zirconia (**b**) subjected to flash sintering (120 V·cm^{-1} applied when the furnace temperature reached 850 °C). (Reprinted from Grasso et al. [28], Copyright (2011) with permission from the Ceramic Society of Japan)

the flash sintering process (Fig. 5.26). Grasso et al. [28] emphasized the significance of the flash sintering technique as an energy-saving method due to the utilization of highly localized heating and short processing times.

Using AC impedance spectroscopy, Park and Chen [29] performed in situ thermometry and measured temperatures exceeding 1500 °C reached within a few

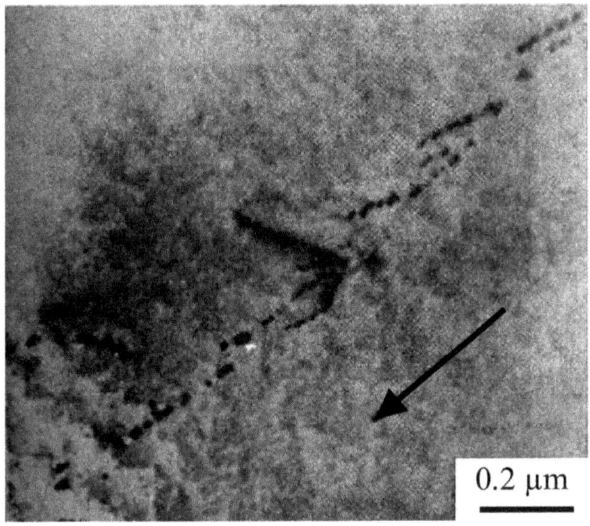

Fig. 5.27 Bright-field electron micrograph of a MgO crystal showing precipitates near a sub-grain boundary. (Reprinted from Narayan [31], Copyright (2013) Acta Materialia Inc., with permission from Elsevier)

seconds in 8 mol.% yttria-stabilized zirconia disks 0.25–0.65 mm thick heated in a furnace at a field intensity of 16 $V{\cdot}mm^{-1}$. They found that the temperature runaway was more pronounced in samples with smaller surface-to-volume ratio. Todd et al. [30] showed that the main electrical and thermal characteristics of flash sintering of 3YSZ can be predicted accurately through the inverse Arrhenius dependence of resistivity and concluded that they are a classic consequence of the negative temperature coefficient of resistivity leading to runaway Joule heating at a constant voltage.

Narayan [31] explained electric field-induced phenomena, such as flash sintering, reduction in the flow stress, and hindered grain growth by a unified mechanism, which includes defect segregation at dislocations and grain boundaries and their selective Joule heating. The segregation effects in applied fields have been observed experimentally. Figure 5.27 shows precipitates with an average size of 10 nm near a sub-grain boundary of a MgO crystal treated in an electric field. The segregation effects were observed only in the samples treated in electric field; further annealing of the crystal without field results in the disappearance of the segregation effects.

Impurities with higher cationic valencies compared with the host atoms increase the electron trapping and generation of defects. The segregation of defects at grain boundaries causes reverse vacancy jumps and slows down the grain growth rate. The segregation effects increase the ionic and electronic transport along the grain boundaries. The AC field was found to be more effective in causing segregation compared with the DC field under similar parameters [31]. These effects in metals are less pronounced because of their higher thermal diffusivity and lower localization of heat compared with ceramics. At high electric fields, ionic and electronic conduction lead to an avalanche, where grain boundaries are selectively heated to melting. Flash sintering of ceramics can thus be understood as a consequence of selective melting of grain boundaries. The enhancement of conductivity with increasing temperature is an avalanche effect in an uncontrolled state and can lead to melting and evaporation of the material between the electrodes. In practice, the current flow is limited to control the process. Narayan argues that Joule heating of the grain boundaries and not of the whole volume of the material should be considered. While in the solid state Joule heating cannot account for high sintering rates observed in flash sintering [26], it can explain the observed heating rates in the presence of liquid phases with high diffusivities (10^{-4}–10^{-5} cm^2 s^{-1}).

Fast volumetric heating leading to flash sintering of a porous compact can also be realized by microwave heating. Flash sintering occurs when the processing conditions are such that facilitate the development of thermal runaway. Bykov et al. [32, 33] suggested that grain boundaries play a key role in the flash sintering process being the regions of high concentrations of defects and impurities. The localized preferential absorption of microwave radiation by the grain boundaries results in the grain-boundary softening or premelting. Thanks to softening, rapid densification of the granular medium having reduced viscosity-grain boundaries can occur via rotation and sliding of the grains, which accommodate their shape by means of fast mass transport through the liquid or quasi-liquid phase. It was shown that the heating rate (the absorbed power) greatly influences the microstructure (Fig. 5.28).

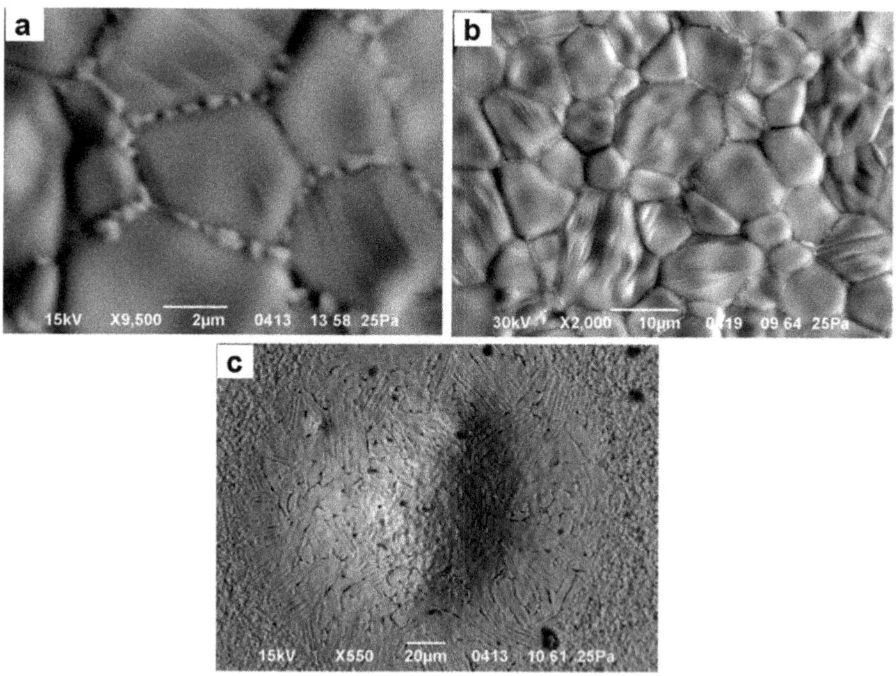

Fig. 5.28 Microstructure of the Yb:$(LaY)_2O_3$ ceramics flash sintered by microwave heating: (**a**) heating at 50 °C·min^{-1} up to 1500 °C, (**b**) heating at 100 °C·min^{-1} up to 1500 °C, (**c**) heating at 7000 °C·min^{-1} up to 1580 °C. (Reprinted from Bykov et al. [32], Copyright (2016) Bykov et al. This article is an open access article distributed under the terms and conditions of the Creative Commons Attribution License http://creativecommons.org/licenses/by/4.0/)

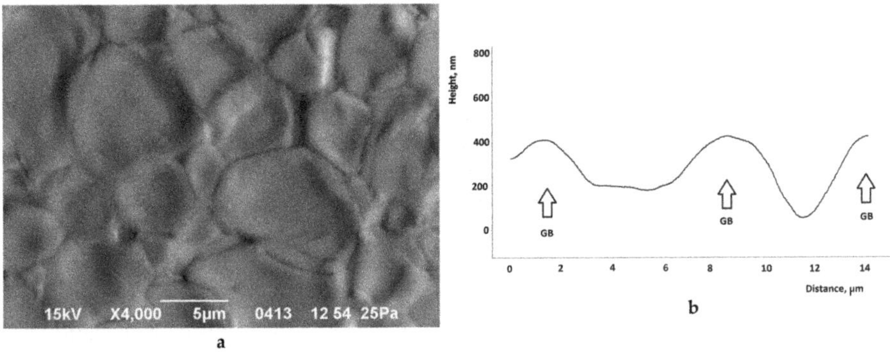

Fig. 5.29 A SEM image of the surface of the sintered Yb:$(LaY)_2O_3$ ceramics flash sintered by microwave heating at 200 °C·min^{-1} up to 1500 °C (**a**), atomic force microscopy profiles of the adjacent grains (**b**). (Reprinted from Bykov et al. [32], Copyright (2016) Bykov et al. This article is an open access article distributed under the terms and conditions of the Creative Commons Attribution License (http://creativecommons.org/licenses/by/4.0/)

At a heating rate of 50 °C·min^{-1}, the grain boundaries showed frozen droplets. At a higher heating rate (100 °C·min^{-1}), these droplets merged and formed grain-boundary layers, which showed a different contrast in comparison with the body of the grains in back-scattered electron microscopy images. At a much higher heating rate (7000 °C·min^{-1}), the regions that experienced melting were larger still being localized.

Bykov et al. [32] analyzed the structure of the specimen's surface and its profile using a combination of scanning electron microscopy and atomic force microscopy. As can be seen in Fig. 5.29, the relief of the sintered compact is uneven, with edges of the grains protruding from the surface. The protruding areas form due to a larger specific volume of the material in the molten state relative to that in the solid state. It was suggested that the molten part of the material fills the triple points of the granular structure and is partially extruded to the surface. The premelting of the particle surface can occur well below the melting point of the bulk material. The elemental analysis of the grains and the intergranular phase in flash-sintered Yb:$(LaY)_2O_3$ showed that the intergranular phase is La- and O-enriched and Y-depleted.

According to Chaim [34], the main problems that make it difficult to investigate the flash sintering mechanisms include the uncertainty of the real temperature at the contacts between the particles (which is difficult to measure, while estimations may vary by hundreds of degrees), the assumption of uniform sintering throughout the specimen volume, and the assumption of the solid-state nature of the flash events. Chaim proposed a model for rapid densification of flash-sintered ceramics [34]. In this model, a liquid film forms by local melting at the inter-particle contacts due to preferred Joule heating followed by thermal runaway. Local densification occurs via particle rearrangement enabled by spreading of the liquid due to local attractive capillary forces. The contacts melt in a random hierarchical manner. The rapid densification is aided by the local increase in the specific volume. The volume densification is closely connected with the percolation effect: at the percolation threshold, a sudden increase in the electrical conductivity and dissipated power is related to the percolation of current through the softened part of the sintered material or liquid [35]. The invasive nature of the melt at the contacts leads to the formation of an advancing front, which forms the percolative pass for the current flow (liquid "circuit"). In composite (multiphase systems), flash events can involve a particular phase. In this regard, study by Candelario et al. [36] should be mentioned, who used yttrium aluminum garnet $Y_3Al_5O_{12}$ (YAG) as a sintering additive in flash sintering of a SiC powder. The YAG additive melted and promoted liquid phase-assisted densification.

A promising approach to studying the flash sintering mechanisms is in situ diffractometry using synchrotron radiation, which allows gaining information on the fast structural changes occurring in the material during the flash sintering process. Reversible transformations (dependent on the presence of electric field) – the formation of new phases and texturing – have been reported [37, 38]. Figure 5.30 shows evidence of a reversible transformation in titania induced by the application of

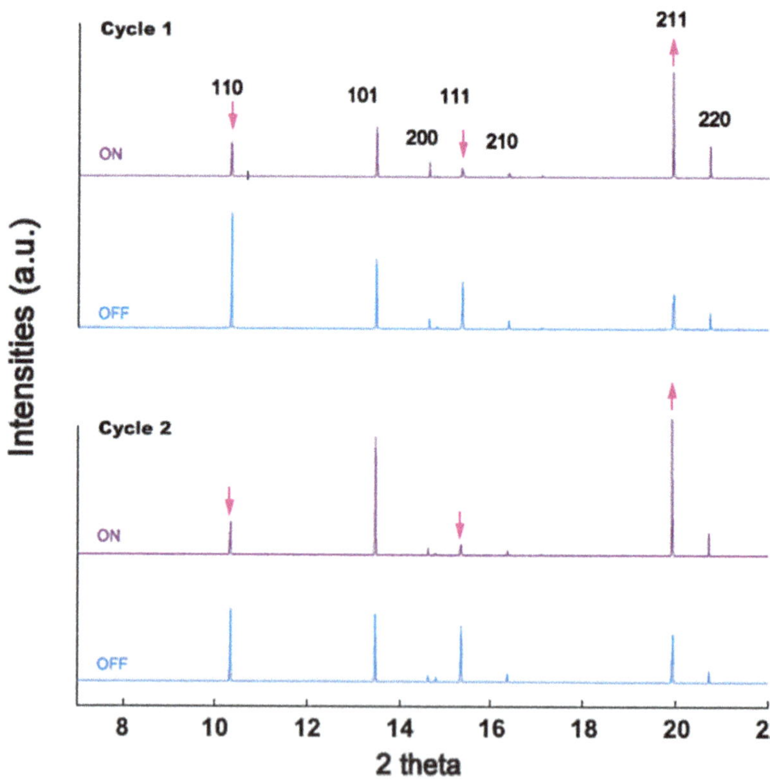

Fig. 5.30 X-ray diffraction patterns of titania compacts experiencing two cycles of a constant flash state under current control: the relative intensities of (211), (110), and (111) reflections change when the electric current is applied relative to the state without applied current. (Reprinted from Jha et al. [38], Copyright (2015) with permission from Elsevier)

electric current. The specimen temperature was measured to be 925 °C using a platinum standard [38]. It was proven that Joule heating was not the cause of the observed changes in the relative intensities of the peaks, as heating of the specimens up to 900–1000 °C without electric field did not lead to any such changes in the diffraction patterns.

McWilliams et al. [18] have reported an unusual phenomenon – the flash sintering behavior of Al alloy AA5083. Upon reaching a certain temperature, the compact consolidated from the alloy powder turned from an insulating state into a conducting state. It was shown that Joule heating alone cannot account for the observed flash sintering behavior. The authors suggested that the dielectric breakdown of the surface oxides present on the powder particles played a significant role in determining the ability of the powder compact to densify in the flash manner. Flash sintering behavior of metallic powders was also observed during processing of polymer/silver mixtures under antenna microwave illumination allowing fast agglomeration and consolidation of silver particles (microwave flash sintering) [39, 40].

An original metal powder flash sintering method using microwaves was elaborated in Ref. [41]. An unusually fast (60 s) thermal and sintering runaway of a Ti-6Al-4V powder was experimentally observed under microwave illumination. The experimental results were compared with the outcomes of the electromagnetic–thermal–mechanical simulation, which is able to predict the cavity microwave distribution, heating, sample densification, and shape distortions. The developed multiphysics model revealed that the specimen's runaway does not result from the intrinsic properties of the sintered material properties but is caused by the resonance phenomenon thermally activated by the surrounding tooling material. In the experiments, a partially densified compact (relative density 40%) was surrounded by a SiC nanopowder. The sample thermal runway was the result of the combination of multiple factors including the resonance phenomenon, the location of the sample in a high magnetic field area, and the confinement of the heat generated in the sample by the nano-SiC powder. The originality of this microwave flash sintering approach is in the presence of a metal thermal runaway that does not originate from the metal properties. The comparison of the microwave and conventional sintering kinetics indicated acceleration of the sintering behavior under microwave heating. The developed microwave flash sintering does not require preheating in a furnace or electric connections (the process is contactless) and can be used for metals and alloys.

Local Joule heating and melting of the inter-particle contacts was suggested as an explanation of a peculiar microstructure development of $Mg_{2.1}Si_{0.487}Sn_{0.5}Sb_{0.01}$ under conditions of flash SPS [42]. A high initial conductivity of the material allowed direct Joule heating of the pre-compacted specimen in an assembly without a die or conductive wrap by pulsed current generated in a SPS facility (Fig. 5.31). The microstructures of the specimen sintered by conventional SPS (sintering in a die with and without a pre-sintering stage) and that obtained by flash SPS conducted without a die differ from each other. Specimens sintered by the SPS in a die were heated at a rate of 100 °C·min^{-1} to 750 °C; the specimen consolidated from the powder directly was held at the maximum temperature for 1 min, while there was no

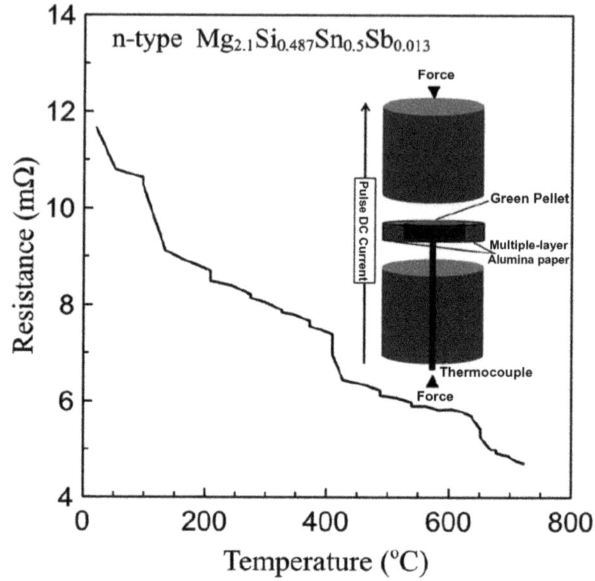

Fig. 5.31 Schematic of an assembly for flash sintering (no die or conductive wrap around the specimen is used, alumina paper covers the cylindrical surface of the specimen to reduce the radiation heat losses) and the resistance change of $Mg_{2.1}Si_{0.487}Sn_{0.5}Sb_{0.013}$ with the temperature during flash sintering under pulsed current in a SPS facility. (Reproduced from Du et al. [42]. This article is licensed under a Creative Commons Attribution 3.0 Unported Licence. https://creativecommons.org/licenses/by/3.0/)

holding at the maximum temperature in the experiment with the pre-sintered compact. The material sintered from the powder by conventional SPS showed no phase separation (Fig. 5.32a). In the material, processed twice by the conventional SPS (Fig. 5.32b), there was mild phase separation: solid solutions with different concentrations formed. The most interesting observation was made when the microstructure of the flash-sintered material was analyzed: a tin-enriched network formed through the microstructure, also engulfing magnesia (MgO) particles present on the surface of the agglomerates of the initial powder (Fig. 5.32c–d). Consequently, during flash sintering, changes in the microstructure, phase composition, and local chemical composition of the material can occur caused by local melting at the particle contacts.

Similar observations in terms of redistribution of elements in solid solutions were made by Corapcioglu et al. [43], who flash sintered $K_{0.5}Na_{0.5}NbO_3$ at 990 °C and 250 V·cm^{-1} for 30 s and found that the distributions of Na and K were not homogeneous. Rather, the distribution resembled a core–shell structure, in which the shell was K-rich and the core was Na-rich. The powder used for sintering did not show any core–shell compositional features. The compositional variations between the shell and the core were explained by grain-boundary melting during flash sintering followed by recrystallization upon cooling. When a flash-sintered sample was heat treated at 1000 °C for 4 h, the core–shell morphology disappeared. So, conventional heat treatment made potassium ions diffuse into the core and sodium ions diffuse from the core to the shell so that the equilibrium stoichiometry of $K_{0.5}Na_{0.5}NbO_3$ was restored.

For porous pre-sintered α-Al_2O_3 subjected to flash sintering at field strengths of 500–1500 V·cm^{-1}, analogies between flash sintering and dielectric breakdown were

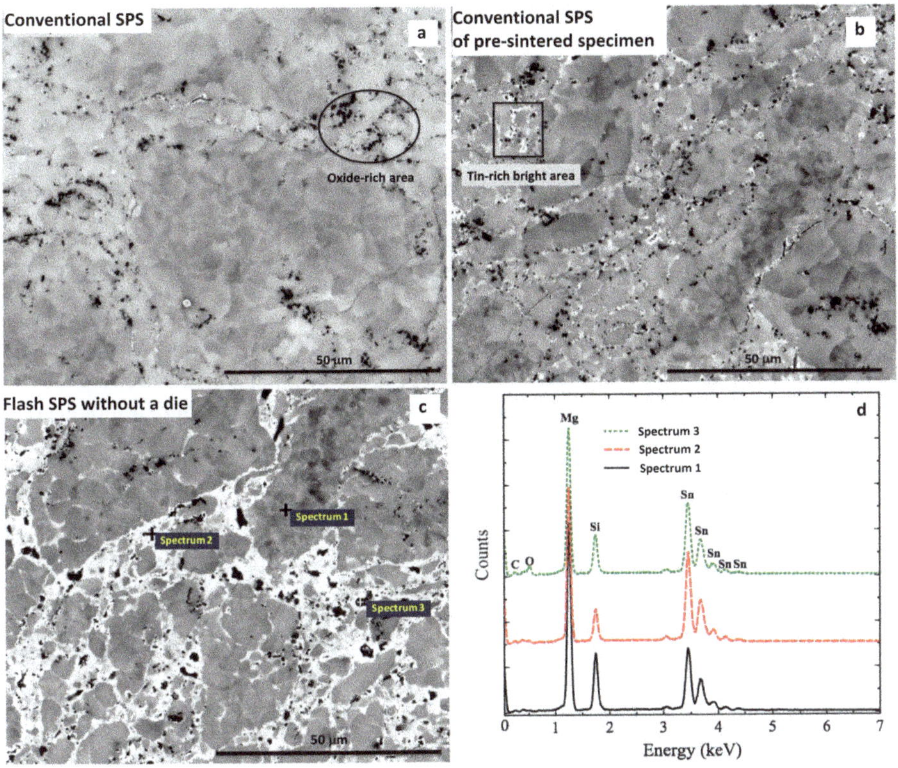

Fig. 5.32 Microstructure of the sintered $Mg_{2.1}Si_{0.487}Sn_{0.5}Sb_{0.013}$ (**a–c**) and spectra (**d**) taken from different points of (**c**); (**a**) conventional SPS from the powder, 750 °C, (**b**) conventional SPS of a pre-sintered compact, 750 °C, (**c**) flash SPS without a die, heating up to 700 °C at 1000 °C·min^{-1}. (Reproduced from Du et al. [42], This article is licensed under a Creative Commons Attribution 3.0 Unported Licence. https://creativecommons.org/licenses/by/3.0/)

found [44]. The authors suggest that elevated temperatures and sample porosity lead to a decrease in the breakdown strength down to values comparable to those at which flash events were observed. The processes of flash sintering and dielectric breakdown were found to be similar in terms of runaway and conductivity evolution. The incubation for flash sintering was compared to the pre-breakdown phenomenon. The relation between the incubation time and the pre-sintering temperature (porosity of the pre-sintered compact) showed that samples with a higher porosity require shorter incubation times. The electrical conductivity of the fully dense samples did not depend on the field strength, while that of the porous samples increased with increasing field strength. Fully dense α-Al_2O_3 materials did not show a flash behavior.

5.3 Materials Densified by Flash Sintering

In SPS, the sintering times are quite short – the total duration of the process is usually from several minutes to several tens of minutes. It is technologically attractive to further shorten the sintering process and reduce the sintering times down to several seconds and increase the energy efficiency of sintering. The greatest limitation of the initially proposed flash sintering approaches is the lack of applicability of flash sintering to a wide range of materials. This limitation has been overcome by Manière et al. [45], who combined SPS and a flash sintering approach and developed a flash (ultra-rapid) spark plasma sintering method (net-shape flash SPS – NSFSPS) applicable to various materials regardless of their electrical resistivity. Powders of metals and insulating ceramics were consolidated to produce compacts with a homogeneous microstructure within sintering times of 8–35 s. Extraordinary fast and homogeneous heating was concentrated in the sample's volume and punches. The controllable flash phenomenon was enabled by the combination of the electric current concentration around the sample and the confinement of the heat generated in this area by the lateral thermal contact resistance (TCR). The NSFSPS approach uses a graphite die electrically insulated by a sprayed boron nitride (BN) layer to concentrate the electric current into the sample, if electrically conductive, or in the nearby graphite foil, if electrically insulating (Fig. 5.33). The graphite die allows an easy control of

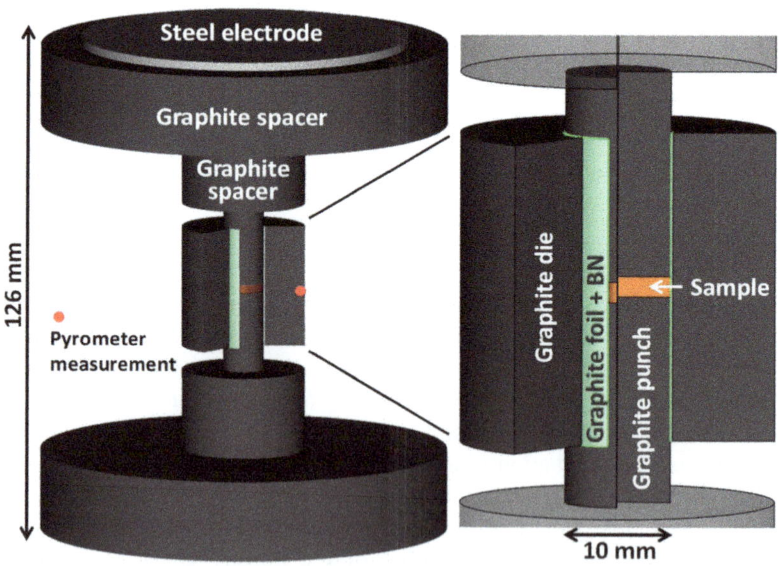

Fig. 5.33 Net-shape flash spark plasma sintering configuration. In order to concentrate the electric current on the sample area, the inner die surface is coated by an electrically insulating boron nitride (BN) spray. (Reprinted from Manière et al. [45], Copyright (2017), Rights managed by Nature Publishing Group. This work is licensed under a Creative Commons Attribution License, http://creativecommons.org/licenses/by/4.0/)

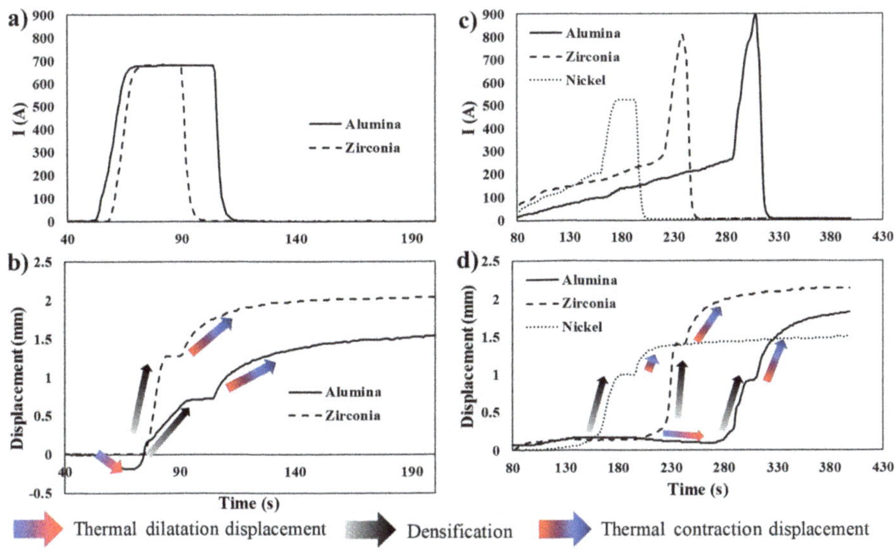

Fig. 5.34 Experimental electric current (**a**, **c**) and displacement (**b**, **d**) curves under constant current mode (**a–b**) and forced mode (**c–d**). (Reprinted from Manière et al. [45], Copyright (2017), Rights managed by Nature Publishing Group. This work is licensed under a Creative Commons Attribution License, http://creativecommons.org/licenses/by/4.0/)

the final shape of the specimen and makes the overall pressure-assisted process more stable. The NSFSPS approach uses graphite tools for high-temperature applications and good thermal shock resistance. This approach is related to the current concentration method described in Ref. [46] but uses electric current patterns different from those of Ref. [46]. Using the NSFSPS approach, nickel, alumina, and yttria-stabilized zirconia powders were sintered in the flash mode [45].

The current cycle was imposed manually in two patterns: constant current mode and forced current mode (Fig. 5.34). In both the "constant current mode" and the "forced mode," the sintering time is shorter than 60 s. An electric current cycle with a forced runaway was realized by manually increasing the current at 70 $A \cdot min^{-1}$ up to the beginning of densification and at 2500 $A \cdot min^{-1}$ further up until the displacement plateau indicating the end of densification was reached. For nickel and alumina, the thermal runaway that appears in traditional flash sintering dealing with materials whose electrical conductivity increases with temperature was artificially imposed by the electric current pattern that reproduced a similar sample's thermal response.

Samples, the fracture surfaces of which are shown in Fig. 5.35, were nearly dense with a residual porosity between 1 and 3% for the ceramic samples and 5% for nickel. For zirconia and alumina, the grain size increased from 37–100 nm to 20 μm in about 10 s. This suggests that fast densification was accompanied by unusually fast grain growth. The comparison of the center/edge images reveals no obvious microstructural differences indicating a very homogeneous process. However, a

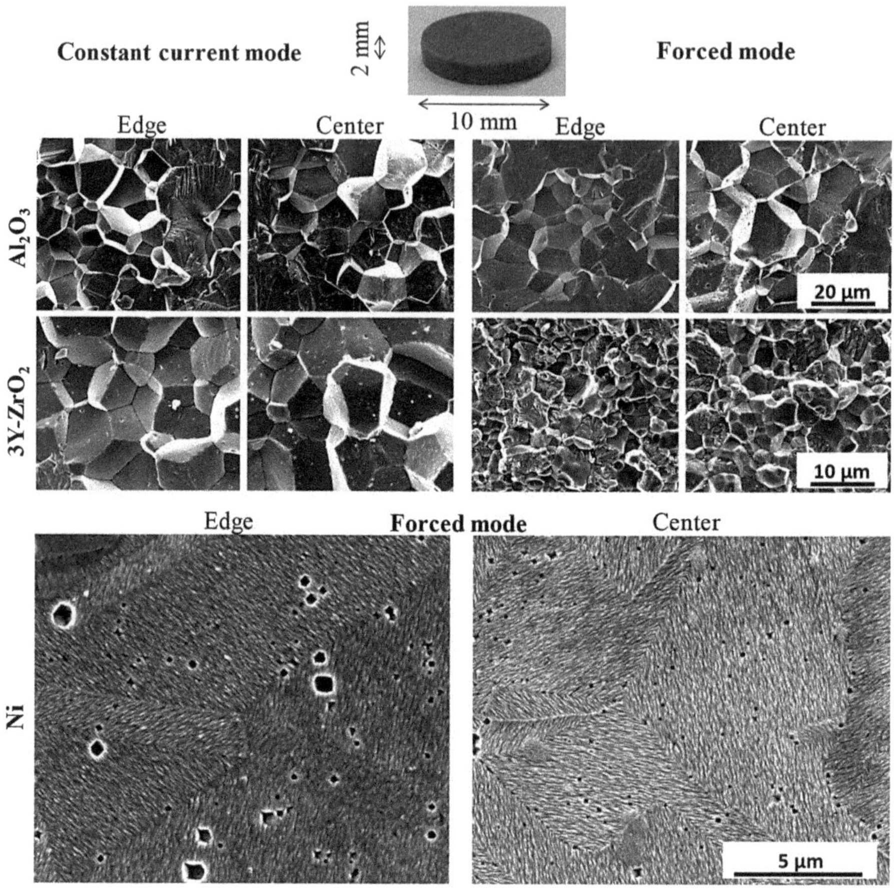

Fig. 5.35 SEM images in the centers and edges of nickel, zirconia, and alumina samples for constant and forced current modes. (Reprinted from Manière et al. [45], Copyright (2017), Rights managed by Nature Publishing Group. This work is licensed under a Creative Commons Attribution License, http://creativecommons.org/licenses/by/4.0/)

noticeable microstructure difference can be seen between the "constant current" mode and the "forced" mode in the zirconia samples. Zirconia becomes electrically conductive at high temperatures and seems to generate some heating instabilities (internal sample temperature runaway and overheating in the sample's area) which cause fast grain growth under the "constant current" mode. In comparison, the alumina sample that possesses a low electrical conductivity at low and high temperatures exhibits nearly no microstructure differences between the two modes. It appears that materials, which are normally impossible to flash sinter using conventional flash methods, are the most stable materials when processed by the NSFSPS method. The TCR that decreases the heat flowing into the lateral punch/die interface is the main factor explaining the high-temperature concentration inside the NSFSPS setup's central column independently of the sample material nature. The heat generated by the central column is contained in the sample, which helps reducing the temperature gradients within the sample. The higher the TCR is, the more homogeneous are the sample temperatures and the more homogeneous are the respective microstructures. The thermal confinement and the short sintering time are favorable conditions to ensure the microstructure homogeneity observed experimentally. The thermal confinement results also from the lateral graphite foil heating. For ceramics samples, a significant amount of the electric current is constrained in the sample/die graphite foil. Under these conditions, the lateral graphite foil can be compared to a susceptor that heats the sample from the edge and considerably stabilizes the sample temperature homogeneity. The forced thermal runaway generates heating rates of 1700 K min^{-1} for nickel and 4300 K min^{-1} for alumina. For zirconia, the transition regime occurs at a sample temperature of 1250 °C, where the heating rate passes from 3400 K min^{-1} to 12000 K min^{-1}. This transition is a result of an intrinsic thermal runaway that is added to the imposed electric current amplification. The differences between NSFSPS and traditional flash sintering should be mentioned. During NSFSPS, the applied pressure accelerates and stabilizes the overall sintering process. A die is used for a better control of the final sample shape of the sample. The NSFSPS method allows extending flash sintering to nearly all materials, controlling sample shape by an added graphite die, and an energy efficient mass production of small and intermediate size objects.

One of the key problems in the processing of ceramics is the requirement of high temperatures needed for sintering and densification. High-temperature sintering equipment is costly, while the processing requires inputs of large amounts of energy. The main advantage of flash sintering is a dramatic reduction of the processing time and temperature, as was illustrated by many examples. Flash sintering has been proven successful for densification of $K_{0.5}Na_{0.5}NbO_3$, a promising lead-free ferroelectric material [43]. A flash sintering process 30 s long at 990 °C was shown to produce a material with a high relative density (94%). In order to achieve the same relative density by conventional sintering, the process should be carried out at 1100 °C for 2 h. Flash sintering at a low furnace temperature of less than 100 °C was reported for $La_{0.6}Sr_{0.4}Co_{0.2}Fe_{0.8}O_3$ under electric fields of 7.5–12.5 V·cm^{-1} [47] – an extraordinary effect caused by a high electrical conductivity of the material.

Materials that have been processed by flash sintering in different laboratories have been classified by Yu et al. [1] using the conductivity mode – ionic conductors, metallic-like conductors, semiconductors, and insulators. Yu et al. [1] see the reason for a limited number of studies of the properties of flash-sintered materials in the small size of the samples dictated by the requirements of the method (application of field and placing samples into a furnace). At the same time, the potential of flash sintering has been proven for different structural and functional materials: lightweight ceramics, materials for solid-state electrochemical devices, ferroelectric materials, and dielectrics, as reviewed by Yu et al. [1]. Recent studies have shown that thermoelectric materials can also be produced by flash sintering: Du et al. [42] found that by switching from conventional SPS in a die to flash SPS without a die, it is possible to produce magnesium silicide stannide materials with altered electrical

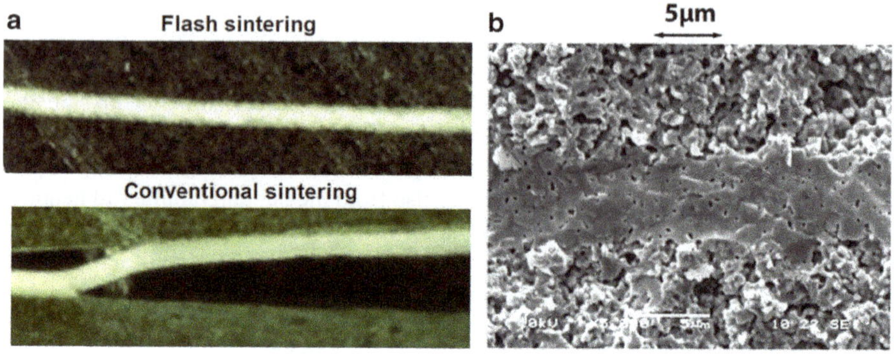

Fig. 5.36 Multilayers NiO–ZrO_2/cubic ZrO_2 sintered by flash and conventional sintering, the bright band in the middle is the electrolyte layer (**a**), the structures of the anode and electrolyte layers sintered and joined by flash sintering (**b**). (Reprinted from Francis et al. [48], Copyright (2013) with permission of John Wiley & Sons)

and thermal conductivities due to local melting-induced phase and structural changes.

Layered structures composed of an anode material (NiO-zirconia) and cubic zirconia electrolyte – structures for solid oxide fuel cell applications – were successfully flash sintered at 915 °C and 150 V·cm^{-1} using as assembly, in which the anode and the electrolyte layers were connected electrically in parallel [48]. The key problem in manufacturing of oxide fuel cells is sintering of the ceramic anode, electrolyte, and cathode layers, as sintering of one layer constrains sintering of the other. Co-sintered layers usually suffer from delamination and defects due to different sintering rates of the layers. In Ref. [48], the flash sintering temperature of the multilayer structure was lower than the temperature at which the anode material sinters when processed separately (1005 °C). The electrolyte as a separate layer sinters at 750 °C. The electrolyte had a high relative density and showed only closed pores, while the anode layer was porous, which is required for solid oxide fuel cells. The flash-sintered multilayers were free from defects and did not show delamination, unlike conventionally sintered multilayers (Fig. 5.36). The experimental results suggested that the behavior of the multilayers during flash sintering was a result of the interaction between the layers, and it was not possible to predict the flash sintering outcome by simply considering the behavior of the individual layers. As flash sintering helped alleviate the problems associated with constrained sintering, it was concluded to be a promising method for fabricating layered structures, such as those required for manufacturing of solid oxide fuel cells.

Studies focusing on comparison of the properties of materials of the same composition produced by sintering in different modes are valuable from the fundamental and practical viewpoints, as they shed light on the mechanisms a certain property is achieved under varied sintering conditions and allow formulating recommendations on the choice of the sintering modes for the material fabrication. Jesus et al. [49] reported the dependences of the grain size and the dielectric constant

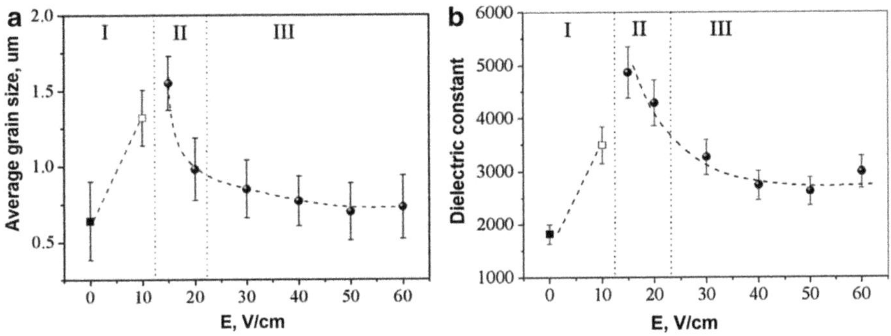

Fig. 5.37 Grain size (**a**) and dielectric constant measured at room temperature at 1 MHz (**b**) of $CaCu_3Ti_4O_{12}$ sintered using different field strength, which determined the sintering mode: conventional-like (region I), field-assisted (region II), flash-dominated (region III). (Reprinted from Jesus et al. [49], Copyright (2016) with permission from Elsevier)

of the sintered $CaCu_3Ti_4O_{12}$ on the field strength, which determined the sintering mode: conventional-like (region I), field-assisted (region II), flash-dominated (region III), as is shown in Fig. 5.37. The specimens were cold-pressed to a density of 55%, suspended on platinum wires, which served as field and current carrying electrodes, and placed into a furnace. The grain size correlated with the dielectric constant showing that under the selected measurement conditions, the dielectric constant of the sintered materials is modulated by the grain-boundary response.

An interesting result of flash sintering was reported in Ref. [50] for the phase transformation in the 3 mol.%Y_2O_3-ZrO_2 system. The powder heat treated at 600 °C contained monoclinic zirconia; conventional sintering at 1100 °C for 1 h led to a partial phase transformation into the tetragonal phase (Fig. 5.38a–b), which showed that prolonged annealing is needed to achieve the formation of a single-phase product from the 3 mol.%Y_2O_3-ZrO_2 mixture. Unlike conventional sintering, flash sintering was very efficient in transforming the mixture into single-phase tetragonal zirconia (Fig. 5.38c–d). However, the effect of electric field during flash sintering on the phase transformation rate remains unclear. Both the bulk and intrinsic grain-boundary conductivity increased with increasing holding time during flash sintering. The estimated grain-boundary thickness δ_{gb} and intrinsic (specific) conductivity σ_{gb}^{sp} of the grain boundary ($\sigma_{gb} = \frac{D_g}{\delta_{gb}}\sigma_{gb}^{sp}$, where D_g is the grain size) are shown in Fig. 5.39. Based on these results, it was concluded that flash sintering created oxygen vacancies in both grains and grain boundaries that accumulated during the holding time.

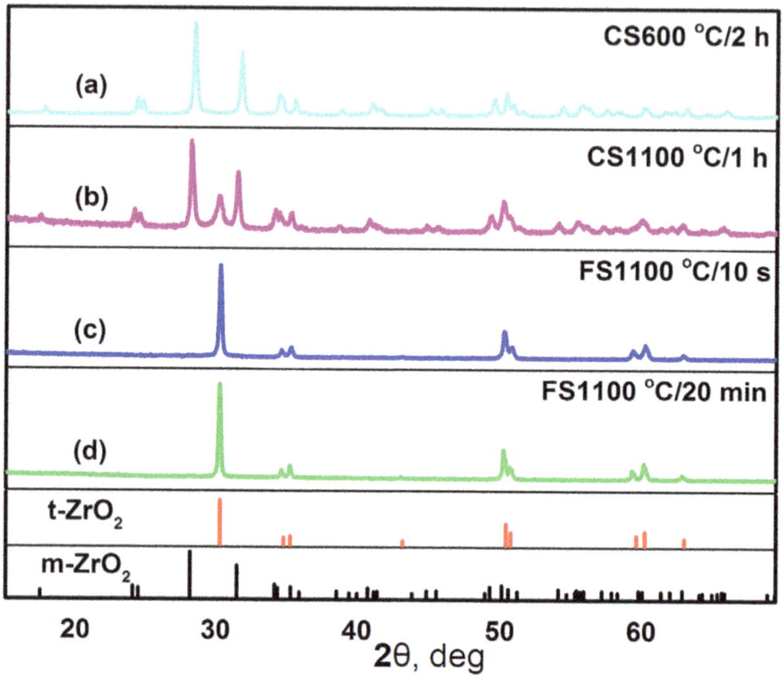

Fig. 5.38 XRD patterns of the 3 mol.%Y_2O_3-ZrO_2 powder heat treated at 600 °C for 2 h (**a**), compact pre-sintered at 1100 °C for 1 h (**b**) and compacts flash sintered at 1100 °C with holding times of 10 s (**c**) and 20 min (**d**). (Reprinted from Liu et al. [50], Copyright (2016) with permission of Elsevier)

5.4 Summary

Flash sintering is a fascinating physical phenomenon, which also offers significant energy saving and shortening of processing times. "Traditional" flash sintering occurs when an electrical potential is applied to a pre-compacted specimen heated in a furnace. The characteristic field strength and power dissipation values in flash sintering are 100–100 V·cm^{-1} and 10–1000 W·cm^{-3}, respectively. "Traditional" flash sintering is accompanied by a sudden increase in the conductivity of the sintered material, while the temperature instability plays a crucial role in the development of flash sintering. Flash sintering can be initiated by arc plasma and microwave radiation. It is now recognized that all types of materials can be flash sintered regardless of their electrical conductivity evolution with temperature if thermal runaway is forced in the sample through tailoring the electric current pattern.

Although possibilities of fast densification of pre-consolidated compacts via flash sintering have been reported for a wide range of materials, structural analyses of the densified materials have been followed by the detailed property testing only in a limited number of studies. The question of specific properties of materials whose

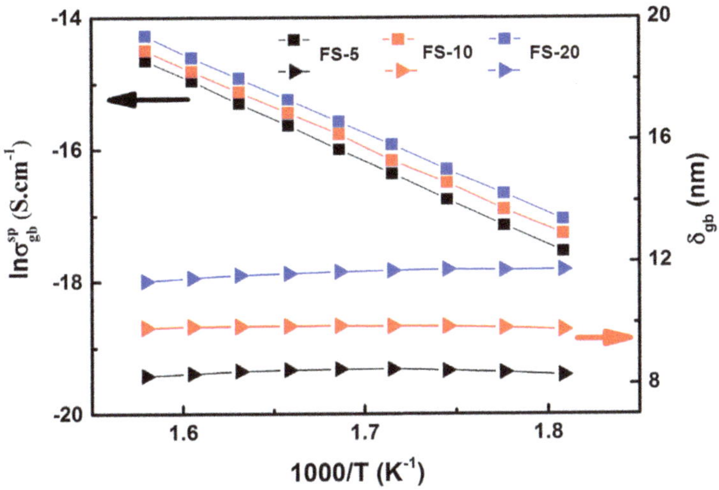

Fig. 5.39 Estimated grain-boundary thickness and intrinsic conductivity of the grain boundary as a function of temperature for flash-sintered 3 mol.% Y_2O_3-ZrO_2 (FS-5, holding time 5 min; FS-10, holding time 10 min; FS-20, holding time 20 min). (Reprinted from Liu et al. [50], Copyright (2016) with permission of Elsevier)

grain boundaries experienced melting as a result of flash sintering is extremely important from the fundamental point of view of the field-induced materials transformations and from a technological perspective of producing materials with advanced properties by energy- and time-saving routes. Until now, most of the flash sintering studies focused on the single-phase materials. At the same time, the role of additives in the processes has been demonstrated, and the behavior of layered structure under flash sintering has been shown to be better compared with conventional sintering in terms of the absence of delamination. For the future research, multiphase materials present interesting and challenging objects to apply the flash sintering approach to, as the phases will most probably differ in electrical conductivity and the character of its evolution with temperature.

By and large, flash sintering emerges as a remarkable way to consolidate powder materials within extremely short periods of time. While the overall process duration may vary due to the presence of the heating and cooling parts of the sintering cycle, the actual time of densification in many cases is less than several seconds. The ability to subject materials to flash sintering has already been demonstrated for a broad variety of materials and for various field-assisted sintering techniques, including the most traditional free pressureless version of flash sintering assisted by an applied DC voltage, flash spark plasma sintering, and flash microwave sintering. While most of flash sintering experiments have been conducted for ceramic materials, it was shown that metals can be flash sintered too, when the "flash" is triggered initially by a ceramic tooling.

From the fundamental research perspective, the underlying physical mechanisms of flash sintering are currently actively debated. While the most active discussion is related to the acceptance or rejection of the thermal runaway phenomena during flash sintering, less attention is paid to the nature of ultra-rapid densification itself. Indeed, it is still not clear, whether the achievement of very high temperature can explain almost instantaneous densification of many material systems.

From the applied research perspective, the main issues are the stability of flash sintering and the related questions of its scalability. Thus, flash sintering is one of the brightest examples of the importance of *the control of non-equilibrium*, which is a major issue for most field-assisted sintering techniques.

References

1. Yu M, Grasso S, Mckinnon R, Saunders T, Reece MJ (2017) Review of flash sintering: materials, mechanisms and modeling. Adv Appl Ceram 116:24–60
2. Cologna M, Rashkova B, Raj R (2010) Flash sintering of nanograin zirconia in < 5 s at 850 degrees C. J Am Ceram Soc 93:3556–3559
3. Francis JSC, Cologna M, Raj R (2012) Particle size effects in flash sintering. J Eur Ceram Soc 32:3129–3136
4. Raj R (2012) Joule heating during flash-sintering. J Eur Ceram Soc 32:2293–2301
5. Saini KK, Sharma CP, Chanderkant SDK, Ravat KB, Chandra S, Tewari SP (1993) Effect of thallium concentration on Tl-2201 superconducting phase by flash sintering process. Physica C 216:59–65
6. Muccillo R, Muccillo ENS (2013) An experimental setup for shrinkage evaluation during electric field-assisted flash sintering: application to yttria-stabilized zirconia. J Eur Ceram Soc 33:515–520
7. Hao X, Liu Y, Wang Z, Qiao J, Sun K (2012) A novel sintering method to obtain fully dense gadolinia doped ceria by applying a direct current. J Power Sources 210:86–91
8. Muccillo R, Kleitz M, Muccillo ENS (2012) Flash grain welding in yttria stabilized zirconia. J Eur Ceram Soc 31:1517–1521
9. Cologna M, Prette ALG, Raj R (2011) Flash-sintering of cubic yttria-stabilized zirconia at 750 degrees C for possible use in SOFC manufacturing. J Am Ceram Soc 94:316–319
10. Prette ALG, Cologna M, Sglavo V, Raj R (2011) Flash-sintering of Co_2MnO_4 spinel for solid oxide fuel cell applications. J Power Sources 196:2061–2065
11. Cologna M, Francis JSC, Raj R (2011) Field assisted and flash sintering of alumina and its relationship to conductivity and MgO-doping. J Eur Ceram Soc 31:2827–2837
12. Zapata-Solvas E, Bonilla S, Wilshaw PR, Todd RI (2013) Preliminary investigation of flash sintering of SiC. J Eur Ceram Soc 33:2811–2816
13. Saunders T, Grasso S, Reece MJ (2016) Ultrafast-contactless flash sintering using plasma electrodes. Sci Rep 6:27222
14. Grasso S, Saunders T, Porwal H, Cedillos-Barraza O, Jayaseelan DD, Lee WE, Reece MJ (2014) Flash spark plasma sintering (FSPS) of pure ZrB_2. J Am Ceram Soc 97(8):2405–2408
15. Olevsky EA, Rolfing SM, Maximenko AL (2016) Flash (ultra-rapid) spark-plasma sintering of silicon carbide. Sci Rep 6:33408
16. Grasso S, Saunders T, Porwal H, Milsom B, Tudball A, Reece M (2016) Flash spark plasma sintering (FSPS) of α and β SiC. J Am Ceram Soc 99(5):1534–1543
17. Grasso S, Kim EY, Saunders T, Yu M, Tudball A, Choi SH, Reece M (2016) Ultra-rapid crystal growth of textured SiC using flash spark plasma sintering route. Cryst Growth Des 16:2317–2321
18. McWilliams B, Yu J, Kellogg F, Kilczewski S (2017) Enhanced sintering kinetics in aluminum alloy powder consolidated using DC electric fields. Metall Mater Trans A 48:919–929

19. Zhang Y, Luo J (2015) Promoting the flash sintering of ZnO in reduced atmospheres to achieve nearly full densities at furnace temperatures of <120 deg C. Scr Mater 106:26–29
20. Hewitt IJ, Lacey AA, Todd RI (2015) A mathematical model for flash sintering. Math Model Nat Phenom 10(6):77–89
21. Dong Y, Chen IW (2016) Thermal runaway in mold-assisted flash sintering. J Am Ceram Soc 99(9):2889–2894
22. Dong Y, Chen IW (2015) Predicting the onset of flash sintering. J Am Ceram Soc 98 (8):2333–2335
23. Dong Y, Chen IW (2015) Onset criterion for flash sintering. J Am Ceram Soc 98 (12):3624–3627
24. Downs JA, Sglavo VM (2013) Electric field assisted sintering of cubic zirconia at 390°C. J Am Ceram Soc 96(5):1342–1344
25. Francis JSC, Raj R (2012) Flash-sinterforging of nanograin zirconia: field assisted sintering and superplasticity. J Am Ceram Soc 95:138–146
26. Raj R, Cologna M, Francis JSC (2011) Influence of externally imposed and internally generated electrical fields on grain growth, diffusional creep, sintering and related phenomena in ceramics. J Am Ceram Soc 94:1941–1965
27. Karakuscu A, Cologna M, Yarotski D, Won J, Francis JSC, Raj R, Uberuaga BP (2012) Defect structure of flash-sintered strontium titanate. J Am Ceram Soc 95:2531–2536
28. Grasso S, Sakka Y, Rendtorff N, Hu CF, Maizza G, Borodianska H, Vasylkiv O (2011) Modeling of the temperature distribution of flash sintered zirconia. J Ceram Soc Jap 119:144–146
29. Park J, Chen IW (2013) In situ thermometry measuring temperature flashes exceeding 1,700°C in 8 Mol% Y_2O_3-stablized zirconia under constant-voltage heating. J Am Ceram Soc 96:697–700
30. Todd RI, Zapata-Solvas E, Bonilla RS, Sneddon T, Wilshaw PR (2015) Electrical characteristics of flash sintering: thermal runaway of joule heating. J Eur Ceram Soc 35:1865–1877
31. Narayan J (2013) A new mechanism for field-assisted processing and flash sintering of materials. Scr Mater 69:107–111
32. Bykov YV, Egorov SV, Eremeev AG, Kholoptsev VV, Plotnikov IV, Rybakov KI, Sorokin AA (2016) On the mechanism of microwave flash sintering of ceramics. Materials 9:684
33. Bykov YV, Egorov SV, Eremeev AG, Kholoptsev VV, Rybakov KI, Sorokin AA (2015) Flash microwave sintering of transparent Yb:$(LaY)_2O_3$ ceramics. J Am Ceram Soc 98(11):3518–3524
34. Chaim R (2016) Liquid film capillary mechanism for densification of ceramic powders during flash sintering. Materials 9:280
35. Chaim R (2017) Particle surface softening as universal behavior during flash sintering of oxide nano-powders. Materials 10:179
36. Candelario VM, Moreno R, Todd RI, Ortiz AL (2017) Liquid-phase assisted flash sintering of SiC from powder mixtures prepared by aqueous colloidal processing. J Eur Ceram Soc 37:485–498
37. Lebrun JM, Morrissey TG, Francis JSC, Seymour KC, Kriven WM, Raj R (2015) Emergence and extinction of a new phase during on–off experiments related to flash sintering of 3YSZ. J Am Ceram Soc 98:1493–1497
38. Jha SK, Lebrun JM, Seymour KC, Kriven WM, Raj R (2016) Electric field induced texture in titania during experiments related to flash sintering. J Eur Ceram Soc 36(1):257–261
39. Perelaer J, Klokkenburg M, Hendriks CE, Schubert US (2009) Microwave flash sintering of inkjet-printed silver tracks on polymer substrates. Adv Mater 21:4830–4834
40. Perelaer J, Jani R, Grouchko M, Kamyshny A, Magdassi S, Schubert US (2012) Plasma and microwave flash sintering of a tailored silver nanoparticle ink, yielding 60% bulk conductivity on cost-effective polymer foils. Adv Mater 24:3993–3998
41. Manière C, Lee G, Zahrah T, Olevsky EA (2018) Microwave flash sintering of metal powders: from experimental evidence to multiphysics simulation. Acta Mater 147:24–34

42. Du B, Gucci F, Porwal H, Grasso S, Mahajan A, Reece MJ (2017) Flash spark plasma sintering of magnesium silicide stannide with improved thermoelectric properties. J Mater Chem C 5:1514–1521
43. Corapcioglu G, Ali Gulgun M, Kisslinger K, Sturm S, Jha SK, Raj R (2016) Microstructure and microchemistry of flash sintered $K_{0.5}Na_{0.5}NbO_3$. J Ceram Soc Japan 124(4):321–328
44. Biesuz M, Luchi P, Quaranta A, Sglavo VM (2016) Theoretical and phenomenological analogies between flash sintering and dielectric breakdown in a-alumina. J Appl Phys 120:145107
45. Manière C, Lee G, Olevsky EA (2017) All-materials-inclusive flash spark plasma sintering. Sci Rep 7:15071
46. Zapata-Solvas E, Gómez-García D, Domínguez-Rodríguez A, Todd RI (2015) Ultra-fast and energy-efficient sintering of ceramics by electric current concentration. Sci Rep 5:8513
47. Gaur A, Sglavo VM (2014) Densification of $La_{0.6}Sr_{0.4}Co_{0.2}Fe_{0.8}O_3$ ceramic by flash sintering at temperature less than 100 °C. J Mater Sci 49:6321–6332
48. Francis JSC, Cologna M, Montinaro D, Raj R (2013) Flash sintering of anode–electrolyte multilayers for SOFC applications. J Am Ceram Soc 96(5):1352–1354
49. Jesus LM, Santos Silva R, Raj R, M'Peko JC (2016) Electric field-assisted flash sintering of $CaCu_3Ti_4O_{12}$: microstructure characteristics and dielectric properties. J Alloys Comp 682:753–758
50. Liu D, Gao Y, Liu J, Wang Y, An L (2016) Effect of holding time on the microstructure and properties of flash sintered Y_2O_3-doped ZrO_2. Ceram Int 42:17442–17446

Chapter 6
Sintering in the Constant Electric Field in the Noncontact Mode and in Magnetic Field

6.1 Sintering in the Constant Electric Field in the Noncontact Mode

A number of reports exist in the literature on the sintering enhancement by an applied field that is not associated with the passage of current through the compact. The specifics of the mechanisms of these types of sintering should apparently include the field-assisted acceleration of diffusion and, possibly, the double-layer effect influencing the specific surface energy [1–3].

Holland et al. [1] suggest that sintering schemes involving electric field in the noncontact mode hold great promise for the future of field-assisted sintering by dramatically reducing processing costs. Newman [2] investigated the effect of electric field strengths up to 7.7 kV cm^{-1} on the porosity of a steel compact using a set-up shown in Fig. 6.1. The following consideration was taken into account: the application of an electric field during sintering, which makes the compact a positive electrode of a capacitance circuit, produces a positive charge on the surface of the compact. Since the vacancies in a metal are negatively charged, the chemical potential of the vacancies at the surface may be lower than in the subsurface layer. As a result, diffusion of vacancies from the pores into the subsurface layer of the compact will occur improving the sintering process. A reduction in the porosity of the 0.2–0.4 mm layer below the surface of a Fe-based alloy compact by as much as 44% was observed. The reduced porosity was confirmed by nitro-carburizing experiments in which carbon diffusion was limited to a depth of only 0.2 mm below the surface. Specimens sintered without an electric field showed carbon diffusion depths over ten times the depth observed in specimens sintered in the applied electric field. The observed decrease in the surface porosity was attributed to a decrease in the chemical potential of vacancies at the charged external surface. The application of this effect in commercial processes is the surface porosity reduction crucial for the corrosion resistance control of metallic parts and metal hardening technologies based

E. A. Olevsky, D. V. Dudina, *Field-Assisted Sintering*,
https://doi.org/10.1007/978-3-319-76032-2_6

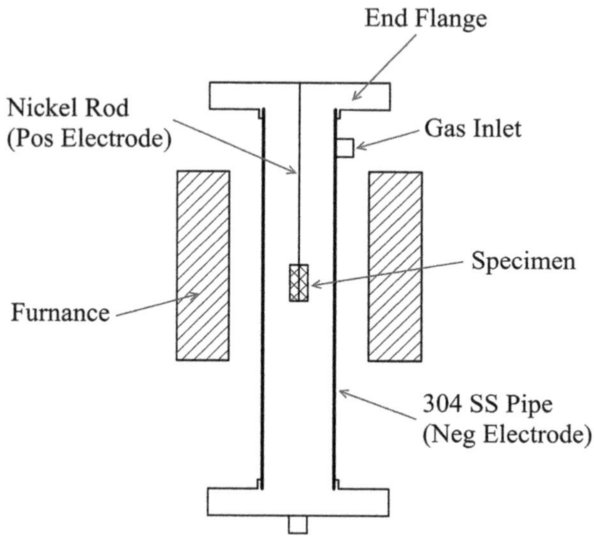

Fig. 6.1 Schematic of a set-up for sintering in the constant electric field (Reprinted from Newman [2], Copyright (2000) with permission from Elsevier)

on surface carburization. The surface modification has to involve a layer of a certain thickness only, leaving the material located deeper non-carburized and ductile.

Holland et al. [1] found that the electric field in the noncontact mode can benefit densification of dielectrics during slow heating, even if it is applied only up to temperatures much lower than those at which any significant diffusion in the sintered material can be expected (Fig. 6.2). A suggestion has been put forward that the electric field can facilitate the removal of surface contaminants from the nanosized powders, such as water and carbon dioxide, which evolve at 400–500 °C upon heating – at temperatures too low for any diffusion activity in the sintered powder material to start. It was pointed out that a great advantage of the noncontact mode is an opportunity of the application of electric fields together with different mechanical loading schemes, not just together with uniaxial pressing as in the commonly accepted scheme of combining pressing and passage of current.

6.2 Sintering in the Constant and Pulsed Magnetic Fields

The conditions of the growth and dissolution of pores in a solid placed in a magnetic field were analyzed by Kornyushin [3]. The magnetic lines around a pore are distorted. If a pore in a one-domain spherical magnetic solid elongates along the direction of the magnetic moment maintaining its volume, the distortion of the magnetic lines will be reduced and so will be the magnetic energy of the system. However, the energy of the system will increase due to an increase in the surface area of the elongated pore. For magnetic materials sintered in a magnetic field, particular pore-size distributions and pore orientations can be expected. These effects will be

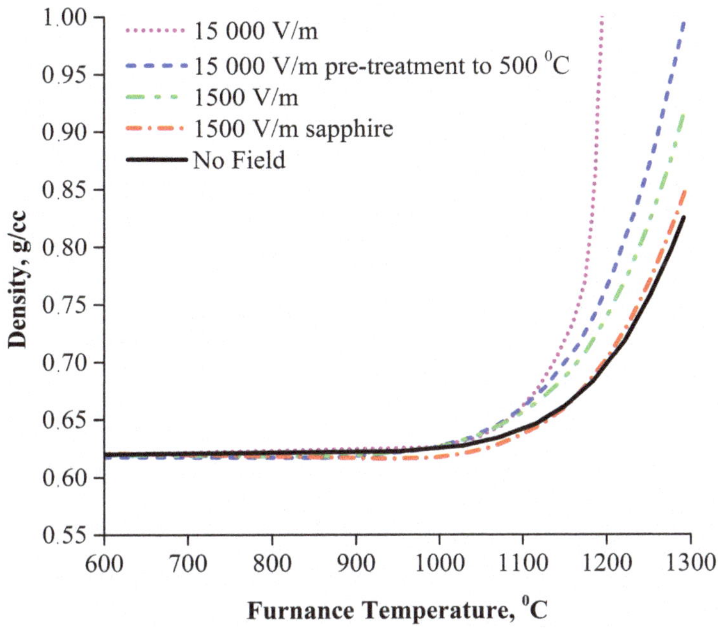

Fig. 6.2 Densification curves of 3YSZ sintered in air with and without electric field (Reprinted from Holland et al. [1], Copyright (2013) with permission from Elsevier)

achieved in materials, for which the Curie temperature and the sintering temperature are comparable. For materials with low Curie temperatures, there will be no influence of magnetic field on sintering. Litvinenko et al. [4] studied the effect of a constant magnetic field (~0.3 T) on the densification of cobalt powders due to the high Curie point of cobalt. The powder compacts were radiantly heated by the heating elements. Although the difference between the remaining porosities of the compact sintered in the magnetic field and the compact sintered without field was slight (65.5% and 68.2%), the resistivity of the compact sintered without magnetic field was almost twice the resistivity of the compact sintered in the presence of magnetic field. The authors explain this effect by a greater perfection of the interparticle contacts formed in the magnetic field.

The influence of a pulsed magnetic field on the density and quality of the sintered compacts has received a more detailed explanation. The contacting particles, when carrying an electric current, form conductors, which experience the action of Ampère's force in the external magnetic field. Such experiments were conducted by Li et al. [5], who combined the SPS processing of iron powders with the action of a pulsed magnetic field and obtained compacts of higher density and increased hardness than in the absence of the magnetic field. The effect of the pulsed magnetic field was concluded to be in the generation of eddy currents, enhancement of mass transport and particle rearrangement by breaking the previously formed bridges.

6.3 Summary

Sintering in the constant electric field in the noncontact mode and in magnetic field presents very rare types of processing and has been reported only in a limited number of publications. Apparently, it can be applicable to a limited number of material systems; however, when applicable, it can produce quite useful results. The specifics of the mechanisms of these types of sintering should include the field-assisted acceleration of diffusion and, possibly, the double-layer effect influencing the specific surface energy.

References

1. Holland TB, Anselmi-Tamburini U, Mukherjee AK (2013) Electric fields and the future of scalability in spark plasma sintering. Scr Mater 69(2):117–121
2. Newman DC (2000) Novel uses of electric fields and electric currents in powder metal (P/M) processing. Mater Sci Eng A 287:198–204
3. Kornyushin YV (1980) Influence of external magnetic and electric fields on sintering, structure and properties. J Mater Sci 15:799–801
4. Litvinenko YM, Solonin SM, Fridman GR (1982) Sintering of loose cobalt, nickel and iron powders in magnetic field. Soviet Powder Metall Metal Ceram 21(1):29–32
5. Li X, Ye Y, Tang Y, Qu S (2010) Effect of pulsed magnetic field on spark plasma sintering of iron-based powders. Mater Trans 51(7):1308–1312

Chapter 7
Microwave Sintering

7.1 Principle of the Method and Microwave Heating Process

Sintering of materials by microwaves is based on the conversion of the energy of microwave electromagnetic field into the thermal form. Over the last decades, methods of materials processing involving the use of microwave energy have been under constant development. In food processing, organic chemistry, pharmaceutical industry, and wood processing, microwave treatment has become an established technology [1–3]. The processes generally rely on microwave absorption in water or organic substances at relatively low temperatures. The advantages of high-temperature microwave processing have been demonstrated in such fields as sintering and joining of ceramic materials [4], inorganic synthesis [5], development of composite materials [6], powder metallurgy [7, 8], industrial and radioactive waste remediation [9], and annealing of implanted semiconductor structures [10, 11]. The development of new microwave processing applications most frequently begins with an attempt to replace the conventional heat source in the existing technology with a microwave source. As a rule, a new microwave technology can only become competitive when it results in a final product that is better significantly than those obtained by conventional technologies.

The development of microwave processing applications relies on the fundamental differences in the processes and their results from the counterpart conventional technologies. An important advantage of microwave heating is volumetric energy absorption in many materials. As opposed to conventional furnace heating by radiant or convective heat flows, microwave heating does not have to fully rely on thermal conductivity. In addition, the power produced by the microwave source can be spent entirely on the heating of the product, without the need to heat thermally massive furnace elements. Therefore, much higher heating rates are achievable with microwaves, which is one of the most important factors in many processes. However, a significant portion of the differences in high-temperature processing observed under

E. A. Olevsky, D. V. Dudina, *Field-Assisted Sintering*,
https://doi.org/10.1007/978-3-319-76032-2_7

microwave versus conventional heating cannot be explained by a different heat deposition pattern. The microwave effects include changes in the activation energy of diffusion [12], anisotropy of diffusion under heating in a linearly polarized microwave field [13] and changes in the solid-state phase transformation temperatures [14]. Apparently, these effects are of a more fundamental nature, associated with the microwave field energy conversion into forms other than the equilibrium thermal motion. The first studies of the use of microwave radiation for high-temperature processing of materials date back to the 1960s [15]. In the field of microwave sintering of ceramics, the pioneering experimental work was done by Tinga et al. [16], Berteaud and Badot [17], Meek et al. [18], and Lynn Johnson [19]. By the end of the 1980s, the feasibility of microwave sintering of oxide ceramic materials, such as Al_2O_3, ZrO_2, and Y_2O_3, had been demonstrated experimentally. It was found that microwave sintering has certain specific features of high potential importance for technology: a significant (50–100 °C) decrease in the process temperature, especially at the intermediate stage of sintering (the shift of kinetic curves of densification is especially pronounced in materials with higher microwave absorptivity [20]) and a reduction in the duration of the high-temperature stage of the sintering process. One of the possible reasons for this is an inverse distribution of porosity at the intermediate stage of sintering due to volumetric heating, which facilitates densification [21].

The achievements and problems in the microwave sintering of ceramics have been summarized in a number of comprehensive reviews, including those by Katz [22], Clark and Sutton [23], Agrawal [24], and Binner and Vaidhyanathan [25]. The demonstrated enhanced processing has induced interest in the use of microwave energy for the sintering of nanostructured ceramics, with a goal to obtain dense materials with the final grain size as close to the initial one as possible. In a study by Binner et al. [26], the two-stage sintering method originally suggested by Chen and Wang for the sintering of nanostructured ceramic materials [27] has been successfully implemented with the use of microwave hybrid heating. The advantages of microwaves in the sintering of advanced metal–ceramic composites and functionally graded materials have been shown by Willert-Porada et al. [28, 29]. Impressive prospects of microwave heating have been demonstrated in the sintering of metal powders, starting from the experiments by Roy et al. [30].

An important role in the development of high-temperature microwave applications is played by the process modeling. While the experimental optimization of microwave processing regimes is labor-intensive and time-consuming, modeling and numerical simulations can facilitate application development significantly. A comprehensive model of a microwave sintering process would pursue obtaining the temporal evolution and spatial distribution of the electromagnetic field in the material, temperature, and variables that reflect the processes in the material, such as stresses, mass fluxes, porosity, grain size, etc. These variables generally depend on the evolution of temperature but may also be influenced by a direct (nonthermal) effect of electromagnetic field. The distribution of the electromagnetic field within the object generally depends on the effective dielectric and magnetic properties of the material, which depend on the temperature and structural variables, such as porosity, the dimensions and properties of the applicator, and the matching conditions between the microwave source, transmission line, and applicator. Similarly, the distribution of temperature within the heated object is determined by 1) the distribution of electromagnetic field within the material; 2) the effective absorption properties, heat capacity, and thermal conductivity of the material (depending on the temperature and porosity); and 3) conditions of heat removal that may involve properties of thermal insulation, emissivity of the material, and convection.

Sintering assisted by microwave heating has found multiple applications in modern material processing technologies. In many important practical cases, the material to be processed by microwaves is placed in a microwave applicator that comprises a cavity resonator. To estimate the electric field in a cavity with a single oscillation mode excited, one can use the expression for the electromagnetic energy stored therein:

$$\frac{1}{2}\varepsilon_0 \int E_0^2 \mathrm{d}V = \frac{P_{\mathrm{inp}} Q}{\omega}, \tag{7.1}$$

where E_0 is the electric field amplitude in the incident wave, ε_0 is the electric constant, ω is the angular frequency, P_{inp} is the power input into the cavity, Q is the quality factor of the oscillation mode, and the integration is over the cavity volume. The cavity Q factor accounts for the losses in the material undergoing heating and the losses in the cavity walls.

The microwave power absorbed per unit volume can be expressed as follows:

$$w = \frac{\omega}{2}\left(\varepsilon_0 \varepsilon'' \vec{E}^2 + \mu_0 \mu'' \vec{H}^2\right), \tag{7.2}$$

where $\vec{E}$ and $\vec{H}$ are the electric and magnetic field strengths and ε'' and μ'' are the dielectric and magnetic loss factors, respectively.

As opposed to conventional furnace heating, in which the energy is delivered by radiant or convective heat flows, the principal feature that distinguishes microwave heating is volumetric energy absorption. Electromagnetic waves in the microwave and millimeter-wave range penetrate most dielectric materials and are absorbed in their volume. The primary mechanism of absorption is excitation of electron oscillation at the frequency of microwave source. The energy of electron oscillation is converted into the energy of lattice vibrations with an equilibrium spectrum via a number of steps (Fig. 7.1) [31].

Electron oscillations induce periodic distortion of lattice potential and thereby modify the spectrum of lattice vibrations driving it away from equilibrium. Then the vibration spectrum equilibrates due to the nonlinear interaction between lattice modes. Heating occurs by direct coupling of the powder compact to the microwaves. The approximate formula for the penetration depth of microwaves into materials, for which $\varepsilon''/\varepsilon' < 0.1$, is [32]

Fig. 7.1 Energy conversion during microwave heating: the electric field E causes oscillation of charged particles and dipole polarization within the solid (**a**); this translates into lattice vibrations (**b**), which are then "thermalized," resulting in the random equilibrium thermal motion (**c**) (Reprinted from Rybakov et al. [31]. Copyright (2013) with permission of John Wiley & Sons)

$$\delta = \frac{\lambda}{2\pi\varepsilon''}\sqrt{\varepsilon'}, \tag{7.3}$$

where λ is the wavelength and ε' and ε'' are the real and imaginary parts of the dielectric constant. Small penetration depths mean that reflection of microwaves from the surface dominates.

Bulk metallic materials and powder compacts, in which the concentration of the conductive particle exceeds the percolation threshold, should reflect the microwave radiation. The penetration of the microwaves is described as skin depth δ determined as

$$\delta = \frac{1}{\sqrt{\pi f \mu \sigma}}, \tag{7.4}$$

where f is the frequency, μ is the material magnetic permeability, and σ is the material electrical conductivity.

7.2 Effective Microwave Dielectric Properties

All electrodynamic calculations require data on the dielectric properties of the materials involved, including the dielectric constant ε' and the imaginary part of the complex dielectric permittivity ε''. In the microwave frequency range, absorption properties of nonmetallic materials vary greatly. The loss factor varies at room temperature from 10^{-4} to 10^{-3} (e.g., in pure alumina and silicon nitride) to 1 and higher (in carbides, borides, some oxides, and intermetallic compounds). Correspondingly, the electromagnetic field penetration depth in the materials varies from meters to fractions of a millimeter.

In the microwave processing practice, most materials are in fact heterogeneous – comprising mixtures of two or more phases. For example, powder compacts contain a solid phase (powder particles) and a gas phase (pores). Composite materials may contain three or more phases. When the characteristic scale of microstructure inhomogeneity is much smaller than the wavelength in the material, the interaction of the electromagnetic field with such a material can be described by averaging-based methods. Within these methods, the material is viewed as homogeneous, possessing the effective properties such as effective dielectric permittivity and/or magnetic permeability. The effective properties are determined by the properties of all components and the microstructure of the material, including the shape of particles and their distribution over size. The effective properties cannot be determined unambiguously by only the volume fractions of the components [33]. In the experiment, nonuniform distribution of components can result in different heat deposition patterns during microwave sintering [34]. Furthermore, microwave absorption properties of ceramic materials of a fixed density can depend on the grain size [35]. However, widely used in practice are approximate methods based on simple models of media with inclusions ("mixing laws"), for which averaged solutions can be obtained analytically. Such models can provide adequate description of real heterogeneous materials for a certain range of component volume fractions.

Bruggeman proposed the so-called effective medium approximation (EMA) to describe the properties of the heterogeneous materials [36]. Within this approximation, the components constituting the heterogeneous material are viewed as inclusions in the "effective" medium whose properties are sought. In the simplest case, the inclusions are assumed to have spherical shape. The following equation can be used to determine the effective dielectric permittivity $\varepsilon_{\mathrm{eff}}$:

$$\sum_i C_i \frac{\varepsilon_i - \varepsilon_{\mathrm{eff}}}{2\varepsilon_{\mathrm{eff}} + \varepsilon_i} = 0, \tag{7.5}$$

where C_i is the volume fraction of the ith component.

An advantage of the EMA is in describing the properties of the heterogeneous material adequately both when the volume fraction of a component tends to zero and when it approaches 100%. This symmetry makes it possible to use this approximation to model processes, in which the volume fractions of the components vary in broad ranges. However, the EMA is not justified in the intermediate range of volume fractions between the abovementioned limiting cases. The EMA is capable of qualitatively describing the phenomenon of percolation that arises in conductive powder compacts with an increase in the particle concentration up to the onset of connectivity between particles. At the percolation threshold, the imaginary part of complex dielectric permittivity, linked with the material's conductivity, increases sharply by orders of magnitude. Therefore, the material, in which the conductive powder volume fraction is above the percolation threshold, should reflect the microwave radiation almost fully. However, in many experiments, microwave heating of such materials has been demonstrated, meaning that the microwave radiation penetrates into such conductive powder compacts. This has been explained by the presence of insulating (e.g., oxide) shells on the conductive particles (Fig. 7.2). The resulting equation for $\varepsilon_{\mathrm{eff}}$ [37, 38] is presented as follows:

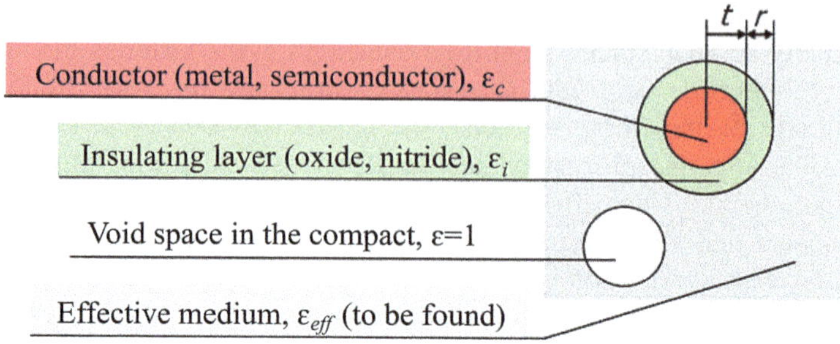

Fig. 7.2 Effective medium approximation for the case of conducting particles with insulating shells (Reprinted from Rybakov et al. [31]. Copyright (2013) with permission of John Wiley & Sons)

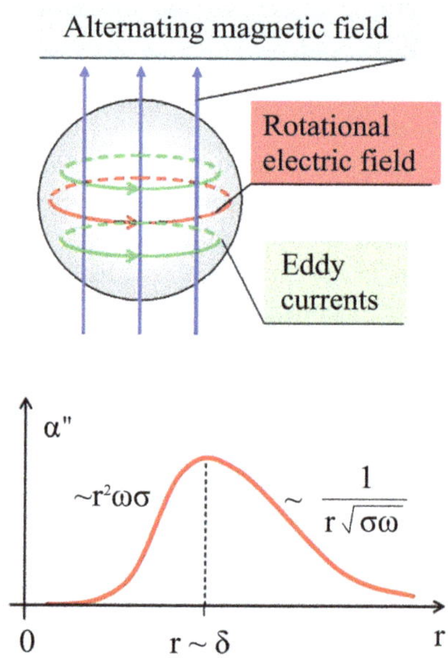

Fig. 7.3 Microwave absorption via magnetic polarization of metal particles and the schematic dependence of absorption on the particle radius (Reprinted from Rybakov et al. [31]. Copyright (2013) with permission of John Wiley & Sons)

$$2\beta\varepsilon_{\text{eff}}^2 + [\beta(3C - 2) + \alpha\varepsilon_i(1 - 3C)]\varepsilon_{\text{eff}} - \alpha\varepsilon_i = 0, \tag{7.6}$$

where C is the volume fraction of solid in the compact, ε_i is the dielectric permittivity of the insulating shells, ε_c is the dielectric permittivity of the conductive material, $\alpha = 2\varepsilon_i + \varepsilon_c + 2(\varepsilon_c - \varepsilon_i)/\xi^3$, $\beta = \varepsilon_i + \varepsilon_c + (\varepsilon_i - \varepsilon_c)/\xi^3$, $\xi = 1 + d$, $d = t/r$, r is the radius of the conductive particle without shell, and t is the shell thickness. The presence of even very thin shells can drastically change the effective dielectric permittivity and explain the microwave penetration into the conductive powder compacts.

Microwave absorption in metal powder particles is predominantly determined by the magnetic polarization of the particles. The alternating magnetic field induces a rotational electric field within the particles, which in turn drives eddy currents that generate heat (Fig. 7.3). The properties of materials containing nonferromagnetic metal particles are, therefore, described by the complex magnetic permeability that depends on the volume fraction of particles, their size, conductivity of metal, and frequency.

Cheng et al. [39] demonstrated that for single-mode applicators, samples compacted from metal powders heat faster when positioned in the magnetic field maximum of the standing wave electromagnetic field distribution in the cavity. This fact is in agreement with the mechanism of microwave absorption in metal particles due to eddy currents.

7.3 Heat Conduction Equation and Materials Parameters

The temperature within the microwave-heated material is governed by the heat conduction equation:

$$c_p \rho \frac{\partial T}{\partial t} - \nabla (\kappa \nabla T) = w, \tag{7.7}$$

where c_p is specific heat capacity, ρ is density, κ is thermal conductivity of the material, and w is local density of the energy released in the material. An analysis of Eq. (7.7) shows that the distribution of temperature in the case of microwave heating is essentially nonuniform, as long as the electromagnetic field is present in the material ($w \neq 0$). For example, in a steady state $\left(\frac{d}{dt} = 0\right)$ with κ not depending on temperature and w uniform, the distribution of T is parabolic with a maximum in the core of the material. This is usually referred to as an inverse temperature profile specific for the microwave heating. If the distribution of the absorbed microwave power, w, is nonuniform, then the temperature distribution follows the power distribution on short timescales, but on longer timescales, it is determined by thermal conduction across the entire workpiece dimensions [40].

The boundary conditions for Eq. (7.7) account for the heat loss at the surface of the material. This heat loss, along with the deposited microwave power, determines the degree of temperature nonuniformity in the material. To improve the uniformity, various means of thermal insulation are widely used in the microwave processing systems. Most commonly, the boundary condition is taken in the simple form:

$$\nabla T|_s = -\alpha\left(T|_s - T_0\right), \tag{7.8}$$

where T_0 is the temperature of the environment away from the material and the coefficient α reflects the quality of thermal insulation that surrounds the material. To account for the heat loss by thermal radiation, whose contribution can be prevailing at higher temperatures, an additional term proportional to T^4 is introduced into Eq. (7.8). In many cases, heat losses by convection in the gas surrounding the sample should also be accounted for [41].

For a powder material undergoing sintering, its parameters used in Eq. (7.7) change with porosity. If the specific heat capacity of air in the pores is neglected, the specific heat capacity, c_p, scales linearly with the density:

$$c_p = (1 - \theta)c_{p0}, \tag{7.9}$$

where θ is the porosity, defined as the volume of pores related to the total volume of the porous material, and c_{p0} is the specific heat capacity of the fully dense material. The thermal conductivity, κ, depends on the material's microstructure. In general, it can be determined within the approximate models in the same way as the dielectric permittivity. For example, within the effective medium approximation, the following equation for the effective thermal conductivity of the porous materials is derived:

$$2\kappa^2 + [\kappa_0(3\theta - 2) + \kappa_a(1 - 3\theta)]\kappa - \kappa_0\kappa_a = 0, \tag{7.10}$$

where κ_0 is the thermal conductivity of the fully dense solid material and κ_a is the thermal conductivity of air in the pores. Calculations of the thermal conductivity within a realistic microstructure model that accounts for the neck formation during sintering have been described in Refs. [42, 43]. For some simple cases of microwave heating, it might be sufficient to model the process based on the heat conduction equation alone, with the distribution of the microwave power considered uniform or prescribed (e.g., exponentially decaying from the material's surface into the bulk, which might be relevant for high-loss materials).

7.4 Self-Consistent Electromagnetic and Thermal Modeling

A self-consistent model of microwave heating would contain an electrodynamic part to determine the electric and (where applicable) magnetic field, $\vec{E}$ and $\vec{H}$, and a thermal part to obtain the temperature, T. The temperature, in turn, determines the values of all material parameters in the coupled model, such as ε, μ, c_p, and κ. With certain simplifying assumptions, in particular regarding the form of the temperature dependencies of the parameters, such a model can be solvable analytically. Otherwise, a simplest numerical implementation of such a model would comprise a one-dimensional slab of material divided into a number of thin layers with constant parameter values. At each iteration, the distribution of electric and magnetic field amplitudes is obtained by the impedance method [31] and then the temperature in each layer is determined by solving Eq. (7.7) by a simple implicit scheme. Before the start of the next iteration, the material parameters in each layer are assigned new values based on the temperature value in the layer.

The combined electromagnetic and thermal models have been a basis for a great number of microwave processing simulations. A semianalytical electromagnetic solution for a cylindrical cavity partially filled with a dielectric material, followed by finite-difference solution of the thermal problem, has been used to simulate the temperature dynamics during microwave heating of alumina [44]. A simple electromagnetic model based on plane-wave absorption in each of the materials present in the applicator has been demonstrated to provide good agreement with experiments. The susceptor material used to improve microwave heating of low-loss materials provides efficient microwave absorption up to some intermediate temperature, then oxidizes, and burns out [45]. The energy balance during the susceptor-assisted microwave sintering is analyzed in Ref. [46] within a plane-wave multilayer electromagnetic/thermal model. More elaborate finite element simulations of the microwave heating of ceramics with temperature-dependent properties based on commercial software such as COMSOL are described in Refs. [47, 48].

An analysis of the coupled electromagnetic and thermal problems in microwave heating, leading to a rich variety of dynamic heating regimes and effects, has attracted a broad interest of researchers. The most widely discussed effect is thermal instability. In a simplest consideration, it results from the microwave-specific nonuniform temperature distribution combined with the rising dependence of microwave absorptivity on temperature. As the deposited power density is proportional to the absorptivity, it is locally higher in the regions with a higher temperature, leading to a higher-temperature growth rate in these regions. Obviously, this gives rise to temperature instability, unless the thermal conduction is strong enough to effectively equalize the temperature over the material. Early models of thermal instability during microwave heating (commonly termed a "thermal runaway") were published by Roussy et al. in the 1980s [49, 50]. The simplest thermal runaway models were based on the thermal conduction equation alone, assuming the electric field amplitude constant and uniform [51, 52]. Such models are capable of obtaining the stability criterion as a function of the rate of material's absorptivity growth with temperature and other material parameters, such as thermal conductivity. Whereas most thermal runaway models consider the temperature dependence of the dielectric loss factor as the primary cause of the instability, it has also been shown that the instability can arise due to the temperature dependence of the real part of the material's dielectric constant [53] or due to the descending temperature dependence of the thermal conductivity [54]. The issue of power or temperature control that would prevent heating instabilities has received considerable attention, resulting in the development of control algorithms for the feedback-loop automatic power control systems [55–57].

7.5 Models of Microwave Sintering

The above-described simulations of microwave heating result in obtaining the distribution of temperature over the ceramic workpiece and the evolution of this distribution. However, a major end result of the sintering process is the densification of a ceramic powder compact. Therefore, in a full-scale model of the sintering process, the obtained distribution and evolution of the electromagnetic field and temperature should be input into a simulation of densification during sintering. Some of the factors, specific to microwave sintering, such as possible inhomogeneity of temperature distribution, deserve mentioning in particular. A prevailing majority of models described sintering in conventional furnaces and, therefore, in many cases did not account for nonuniform temperature distributions. A continuum sintering model that attempts to analyze the influence of the temperature nonuniformity associated with microwave heating has been suggested by Rybakov and Semenov [58]. The powder material undergoing sintering is considered as a viscoelastic medium characterized by shear and bulk viscosities, and its densification is primarily governed by an intrinsic sintering stress. Within such a model, the effect of nonuniform heating manifests itself via thermoelastic stresses and nonuniform effective viscosity of the porous ceramic material that depends on the temperature. The results of the analysis suggest that the nonuniform, temperature-dependent viscosity of the ceramic material can be the cause for nonuniform localized densification, whereas the effect of thermoelastic stresses induced by volumetric microwave heating is limited.

The attempts to combine the electromagnetic and thermal modeling of microwave heating with the simulation of temperature-dependent densification are not numerous. In the simplest case, a fitting equation that relates the density and the temperature is used at the final stage of the simulation. As a result, the simulation output is translated from temperature to density terms. Examples of such simulations can be found in Refs. [59–62]. A limitation of such simulations is that they do not reflect changes in the dimensions and shape of the sintered objects. Also, the changes in the material's properties, such as the dielectric permittivity and heat conductivity, with the evolving porosity are often not taken into account. Another simplified approach is based on the use of master sintering curves proposed by Su and Johnson [63]. These curves are the experimentally obtained dependencies of the relative density versus the logarithm of the temperature evolution function:

$$\theta(t, T(t)) \equiv \int_0^t \frac{1}{T} \exp\left(-\frac{E_a}{RT}\right) \mathrm{d}t',$$

where the value of the activation energy, E_a, is picked to obtain the best fit to the experimental data. Using the master sintering curves, it is possible to obtain the density as a function of the local temperature evolution. A one-dimensional simulation that combines all three modeling stages, viz., finding the absorbed microwave power by an impedance method, finding the temperature distribution by a finite-difference method, and finding the shrinkage of each layer in the mesh by the master sintering curve method, has been described in Refs. [59, 64] for the microwave sintering of metal–ceramic functionally graded materials with a composition spatially varying from the pure ceramic to the pure metal. The master sintering curves for the intermediate concentrations of the composite material were interpolated based on the curves for the pure metal and the pure ceramic. At each iteration, the mesh spacing was adjusted according to the calculated local densification. However, it should be noted that the master sintering curve approach, in its originally suggested form [63], assumes a fixed relationship between the average grain size of the sintered material and its relative density. This assumption significantly narrows down the applicability area of the master sintering curve concept. In addition, it contradicts a major idea of sintering optimization, viz., maximum densification achieved along with the controllable grain size evolution, which has been one of the incentives driving the implementation of the microwave sintering techniques.

A more rigorous approach is to incorporate the nonuniform temperature distribution resulting from the microwave heating simulation into continuum sintering models. In a numerical study of microwave sintering [65], the distribution of microwave field and temperature is obtained for a one-dimensional slab of a ceramic material. The porosity and grain size are computed using a continuum model applicable for late stages of sintering. The material constants are assumed to depend on porosity; however, whereas the complex dielectric permittivity is calculated using the Maxwell–Wagner mixture rule, the thermal conductivity (as well as the specific heat capacity) is simply proportional to the values for the solid material and air weighed with their respective volume fractions. The coupled system of nonlinear equations is solved numerically. The authors choose the microwave heating regime in which the temperature distribution in the ceramic material is (almost) uniform and compare the microwave process with the conventional one in which the temperature is assumed to grow uniformly at the same heating rate. Naturally, no difference is observed in the obtained evolution of porosity and grain size. This result is trivial because neither the differences in the temperature distribution and/or evolution nor the microwave nonthermal effects have been taken into account in this comparison.

A simulation of microwave sintering of solid oxide fuel cell materials [66] utilizes a continuum sintering model with some empirical parameters and takes into account the evolution of the powder particle size distribution. The electromagnetic field is assumed to be decaying exponentially from the surface into the bulk of the material. The effective diffusivity under microwave heating is assumed to have a lower activation energy due to the nonthermal microwave influence, and, in addition, the microwave absorptivity is related to the specific surface area of the powder (that varies in the course of simulation due to grain coarsening). As a result of numerical modeling in the 2D geometry, the density and particle size evolution with time is obtained along with the changes in the dimensions of the samples.

A full-scale comprehensive microwave sintering simulation that combines the numerical solution of Maxwell equations, thermal conduction equation, and the evolution equations for the relative density and grain size has been developed by Riedel and Svoboda [67]. The finite-difference electromagnetic simulation is implemented for the oscillation modes of the cylindrical cavity applicator with a zero azimuthal index. To save the computation time, the electromagnetic simulation is repeated once in a hundred time steps for the thermal conduction and densification/grain growth simulation. The cavity tuning (that changes due to the evolution of the ceramic workpiece properties) is adjusted automatically via the frequency shift. The numerical model is capable of describing the influence of the sintering process on the evolution of thermal instabilities during microwave heating. As a final result of the simulation, density and grain size distributions are obtained for the sintered product. Being a comprehensive example of the electromagnetic–thermal–mechanical-coupled solution of a microwave sintering problem, this modeling framework is lacking consideration of the impact of the microwave nonthermal effects on the sintering process. Incorporation of these effects is one of the important tasks for the multiphysics models of microwave sintering.

7.6 Experimental Evidence of Microwave Nonthermal Effects

Microwave sintering has become an important field for investigations of direct electromagnetic field contributions into mass transfer. From a physical standpoint, one should distinguish between "microwave effects" of thermal and nonthermal origin. The problem of nonthermal microwave effects has become one of the most controversial issues in the literature on high-temperature microwave processing of materials. In essence, it is related to the principal difference between the microwave and conventional processes that lies in the nature of their energy sources. In the microwave processes, a coherent electromagnetic field drives an oscillatory motion of the charged particles. In the conventional processes, the energy exchange between the furnace and material is mostly provided by means of quasi-equilibrium thermal electromagnetic radiation that has a continuous spectrum, predominantly in the infrared range. The process of thermal radiation absorption in the material is usually viewed as excitation of oscillations that have an equilibrium spectrum (described by the Maxwell–Boltzmann statistics). As most materials are not transparent for the infrared radiation, its absorption occurs mostly at the external surface of the solid, from where heat is transferred to the bulk by thermal conduction. It is sufficient to describe the state of the solid statistically, i.e., assume that motion of all forms can be described by one variable, the temperature. Changes in the temperature of the solid mean slow evolution of the spectrum, which at all times remains quasi-equilibrium.

The temperature field dynamics data are commonly used as a basis for comparisons between different processes, including those utilizing microwaves. In principle, the term "microwave effect" or, more strictly, "microwave field nonthermal effect" should be reserved for the deviations in the microwave processes from the conventional ones that occur given the identical temperature dynamics, $T(r, t)$ in these processes. In practice, at least two general problems challenge the identification of the nonthermal effects. One of them is inaccurate or incomplete temperature measurements [68]. The second problem is that it is impossible to devise a pair including a microwave and a conventional process that would have identical temperature fields. In particular, due to volumetric energy deposition and surface heat loss, it is not possible to implement a microwave process with a uniform temperature distribution in the material undergoing processing. Unless there is no microwave absorption within the material, such as in the susceptor-assisted microwave heating, temperature gradients are an inherent attribute of microwave heating, and the effects caused by them will always contribute to differences in the process flow. Possible mechanisms of thermal gradients influence are, for example, thermoelastic stresses [69] and thermal diffusion [70, 71]. However, both experimental [20, 72] and theoretical [58] studies suggest that temperature gradients are generally not the main reason for enhanced mass transport rates in microwave processing. The same can be said about the microscopic temperature gradients that were speculated to arise due to localized microwave heating of grain boundaries, whose absorption properties are higher than those of the bulk [73]. The thermal conduction equilibrates the temperature efficiently, and it has been argued that no significant temperature differences can arise between the grain boundary and the bulk [74].

The microwave field nonthermal effect is associated with non-equilibrium excitations in the solid. In contrast to thermal radiation, electromagnetic waves in the microwave and millimeter-wave range penetrate most dielectric materials and are absorbed in their volume. The primary mechanism of absorption is excitation of electron oscillation at the frequency of the microwave source. This means that all absorbed energy is for a certain time contained in non-equilibrium modes. The fraction of energy stored in the non-equilibrium excitations is determined by the time constant of the process. For a uniform solid, it is generally negligible. The presence of defects, however, can increase the lifetime of the non-equilibrium excitations. Probably, the most obvious example is vacancy diffusion, which can convert a portion of the microwave field energy into the energy of ion oscillatory migration on the lattice. In particular, rotation of bivacancy dipoles has been shown to modify the microwave absorption behavior of ceramics [75].

The motion of charged particles under the action of electromagnetic field can lead not only to heating. In principle, the field can influence interactions between the particles and/or the transport phenomena. The estimates suggest that under the conditions of microwave processing of materials, the field strength is usually too low to cause noticeable changes in the interaction of particles, i.e., a direct influence of the electromagnetic field on the rates of chemical reactions should be negligible [76]. Therefore, the influence on the transport phenomena should be viewed as the main mechanism of nonthermal microwave effect. In this manner, the microwave field should be also capable of influencing chemical and phase transformations in solids where they are often limited by mass transport [77].

There are many experimental observations that suggest a nonthermal influence of the microwave fields on mass transport. One of the first, yet still not explained in full, was the enhancement of oxygen diffusion in sapphire crystals heated in a 28 GHz millimeter-wave furnace, as reported by Janney et al. [12]. The evolution of the spatial distribution of oxygen tracer isotope concentration was investigated within a layer with a thickness of the order of 1 μm, in which the temperature could be considered constant. A 40% decrease in the apparent activation energy for diffusion was observed under millimeter-wave heating, as compared with conventional heating. Similarly, the activation energy for grain growth during millimeter-wave annealing of sintered fine-grain alumina was found to be 20% lower compared with conventional heating [78].

The microwave influence on atom diffusion has been reported in a number of studies. A half-order of magnitude increase in the pre-exponential factor has been observed in the microwave synthesis of titanium carbide by carbothermal reduction in TiO_2 [79], whereas the activation energies were the same for the microwave and conventional processes within the experimental error. A very similar result was obtained in a study of the synthesis of lead titanate from lead and titanium oxides, where the process of reaction front propagation is limited by bulk diffusion [80].

The activation energy has also been found to be the same for conventional and microwave heating in the process of interdiffusion in the semiconductor heterostructure system InGaAs [81] although the pre-exponential factor in this case was lower for microwave heating than for conventional heating. In the studies of creep deformation of nanostructured alumina-based ceramics under microwave heating [82, 83], it has been found that the apparent activation energy of creep (which is based on grain-boundary diffusion) is lower than the values for the same ceramic compositions heated conventionally.

Lee et al. [84] studied the microwave influence on the annealing process of thin amorphous silicon films. Due to a small thickness of the films, the temperature variations were also negligible. It has been found that microwave annealing not only accelerates crystallization in the films but also enhances the process of hydrogen removal, which is known to be diffusion-controlled under conventional heating. Rowley et al. [85] discovered that microwaves enhance the process of bulk oxygenation of dense YBCO ceramics. The temperature in these experiments was independently verified by measuring the equilibrium oxygen content, which is highly temperature-sensitive. The microwave influence on the diffusion healing of cracks was analyzed on thin plates of polycrystalline alumina [86]. The diffusivity was found to increase monotonically with the microwave power. However, in contrast to Ref. [12], the activation energy for diffusion under microwave heating was greater than under conventional heating.

The influence of millimeter-wave annealing on interdiffusion in KCl–KBr diffusion couples was studied in Ref. [87]. It should be noted that the processes considered in the above-listed studies [84–87] were not purely volumetric but included the surface or interface phenomena; therefore, their rates depend not only on the volume diffusivity but also on the boundary conditions that include the exchange kinetic coefficients. The microwave effect on the surface kinetics was studied [88] for the transport of oxygen in partially stabilized zirconia. It has been found that whereas microwave heating does not influence the volume diffusion coefficient, the surface exchange kinetic coefficients are higher by about an order of magnitude under microwave heating compared with conventional heating.

A significant body of evidence for nonthermal effects has been obtained in the experiments on microwave sintering. A reduction in the apparent activation energy of densification was observed by Bykov et al. [89] for the millimeter-wave sintering of Si_3N_4 ceramics. By varying the heating rate, it was demonstrated that the magnitude of this reduction correlated with the absorbed microwave power, and this could not be attributed to the effect of temperature gradients. A detailed investigation of the nonthermal effects during microwave sintering was accomplished by Wroe and Rowley [90] via studying the dependency of the densification rate on the microwave power. The identity of the temperature–time schedules was ensured by simultaneously using microwave and conventional heating each having a controlled power. The results proved that the observed enhancement of densification correlates with the microwave power and not the temperature and is, therefore, of a nonthermal nature. A similar approach was further developed by investigating the sintering kinetics under hybrid heating with a varying proportion of microwave/conventional power while maintaining identical temperature–time schedules [20, 91]. The results of the experiments have demonstrated that the achieved density for a given temperature is increased by increasing the microwave power and the effect is more pronounced when the ceramic material possesses higher microwave absorptivity. The temperature gradients were assessed and found too small to be the cause of the observed effect.

The results of a comparative study of the closed porosity evolution upon microwave and conventional sintering accomplished by Willert-Porada [92] are of special importance for understanding the mechanism of the nonthermal microwave effect. It is known that the size of large pores surrounded by a large number of grains cannot be reduced by the capillary forces [93]. Therefore, the changes in the evolution of the closed porosity under microwave heating observed in Ref. [92] are not only evidence of the changes in mass transport but also suggest that microwaves produce a new driving force for this process. Similar conclusions can be drawn from the studies of the microwave influence on the formation and decomposition of solid solutions [94–97].

A direct observation of nonthermal microwave mass transport enhancement has been made in the experiments on microwave heating of thin amorphous alumina membranes with a regular structure of cylindrical pores with a diameter of tens of nanometers [98]. Small thickness of the membranes (10–20 μm) eliminated the influence of temperature gradients. The observed enhancement of the closure process of the pores of such a small diameter suggests that microwave nonthermal effects can be strong enough to compete with very high capillary stresses typical for such pores. Furthermore, a correlation of the pore closure rates with the electric field strength has been demonstrated [99].

One of the most convincing demonstrations of the nonthermal nature of the microwave-enhanced mass transport is the study of the orientation dependence of diffusion during heating by a linearly polarized microwave field. It has been shown by Whittaker [13] that the rate of diffusion mass transport in sintered diffusion couples is much higher along the electric field vector than across it. In a more recent study by Link et al. [100, 101], the anisotropy of pores formed in ceramics under microwave heating has been demonstrated. It has been shown that the pores are prevailingly elongated across the electric field vector. In addition, there have been a number of studies demonstrating that the temperature of solid-state phase transformations is changed under microwave heating [14, 91, 102–104]. The effect on phase transformations has been shown to depend on the microwave electric field strength [105]. Also, decrystallization of ferrite materials [106] and titanium oxide [107] has been observed as a result of heating in a magnetic field maximum of a single-mode microwave cavity.

7.7 Models of Microwave Nonthermal Effects in Solids

The experimental observation of the nonthermal effects involved in microwave sintering has stimulated development of a number of hypotheses and models attempting to explain them by suggesting a specific physical mechanism. Initially, a microwave influence on the transport coefficients in the solid was proposed [108–110]. However, the electric field strength required for the effect to be noticeable exceeds the values that are typical for the microwave processing by orders of magnitude [111]. There have been attempts to develop models of the microwave heating processes based on thermodynamic considerations, accounting for the extra term in the free energy associated with microwave-induced polarization [112] or postulating the change in the vacancy formation energy under microwave field [113]. A generalized approach to the description of microwave heating accounting for irreversible processes within the framework of non-equilibrium thermodynamics has been presented in Ref. [114]; however, no quantitative assessment of the microwave effects' significance has been accomplished.

Due to the perturbations of alternating drift vacancy flows near pore surfaces that have limited permeability for vacancies, the averaged product of the electric field and the charge density is nonzero, which leads to nonlinear "rectification" of the vacancy currents and gives rise to steady-state driving force for the vacancy flows that is proportional to $|E|^2$ [115]. The nature of this force is similar to the ponderomotive force known in plasma physics and electronics.

The origin of ponderomotive forces is the interaction of the microwave field with the effective electric charges of vacancies. In ionic crystals, vacancies can be treated as isolated point defects with effective electric charges equal to the magnitude and opposite to the sign of the corresponding ions [116]. For the modeling of electromigration and diffusion processes, it is convenient to deal with the migration of vacancies, bearing in mind that real mass transfer is provided by ion migration [117]. The combination of diffusion and electromigration fluxes for any type of vacancies can be described as

$$\vec{J} = -D\vec{\nabla}C_v + DC_v\frac{q\vec{E}}{kT}, \tag{7.11}$$

where C_v, D, and q are the vacancy concentration, diffusivity, and effective electric charge, respectively. The first term describes diffusion and the second one represents electromigration. Due to oscillations of the electric field during microwave sintering, the average electromigration mass transfer is, in general, equal to zero.

Different hypotheses about the physical nature of this rectification can be put forward. For the typical conditions of microwave sintering, the amplitude of the drift part of the flux exceeds the diffusion part by 2–3 orders of magnitude. However, the drift flux oscillates at the microwave frequency and needs to be "rectified" through a mechanism that transforms oscillatory vacancy motion into its monotonic drift. Such nonlinear rectification can be caused by perturbations of the vacancy flow near pore surfaces that have limited permeability for vacancies. Rybakov et al. [118] and Olevsky et al. [119] suggested that there exists a very steep gradient of the surface diffusivity with distance from the surface. It is known that surface diffusivity is active not just at the surface of a powder particle, but also within several atomic layers under the surface. At the distance δ_s, the diffusivity along the surface is changed from the typical values of the surface diffusivity to the typical values of volume (a four- to five-order decrease in the diffusivity depending on the temperature and the material).

The oscillating component of an electric field, normal to the free surface moves positive and negative vacancies of a ceramic material to the free surface. The concentration of vacancies builds up in a thin layer near the surface when the electric field drives them toward the surface and/or is depleted during the next half-period when the field changes its direction and drives them from the surface into the bulk. Because of the diffusivity difference, the flux of the non-equilibrium vacancies along a free surface under the influence of the tangential electric field component is much faster than the flux of the opposite-sign vacancies in the bulk of the material in the vicinity of a free surface.

While the normal component of the electric field moves positive vacancies to the surface, the tangential component shifts them to the contact. The same is valid for the negative vacancies during the next half-period of the oscillation. As a result of the in-phase motion of vacancies with the electric field, there is a systematic "ponderomotive" material transport along the surface. Vacancy flux means matter flux in the opposite direction. When vacancies move to the contact, the matter flux is directed from the contact to the surface. In modeling performed in Refs. [118, 119], the mobility of vacancies inside the body is considered negligible in comparison with the surface mobility. This makes the results of those calculations the upper bound for the electromigration mass transfer. The important detail of this rectification mechanism is the existence of non-equilibrium vacancies at the surface during oscillations.

The existence of an equilibrium surface concentration of vacancies means that vacancies do not immediately disappear when approaching the surface. It is reasonable to assume that the rate of vacancy annihilation at the surface is proportional to the deviation of vacancy concentration from the equilibrium concentration [111]. If the frequency of the electromagnetic radiation is high enough, such as in the case of microwave sintering, then the complete annihilation of the excess free surface vacancies does not occur during the period of the oscillations and some drift of non-equilibrium vacancies takes place. Experimentally, the existence of this material transfer was observed during microwave heating of a material with spherical pores [120]. In general, this effect can be readily detected on any flat surface during microwave heating: after being heated in a polarized microwave field, any flat surface must gain some slope due to the "rectified" mass transfer. This effect can be a subject for the future experiments. At the initial stage of sintering, the very geometry of the powder material leads to the polarization and intensification of the microwave radiation in the vicinity of inter-particle contacts [121]. Figure 7.4 shows

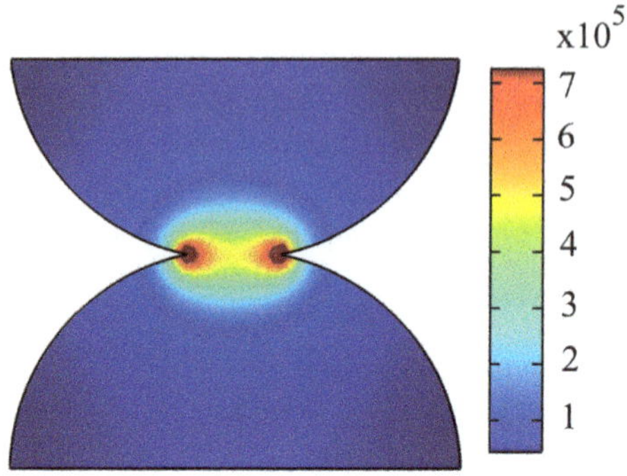

Fig. 7.4 Electric field concentration at the neck between the particles and streamlines of the electric field near the particles

the intensification of the electric field at the contact between the particles when the average amplitude of the external electric field is assumed to be 10^5 Vm^{-1}. As a result, the ponderomotive material transport significantly influences the contact growth during sintering.

It is known from classical electrodynamics that when the electric field vector is perpendicular to the bisecting plane of a dihedral angle, the field strength increases without limit upon approaching the angle vertex, whereas when the field vector lies within this plane, it vanishes. It has been argued [122] that for the former case of field orientation, the mass flow resulting from the ponderomotive effect is directed from the grain boundary into the pore, and for the latter case, the direction of the mass flow is reverse. While the grain boundaries in a ceramic sample undergoing microwave sintering are oriented randomly with respect to the E-field vector, it is important that the former effect is much stronger, due to a higher field strength, and, therefore, the net contribution of the ponderomotive effect into the pore closure is positive.

In accordance with these considerations, Rybakov et al. [118] have accomplished a numerical simulation of the pore closure due to the microwave ponderomotive effect in ceramics. The ponderomotive effect on densification during sintering has been analyzed in terms of a continuum sintering model [123–125]. In order to demonstrate the most significant aspects of the effect, the model space comprised a regular rectangular array of rhombic pores [126]. A representative element of the model space is shown in Fig. 7.5.

The values obtained for the normal and tangential E-field components on the pore surface are utilized to compute the 2D area displaced from the grain boundary into the pore in one time step. Then the vertical shrinkage of the model space is calculated, and the mesh is modified accordingly to account for densification. The obtained results made it possible to describe the evolution of porosity. A comparison of the conventional sintering process (free sintering) and that assisted by microwaves was made based on the calculation of the ratio of the characteristic sintering times for these two processes:

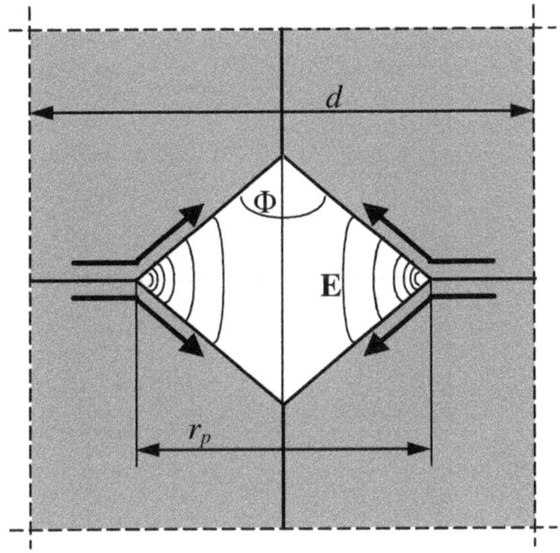

Fig. 7.5 A representative element of the model space: a rhombic pore surrounded by four ceramic grains. The E-field distribution within pores (lines) and preferred directions of the mass flow (arrows) are shown schematically (Preprinted from Rybakov et al. [118]. Copyright (2012), with permission from Elsevier))

$$\frac{\tau_{MW}}{\tau_{FS}} \cong \frac{C_{v0}\alpha\delta_{gb}}{d^2\varepsilon_0 E_0^2}, \tag{7.12}$$

where C_{v0} is the concentration of vacancies, α is the coefficient of surface tension, δ_{gb} is the grain-boundary thickness, d is the grain size, and E_0 is the field strength.

The ratio of the characteristic times of the two sintering processes was calculated to be close to 1 taking $C_{v0}=10^{-5}$, $\alpha=1$ J m^{-2}, $\delta_{gb}=10^{-9}$ m, $d=10^{-6}$ m and $E_0=3\times10^4$ V m^{-1}, which means that under the chosen conditions, the contribution of the microwave ponderomotive effect to sintering is significant. For the detailed mechanism of pore closure due to the ponderomotive effect of microwave field, the reader is referred to Ref. [31].

The demonstrated effect will remain even if the greater part of the pore surface is curved instead of flat, as long as the dihedral angles that are responsible for E-field enhancement are present near the grain-boundary edges. If the pore shape, assumed to remain constant for the sake of simplicity, is allowed to change during the sintering process, the ponderomotive effect will result in the removal of the material from the regions, where the field is enhanced, and its accumulation in other corners of the pore. Therefore, in a linearly polarized microwave electric field, the pores will elongate across the electric field vector, in exact agreement with the experimental observations [120].

In the course of sintering, the pore surfaces will tend to smoothen due to surface diffusion, and the pore shape will change from faceted to ellipsoid. The electric field will still be enhanced in this configuration, but not to infinite strength. Therefore, the resulting ponderomotive contribution to densification will be most pronounced in the initial stages of sintering and decrease toward the end of the sintering process. This is in agreement with most comparative studies of microwave vs. conventional

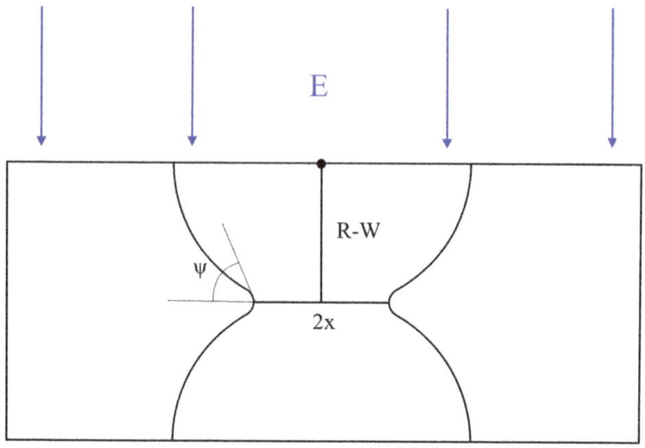

Fig. 7.6 Unit cell of the powder material used for modeling in Olevsky et al. [119]

sintering, which report the most drastic differences in kinetics in the early stages of densification. However, it should also be noted that the ponderomotive effect can be relevant in microwave sintering of nanocrystalline materials, in which the pore surfaces tend to retain a faceted shape reflecting the material crystalline structure [127].

Theoretical considerations related to the action of ponderomotive forces were further developed by Olevsky et al. [119], who performed assessment of the contributions of the ponderomotive effect to the kinetics of single contact growth during sintering. The direction of vacancy drift depends on the direction of the electric field at the surface. For example, streamlines in Fig. 7.6 show the directions of the considered electric field near the contact of two dielectric particles. It is clear from the geometry of streamlines that if the normal component of the electric field is directed outside the particle in the vicinity of the contact, then its tangential component is directed to the contact. While the normal component of the electric field moves positive vacancies to the surface, the tangential component shifts them to the contact. The same is valid for the negative vacancies during the next half-period of oscillations. As a result of the in-phase motion of vacancies with the electric field, there is a systematic "ponderomotive" material transport along the surface.

In the modeling of microwave sintering, it should be found in what form the macroscopic electromagnetic field manifests itself at the microscopic level of powder particles. For this purpose, the homogenization theory for periodic media can be used [128]. The geometry of the unit cell of a powder material used by Olevsky et al. [119] is shown in Fig. 7.6. It consists of two spherical particles with a single place of contact between them and some space around these particles. It was assumed that the external electric field has only one component directed along the center-to-center line of the particles. The formulation of the microwave heating problem for this unit cell followed the approach proposed in Ref. [129]. The modeling of the diffusion processes in the designed unit cell is based on the approach developed earlier [130]. Olevsky et al. [119] consider grain-boundary diffusion and surface diffusion to be the main mechanisms of the matter redistribution. This is the case for the sintering of fine powders and nanopowders. From the symmetry considerations, it is clear that the middle line at the particle surface between the different necks is the line of zero matter transport across it and the diffusion matter transport is equal to zero at the intersection of the particle boundaries with the boundaries of the unit cell. The model describes the consolidation processes only for the first stage of sintering with porosity higher than 0.1–0.15 when single necks between particles are discernible [131]. The manifestations of the ponderomotive effects are the most pronounced at this stage.

In Ref. [119], the relative rate of the grain-boundary diffusion and surface diffusion was estimated through the following parameter:

$$\xi = \frac{\delta_g D_g}{\delta_s D_s}, \tag{7.13}$$

where δ_s is the thickness of the surface layer, in which diffusion takes place, D_s is the surface diffusivity, δ_g is the grain-boundary thickness, and D_g is the grain-boundary diffusivity.

The parameter ξ is an indicator of the process non-equilibrium. If ξ is small (ξ is equal to or less than 0.01), then the rapid matter redistribution by surface diffusion keeps the free surfaces of the particles close to their equilibrium positions through the whole sintering process. For larger ξ values, the matter redistribution is in substantial non-equilibrium. In general, the full spectrum of sintering conditions in the present model can be described through the variations of the two independent parameters: ξ and ψ. The dihedral angle ψ is assumed to be constant and equal to 0.8π in all the calculations, which lies within the typical range for ceramic materials [132]. The parameter ξ was taken equal to 0.001. The utilized numerical approach was based on the method of lines where all the derivatives along the surfaces in the equation for matter fluxes are replaced by their finite-difference approximations. An implicit numerical scheme has been used for the approximation of the time derivatives. The details of the numerical approach for diffusion modeling can be found in Ref. [130]. The numerical calculations allowed the prediction of the evolution of the distances between particles, neck radii, and particle shapes for different values of ξ. Modeling of the contact formation assumed parallel calculations of the electric field evolution around the particles and of the mass transfer produced by diffusion and electromigration. The average electromigration matter flux along the particle surface

$$J_E = \frac{D_s \Omega}{kT} \frac{\varepsilon_0 \varepsilon''^2}{\varepsilon'^2 + \varepsilon''^2} E_\tau^{(\mathrm{out})} E_n^{(\mathrm{out})}, \tag{7.14}$$

where D_s is the surface diffusivity, Ω is the atomic volume, ε' and ε'' are the real and imaginary parts of the dielectric constant ε, and $E_\tau^{(\mathrm{out})}$ and $E_n^{(\mathrm{out})}$ are the tangential and normal components of the electric field outside the particles.

Unlike the traditional surface diffusion, the electromigration surface flux does not depend on the thickness of the surface layer. The electromigration flux is the surface matter flux that modifies the common surface diffusion matter flux. The vacancy flux J_E is directed along the surface to the place of contact between the particles. It is important to note the specifics of the electric field distribution at the particle surfaces. The results of electric field calculations at the particle surfaces during sintering demonstrate that the electromigration flux reaches considerable values only in the vicinity of the triple line of contact. The matter flux in the triple point is directed from the neck to the surface of the particles, and because of the flux continuity condition, the same flux must be induced from the contact to the triple point. Therefore, the matter transport from the grain boundary to the surface increases. As a result, the volume shrinkage during microwave sintering exceeds that taking place during conventional sintering. The increase in the shrinkage rate can be estimated through the fictional additional compressive stress at the contact between particles that corresponds to the additional electromigration matter flux:

$$p = \frac{\varepsilon_0 \varepsilon''^2}{\varepsilon'^2 + \varepsilon''^2} \max\left|E_\tau^{(\text{out})} E_n^{(\text{out})}\right|_\Gamma \frac{X}{\xi \delta_g}, \tag{7.15}$$

where X is the neck radius and $\max\left|E_\tau^{(\text{out})} E_n^{(\text{out})}\right|_\Gamma$ is calculated at the particle boundaries. It can be seen from the equation above that the additional pressure is proportional to the squared surface conductivity through ε'' and it is inversely proportional to the diffusivity ratio ξ. The additional pressure p as a function of the neck radius is shown in Fig. 7.7 [119]. It can be seen that in some cases the additional pressure is of the same order of magnitude or exceeds the values of the sintering stress observed in the experiments on traditional sintering (0.1–10 MPa) [131]. The

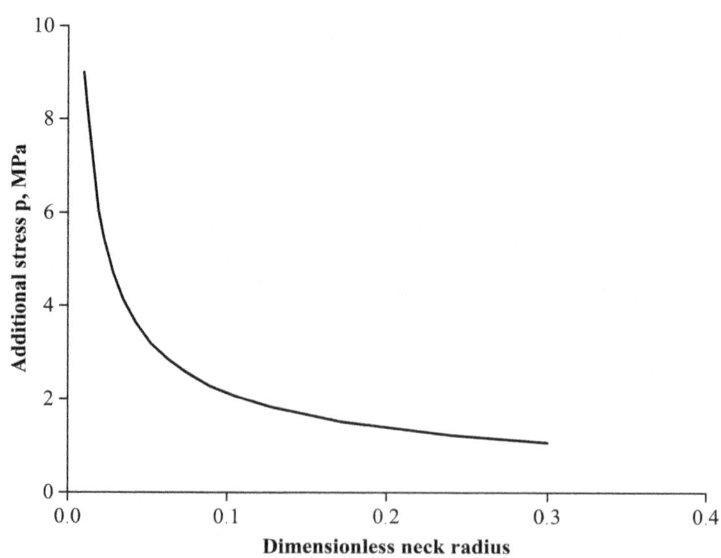

Fig. 7.7 Additional compressive stress at the contact between the particles as a function of the dimensionless neck radius [119]

parameter ξ is a function of temperature. If the activation energy of grain-boundary diffusion exceeds the activation energy of surface diffusion, which is usually the case, then ξ exponentially increases with the temperature. The calculations show that the product $\left|E_\tau^{(\mathrm{out})} E_n^{(\mathrm{out})}\right| X$ decreases with the neck radius. Therefore, the effect of the nonthermal mass transfer during microwave sintering is more pronounced for comparatively low temperatures (for small ξ) at the initial sintering stage.

The compressive stress is proportional to the intensity of the electric field at the inter-particle neck and inversely proportional to the ratio of the grain-boundary and surface diffusivities. Based on these modeling results, it can be concluded that the electromigration matter transport can substantially accelerate shrinkage during microwave sintering in comparison with conventional sintering.

The account for the quasi-electrostatic microwave field intensification at the dihedral angle on the pore surface near grain-boundary edge has made it possible to explain enhanced closure of thermodynamically stable faceted pores. The ponderomotive effect can be observed in materials, in which there is a small fraction of defects (such as vacancies) that are mobile and possess a significantly higher electric susceptibility compared with an average site in the crystalline lattice. A necessary condition for the ponderomotive effects is the presence of macroscopic structural nonuniformities in the material, such as pore surfaces and/or grain boundaries. The contribution of the microwave electromagnetic field effect to the densification rate is calculated starting from a model of the ponderomotive effect in the vicinity of an individual pore and then averaging over an array of such pores. The following issues are critical for the effect to be pronounced during sintering: (1) pore surfaces and grain boundaries possess limited permeability for the high-frequency vacancy flows; (2) the electric field is intensified quasi-electrostatically in the vicinity of dihedral angles at pore–grain-boundary junctions; and (3) significant microwave absorption is associated with the transport of mobile vacancies (or other mobile defects) in the material. As a result, it has been demonstrated that the role of the ponderomotive effect in enhanced densification can be highly significant under realistic microwave sintering conditions. Most experimentally observed nonthermal effects in high-temperature microwave processing discussed above are consistent with the ponderomotive model. These effects are more pronounced in the materials with higher ionic conductivity (hence vacancy mobility); the effects scale with the absorbed microwave power (or electric field square); the affected mass transport mechanisms are surface and grain-boundary diffusion, but rarely volume diffusion; in the case of linear polarization of the field, there is an anisotropy of the effect.

7.8 Grain Growth During Microwave Sintering

Although experimental efforts have been focused on the analyses of the kinetics of grain growth during microwave sintering [78], no reliable model is available yet to describe this important aspect of the microwave-assisted processing. Some studies included conventional models of grain growth for the modeling of the coupled densification – microstructure evolution kinetics during microwave sintering [65, 67]. No electromagnetic field effects have been included in this kind of studies.

Raj et al. [133] put forward a concept of microwave sintering densification and grain growth affected by the overheating of the grain-boundary regions; this issue was also analyzed by Johnson [74] when, in the beginning of the 1990s, the specifics of microwave plasma sintering were actively discussed. Overall, this important aspect of microwave sintering is clearly underdeveloped. At the same time, its importance cannot be underestimated, as one of the frequently claimed expected advantages of the microwave sintering is the grain size retention.

7.9 Selected Examples of Materials Consolidated by Microwave Sintering

This section discusses experimental approaches to improve the structural uniformity of the sintered samples and presents examples of the microstructure evolution of microwave sintered materials and practical applications of microwave sintering.

Holcombe and Dykes [134] showed that "casketing" – surrounding samples that are microwave susceptors themselves with coarse dense granules of materials transmitting infrared radiation – is important for obtaining uniform microstructures and avoid cracking related to thermal stresses. Granules of yttria, alumina, or magnesia can be used for this purpose. The presence of granules in contact with the sample minimizes the temperature difference between the sample's surface and its interior for microwave-absorbing materials, as heat that radiates away from the surface is reduced. Using casket arrangements, crack-free LiH and TiB_2-based pellets were obtained. Without casketing, crack in the sintered pellets were unavoidable.

The temperature homogeneity can also be improved by hybrid heating, in which case an external heat flux comes from a susceptor to the sample coupled to microwaves. As has been shown theoretically by Manière et al. [135, 136], during direct microwave heating, materials dissipating microwave power show inherent heating instability, which may cause non-reproducible experimental results. In contrast to hybrid heating, direct heating configurations favor the formation of hot spots resulting in nonhomogeneous densification and shape distortions of the specimen (Fig. 7.8).

Hybrid heating configurations reduce thermal inhomogeneity and lead to densification homogenization. Consequently, with the use of hybrid heating, densification

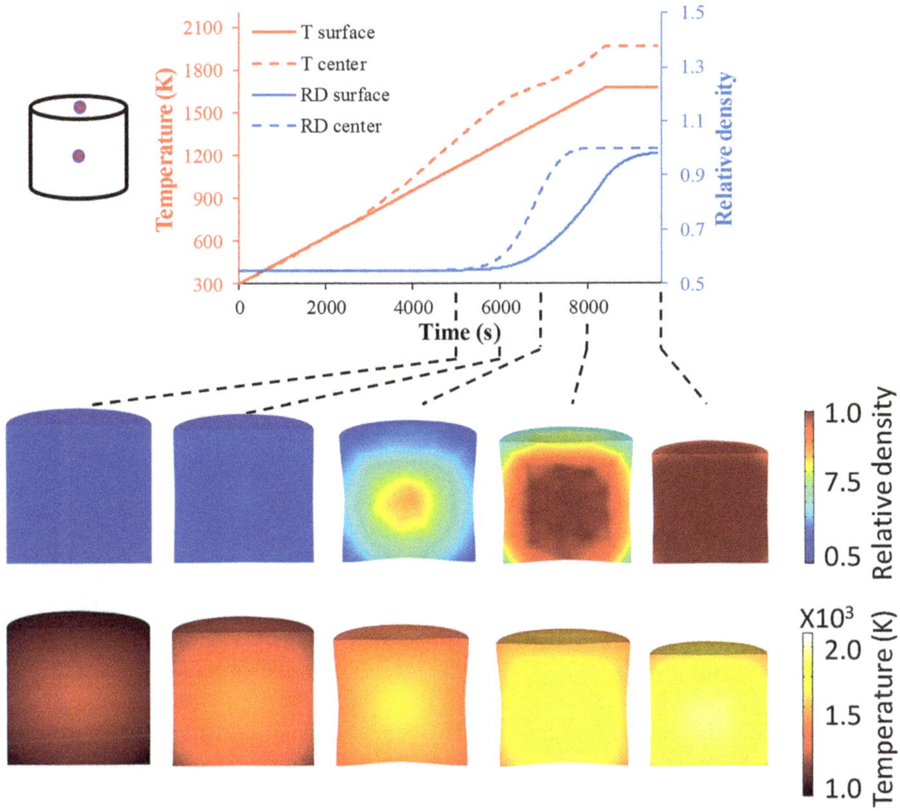

Fig. 7.8 Fully coupled electromagnetic–thermal–mechanical finite element modeling of relative density (RD) and temperature fields for the direct 3Y-ZrO_2 sample heating configuration (initial sample diameter is 7 mm). (Reprinted from Manière et al. [136], Copyright (2017) with permission of John Wiley & Sons)

of the specimen is achieved without shape distortions. Charmond et al. [137] observed differences in the grain size between the top, center, and bottom of yttria-stabilized tetragonal zirconia compacts obtained by direct heating microwave sintering.

Agarwal [138] reported successful microwave sintering of a number of pure metals and steels. An important result is a possibility of sintering refractory metals by microwaves at much lower temperatures than those required in conventional sintering. Thus, a molybdenum powder was sintered by microwaves at 1600 °C with a holding time of 1 min to a relative density of 98% while retaining the grain size in the submicron range. Anklekar et al. [139] found that the sintered density values of the conventionally sintered and microwave sintered copper and nickel steel parts were very close to each other. However, mechanical properties of the microwave sintered parts showed a marked improvement over those of the conventionally sintered parts. This was explained by the different microstructures and porosity

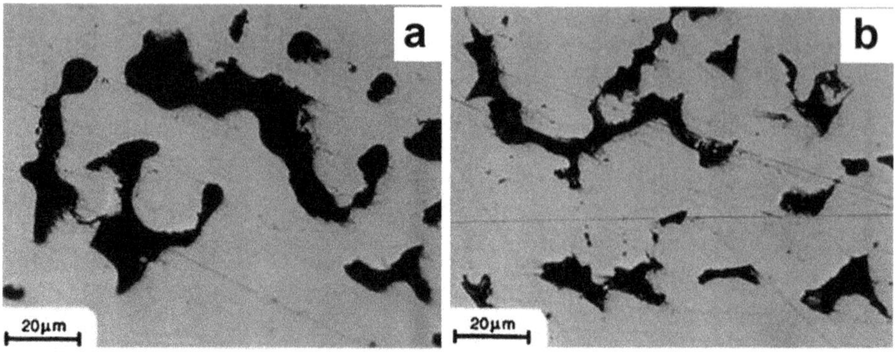

Fig. 7.9 Optical micrographs of FC-0208 (Fe-2Cu-0.8C) sintered at 1260 °C for 20 min in a reducing atmosphere: (**a**) microwave sintering, (**b**) conventional sintering (Reprinted from Anklekar et al. [139]. Copyright (2005) The Institute of Materials, Minerals and Mining, by permission of Taylor & Francis Ltd., www.tandfonline.com, on behalf of The Institute of Materials, Minerals and Mining)

distributions, those of the microwave sintered parts favoring better mechanical properties. Optical images of the sintered FC-0208 (Fe-2Cu-0.8C) alloy (Fig. 7.9) showed the presence of interconnected and isolated pores having distinct characteristics depending on the sintering method: in the microwave sintered part, the pores are smooth and rounded having very few stress concentration regions, while in the conventionally sintered part, the pores have sharp edges and triangular or wedge shapes with a lot of stress concentration regions. As was mentioned above, the density difference between these sintered parts was only minor: 6.94 $g{\cdot}cm^{-3}$ for the microwave and 6.90 $g{\cdot}cm^{-3}$ for the conventionally sintered material The toughness parameter of the sintered parts was determined on cylindrical tubular samples as a measure of ductility of the materials by using the following equation: $K=P(D-t)/Lt^2$, where P is the crushing strength or the minimum load for failure, L is the length of the cylindrical tubular sample, D is the outer diameter of the cylindrical tubular sample, and t is the wall thickness of the cylindrical tubular sample. The toughness parameter for materials, whose microstructures are shown in Fig. 7.8, was 637 MPa and 475 MPa for the microwave and conventionally sintered material, respectively. In addition to the pore shape differences, the microstructures of the microwave and conventionally sintered parts showed a difference in the character of distribution of copper – the alloying element – in the iron matrix. An altered distribution of alloying elements is believed to be another factor contributing to better mechanical properties of the microwave sintered alloyed steel parts.

The microwave sintering-induced dissolution–precipitation mechanism of grain growth was addressed by Bao and Yi [140]. They reported a lower activation energy for grain growth of tungsten carbide WC in WC–8Co cemented carbides processed by microwave sintering: the activation energy for conventionally sintered materials was found to be twice as high as that determined for microwave sintered cemented carbides. Based on these results, it was concluded that different mechanisms operate in these two sintering processes. It was suggested that local hot spots form in the microwave sintered WC–8Co and are associated with cobalt particles and particles of residual carbon. Within these hot spots, the cobalt binder experiences melting allowing for local rearrangement of the carbide grains. Furthermore, tungsten and carbon tend to dissolve into molten cobalt. Both particle rearrangement and the dissolution–precipitation process during microwave sintering lead to a faster grain growth as compared with conventional sintering. Despite the fact that grain growth in microwave sintered WC–8Co was more rapid than in conventionally sintered counterparts, grains in the microwave sintered materials were smaller due to a significantly shorter sintering time.

Growth of inter-particle necks during microwave sintering of metals was studied by Demirskyi et al. [141]. They applied a classical sphere-to-sphere approach to the analysis of the neck formation between loosely packed copper particles subjected to microwave treatment. A monolayer of spherical copper particles was placed on a microwave-transparent mullite substrate, which was then placed on an insulator. No additional susceptors were used in the experimental setup, which ensured direct microwave heating only. The neck size was measured both on fractured necks and contacting particles using scanning electron microscopy images. Similarities between the microwave and conventional sintering processes were found for long sintering times. However, during the initial stage of microwave sintering, anomalous neck growth was revealed. Experimentally determined large n values in $(x/a)^n \sim t$ relation, where x is the inter-particle neck diameter, a is the particle diameter, and t is the time, are indicative of nonthermal effects operating during microwave sintering. Indeed, in experiments on microwave sintering of copper, n was equal to 11.2 at 800 °C. This is a high value, considering $n = 7$ for the classic neck growth by surface diffusion. After the neck formation was complete, the kinetics of neck growth during microwave sintering became similar to that found for conventional sintering. Evidence of the formation of a liquid phase at inter-particle contacts was detected in microwave sintered nickel [141].

In a comparative study of the behavior of Distaloy AE in conventional and microwave sintering processes [142], it was found that microwave sintered samples obtained below a certain temperature (1050 °C) showed a better ductility and a higher strength than their conventionally sintered counterparts obtained at the same temperature and soaked for the same time. However, there were no significant differences between the properties of the samples sintered by the two methods at higher temperatures.

As microwaves are absorbed by metallic powders and are reflected by bulk metals, it is possible to use microwave power to braze metals and alloys using a powder brazing layer, which can be selectively heated by microwaves [138]. Using this approach, steel and cast iron were brazed by microwaves, the required treatment time being only 2–3 min.

Successful consolidation of metallic powders by microwave sintering allowed extending its applications to metal matrix composites. Thus, hybrid heating microwave sintering was combined with hot extrusion to produce rod-shape samples of aluminum matrix composites reinforced with Si_3N_4 nanoparticles [143] and magnesium matrix composites reinforced with SiO_2 nanoparticles [144]. The microwaves were absorbed internally by the sample, and the radiant heat was also provided externally from SiC susceptors. Microwave sintering was used to produce bulk Al-$Ni_{50}Ti_{50}$ composites, in which the amorphous structure of the $Ni_{50}Ti_{50}$ reinforcing particles was preserved after sintering [145].

Densification during microwave sintering can be improved by preliminary ball milling of the powder, as has been proven for a tungsten powder [146]. The compacts were sintered from the as-received powder and the powder obtained by high-energy ball milling of the as-received powder. The compact sintered from the as-received powder by microwaves at 1800 °C with a soaking time of 1 h was 85% dense, while that sintered from the milled powder under the same conditions was 93% dense. A higher relative density and a higher hardness achieved in the compact obtained by sintering of the ball milled powder were attributed to a larger specific surface area of the milled powder relative to the as-received powder and accumulation of strain during milling.

The influence of the powder structure on its behavior in the microwave field was also addressed in a study by Fang et al. [147], who reported the microwave treatment effect on alumina samples of different crystallinity. The samples to be sintered were obtained by thermal treatment of compacts made of a X-ray amorphous powder (precursor); treatment was conducted at 800–1500 °C. When annealed at 800 and 1000 °C, the powder was still poorly crystallized and contained γ-alumina; traces of α-alumina appeared after annealing at 1000 °C. Powders preheated up to 1200 °C and above and sintered samples were single-phase α-Al_2O_3. The effect of microwaves on densification was evaluated by comparing the results with those obtained by conventional sintering. Sintering was conducted for 10 min at 1500 °C. Densification of alumina in the microwave field was higher than that achieved in the conventional process at the same temperature. The maximum enhancement of 60% was observed for the X-ray amorphous alumina and decreased with an increase in its crystallinity. These observations were explained by a higher microwave absorption of the X-ray amorphous alumina compared with well-crystallized α-Al_2O_3.

Microwave sintering was successfully used to sinter B_6O monolithic ceramics [148], which is prone to decomposition – oxygen loss and formation of nonstoichiometric B_6O_x. Sintering by microwaves was chosen as a means to reduce the sintering temperature and avoid undesirable decomposition. Cold-pressed and then cold isostatically pressed compacts with a relative density of 45% were subjected to microwave treatment. The compacts were placed in a BN crucible with larger dimensions than those of the compacts and were surrounded by a mixture of hexagonal BN and tungsten powder. The mixture of BN and W powders acted as a microwave susceptor. Microwave sintering of compacts cold-pressed from a powder with 500 nm crystallites at 1850 °C for 5 min resulted in the formation of compacts with a relative density of more than 98% and grains with a size of 1.4 μm. These sintering conditions allowed obtaining the best flexural strength to toughness ratio in B_6O (400 MPa and 2.8 MPa·$m^{1/2}$, respectively).

In the presence of grain growth inhibitors, the advantage of microwave sintering over conventional sintering is less pronounced than in the case of materials, in which grain growth is not inhibited by additives. In microwave sintered alumina containing a ZrO_2 additive (5–15 vol.%), grain growth was governed by the additive, and only a slight difference was observed between the grain sizes of microwave and conventionally sintered compacts [149]. Compacts sintered by the two methods at the same temperature (1400 °C) were compared: the microwave sintered materials had a hardness of 18–19 GPa, and the conventionally sintered materials had a hardness of 16–17 GPa. This difference in the hardness was due to slightly smaller grains and higher relative densities of the microwave sintered materials. A shorter sintering cycle was an advantage of the microwave method for the Al_2O_3-ZrO_2 ceramics: the holding times were 10 min and 2 h in the microwave and conventional sintering, respectively.

In the area of materials for medicine, microwave sintering can be used for sintering of porcelains [150] and zirconia [151] for dental applications. Medeiros et al. [150] compared results of sintering by hybrid microwave heating with those obtained by conventional sintering in vacuum. Materials obtained by these two methods showed close values of porosity, flexural strength, and surface microhardness. Monaco et al. [151] suggested microwave sintering as a sintering method ensuring a more uniform microstructure of ZrO_2-based ceramics than conventional sintering. The latter can result in the formation of final components with microstructural flaws, which are highly undesirable in dental reconstruction. Microwave sintering in a single-mode applicator allowed reducing the sintering temperature from 1450 to 1200 °C and shortening the soaking time from 600 to 6 min to produce ZrO_2-based ceramics with an apparent porosity of 0.01%. An advantage of microwave sintering is also in hindering grain growth undesirable from the viewpoint of mechanical strength of the sintered components. Fast fabrication of zirconia components by microwave sintering makes it possible to introduce them in chairside treatments [151].

Microwave sintering was suggested as an efficient method of immobilizing radioactive soil waste [152]. A way to immobilize radioactive nuclides is to incorporate them into a glassy structure. It was found that soil samples experience vitrification upon high-temperature sintering, microwave sintering being more efficient enabling faster transformation of the crystalline material into a glass phase. Figure 7.10 shows XRD patterns of the soil sintered by conventional and microwave sintering at different temperatures. As the temperature during the microwave treatment increased, the transformation degree of the crystalline phases into a glass increased. The soil sample microwave sintered at 1300 °C for 30 min was fully vitrified, while the conventionally sintered material sintered at the same temperature for 2 h was still partially crystalline. Neodymium oxides Nd_2O_3 was used as a model contaminant, as the Nd^{3+} ion has an ionic radius close to those of actinides found in the contaminated soil and has the same valence with the actinide ions. The selected microwave sintering conditions were successfully used to produce glassy Nd-containing phases up to 25 wt.% of Nd_2O_3. The glassy materials demonstrated very low neodymium leaching rates, indicating that microwave sintering is a promising method of the safe disposal of radioactive soil.

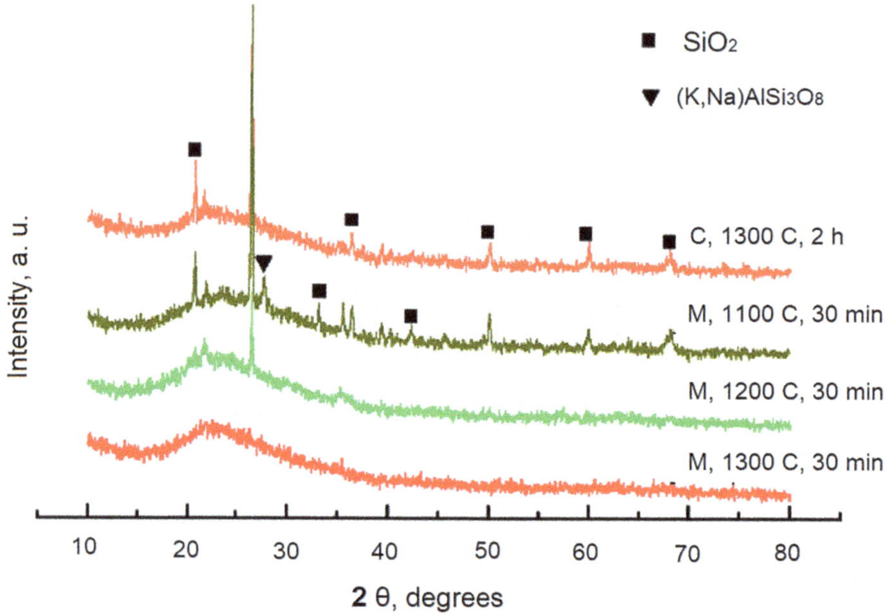

Fig. 7.10 XRD patterns of the soil sintered by conventional sintering at 1300 °C and microwave sintering at 1100, 1200, and 1300 °C. "M" denotes microwave sintering, "C" – conventional sintering (Reprinted from Zhang et al. [152]. Copyright (2017), with permission from Elsevier)

More examples of microwave sintering applications for the processing of metals, ceramics, and composite materials can be found in reviews by Oghbaei and Mirzaee [153], Agrawal [154], and Zhao and Chen [155].

7.10 Summary

Microwave sintering is a cross-disciplinary area of technology that requires expertise in electromagnetism, thermal engineering, and materials science. The optimization of microwave sintering requires extensive process modeling. While both the fundamentals and numerical calculation methods are well established for the electromagnetic and thermal modeling, extensive computation resources are required for the simulation of large multimode microwave cavity applicators. In addition, the dielectric property measurements in the entire range of the process temperatures are important in explaining the microwave absorption characteristics and in the validation of the modeling results, as this will be crucial for understanding and gaining industrial confidence on control procedures. The principal direction of the research in the modeling of microwave sintering can be seen in the development of multiphysics simulation frameworks that would combine the electromagnetic and thermal solutions with the advanced sintering models, taking into account microwave sintering-specific constitutive equations. This will make it possible to predict and optimize density distribution and shrinkage in the final sintered products. Ultimately, an effective microwave sintering process should enable the manufacturing of components with controllable structure (such as grain size) characteristics and macroscopic shape. The solution of this kind of problems requires multiscale approaches and special optimization techniques. Incorporating further advances into the studies of the nonthermal effects of the microwave electromagnetic field on mass transport phenomena, the comprehensive modeling should result in the development and industrial mastering of a microwave sintering technology, capable of producing high-quality materials with unique microstructures and properties.

References

1. Zong L, Zhou S, Sgriccia N, Hawley MC, Kempel LC (2003) A review of microwave-assisted polymer chemistry (MAPC). J Microw Power Electromagn Energy 38(1):49–74
2. Venkatesh MS, Raghavan GSV (2004) An overview of microwave processing and dielectric properties of agri-food materials. Biosyst Eng 88(1):1–18
3. Vongpradubchai V, Rattanadecho P (2009) The microwave processing of wood using a continuous microwave belt drier. J Chem Eng Process: Process Intensification 48(5):997–1003
4. Clark DE, Folz DC, Folgar CE, Mahmoud MM (eds) (2005) Microwave solutions for ceramic engineers. American Ceramic Society, Westerville
5. Rao KJ, Vaidhyanathan B, Ganguli M, Ramakrishnan PA (1999) Synthesis of inorganic solids using microwaves. Chem Mater 11(4):882–895
6. Makino Y, Ohmae T, Setsuhara Y, Miyake S, Sano S (1999) Sintering of $Al_2O_3 - ZrO_2$ composites using millimeter-wave radiation. Key Eng Mat 161–163:41–44
7. Gupta M, Leong EWW (2007) Microwaves and metals. Wiley Asia, Singapore
8. Mondal A (2011) Microwave sintering of metals. Lambert Academic Publishing, Saarbrücken
9. Oda SJ (1992) Microwave remediation of hazardous waste: a review. In: Beatty RL, Sutton WH, Iskander MF (eds) Microwave processing of materials III, Materials Research Society Symposium Proceedings, vol 269. Materials Research Society, Pittsburgh, pp 453–464
10. Zhang SL, Buchta R, Sigurd D (1994) Rapid thermal processing with microwave heating. Thin Solid Films 246(1–2):151–157
11. Alford TL, Thompson DC, Mayer JW, David Theodore N (2009) Dopant activation in ion implanted silicon by microwave annealing. J Appl Phys 106:114902
12. Janney MA, Kimrey HD, Allen WR, Kiggans JO (1997) Enhanced diffusion in sapphire during microwave heating. J Mater Sci 32:1347–1355
13. Whittaker AG (2005) Diffusion in microwave-heated ceramics. Chem Mater 17:3426–3432
14. Robb GR, Harrison A, Whittaker AG (2002) Temperature-resolved, in-situ powder X-ray diffraction of silver iodide under microwave irradiation. Phys Chem Comm 5:135–137
15. Osepchuk JM (1984) A history of microwave heating applications. IEEE Trans Microwave Theory Tech 32(9):1200–1224
16. Tinga WR, Voss WAG (1968) Microwave power engineering. Academic Press, New York
17. Berteaud AJ, Badot JC (1976) High temperature microwave heating in refractory materials. J Microw Power 11(4):315–320
18. Meek TT, Holcombe CE, Dykes N (1987) Microwave sintering of some oxide materials using sintering aids. J Mater Sci Lett 6(8):1060–1062
19. Lynn Johnson D (1991) Microwave and plasma sintering of ceramics. Ceram Int 17:295–300

20. Wang J, Binner JGP, Vaidhyanathan B, Joomun N, Kilner J, Dimitrakis G, Cross TE (2006) Evidence for the microwave effect during hybrid sintering. J Amer Ceram Soc 89 (6):1977–1984
21. Birnboim A, Gershon D, Calame J, Birman A, Carmel Y, Rodgers J, Levush B, Bykov Y, Eremeev A, Holoptsev V, Semenov V, Dadon D, Martin P, Rosen M, Hutcheon R (1998) Comparative study of microwave sintering of zinc oxide at 2.45, 30 and 83 GHz. J Amer Ceram Soc 81:1493–1501
22. Katz JD (1992) Microwave sintering of ceramics. Annu Rev Mater Sci 22:153–170
23. Clark D, Sutton WH (1996) Microwave processing of materials. Annu Rev Mater Sci 26:299–331
24. Agrawal DK (1998) Microwave processing of ceramics: a review. Curr Opin Solid State Mater Sci 3(5):480–486
25. Binner JGP, Vaidhyanathan B (2004) Microwave sintering of ceramics: what does it offer? Key Eng Mater 264–268:725–730
26. Binner JGP, Annapoorani K, Paul A, Santacruz I, Vaidhyanathan B (2008) Dense nanostructured zirconia by two stage conventional/hybrid microwave sintering. J Eur Ceram Soc 28:973–977
27. Chen LW, Wang XH (2000) Sintering dense nanocrystalline oxide without final stage grain growth. Nature 404:168–171
28. Willert-Porada M, Borchert R (1997) Microwave sintering of metal-ceramic FGM. In: Shiota I, Miyamoto Y (eds) Functionally graded materials. Elsevier, Amsterdam, pp 349–354
29. Gerdes T, Willert-Porada M (1994) Microwave sintering of metal-ceramic and ceramic-ceramic composites. In: Iskander MF, Sutton WH, Lauf RJ (eds) Microwave processing of materials IV, Materials Research Society Symposium Proc, vol 347. Materials Research Society, Pittsburgh, pp 531–538
30. Roy R, Agrawal D, Cheng J, Gedevanishvili S (1999) Full sintering of powdered-metal bodies in a microwave field. Nature 399:668–670
31. Rybakov KI, Olevsky EA, Krikun EV (2013) Microwave sintering: fundamentals and modeling. J Amer Ceram Soc 96(4):1003–1020
32. Troch PA, Vandersteene F, Su Z, Hoeben R, Wuethrich M (1996) Estimating microwave observation depth in bare soil through multi-frequency scatterometry. Proc 1st EMSL User Workshop. Ispra, Italy, SAI, JRC
33. Hashin Z, Shtrikman S (1962) A variational approach to the theory of the effective magnetic permeability of multiphase materials. J Appl Phys 33(10):3125–3131
34. Ishizaki K, Battabyal M, Pittini YY, Nicula R, Vaucher S (2010) Microwave sintering explored by X-ray microtomography. In: Bordia RK, Olevsky EA (eds) Advances in sintering science and technology, ceramic transactions, vol 209. The American Ceramic Society, Westerville, pp 211–217
35. Penn SJ, Alford NM, Templeton A, Wang X, Xu M, Reece M, Schrapel K (1997) Effect of porosity and grain size on the microwave dielectric properties of sintered alumina. J Amer Ceram Soc 80(7):1885–1888
36. Bruggeman DAG (1935) Berechnung Verschiedener Physikalischer Konstanten von Heterogenen Substanzen, I. Dielektriziätskonstanten und Leitfähigkeitender Mischkörper aus Isotropen Substanzen. Ann Phys-Berlin Series 5(24):636–679
37. Rybakov KI, Semenov VE, Egorov SV, Eremeev AG, Plotnikov IV, Bykov YV (2006) Microwave heating of conductive powder materials. J Appl Phys 99:023506
38. Buchelnikov VD, Louzguine-Luzgin DV, Xie G, Li S, Yoshikawa N, Sato M, Anzulevich AP, Bychkov IV, Inoue A (2008) Heating of metallic powders by microwaves: experiment and theory. J Appl Phys 104:113505
39. Cheng J, Roy R, Agrawal D (2002) Radically different effects on materials by separated microwave electric and magnetic fields. Mater Res Innovat 5:170–177
40. Bhattacharya M, Basak T (2008) Generalized scaling on forecasting heating patterns for microwave processing. AIChE J 54(1):56–73

41. Egorov SV, Rybakov KI, Semenov VE, Bykov YV, Kanygina ON, Kulumbaev EB, Lelevkin VM (2007) Role of convective heat removal and electromagnetic field structure in the microwave heating of materials. J Mater Sci 42:2097–2104
42. Birnboim A, Olorunyolemi T, Carmel Y (2001) Calculating the thermal conductivity of heated powder compacts. J Amer Ceram Soc 84(6):1315–1320
43. Olorunyolemi T, Birnboim A, Carmel Y, Wilson OC, Lloyd IK (2002) Thermal conductivity of zinc oxide: from green to sintered state. J Amer Ceram Soc 85(5):1249–1253
44. Tian YL, Black WM, Sa'adaldin HS, Ahmad I, Silberglitt R (1995) Dynamic model for electromagnetic field and heating patterns in loaded cylindrical cavities. In: Clark DE, Folz DC, Oda SJ, Silberglitt R (eds) Microwaves: theory and application in materials processing III, ceramic transactions, vol 59. The American Ceramic Society, Westerville, pp 261–268
45. Peng H, Tinga WR, Sundararaj U, Eadie RL (2003) Microwave sintering process model. J Microw Power Electromagn Energy 38(4):243–258
46. Lasri J, Ramesh PD, Schachter L (2000) Energy conversion during microwave sintering of a multiphase ceramic surrounded by a susceptor. J Amer Ceram Soc 83(6):1465–1468
47. Duan Y, Sorescu DC, Johnson JK (2006) Finite element approach to microwave sintering of oxide materials. Proc COMSOL Users Conference, Boston
48. Santos T, Valente MA, Monteiro J, Sousa J, Costa LC (2011) Electromagnetic and thermal history during microwave heating. Appl Therm Eng 31(16):3255–3261
49. Roussy G, Chenot P, Colin P, Thiebaut JM (1980) Control of microwave-heating of granular materials. Thermochim Acta 41(2):225–236 (in French)
50. Roussy G, Mercier J (1985) Temperature runaway of microwave heated materials: study and control. J Microw Power 20(1):47–51
51. Roussy G, Bennani A, Thiebaut JM (1987) Temperature runaway of microwave irradiated materials. J Appl Phys 62(4):1167–1170
52. Coleman CJ (1991) On the microwave hotspot problem. J Aust Math Soc Series B-Applied Mathematics 33:1–8
53. Fliflet AW (2008) Self-consistent electromagnetic-thermal model for calculating the temperature of a ceramic cylinder irradiated by a high-power millimeter-wave beam. IEEE Trans Plasma Sci 36(3):582–590
54. Parris PE, Kenkre VM (1997) Thermal runaway in ceramics arising from the temperature dependence of the thermal conductivity. Phys Status Solidi B 200(1):39–47
55. Alliouat M, Lecluse Y, Massieu J, Mazo L (1990) Control algorithm for microwave sintering in a resonant system. J Microw Power Electromagn Energy 25(1):25–31
56. Liu B, Marchant TR (2002) The occurrence of limit-cycles during feedback control of microwave heating. Math Comput Model 35(9–10):1095–1118
57. Beale GO, Arteaga FJ, Black WM (1992) Design and evaluation of a controller for the process of microwave joining of ceramics. IEEE Trans Ind Electron 39(4):301–312
58. Rybakov KI, Semenov VE (1996) Densification of powder materials in nonuniform temperature fields. Phil Mag A 73(2):295–307
59. Zharova NA, Rybakov KI, Semenov VE, Egorov SV (2001) Computer simulation of millimeter-wave sintering of ceramic and composite materials. In: Clark DE, Sutton WH, Lewis DA (eds) Microwaves: theory and application in material processing V, ceramic transactions, vol 111. The American Ceramic Society, Westerville, pp 11–18
60. Birnboim A, Carmel Y (1999) Simulation of microwave sintering of ceramic bodies with complex geometry. J Amer Ceram Soc 82(11):3024–3030
61. Egorov SV, Zharova NA, Bykov YV, Semenov VE (2006) Microwave sintering of large-size ceramic workpieces. In: Willert-Porada M (ed) Advances in microwave and radio frequency processing, Proc 8th Int Conf Microwave and High-Frequency Heating. Springer, Berlin–Heidelberg–New York, pp 577–582
62. Bouvard D, Charmond S, Carry CP (2010) Finite element modelling of microwave sintering. In: Bordia RK, Olevsky EA (eds) Advances in sintering science and technology, ceramic transactions, vol 209. The American Ceramic Society, Westerville, pp 173–180

63. Su H, Johnson DL (1996) Master sintering curve: a practical approach to sintering. J Amer Ceram Soc 79(12):3211–3217
64. Bykov YV, Egorov SV, Eremeev AG, Plotnikov IV, Rybakov KI, Semenov VE, Sorokin AA, Holoptsev VV (2012) Fabrication of metal-ceramic functionally graded materials by microwave sintering. Inorg Mater Appl Res 3(3):261–269
65. Chatterjee A, Basak T, Ayappa KG (1998) Analysis of microwave sintering of ceramics. AIChE J 44(10):2302–2311
66. Darcovich K, Whitfield PS, Amow G, Shinagawa K, Miyahara RY (2005) A microstructure based numerical simulation of microwave sintering of specialized SOFC materials. J Eur Ceram Soc 25:2235–2240
67. Riedel R, Svoboda J (2006) Simulation of microwave sintering with advanced sintering models. In: Willert-Porada M (ed) Advances in microwave and radio frequency processing. Springer, Berlin, pp 210–216
68. Rothman SJ (1994) Critical assessment of microwave-enhanced diffusion. In: Iskander MF, Lauf RJ, Sutton WH (eds) Microwave processing of materials IV, Materials Research Society Symposium Proceedings, vol 347. Materials Research Society, Pittsburgh, pp 9–18
69. Beruto D, Botter R, Searcy AW (1989) Influence of temperature gradients on sintering: experimental tests of a theory. J Amer Ceram Soc 72:232–235
70. Young RM, McPherson R (1989) Temperature-gradient-driven diffusion in rapid-rate sintering. J Amer Ceram Soc 72:1080–1081
71. Olevsky E, Froyen L (2009) Impact of thermal diffusion on densification during SPS. J Amer Ceram Soc 92S:122–132
72. Bykov YV, Eremeev AE, Holoptsev VV (1994) Experimental study of the non-thermal effect in microwave sintering of piezoceramics. In: Iskander MF, Lauf RJ, Sutton WH (eds) Microwave processing of materials IV, Materials Research Society Symposium Proceedings, vol 347. Materials Research Society, Pittsburgh, pp 585–590
73. Meek TT (1987) Proposed model for the sintering of a dielectric in a microwave field. J Mater Sci Lett 6:638–640
74. Johnson DL (1991) Microwave heating of grain boundaries in ceramics. J Amer Ceram Soc 74:849–850
75. Endicott MR, Kenkre VM, Kus M (1994) Theory of a confinement effect of dipole rotations resulting in saturation in microwave heating of ceramics. Phys Stat Sol (b) 184(1):99–111
76. Stuerga DAC, Gaillard P (1996) Microwave athermal effects in chemistry: a myth's autopsy. J Microw Power Electromagn Energy 31:87–113
77. Booske JH, Cooper RF (2006) How the coupling of microwave and RF energy in materials can affect solid state charge and mass transport and result in unique processing effects. In: Willert-Porada M (ed) Advances in microwave & radio frequency processing. Springer, Berlin, pp 461–471
78. Janney MA, Kimrey HD, Schmidt MA, Kiggans JO (1991) Grain growth in microwave-annealed alumina. J Amer Ceram Soc 74(7):1675–1681
79. Binner JGP, Hassine NA, Cross TE (1995) The possible role of the pre-exponential factor in explaining the increased reaction rates observed during the microwave synthesis of titanium carbide. J Mater Sci 30:5389–5393
80. Eremeev AG, Plotnikov IV, Rybakov KI, Bykov YV, Rachkovskii AI (2007) Comparative study of diffusion rates during lead titanate synthesis under microwave and conventional heating. In: Silaghi AM, Gordan IM (eds) Proc 11th Int Conf Microwave and high frequency heating. Editura Universitatii din Oradea, Oradea, Romania, pp 232–235
81. Bykov YV, Eremeev AG, Zharova NA, Plotnikov IV, Rybakov KI, Drozdov MN, Drozdov YN, Skupov VD (2003) Diffusion processes in semiconductor structures during microwave annealing. Radiophys Quantum Electron 46(8–9):749–755
82. Kimura H, Yamazaki Y (2006) Millimeter wave thermo-mechanical processing for bulk nanocrystalline ceramics. In: Proc International Microwave Power Institute's 40th Annual Symposium, pp 52–55

83. Egorov SV, Eremeev AG, Plotnikov IV, Sorokin AA, Bykov YV, Chuvil'deev VN, Gryaznov MY, Shotin SV (2008) Plastic deformation of ultrafine alumina ceramics under microwave heating. In: Proc. Global Congress on Microwave energy applications. Japan Society of Electromagnetic Wave Energy Applications, Tokyo, Japan, pp 65–68
84. Lee JN, Choi YW, Lee BJ, Ahn BT (1997) Microwave-induced low-temperature crystallization of amorphous silicon thin films. J Appl Phys 82:2918–2921
85. Rowley AT, Wroe R, Vazquez-Navarro D, Lo W, Cardwell DA (1997) Microwave-assisted oxygenation of melt-processed bulk $YBa_2Cu_3O_{7-\delta}$ ceramics. J Mater Sci 32:4541–4547
86. Wilson DA, Lee KY, Case ED (1997) Diffusive crack-healing behavior in polycrystalline alumina: a comparison between microwave annealing and conventional annealing. Mater Res Bull 32:1607–1616
87. Get'man OI, Panichkina VV, Radchenko PY, Samelyuk AV, Skorokhod VV, Eremeev AG, Plotnikov IV, Matsokin VP (2008) Effect of microwave heating on diffusion in KCl–KBr single crystals. Powd Metall Metal Ceram 47(11–12):660–668
88. Joomun NI, Kilner JA, Wang J, Vaidhyanathan B, Binner JGP (2003) Microwave hybrid annealing and its effect on oxygen diffusion in yttria-stabilized zirconia polycrystals. In: Proc 9th Int Conf on MW & HF Heating, Loughborough, UK, pp 405–408
89. Bykov Y, Eremeev A, Holoptsev V (1996) Influence of specific absorbed microwave power on activation energy of densification in ceramic materials. In: Iskander MF, Kiggans JO, Bolomey JC (eds) Microwave processing of materials V, Materials Research Society Symposium Proceedings, vol 430. Materials Research Society, Pittsburgh, pp 385–390
90. Wroe R, Rowley AT (1996) Evidence for a non-thermal microwave effect in the sintering of partially stabilized zirconia. J Mater Sci 31:2019–2026
91. Binner J, Vaidhyanathan B, Wang J, Price D, Reading M (2008) Evidence for non-thermal microwave effects using single and multimode hybrid conventional/microwave systems. J Microw Power Electromagn Energy 42(2):47–63
92. Willert-Porada M (1997) A microstructural approach to the origin of 'microwave effects' in sintering of ceramics and composites. In: Clark DE, Sutton WH, Lewis DA (eds) Microwaves: theory and application in materials processing IV, ceramic transactions, vol 80. The American Ceramic Society, Westerville, pp 153–164
93. Lange FF (1984) Sinterability of agglomerated powders. J Amer Ceram Soc 67(2):83–89
94. Willert-Porada M (1996) Microwave effects on spinodal decomposition. In: Iskander MF, Kiggans JO, Bolomey JC (eds) Microwave processing of materials V, Materials Research Society Symposium Proceedings, vol 430. Materials Research Society, Pittsburgh, pp 403–409
95. Li J, Huang Y, Xie Z (1996) Microwave interaction with ceramics and its application to spinodal decomposition. In: Koumoto K, Sheppard LM, Matsubara H (eds) Mass and charge transport in ceramics, ceramic transactions, vol 71. The American Ceramic Society, Westerville, pp 259–268
96. Get'man OI, Panichkina VV, Radchenko PY, Skorokhod VV, Andreeva MG, Eremeev AG, Kholoptsev VV (2009) Diffusion processes and structurization in microwave sintering of $BaTiO_3$–$SrTiO_3$ and Al_2O_3–Cr_2O_3 powder systems with complete miscibility. Powd Metall Metal Ceram 48(5–6):279–289
97. Thakur OP, Prakash C, Agrawal DK (2002) Dielectric behavior of $Ba_{0.95}Sr_{0.05}TiO_3$ ceramics sintered by microwave. J Mater Sci Eng B 96:221–225
98. Bykov YV, Egorov SV, Eremeev AG, Rybakov KI, Semenov VE, Sorokin AA, Gusev SA (2001) Evidence for microwave enhanced mass transport in the annealing of nanoporous alumina membranes. J Mater Sci 36:131–136
99. Egorov SV, Eremeev AG, Rybakov KI, Semenov VE, Sorokin AA, Gusev SA (2004) Microwave intensity-dependent mass transport enhancement in nanostructured alumina membranes. In: Folz DC, Booske JH, Clark DE, Gerling JF (eds) Microwave and radio frequency applications (Proc Third World Congress on Microwave and radio frequency applications). The American Ceramic Society, Westerville, pp 167–174

100. Link G, Miksch S, Takayama S, Thumm M (2006) Anisotropic sintering in polarized microwave fields – Evidence for non-thermal microwave effects. In: Proc. Joint 31st Int Conf Infrared Millimeter Waves and 14th Int Conf Teraherz Electronics, IRMMW-THz 2006, p 285
101. Rybakov KI, Semenov VE, Link G, Thumm M (2007) Preferred orientation of pores in ceramics under heating by a linearly polarized microwave field. J Appl Phys 101:084915
102. Jones M, Valecillos MC, Hirao K, Brito ME, Toriyama M (2006) Sintering behaviour and mechanical properties of microwave sintered silicon nitride. In: Willert-Porada M (ed) Advances in microwave and radio frequency processing. Springer, Berlin, pp 562–569
103. Jung YM, Kim SW (2008) Effect of magnesium addition on the phase transformation of α-alumina prepared from route of ammonium aluminum carbonate hydroxide. Solid State Phenom 135:139–142
104. Vaidhyanathan B, Saremi-Yarahmadi S, Wijayanth KGU (2011) Fabrication of nanostructured α-Fe_2O_3 films for solar-driven hydrogen generation using hybrid heating. In: Mathur S, Widjaja S, Singh D (eds) Nanostructured materials and nanotechnology V, ceramic engineering and science proceedings, vol 32(7). Wiley, Hoboken, pp 11–22
105. Rybakov KI, Eremeev AG, Egorov SV, Bykov YV, Pajkic Z, Willert-Porada M (2008) Effect of microwave heating on phase transformations in nanostructured alumina. J Phys D Appl Phys 41:102008
106. Roy R, Peelamedu R, Grimes C, Cheng J, Agrawal D (2002) Major phase transformations and magnetic property changes caused by electromagnetic fields at microwave frequencies. J Mater Res 17(12):3008–3011
107. Roy R, Fang Y, Cheng J, Agrawal D (2005) Decrystallizing solid crystalline titania, without melting, using microwave magnetic fields. J Amer Ceram Soc 88(6):1640–1642
108. Kashcheev VA, Poluektov PP (1991) The use of alternating electric field for the stimulation of diffusion flow of charged impurities. Sov Tech Phys Lett 17:577
109. Booske JH, Cooper RF, Dobson I (1992) Mechanisms for nonthermal effects on ionic mobility during microwave processing of crystalline solid. J Mater Res 7(2):495–501
110. Bokhan YI (1992) Diffusion of charged impurities in high-frequency field. Sov Tech Phys Lett 18:339
111. Freeman SA, Booske JH, Cooper RF (1995) Microwave field enhancement of charge transport in sodium chloride. Phys Rev Lett 74:2042–2045
112. Bokhan YI, Shkrob IA (1994) Synthesis in an RF field of a ceramic material with a structural phase transition. Tech Phys Lett 20(6):439
113. Hao HS, Xu LH, Huang Y, Zhang XM, Xie ZP (2009) Kinetics mechanism of microwave sintering in ceramic materials. Sci China – Series E Technol Sci 52(9):2727–2731
114. Bergese P (2006) Specific heat, polarization and heat conduction in microwave heating systems: a nonequilibrium thermodynamic point of view. Acta Mater 54:1843–1849
115. Rybakov KI, Semenov VE, Freeman SA, Booske JH, Cooper RF (1997) Dynamics of microwave-induced currents in ionic crystals. Phys Rev B 55(6):3559–3567
116. Kittel C (1996) Introduction to solid state physics, 7th edn. Wiley, New York
117. Manning JR (1968) Diffusion kinetics for atoms in crystals. Van Nostrand-Reinhold, Princeton
118. Rybakov KI, Olevsky EA, Semenov VE (2012) The microwave ponderomotive effect on ceramic sintering. Scr Mater 66:1049–1052
119. Olevsky EA, Maximenko AL, Grigoryev EG (2013) Ponderomotive effects during contact formation in microwave sintering. Modelling Simul Mater Sci Eng 21:055022
120. Rybakov KI, Semenov VE, Link G, Thumm M (2007) Preferred orientations of pores in ceramics under heating by a linearly polarized microwave field. J Appl Phys 101:084915
121. Calame JP, Rybakov KI, Carmel Y, Gershon D (1997) Electric field intensification in spherical neck ceramic microstructures during microwave sintering. In: Clark DE et al (eds) Microwaves: theory and application in materials processing IV, ceramic transactions, vol 80. The American Ceramic Society, Westerville, pp 135–142

122. Booske JH, Cooper RF, Freeman SA, Rybakov KI, Semenov VE (1998) Microwave ponderomotive forces in solid state ionic plasmas. Phys Plasmas 5(5):1664–1670
123. Olevsky EA (1998) Theory of sintering: from discrete to continuum. Mater Sci Eng R 23 (2):41–100
124. Olevsky EA, Tikare V, Garino T (2006) Multi-scale study of sintering: a review. J Amer Ceram Soc 89(6):1914–1922
125. Olevsky EA, Molinari A (2000) Instability of sintering of porous bodies. Int J Plasticity 16:1–37
126. Rybakov KI, Semenov VE (1999) In: Vincenzini P (ed) Ceramics: getting into the 2000's – part C (Proceedings of the 9th Cimtec – World Ceramic Congress). TechnaSrl, Faenza, pp 397–404
127. Ding L, Davidchack RL, Pan J (2009) A molecular dynamics study of sintering between nanoparticles. Comput Mater Sci 45:247–256
128. Bachvalov NS, Panasenko GP (1989) Homogenization: averaging processes in periodic Media. Kluwer, Dordrecht
129. Bardzokas DI, Zobnin AI (2005) Mathematical modeling of physical processes in composite materials with periodical structure, Editorial URSS, Moscow, Russia, 336 p
130. Maximenko A, Olevsky E (2004) Effective diffusion coefficients in solid-state sintering. Acta Mater 52:2953–2963
131. Rahaman MN (1995) Ceramic processing and sintering. Marcel Dekker Inc, New York
132. Hague DC, Mayo MJ (1999) Sinter-forging of nanocrystalline zirconia: II. Simulation J Amer Ceram Soc 82:545–555
133. Raj R, Cologna M, Francis JSC (2011) Influence of externally imposed and internally generated electrical fields on grain growth, diffusional creep, sintering and related phenomena in ceramics. J Amer Ceram Soc 94(7):1941–1965
134. Holcombe CE, Dykes NL (1990) Importance of "casketing" for microwave sintering of materials. J Mater Sci Lett 9:425–428
135. Manière C, Zahrah T, Olevsky EA (2017) Inherent heating instability of direct microwave sintering process: sample analysis for porous 3Y-ZrO_2. Scr Mater 128:49–52
136. Manière C, Zahrah T, Olevsky EA (2017) Fully coupled electromagnetic-thermal-mechanical comparative simulation of direct vs hybrid microwave sintering of 3Y-ZrO_2. J Amer Ceram Soc 100(6):2439–2450
137. Charmond S, Carry CP, Bouvard D (2010) Densification and microstructure evolution of Y-Tetragonal Zirconia polycrystal powder during direct and hybrid microwave sintering in a single-mode cavity. J Eur Ceram Soc 30:1211–1221
138. Agrawal D (2006) Microwave sintering, brazing and melting of metallic materials, Sohn International Symposium "Advanced processing of metals and materials volume 4 – New, improved and existing technologies: Non-ferrous materials extraction and processing. In: Kongoli F, Reddy RG (eds) TMS (The Minerals, Metals & Materials Society), pp 183–192
139. Anklekar RM, Bauer K, Agrawal DK, Roy R (2005) Improved mechanical properties and microstructural development of microwave sintered copper and nickel steel PM parts. Powder Metall 48(1):39–46
140. Bao R, Yi J (2014) Densification and alloying of microwave sintering, WC–8wt.%Co composites. Int J Refract Met Hard Mater 43:269–275
141. Demirskyi D, Agrawal D, Ragulya A (2010) Neck growth kinetics during microwave sintering of copper. Scr Mater 62:552–555
142. Schmidt J, Schubert T, Weißgärber T, Kieback B (2004) Microwave assisted sintering of metallic materials, Proc. Euro PM2004
143. Reddy Matli P, Ubaid F, Abdul Shakoor R, Parande G, Manakari G, Yusuf M, Mohamed Amer Mohamed A, Gupta M (2017) Improved properties of Al–Si_3N_4 nanocomposites fabricated through a microwave sintering and hot extrusion process. RSC Adv 7:34401–34410

144. Parande G, Manakari V, Meenashisundaram GK, Gupta M (2016) Enhancing the hardness/compression/damping response of magnesium by reinforcing with biocompatible silica nanoparticulates. Int J Mater Res 107:1091–1099
145. Penchal Reddy M, Ubaid F, Shakoor RA, Mohamed AMA, Madhuri W (2016) Structural and mechanical properties of microwave sintered Al-$Ni_{50}Ti_{50}$ composites. J Science: Adv Mater Devices 1:362–366
146. Prabhu G, Chakraborty A, Sarma B (2009) Microwave sintering of tungsten. Int J Refract Met Hard Mater 27:545–548
147. Fang Y, Cheng J, Agrawal DK (2004) Effect of powder reactivity on microwave sintering of alumina. Mater Lett 58:498–501
148. Demirskyi D, Vasylkiv O (2016) Microstructure and mechanical properties of boron suboxide ceramics prepared by pressureless microwave sintering. Ceram Int 42:14282–14286
149. Benavente R, Salvador MD, Penaranda-Foix FL, Pallone E, Borrell A (2014) Mechanical properties and microstructural evolution of alumina–zirconia nanocomposites by microwave sintering. Ceram Int 40:11291–11297
150. Danielle Sales Cunha Medeiros F, Menezes RR, Neves GA, Navarrode Lima Santana L, Sivini Ferreira H, Silva Guedes de Lima D, Jackson Guedes de Lima S (2015) Microwave-assisted sintering of dental porcelains. Ceram Int 41:7501–7510
151. Monaco C, Prete F, Leonelli C, Esposito L, Tucci A (2015) Microstructural study of microwave sintered zirconia for dental applications. Ceram Int 41:1255–1261
152. Zhang S, Shu X, Chen S, Yang H, Hou C, Mao X, Chi F, Song M, Lu X (2017) Rapid immobilization of simulated radioactive soil waste by microwave sintering. J Hazard Mater 337:20–26
153. Oghbaei M, Mirzaee O (2010) Microwave versus conventional sintering: a review of fundamentals, advantages and applications. J Alloys Compd 494:175–189
154. Agrawal D (2006) Microwave sintering of ceramics, composites and metallic materials, and melting of glasses. Trans Indian Ceram Soc 65(3):129–144
155. Zhao Y, Chen J (2008) Applications of microwaves in nuclear chemistry and engineering. Prog Nuclear Energy 50:1–6

Chapter 8
Induction Heating Sintering

8.1 Principle of Induction Heating Sintering

Induction heating is heating by eddy currents created in an electrically conductive material by an alternating magnetic field. Similar to Eq. (7.4), the penetration of the eddy currents D is determined as

$$D = \frac{1}{\sqrt{\pi f \mu \sigma}}, \tag{8.1}$$

where f is the frequency, μ is the material magnetic permeability, and σ is the material electrical conductivity.

Induction sintering was suggested as a less costly process compared with conventional sintering [1], as the induction units are smaller, simpler, and easier to maintain than conventional sintering furnaces. As induction heating is fast, it can be conducted without a protective atmosphere. Two schemes are possible that allow utilizing the effect of induction heating for the sintering of powders. In the first scheme, a much more common one, a conductive container or a die, playing the role of a susceptor, is heated by the eddy currents, while the powder in the die is heated through radiation and thermal conduction [2–4]. This scheme is usually used in medium- and high-frequency processes. In the second scheme, the eddy currents are induced directly in the porous compact to be sintered, which is placed inside a coil carrying a high-frequency alternating current [5]. As significant induced currents can only be generated in conductors, direct induction sintering requires powder compacts having high electrical conductivity in the beginning of the process. A feature of the induction sintering technique is the cold environment of the heated compact [5].

Both schemes of induction heating sintering provide fast heating and fast cooling (the latter happens when the coil current is switched off), dramatically shorten the sintering time, and ensure energy efficiency. For successful induction sintering without a susceptor, the compact should be electrically conductive and of a simple shape to ensure uniform heating. For parts with irregular cross-sections, coils of

E. A. Olevsky, D. V. Dudina, *Field-Assisted Sintering*,
https://doi.org/10.1007/978-3-319-76032-2_8

complicated design may be necessary. A dramatic reduction of the sintering time can be seen from an example given by Hermel et al. [2]: an iron compact induction sintered for 12 min reaches the strength of the compact conventionally sintered for 60 min at the same temperature.

The energy is efficiently transferred during induction sintering if the following requirements are satisfied [2]:

- The penetration depth δ should be smaller than the thickness d of the part: $2\delta < d$.
- The cross-sections of the coil and the sintered part should be similar with small gaps between them.
- Undesirable thermal gradients should be avoided by selecting optimal heating rates and penetration depths.

Direct induction sintering is used less frequently than induction sintering with a susceptor, as direct sintering is conducted without a die and proceeds, therefore, without the application of pressure.

Conta [6] reported the existence of an incubation period at the beginning of induction heating of unsintered steel powder compacts at 3 kHz. This incubation period was due to poor coupling of the metal compact to the induction coil and, as a result, slow and inefficient heating. Poor coupling was caused by a very high initial resistance of the material. For eliminating the incubation period problem, preheating of the material within the reference depth was suggested by either radiation heating or radio-frequency (RF) induction heating. Preheating resulted in a more than an order of magnitude reduction in the resistivity and considerable net energy savings during the subsequent induction heating to the sintering temperature.

Induction heating can be conducted at high rates – from 100 $°C \cdot min^{-1}$ [7, 8] to 1400 $°C \cdot min^{-1}$ [9]. As the alternating current is switched off, cooling of the induction sintered samples occurs at 500–600 $°C \cdot min^{-1}$ [10, 11]. The temperature during induction sintering is usually controlled by a pyrometer focused on the wall of the die. In direct induction sintering, the pyrometer is focused on the specimen's surface.

Fast densification is characteristic of induction sintering. Kim et al. [10] reported the formation of nearly fully dense WC–Co by induction heating sintering within 1 min. Figure 8.1 shows the evolution of the specimen's displacement and the temperature of the surface of the graphite die (susceptor) with the heating time during induction heating sintering of WC–15vol.%Co at a pressure of 60 MPa. When the powder was heated up to 950 °C, no significant shrinkage was observed. As the temperature was raised to 1180 °C, fast densification occurred resulting in the formation of a compact with a relative density of 99.4%. Accelerated densification by induction heating was suggested to be caused by a combination of electrical discharge, resistance heating, and pressure effects. A possibility of localized heating at inter-particle contacts causing surface melting and oxide breakdown was also considered.

Interestingly, in the same study [10], WC–15vol.%Co materials with a relative density reaching 96.6% were obtained by presureless induction sintering. In those experiments, compacts with a green density of 52% were subjected to pressureless sintering for 150–500 s at the maximum temperature of 1150 °C. Figure 8.2

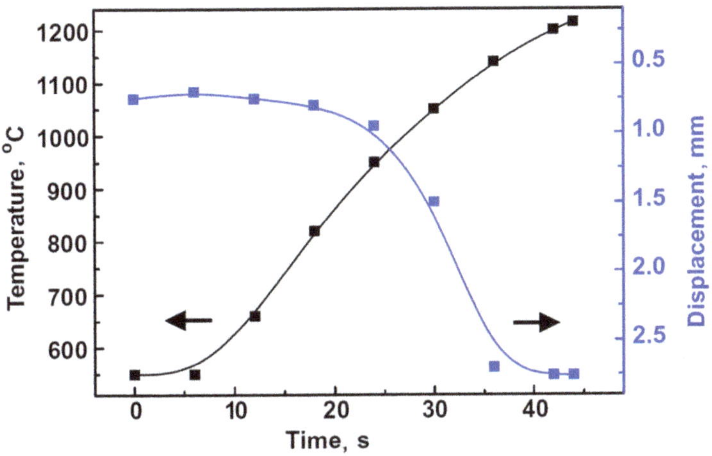

Fig. 8.1 Variations of the temperature and displacement with the heating time during induction sintering of WC–15vol.%Co materials at a pressure of 60 MPa. (Reprinted from Kim et al. [10]. Copyright (2004) with permission from Elsevier)

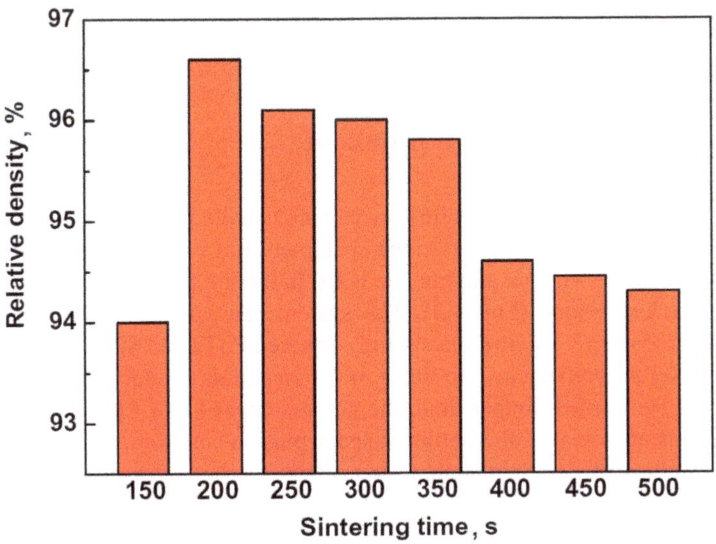

Fig. 8.2 Variation of the relative density of the WC–15vol.%Co materials with the sintering time during pressureless induction sintering. (Reprinted from Kim et al. [10]. Copyright (2004) with permission from Elsevier)

demonstrates the effect of the sintering time on the relative density of the WC–15vol.%Co materials. When the heating time was 200 s, a density of about 96.6% was reached. Rather unexpectedly, when the sintering time was increased, the relative density of the composite decreased. This effect remained unexplained and, therefore, requires further investigation.

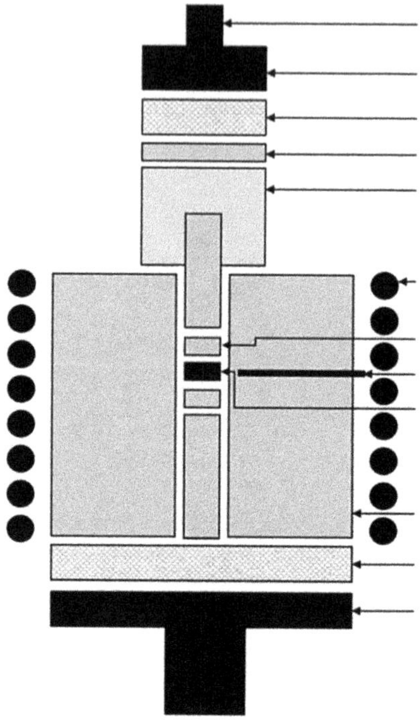

Fig. 8.3 Schematic of induction hot press according to [12]. (Reprinted from LaLonde et al. [12]. Copyright (2011) with the permission of AIP Publishing)

8.2 Induction Sintering Equipment

In material development studies, the powders are usually placed in a conductive die, and a uniaxial pressure is applied to the specimen. LaLonde et al. [12] described a hot-press system, in which a graphite die is RF induction heated, while the sample is pressed in a load frame (Fig. 8.3). The system can operate under vacuum or a protective atmosphere. Having successfully sintered PbTe and SiGe thermoelectrics, which showed properties close to those of the materials produced by conventional hot pressing for a much longer time, the authors argued that RF induction heating sintering presents a viable alternative to spark plasma sintering.

In the studies of direct induction sintering of nickel, Guyon et al. [5] used a facility shown schematically in Fig. 8.4. Induction sintering in a heat-resistant glass tube was also used by Çivi et al. [13] to sinter iron-based parts.

8.3 High Heating Rates in Induction Heating Sintering

Guyon et al. [5] found that heating of 60% dense nickel compacts cold-pressed at 450 MPa before sintering by eddy currents for 70 s up to 1100 °C (heating rate 900 $°C \cdot min^{-1}$) without the dwell time at the maximum temperature allows reaching a relative density of 93%. When slow heating was used, the rim areas of the compact

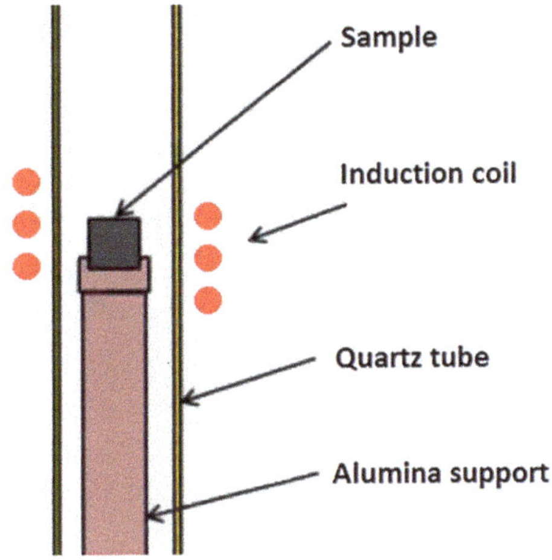

Fig. 8.4 Schematic of induction sintering setup with direct heating of the sample according to [5]. (Reprinted from Guyon et al. [5]. Copyright (2014) The Institute of Materials, Minerals and Mining, by permission of Taylor & Francis Ltd., www.tandfonline.com, on behalf of The Institute of Materials, Minerals and Mining)

were dense, while the center of the compact remained porous. Upon fast heating, a uniform heating was obtained. It was concluded that nickel compacts with high densities, small grain sizes, and homogeneous microstructures can be obtained by direct induction sintering with processing times much shorter than those used in conventional sintering. In the area of direct induction sintering, the work conducted by Guyon et al. is remarkable in the sense that densification enhancement during induction sintering was discussed as caused by enhancement of diffusion under the electric current. The same powder was sintered by conventional sintering and induction heating sintering. During conventional sintering conducted by heating the compacts at different heating rates up to 1000 °C, at a given temperature, the relative density was higher when the heating rate was lower, and the temperature corresponding to the maximum densification rate increased with increasing heating rate. These observations showed that the powder experienced solid-state sintering. From the determined value of the activation energy, it was concluded that the process is controlled by grain-boundary diffusion. The master sintering curve approach was used to compare the results of induction sintering with those of conventional sintering. The master sintering curves were extrapolated to the thermal cycles used in induction sintering. In order to quantify the specific effect of induction heating on densification, the densities predicted by the master sintering curve approach adjusted from the conventional sintering data were calculated. At a high heating rate (900 $°C{\cdot}min^{-1}$), the density predicted by the master sintering curve approach above 900 °C was always lower than the density obtained in the induction sintering experiments. Based on these results, the authors assumed that a favorable effect of induction heating on densification was due to a direct influence of the electric field/current or an indirect influence of the heating method through high heating rates. Compacts obtained by induction sintering at a heating rate of 25 $°C{\cdot}min^{-1}$ showed microstructural differences between the core and the edge regions. Optical images of the cross-sections of the nickel compacts induction sintered at heating rates of 25 $°C{\cdot}min^{-1}$ and 900 $°C{\cdot}min^{-1}$ (heating up to

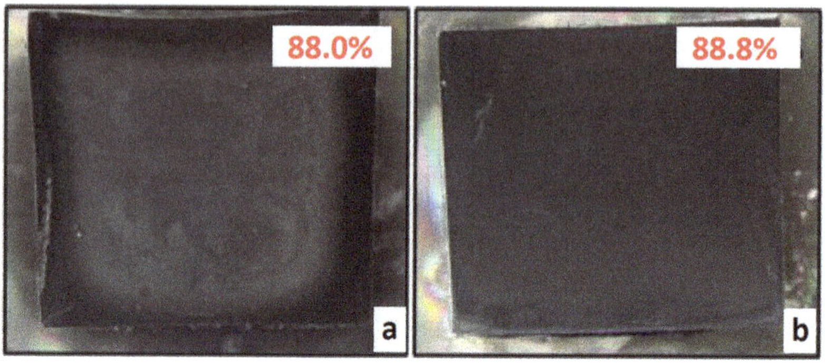

Fig. 8.5 Optical images of the cross-sections of nickel induction sintered at a heating grate of 25 °C·min^{-1} (**a**) and 900 °C·min^{-1} (**b**). Heating up to 1000 °C, no dwell time; the relative density values are given in the upper right corners of the images. Dark areas are denser than light areas. (Reprinted from Guyon et al. [5]. Copyright (2014) The Institute of Materials, Minerals and Mining, by permission of Taylor & Francis Ltd., www.tandfonline.com, on behalf of The Institute of Materials, Minerals and Mining)

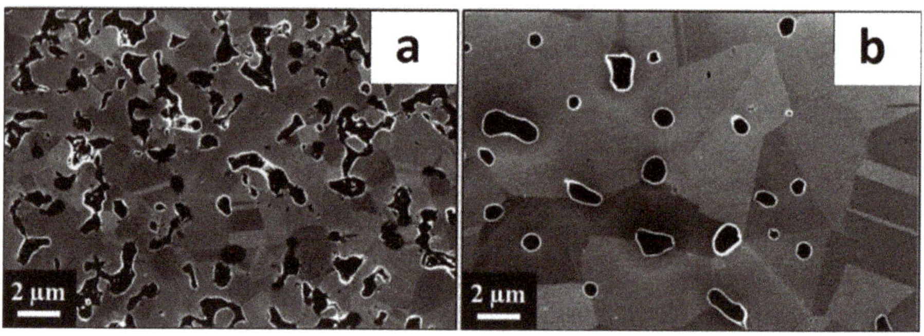

Fig. 8.6 SEM images of the core (**a**) and edge (**b**) of the nickel compact induction sintered at a heating rate of 25 °C·min^{-1}, heating up to 1000 °C, no dwell time. (Reprinted from Guyon et al. [5]. Copyright (2014) The Institute of Materials, Minerals and Mining, by permission of Taylor & Francis Ltd., www.tandfonline.com, on behalf of The Institute of Materials, Minerals and Mining)

1000 °C, no dwell time) are shown in Fig. 8.5. SEM images of the core and the edge of the nickel compact induction sintered at a heating rate of 25 °C·min^{-1} confirm that the former is denser than the latter (Fig. 8.6). The thickness of the dense shell is 1–2 mm, which is close to the penetration depth of the eddy currents during induction sintering. This suggests that inter-particle contacts directly heated by eddy currents sinter earlier than the contacts in the central part of the specimen, which are heated only by conduction. Faster sintering of the shell of the compact may also hinder densification of the central part, the shell acting as an envelope. Uniform sintering of the compact in experiments conducted at a high heating rate was explained by a high temperature of the sintering start, at which the electrical conductivity is lower, the penetration depth of the eddy currents is greater, and the viscosity of the material is lower weakening the "envelope" effect of the faster sintered shell.

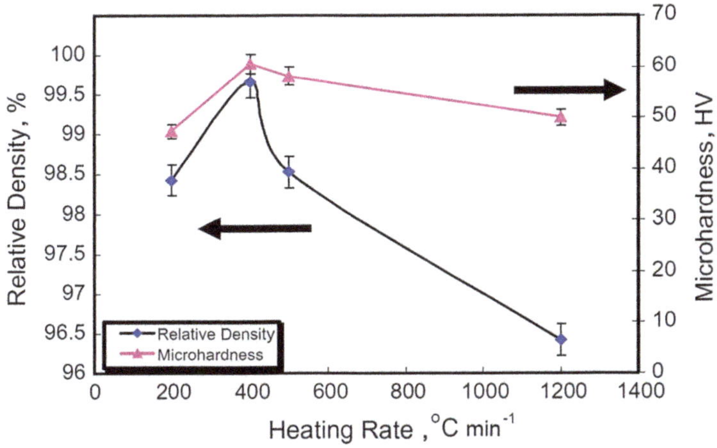

Fig. 8.7 Relative density and microhardness of Mg–HAp composites containing 1 wt.% of HAp induction sintered at 500 °C and 50 MPa vs. heating rate. (Reprinted from Khalil and Almajid [11]. Copyright (2011) with permission from Elsevier)

The effect of the heating rate on the sintering outcome was also investigated for induction sintering with a susceptor [11]. Heating rates that are too high were found to facilitate the formation of compacts with a low relative density. Magnesium–hydroxyapatite (Mg–HAp) composites containing 1 wt.% of HAp were induction sintered at 500 °C and 50 MPa. At first, the relative density and microhardness increased with increasing heating rate and reached maximum at a heating rate of 400 $°C{\cdot}min^{-1}$. At higher heating rates, compacts with lower relative densities and microhardness were obtained (Fig. 8.7). The authors suggested that as the Mg–HAp nanopowders were heated from the graphite mold and punches, at high heating rates, the layers adjacent to the tooling surface were sintered to nearly full density, while the compact's interior was not densified. An increase in the heating rate resulted in an increase in the non-densified volume, and the average relative densities and microhardness of the compacts decreased. The crystallite size was also greatly influenced by the heating rate, as can be seen in Fig. 8.8. At high heating rates, materials with very fine crystallites were obtained. At the same time, the materials retained porosity, which weakened their mechanical strength and lowered their hardness. Compacts produced at a heating rate of 400 $°C{\cdot}min^{-1}$ showed the best mechanical properties in this series of tests (Figs. 8.7 and 8.9).

In a study by Kim et al. [9], experiments on induction sintering were set in a different manner, and conclusions derived from them should not be analyzed together with the results obtained in Ref. [11]. Binderless tungsten carbide WC was sintered at the output of the total power ranging from 60% to 90%. As the output of the total power was increased from 60% to 90%, the heating rate increased from 950 to 1400 $°C{\cdot}min^{-1}$. The samples were heated until a plateau on the displacement–time curve was reached (Fig. 8.10). In these series of experiments, the maximum temperature was not constant, and this is the main difference between this study and investigations conducted in Ref. [11]. An increase in the output of the total power

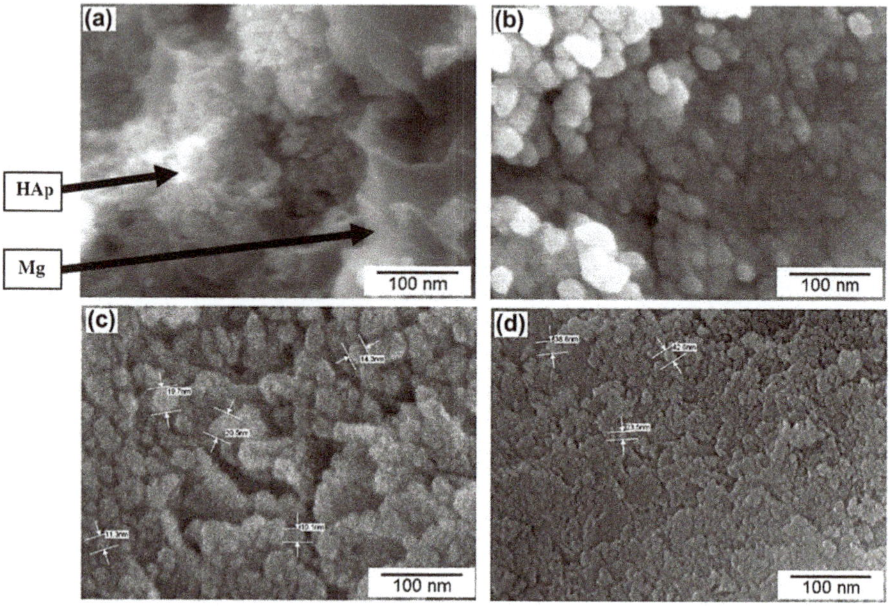

Fig. 8.8 Fracture surface of Mg–HAp composites containing 1 wt.% of HAp induction sintered at 500 °C and 50 MPa using different heating rates: (**a**) 200 °C·min^{-1}, (**b**) 400 °C·min^{-1}, (**c**) 500 °C·min^{-1}, (**d**) 1200 °C·min^{-1}. (Reprinted from Khalil and Almajid [11]. Copyright (2011) with permission from Elsevier)

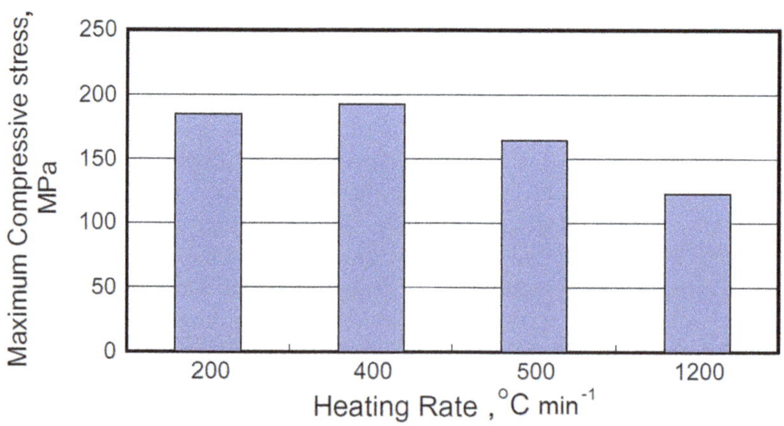

Fig. 8.9 Maximum compressive stress of Mg–HAp composites containing 1 wt.% of HAp induction sintered at 500 °C and 50 MPa using different heating rates. (Reprinted from Khalil and Almajid [11]. Copyright (2011) with permission from Elsevier)

from 60% to 90% led to an increase in the maximum temperature, shortening of the time required to reach a plateau on the displacement–time curve (Fig. 8.11). The relative density of the compacts increased with the output of the total power (and, hence, with the heating rate).

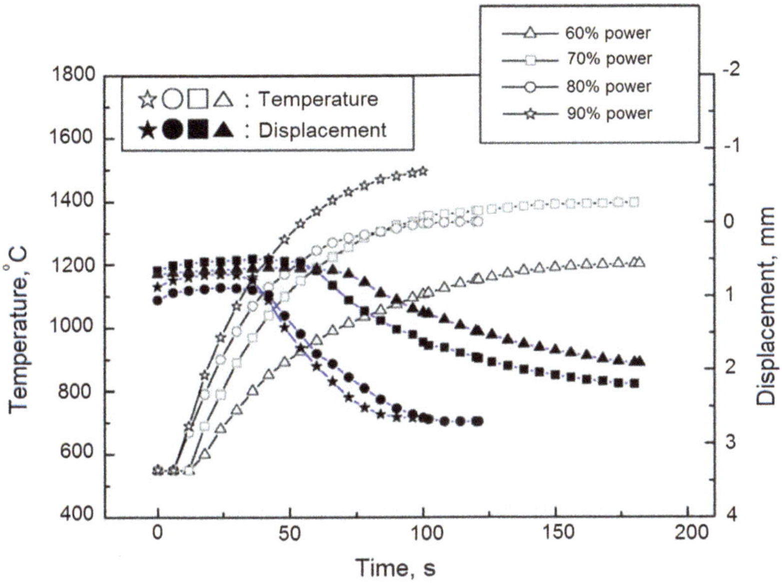

Fig. 8.10 Variation of temperature and displacement with time during induction sintering of binderless WC at 60 MPa. (Reprinted from Kim et al. [9]. Copyright (2006) with permission from Elsevier)

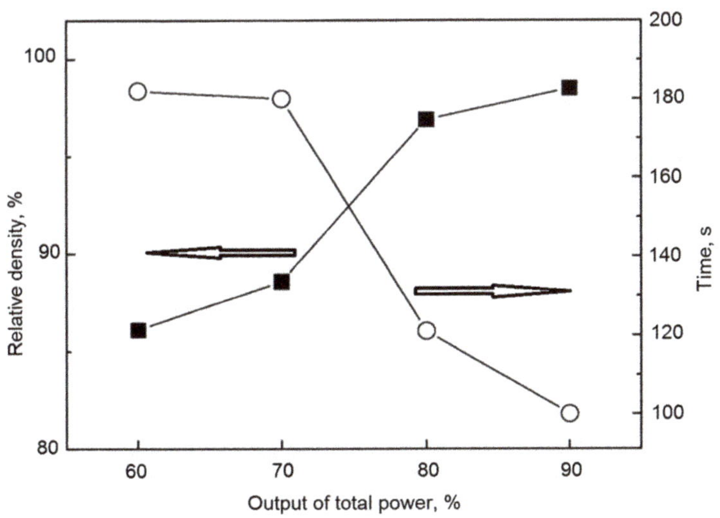

Fig. 8.11 Relative density and time required to reach the plateau on the displacement–time curve vs. output of the total power during induction sintering of binderless WC at 60 MPa. (Reprinted from Kim et al. [9]. Copyright (2006) with permission from Elsevier)

8.4 Selected Examples of Materials Processed by Induction Heating Sintering

Induction sintering has been proven to be an efficient method for the production of nanostructured/fine-grained materials by rapid consolidation of powders – metals [14], single-phase ceramics [9, 15–18], and composites [10, 11, 19–35] – as well as for the preparation of bulk materials containing metastable phases, which otherwise (upon conventional sintering) would have transformed into the corresponding stable phases with a concomitant deterioration of properties of the composites [7, 8, 36, 37].

Kim et al. [19] sintered WC–10wt.%Co by induction sintering in a graphite die and produced a compact 99.5% dense with an average grain size of 260 nm within 1 min. The sintered WC–10wt.%Co materials showed a high fracture toughness (13 $MPa{\cdot}m^{1/2}$). Successful sintering in a short time maintaining a small grain size and enhanced properties of the WC–Co sintered materials was suggested to be the result of not only high heating rates but also of the intrinsic current effects. An explanation for increased fracture toughness was suggested based on an increased solubility of tungsten carbide in cobalt induced by electric current.

Rapid heating and short overall sintering times were used as benefits to sinter materials containing metastable phases – metallic glasses – avoiding their crystallization. The amorphous state of the metallic glass should be preserved during consolidation of the powder mixtures, if the glass is introduced to reinforce a metal or a metallic alloy [7, 8, 36, 37]. Figure 8.13 shows the microstructure and the XRD pattern of the magnesium alloy matrix composite reinforced with particles of $Zr_{57}Nb_5Cu_{15.4}Ni_{12.6}Al_{10}$ metallic glass [7]. The bright particles in the image (Fig. 8.12a) are particles of the metallic glass, while the dark background is the magnesium alloy AZ91. Above their glass transition temperatures, metallic glasses enter a Newtonian flow in the supercooled liquid region ΔT_x between the glass transition T_g and the crystallization T_x temperatures. In the supercooled liquid, the viscosity of metallic glasses drops drastically, and they become liquid-like. In order to benefit from this behavior of metallic glass during sintering and use it as a reinforcement in metal matrix composites, the sintering temperature T_{sint} was selected to satisfy the following:

$$T_g < T_{sint} < T_{sol},$$

where T_{sol} is the solidus temperature of the matrix alloy. Fast heating (at a rate of 100 $°{\cdot}min^{-1}$) and a short holding time (2 min) at the maximum temperature allowed preserving the amorphous state of the metallic glass, as was confirmed by the XRD analysis – a halo corresponding to the metallic glass phase was observed at $2\theta \approx 37°$ (Fig. 8.12b). In a similar manner, the amorphous state of the $Cu_{54}Zr_{36}Ti_{10}$ metallic glass was preserved during consolidation of an aluminum alloy matrix composite with metallic glass as a reinforcement (Fig. 8.13) [8]. The energy-dispersive spectroscopy (EDS) data (Fig. 8.13a) confirm the chemical composition of the reinforcing particles

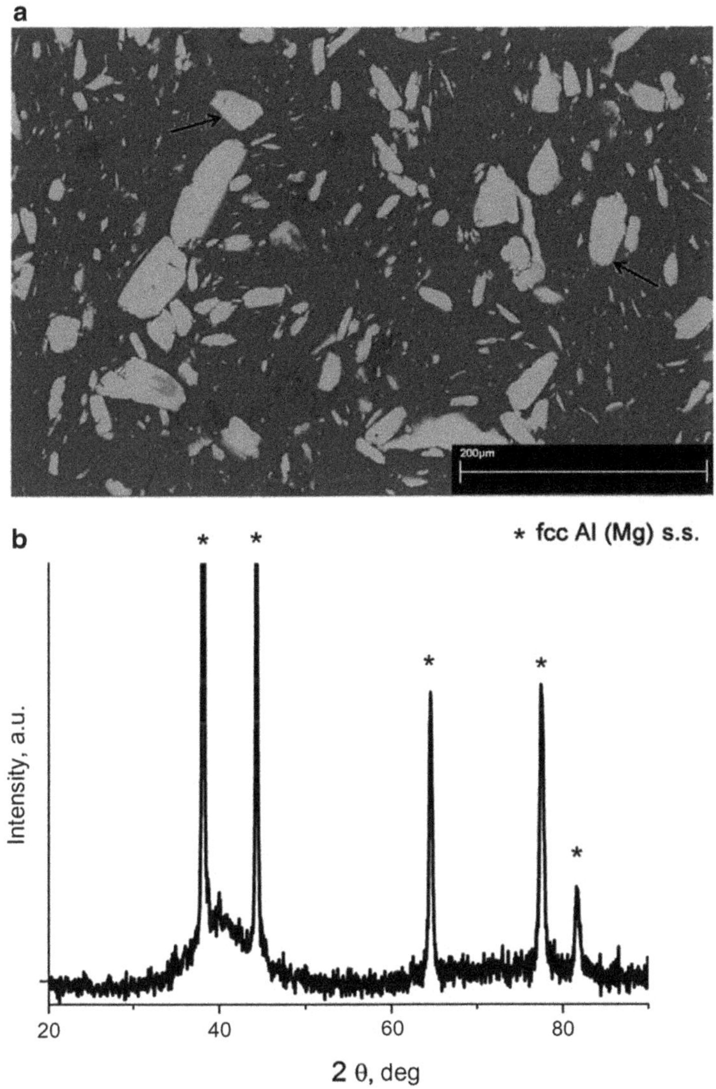

Fig. 8.12 Microstructure (**a**) and XRD pattern (**b**) of the Mg alloy AZ91–15vol.% $Zr_{57}Nb_5Cu_{15.4}Ni_{12.6}Al_{10}$ composite induction sintered at 440 °C. The bright particles are particles of metallic glass (marked by the arrows). (Reprinted from Dudina et al. [7]. Copyright (2009) with permission from Elsevier)

(spectrum I) and the matrix (spectrum II), while a halo present in the XRD pattern (Fig. 8.13b) indicates the presence of the amorphous phase in the induction sintered composite. Attractive mechanical properties in compression were obtained in the sintered composites, thanks to the introduction of the metallic glass reinforcement and preservation of its amorphous structure in the sintered state. Crystallization of

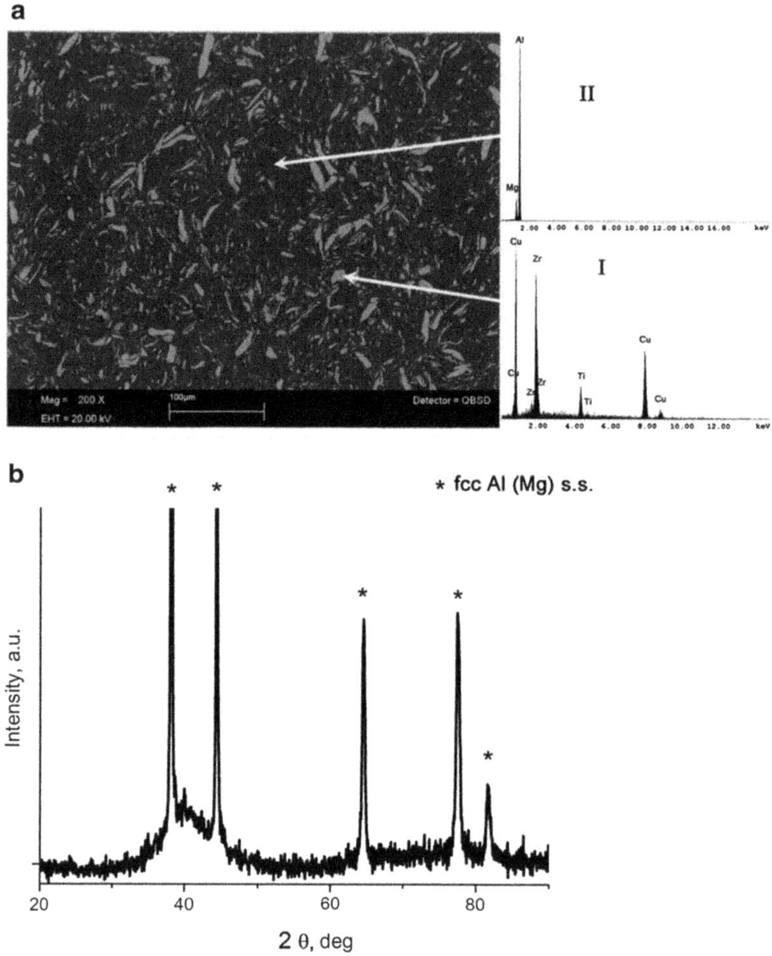

Fig. 8.13 Microstructure (**a**) and XRD pattern (**b**) of the Al alloy 520.0–15 vol.% $Cu_{54}Zr_{36}Ti_{10}$ composite induction sintered at 447 °C. The bright particles are particles of metallic glass. (Reprinted from Dudina et al. [8]. Copyright (2010) with permission from Elsevier)

metallic glass is undesirable during consolidation as it leads to the formation of a mixture of brittle intermetallics. The yield strength of the AZ91–15vol.% $Zr_{57}Nb_5Cu_{15.4}Ni_{12.6}Al_{10}$ induction sintered composite was 330 MPa, which is 2.4 times greater than the yield strength of the matrix alloy AZ91 obtained by casting. The yield strength of the Al520.0–15vol.%$Cu_{54}Zr_{36}Ti_{10}$ composite was 580 MPa, which is three times greater than the yield strength of the matrix alloy Al520.0 obtained by casting. The sintered composites showed appreciable plasticity: the deformation at fracture was 14% for the Al520.0–15vol.%$Cu_{54}Zr_{36}Ti_{10}$ composite and 11% for the AZ91–15vol.%$Zr_{57}Nb_5Cu_{15.4}Ni_{12.6}Al_{10}$ composite.

Induction sintering was shown to be efficient for the production of bulk nanostructured aluminum [14]. Nanostructured aluminum powders (crystallite size 23 nm) obtained by cryomilling were consolidated by induction sintering in vacuum at a pressure of 50 MPa. The crystallite size of the consolidated aluminum was below 100 nm. Comparative compression tests showed that the yield strength of the aluminum obtained by induction sintering of the cryomilled powders – 270 MPa (sintering temperature 500 °C, heating rate 800 $°C{\cdot}min^{-1}$, holding time at the maximum temperature 3 min, crystallite size 50 nm) – is twice the yield strength of aluminum alloy 1050 obtained by equal-channel angular pressing and five times the yield strength of the aluminum obtained by induction sintering of the as-received (non-milled) powders.

In steel induction sintered from a commercial steel powder blend, finer grains, a smaller perlite spacing, higher concentrations of carbides at the grain boundaries, and a higher hardness of the ferrite phase compared with conventionally sintered specimens were found [1]. A greater amount of perlite and a higher hardness of ferrite indicated that a greater degree of alloying occurred during induction sintering than during conventional sintering. It was suggested that more pronounced alloying was caused by high localized temperatures developing at the particle contacts as regions of high resistance in the induction sintered specimens. Induction sintering was, therefore, proposed as a method, suitable for both sintering of powders and alloying between the components of an alloy.

As in other sintering techniques, preliminary ball milling leading to refinement of the crystalline structure reduces the onset temperature of induction sintering. The shrinkage of $TiC_{0.5}N_{0.5}$ samples obtained from the powders milled for different times with increasing temperature is shown in Fig. 8.14 [17]. The onset sintering temperature was significantly reduced by applying preliminary ball milling: while shrinkage of the sample obtained from the as-received powder started at 1170 °C, that of the sample obtained from the powder milled for 10 h was observed at 820 °C [17]. Compacts with finer grains were obtained from the powders milled for longer durations: the crystallite sizes of the compacts sintered from the powders milled for 0, 1, 4, and 10 h were 227, 201, 196, and 158 nm, respectively. Along with a reduction in the sintering temperature, ball milling of the powder prior to induction sintering resulted in higher relative densities, higher hardness, and higher fracture toughness of the sintered $TiC_{0.5}N_{0.5}$.

Similar trends for the relative density and hardness were observed by Shon et al. [15] in induction sintered titanium carbide obtained from the ball milled powders. The TiC powder ball milled for 1, 4, and 10 h had a crystallite size of 45, 29, and 25 nm, respectively. Induction sintering of the ball milled powders in a graphite die at 80 MPa for 4 min resulted in the formation of TiC compacts with a crystallite size of 295, 114, and 73 nm when the powders milled for 1, 4, and 10 h were used, showing that although grain growth does occur during induction sintering, it is not pronounced and bulk nanostructured ceramics can be obtained. As the milling time was increased from 1 to 10 h, the relative density of the compacts increased from 97% to 99%, while their Vickers hardness increased from 11.8 to 25.7 GPa. Unlike induction sintered $TiC_{0.5}N_{0.5}$ [17], whose fracture toughness increased with the

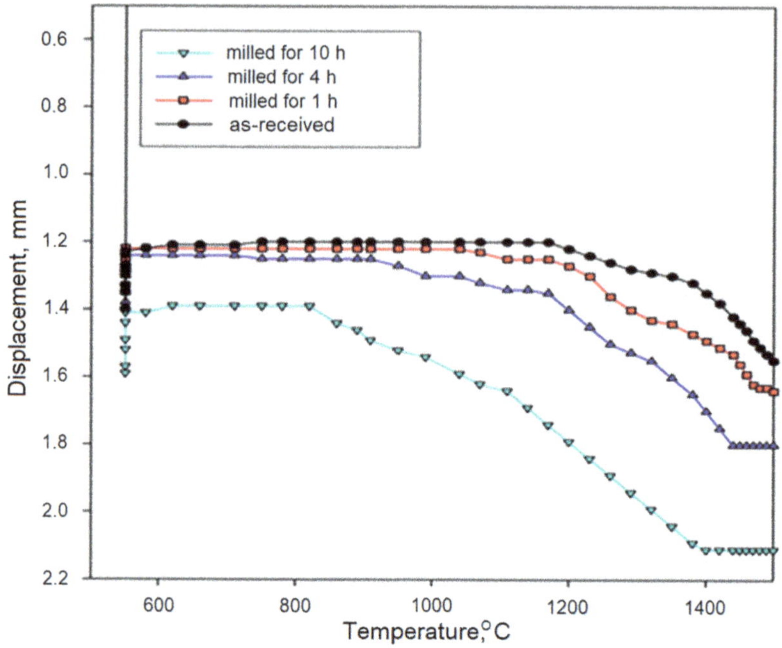

Fig. 8.14 Shrinkage of $TiC_{0.5}N_{0.5}$ samples obtained from the powders milled for different times with increasing temperature. (Reprinted from Kim et al. [17]. Copyright (2012) with permission from Elsevier)

milling time, the fracture toughness of TiC slightly decreased with the milling time, still remaining at a reasonably high level. A reduction in the fracture toughness was explained by a small crystallite size of the sintered compacts remaining after induction sintering.

The capabilities of induction sintering can be extended to reactive systems. Inducing synthesis of new phases by induction heating – conducting reactive induction sintering – is worthwhile in the powder mixtures capable of fast chemical transformations into solid-state products. The fabrication of intermetallic composites in Ti–Al [38] and Ti–Ni–Al [39] systems and the reactive deposition of Ni–Al coatings on mild steel [40] using induction heating have been reported.

8.5 Summary

Induction heating sintering is realized using eddy currents induced directly in the conductive specimens or in the susceptor. An advantage of schemes with a susceptor is the possibility of applying mechanical pressure during induction heating to facilitate sintering. For direct induction sintering, the powder compacts should possess a high electrical conductivity in the beginning of the process. Due to poor coupling of metal compacts to the induction coil, there exists an incubation period at the beginning of induction heating. Both heating schemes enable fast heating and dramatically shorten the sintering time relative to conventional sintering. In an induction sintering run, the total duration of heating and holding at the maximum temperature (if any) can be a few minutes only. The heating rate in induction sintering plays a significant role influencing the microstructure uniformity and final densities. Successful consolidation of ceramics, metals, and composites with nanosized grains has been reported. Induction sintering has also proven efficient for consolidation of composites containing metastable phases – metallic glasses. Explanations of the observed fast densification provided so far together with the experimental data are based on a beneficial combination of resistance heating of the sample, localized heating at the inter-particle contacts, and pressure effects (for the pressure-assisted processes). A possibility of surface melting and oxide breakdown at the inter-particle contacts has also been suggested as favoring rapid densification. In the future studies, specifics of mass transport enhancement by eddy currents need to be elaborated. The development of a coupled electromagnetic–thermal–mechanical modeling framework taking into account the specifics of mass transport during induction heating sintering is a major direction for the future studies. This direction should open up the predictive capabilities and allow for the optimization of induction heating sintering enabling the *controlled non-equilibrium* conditions needed for the successful outcomes of most field-assisted sintering techniques. Another promising direction in induction sintering-related research is controlled preparation of materials with gradient structures, including those forming as a result of reactive sintering.

References

1. Hirschho JS, Samat M, Maxwell GM (1970) Induction sintering has potential for powder metal parts. Metal Progress 97:135–138
2. Hermel W, Leitner G, Krumphold R (1980) Review of induction sintering: fundamentals and applications. Powder Metall 3:130–135
3. Schatt W, Wieters KP (1997) Powder metallurgy. Processing and materials. EPMA, Shrewsbury, 492 p
4. German RM (1996) Sintering theory and practice. Wiley, New York, 568 p
5. Guyon A, Bouvard D, Chaix JM, Carry CP (2014) Densification and microstructure changes of micron size nickel powder during direct induction sintering. Powder Metal 57:54–60
6. Conta RL (1977) A process for improved induction heating of powder metal compacts. IEEE Trans Ind Appl IA-13(4):330–334
7. Dudina DV, Georgarakis K, Li Y, Aljerf M, LeMoulec A, Yavari AR, Inoue A (2009) A magnesium alloy matrix composite reinforced with metallic glass. Comp Sci Tech 69:2734–2736
8. Dudina DV, Georgarakis K, Li Y, Aljerf M, Braccini M, Yavari AR, Inoue A (2010) Cu-based metallic glass particle additions to significantly improve overall compressive properties of an Al alloy. Composites Part A 41:1551–1557
9. Kim HC, Yoon JK, Doh JM, Ko IY, Shon IJ (2006) Rapid sintering process and mechanical properties of binderless ultra fine tungsten carbide. Mater Sci Eng A 435–436:717–724

10. Kim HC, Oh DY, Shon IJ (2004) Sintering of nanophase WC-15vol.%Co hard metals by rapid sintering process. Int J Refractory Metals & Hard Mater 22:197–203
11. Khalil KA, Almajid AA (2012) Effect of high-frequency induction heat sintering conditions on the microstructure and mechanical properties of nanostructured magnesium/hydroxyapatite nanocomposites. Mater Des 36:58–68
12. LaLonde AD, Ikeda T, Snyder GJ (2011) Rapid consolidation of powdered materials by induction hot pressing. Rev Sci Inst 82:025104
13. Çivi C, Tahralı N, Atik E (2014) Reliability of mechanical properties of induction sintered iron based powder metal parts. Mater Des 53:383–397
14. El-Danaf EA, Soliman MS, Almajid AA, Khalil KA (2013) Mechanical characterization of cryomilled Al powder consolidated by high-frequency induction heat sintering. Adv Mater Sci Eng, article ID 397351, 10 p
15. Shon IJ, Kim BR, Doh JM, Yoon JK (2010) Consolidation of binderless nanostructured titanium carbide by high-frequency induction heated sintering. Ceram Int 36:1797–1803
16. Shon IJ, Ko IY, Chae SM, Na K (2011) Rapid consolidation of nanostructured $TaSi_2$ from mechanochemically synthesized powder by high frequency induction heated sintering. Ceram Int 37(2):679–682
17. Kim W, Suh CY, Roh KM, Lim JW, Lee S, Du SL, Shon IJ (2013) High-frequency induction heated sintering of high-energy ball milled $TiC_{0.5}N_{0.5}$ powders and mechanical properties of the sintered products. Ceram Int 39(1):585–591
18. Shon IJ (2016) High-frequency induction sintering of B_4C ceramics and its mechanical properties. Ceram Int 42:19406–19412
19. Kim HC, Shon IJ, Munir ZA (2005) Rapid sintering of ultra-fine WC-10 wt% co by high-frequency induction heating. J Mater Sci 40:2849–2854
20. Kim HC, Shon IJ, Yoon JK (2006) Rapid sintering of ultrafine WC-Ni cermets. Int J Refractory Metals Hard Mater 24:427–431
21. Kim SW, Khalil KA (2006) High-frequency induction heat sintering of mechanically alloyed alumina–yttria-stabilized zirconia nano-bioceramics. J Am Ceram Soc 89(4):1280–1285
22. Khalil KA, Kim SW (2006) Effect of processing parameters on the mechanical and microstructural behavior of ultra-fine Al_2O_3-(ZrO_2+8%molY_2O_3) bioceramic, densified by high frequency induction heat sintering. Int J Appl Ceram Technol 3(4):322–330
23. Montasser D (2010) Microstructure and mechanical properties of biocompatible high density Ti–6Al–4V/W produced by high frequency induction heating sintering. Mater Des 31:3964–3970
24. Khalil KA, Kim SW, Kim HY (2007) Observation of toughness improvements of the hydroxyapatite bioceramics densified by high-frequency induction heat sintering. Int J Appl Ceram Technol 4(1):30–37
25. Khalil KA, Kim SW, Dharmaraj N, Kim KW, Kim HY (2007) Novel mechanism to improve toughness of the hydroxyapatite bioceramics using high-frequency induction heat sintering. J Mater Proc Technol 187–188:417–420
26. Khalil KA, Kim SW (2007) Mechanical wet-milling and subsequent consolidation of ultra-fine Al_2O_3-(ZrO_2+3%mol.Y_2O_3) bioceramics by using high-frequency induction heat sintering. Trans Nonferrous Met Soc China 17:21–26
27. Khalil KA, Kim SW, Kim HY (2007) Consolidation and mechanical properties of nanostructured hydroxyapatite–(ZrO_2+3mol.%Y_2O_3) bioceramics by high-frequency induction heat sintering. Mater Sci Eng A 456:368–372
28. Woo KD, Kim BR, Kwon EP, Kang DS, Shon IJ (2010) Properties and rapid consolidation of nanostructured TiC-based hard materials with various binders by a high-frequency induction heated sintering. Ceram Int 36(1):351–355
29. Shon IJ (2016) Rapid consolidation of nanostructured WC-FeAl hard composites by high-frequency induction heating and its mechanical properties. Int J Refractory Metals Hard Mater 61:185–191

30. Kim HC, Park HK, Jeong IK, Ko IY, Shon IJ (2008) Sintering of binderless WC–Mo_2C hard materials by rapid sintering process. Ceram Int 34:1419–1423
31. Shon IJ (2016) Enhanced mechanical properties of the nanostructured AlN-graphene composites rapidly sintered by high-frequency induction heating. Ceram Int 42:16336–16342
32. Shon IJ, Jeong IK, Ko IY, Doh JM, Woo KD (2009) Sintering behavior and mechanical properties of WC–10Co, WC–10Ni and WC–10Fe hard materials produced by high-frequency induction heated sintering. Ceram Int 35:339–344
33. Kim HC, Kim DK, Woo KD, Ko IY, Shon IJ (2008) Consolidation of binderless WC–TiC by high frequency induction heating sintering. Int J Refractory Metals Hard Mater 26:48–54
34. Shon IJ, Kang HS (2011) Properties and fast low-temperature consolidation of nanocrystalline Ni–ZrO_2 composites by high-frequency induction heated sintering. J Alloys Compd 509 (6):2964–2969
35. Shon IJ, Kim TW, Doh JM, Yoon JK, Park SW, Ko IY (2011) Mechanical synthesis and rapid consolidation of a nanocrystalline $3.3Fe_{0.6}Cr_{0.3}Al_{0.1}$–$Al_2O_3$ composite by high frequency induction heating. J Alloys Compd 509(2):L7–L10
36. Dudina DV, Georgarakis K, Yavari AR (2012) Metal matrix composites reinforced with metallic glass particles: state of the art. In: Metal Matrix Composites, Paulo Davim J (ed), Nova Science Publishers, Inc., USA, p 1–30
37. Aljerf M, Georgarakis K, Louzguine-Luzgin D, Le Moulec A, Inoue A, Yavari AR (2012) Strong and light metal matrix composites with metallic glass particulate reinforcement. Mater Sci Eng A 532:325–330
38. Fu EKY, Rawlings RD, McShane HB (2001) Reaction synthesis of titanium aluminides. J Mater Sci 36(23):5537–5542
39. Widodo Besar Riyadi T, Zhang T (2014) Combustion synthesis of Ti_3Al composite by induction heating, Proc 3rd applied science for technology innovation, ASTECHNOVA 2014, International Energy Conference, Yogyakarta, Indonesia, 13-14 August 2014, pp 18–23
40. Lee HY, Kim TJ, Cho YJ (2010) Effects of induction heating conditions on NiAl based intermetallic compound coating. J Korean Inst Metals Mater 48(2):141–147

Chapter 9
Magnetic Pulse Compaction

9.1 Principles of MPC

Magnetic pulse compaction (MPC) belongs to a class of compaction methods, in which the mechanical load is applied in a (quasi) dynamic manner [1]. Along with magnetic pulse compaction, this class of compaction methods includes forging and explosive compaction. A distinctive feature of these methods is the concentration of high energy in a small volume. MPC is suitable for producing dense and porous compacts. In a non-densified state, all powders have a very poor electrical conductivity. Although magnetic field can induce eddy currents in each powder particle, they are confined in the particle itself due to an extremely high inter-particle resistance. The particles experience constriction; however, there is no mutual displacement of the particles, i.e., there is no densification. This is the reason why direct pressing of powders is not possible in many cases by applying a magnetic field to a sample composed of loose particles. In order to efficiently convert the energy of the magnetic field into the mechanical energy and gain the required pressures of the powder sample, shells made of a conductive metal (copper, aluminum) are used as containers shaping the compact and transferring pressure to the powder on a macro-level. Electrically conductive plates can also be used to transfer pressures in a uniaxial mode.

The first experiments on powder compaction using a pulse magnetic field were conducted by Sandstrom in 1964 [2]. The powder in a tube was compressed by a pulse magnetic field produced by an eigencurrent (Z-pinch schematics). The compaction occurs due to the action of the repulsive forces between the electric currents passing in the opposite directions in the coaxial conducting tubes (electrodynamic pressing). In this geometry, the shell is compressed and the powder that it contains is compacted. If the central electrode is placed inside a conducting tube, the latter can be forced to expand and thus transfer pressure in the radial direction to a tube-shaped compact [1]. Electrodynamic MPC setups were not capable of achieving high densities of the compacts due to high inductances of the current leads and other parts of the setup compared to the working elements.

E. A. Olevsky, D. V. Dudina, *Field-Assisted Sintering*,
https://doi.org/10.1007/978-3-319-76032-2_9

Induction MPC is based on the interaction of the pulsed magnetic field with the magnetic field of eddy currents induced in an electrically conductive element – a cylindrical shell in radial compaction and a plate in uniaxial compaction. In uniaxial compaction, the principle of magnetic hammer is realized. A plate made of a conductive material is repelled from a flat inductor when a pulse of high current passes through the latter. The concentrator transfers the mechanical force to a punch, which densifies the powder in a die. This method is used to make washers and parts of complex shapes but limited heights. Barbarovich [3] applied the pulse theory to determine the velocity of the moving plate taking into account the experimentally observed facts that the motion of a part under the action of a pulsed magnetic field does not begin at once but a certain time after the beginning of the discharge. For massive parts, this usually occurs after the first current maximum, the time being needed for the energy to build up. The lower is the weight of the plate, the higher is the velocity that it develops. In radial compaction, magnetic field compresses a copper or an aluminum tube (shell) containing the powder to be compacted. The compression of the shell occurs when it is placed in a cylindrical inductor (Θ-pinch method) or when a pulse of current passes directly through the shell (Z-pinch method). In the Θ-pinch method, the magnetic field of the inductor interacts with the eddy currents induced in the conductive shell.

When a cylindrical electrically conducting shell is radially compressed, in the gap between the working winding of the inductor and the surface of the shell, a strong magnetic field appears for a short period of time. Due to the pulse nature of the process and a skin effect, the field does not penetrate inside the conductive shell. The outer lateral surface of the cylindrical shell experiences pressure equal to the volume density of the magnetic field [4]:

$$p = W_{\mathrm{m}} = \frac{1}{2}\mu_0\mu_{\mathrm{r}}H^2, \tag{9.1}$$

where μ_{r} is the relative magnetic permeability.

The resultant pressure on the powder compact is equal to the difference between the pressures on the outer and inner surfaces of the shell:

$$p = W_{\mathrm{m}} = \frac{1}{2}\mu_0\mu_{\mathrm{r}}\left(H_1^2 - H_2^2\right), \tag{9.2}$$

where H_1 is the magnetic field strength in the gap between the shell and the inductor and H_2 is the magnetic field strength at the inner surface of the shell. The value of the pressure counteracting the pressure on the outer surface of the shell depends on the frequency of the current, the rate of change of the magnetic flux, and the electrical conductivity of the shell material. At high current frequencies, a reduction in the magnetic field strength in the direction from the surface to the interior of the conducting material can be rather significant. The electrical conductivity of the shell affects the penetration depth of the magnetic field into the shell. The penetration of the magnetic field to the inner surface of the shell would weaken the result of compression. The penetration depth of the magnetic field can be calculated as follows:

$$\delta = \frac{1}{\sqrt{\pi f \mu \sigma}}, \tag{9.3}$$

where f is the frequency, μ is the material magnetic permeability, and σ is the material electrical conductivity. The thickness of the shell should be equal to or greater than the field penetration depth. When a copper shell is compressed using a 10^4 Hz frequency current in the inductor, the field penetration depth δ is 0.7 mm. For the thickness of the shell equal to the field penetration depth, the field strength decreases to 37% of its initial value. When the shell is three times thicker than the penetration depth, the field strength at the inner wall of the shell is only 5%. The maximum density of the compacts is obtained when the shell thickness is close to the field penetration depth. The deformation of thicker shells causes higher energy losses and less energy is spent for the compaction process itself. However, shells that are too thin are prone to failure during MPC. After MPC, the shell has to be removed by cutting and unfolding or chemical dissolution. If reuse of the shell is possible by a certain technical approach, the manufacturing costs can be dramatically reduced.

When the skin effect is significant, the pressure can be estimated as

$$p(t) = \frac{1}{2\mu_0\mu_a} B_0^2 e^{-2\alpha t} \sin^2 \omega t, \tag{9.4}$$

where B_0 is the amplitude of the magnetic induction and α is the attenuation coefficient.

The electrical conductivity of the shell significantly influences the magnetic pressure. The higher is the electrical conductivity of the shell, the higher is the induced current, and, as a consequence, the higher are the magnetic induction and magnetic pressure.

In order to assess the kinetic characteristics of the MPC process based on the use of cylindrical shells, the displacement of the shell should be analyzed under the action of a pulse magnetic pressure. The equation of motion of an elementary element of the shell of external radius r_e and wall thickness δ_0 ($\delta_0 \ll r_e$) can be written in the following form:

$$\rho \frac{d^2u}{dt^2} = F_{vm} - F_{vd}, \tag{9.5}$$

where ρ is the density of the material of the shell, u is the distance, F_{vm} is the volume density of the force induced by the magnetic field, and F_{vd} is the volume density of the counter force (shell resistance to deformation). If the penetration of the magnetic field to the inner surface of the shell can be neglected, we obtain

$$\rho\delta \frac{d^2u}{dt^2} = p(t) - p_{vd}, \tag{9.6}$$

where p_{vd} is the counter pressure, which is constant for an ideal ductile material and depends on the deformation for real materials (rapidly increases and then remains

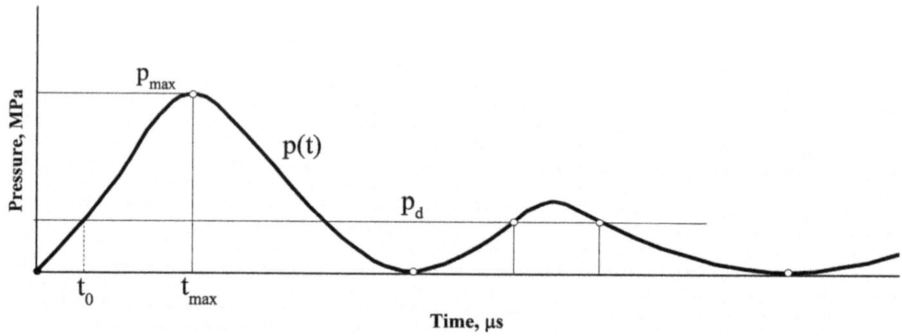

Fig. 9.1 Evolution of magnetic pressure $p(t)$ with time at $p_d = \text{const}$. (Drawn using data of Ref. [1])

practically constant as the deformation increases for cold-worked metals and gradually increases for annealed metals). Figure 9.1 shows the evolution of the magnetic pressure $p(t)$ with time at $p_d = \text{const}$ [1].

Due to high displacement rates of the shell and powder particles, inertia forces play a significant role. Adiabatic compression of a material induces a number of thermal effects in the compact. The heat is also transferred from the shell carrying the eddy currents. The presence of pores in the material that is to be compressed in the MPC is responsible for an increase in the fraction of the thermal energy dissipated into the material, which is related to local deformation and adiabatic compression of air filling the compressed pores.

In MPC, the pressure increases more slowly than in explosive compaction, which is characterized by a rapid pressure increase during a short pulse of 1–10 μs. In MPC, the pulse duration can be varied by adjusting the experimental parameters. This eliminates the unloading wave which could otherwise have followed the compression wave and prevents crack formation in the compacted materials. The pulse duration in MPC ranges usually from tens to hundreds of μs. Unlike the dynamic pressing, MPC generates an acclivous leading edge of the compression wave, which excludes the emergence of the next discharge wave and avoids exfoliation of the compacts.

9.2 Equipment for MPC

At present, the MPC method uses two technological schemes, which are based on uniaxial and radial compaction. Figure 9.2 shows a magnetic pulse press for flat uniaxial compression. As the magnetic field in the plate is completely damped, an analogy may be drawn between the electromagnetic forces acting on an electrically conducing surface and a gas exerting a pressure on it.

The main requirements for conducting a successful MPC process are a high electrical conductivity of the shell, a short distance between the inductor and the accelerated part, a high energy of the magnetic field, and a high rate of change of the magnetic flux.

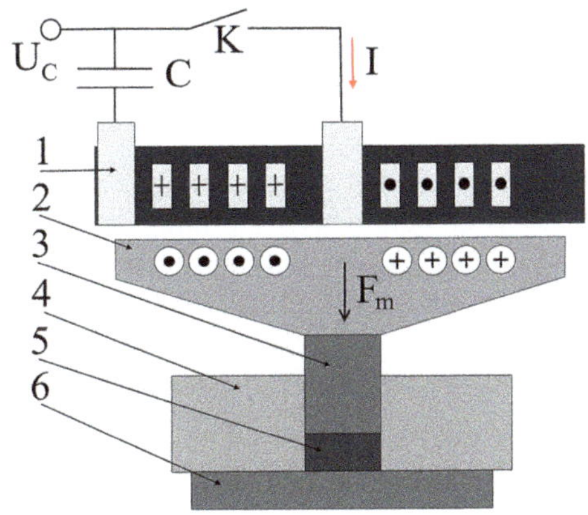

Fig. 9.2 Magnetic pulse press for flat uniaxial compression: 1, inductor; 2, metallic plate (concentrator); 3, punch; 4, die; 5, powder; 6, base. (Drawn using data of Refs. [5, 6])

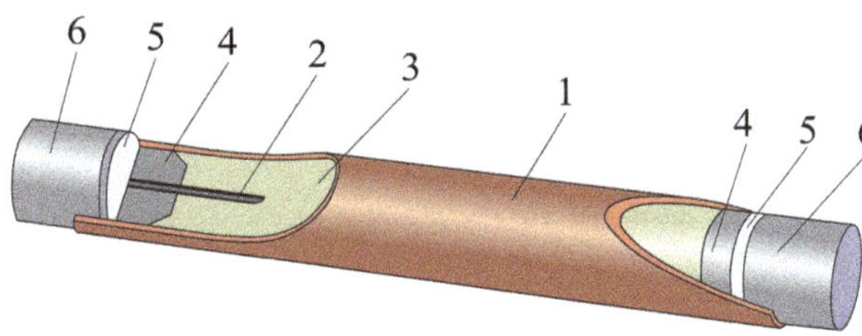

Fig. 9.3 An assembly for radial MPC [7]: 1, copper shell; 2, rod; 3, powder; 4, bushing; 5, gasket; 6, plug. (Drawn using data of Ref. [7])

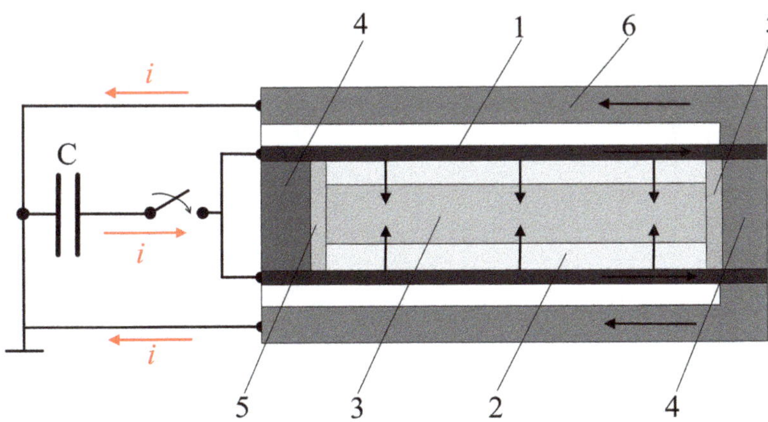

Fig. 9.4 Principle scheme of magnetic pulse compaction of tubes: 1, copper shells; 2, powder; 3, steel rod; 4, plugs; 5, gaskets; 6, reverse conductor. (Reprinted from Ivanov et al. [8] http://www.iiss.sanu.ac.rs/download/vol37_1/vol37_1_06.pdf, this article is available under the terms of the Creative Commons Attribution License (CC BY), https://creativecommons.org/licenses/by/4.0/)

The radial MPC (Fig. 9.3) is suitable for making tubes and rods several tens of centimeters long. Figure 9.4 is suitable for both Θ-pinch and Z-pinch compaction schemes.

The technology of MPC is promising for consolidating long-length parts. For ductile materials, MPC allows obtaining nearly fully dense compacts without a heating stage. Powders of materials that lack ductility can be compacted by MPC to densities exceeding 0.8 of theoretical density, while full densification can be achieved by the subsequent sintering. Due to a strained state of the material and high concentrations of defects in the particles, the required sintering temperatures are reduced such that a nanostructure of the material can be preserved in the sintered state. Both conductive and non-conductive materials can be compacted. The magnetic pulse compaction approaches have been successfully used for the pressure treatment of nanosized powders [7–10].

The powder compaction involves several distinct stages [11]: 1) packing (particle rearrangement), 2) elastic deformation at contact points and 3) plastic deformation at contact points of ductile particles and fragmentation of brittle ones. In the powder just loaded into a shell or a die, groups of particles become locked by friction. As the punch moves down toward the powders, the particles move to lower positions until further movement is not possible without particle deformation. At the end of the first stage, the particles form a rigid system. During the second stage, sliding between the particles and elastic compression at the contact points result in further densification of the powder. During the final stage, the material around contact points is subject to plastic deformation and flow. At this stage of compaction, there is a significant increase in resistance to densification due to strain hardening (cold working). Figure 9.5 demonstrates that multiple impacts are more effective as long as no delamination effects appear [11], which agrees well with Ref. [12].

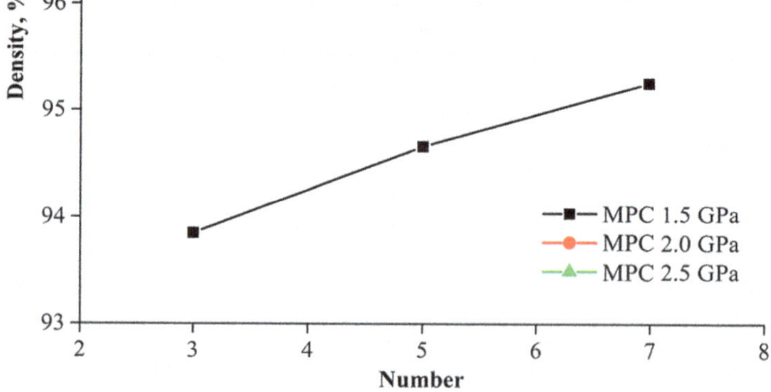

Fig. 9.5 The density of compacts after multiple pressing cycles. (Reprinted from Park et al. [11], Copyright (2012) with permission from Elsevier)

9.3 Modeling of Uniaxial and Radial MPC

Despite a rather long history of the experimental research, the systematic theoretical description of MPC started only in the 2000s. Reference [13] describes the first attempt of the modeling of the radial compression of a powder, taking into account the dynamic effects that appear when the can is accelerated by an external pulse magnetic field. The so-called inertia effect is noted in this work, when due to the inertia of the mechanical system (the can and the powder), a pressure level can be achieved that is significantly higher than the stress applied to the can by the magnetic field. It should be noted that the dynamics of the electric circuit has not been considered (the magnetic field pulse was defined by a decaying sinusoid).

The theoretical description of uniaxial pressing was developed by Boltachev et al. [14, 15] by assuming that the punch (impactor) acceleration by the magnetic field and the pressing of a powder occur independently; i.e., first, the acceleration of the impactor (which is not in direct contact with the powder) occurs, and after that the impactor comes into contact with the powder, and the powder compression occurs.

A theoretical model for radial MPC was developed by Boltachev et al. [9, 16], who found a closed solution of both the equations of the dynamics of a mechanical deformed system (a powder and a shell) and the equations of the dynamics of an electric circuit, which generates the magnetic field. By simultaneously solving the differential equations, it is possible to bring the theoretical models to the level of exact quantitative predictions and fully characterize the process. The powder compaction was described within a continual approach of plastically hardened porous body using empirical laws of hardening, in which the free parameters were determined from the experimentally obtained adiabatic curves of uniaxial compaction, examples of which are shown in Fig. 9.6 [17].

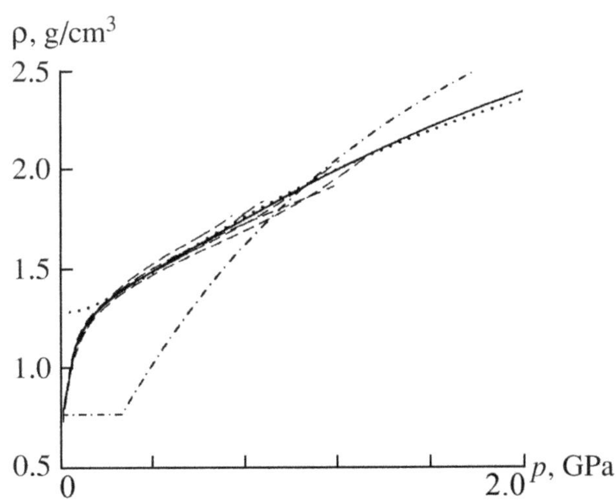

Fig. 9.6 Adiabatic compression curves for an alumina-based nanopowder produced by electric explosion: dash-dotted line is a theoretical case with the constant yield stress of 2 GPa, dotted line is the theory with a parabolic hardening law, solid line is the case with a modified hardening function, and dashed lines are experimental curves. (Reprinted from Boltachev et al. [17], Copyright (2007) with permission of Springer)

The theoretical modeling demonstrated that a satisfactory description of the radial compaction of powder in accordance with the Z-pinch mode can be achieved based on the approximation of a pronounced skin effect. At the same time, the simulation of the densification processes in the Θ-pinch mode requires the consideration of the diffusion of the magnetic field in the conductive material (shell, solenoid) and of their heating, largely due to the passage of electric current.

The complete system of equations, which takes into account all these factors in relation to the movement of a hollow shell, was formulated in Ref. [18]. It should be noted that for a fixed shell and neglecting heating of the material, the problem of the diffusion of the magnetic field is described by a linear differential equation, and its solution is possible in a closed form [19, 20]. Such an analysis, in particular, enables the determination of the scope of the problem parameters (size of the shell, the amplitude, and pulse width of the external magnetic field), when the shell's volume can be expanded by the residual magnetic field present in its cavity. This effect is seen as a possible way of a noncontact removal of spent shells from the compressed products. Developed in Ref. [18], the theoretical model was generalized in Ref. [16] to describe the motion of a shell in the presence of a densifying powder inside it. The analysis of pulsed processes of radial compaction based on the simultaneous solution of differential equations defining the dynamics of the electric circuit and the system being deformed, was carried out. This allowed studying the influence of various experimental setup parameters on the compaction process. Inertia effects in compaction dynamics were investigated, and the "resonance" compaction conditions, corresponding to the maximal usage of the inertial properties of the "powder + shell" system being deformed, were revealed. Thus, a theory of powder pressing for the Θ-pinch mode, sufficient for reliable reproduction of the known experimental data and, as a consequence, for the confident prediction of these processes, was created.

The correctness of the theoretical model was confirmed by a direct comparison of the calculated data with the existing experimental data on the nanopowder compact final density (green density). The theoretical model (taking into account the magnetic field diffusion) was developed for the pressing by relatively thin-walled conducting shells with a thickness that is comparable to or less than the magnetic field skin layer characteristic thickness. In particular, within the model, the cylindrical shell expansion caused by the diffusion of the magnetic field inside is studied. This effect was experimentally confirmed and is considered as one of the options of the shell detachment from the compact [16]. In general, the developed theoretical models allow performing a reliable prediction of the processes of nanopowder MPC and choosing the optimal conditions for manufacturing compacts with the required characteristics. It has been shown that the theory of plastically hardened porous body allows us to correctly characterize the nanopowder compaction process at a macroscopic level. Tuning of the resonant conditions is possible by varying both the external action characteristics (electric circuit parameters) and the inertia properties of the mechanic system being deformed. Within the Θ-pinch theory taking into account the magnetic field diffusion, complete agreement is achieved between the theoretical calculations and the experimental data on powder compact final densities.

Mathematical models have not only enabled determining the features of the compaction processes, but also helped discover a phenomenon of the shell expansion at the end of the pulse previously not known from the experiments but very useful for the procedure of detaching the shell from the compact.

The mathematical models and calculation methods elaborated for the radial MPC were successfully transferred to a more complex object for modeling – uniaxial MPC. In Ref. [10], the size effect was considered, which is one of the most important effects for any type of the compaction processes. In that publication, the size effect was related to the size of the powder particles. Calculation performed for particles ranging from 10 to 100 nm taking into account Hertz elastic forces, tangent friction forces, and dispersion forces of attraction have shown that the latter make fine powders more difficult to densify than coarse powders.

A mathematical model of uniaxial MPC of powders that includes a consistent solution describing the dynamics of a pulse magnetic field and the dynamics of the mechanical system (the moving pressing parts + compacted powder), was presented by Olevsky et al. [21]. The model takes into account a number of important factors such as dissipative losses at the friction contact boundaries, the elastic properties of the pressing device, and the inertia properties of the moving parts of the press. The constitutive behavior of the compacted powder is described in the framework of the theory of plastic porous bodies with hardening [17, 22–27]. To determine the number of parameters of the theory (the law of hardening of a powder body, the coefficients of friction, the elastic characteristics of the press, etc.), experimental research on pressing of two different nanosized powders of alumina has been conducted. It was shown that the developed theoretical model can reliably reproduce the results of the conducted experiments: the time cyclograms of the electric current passing through the inductor and of the force (pressure) applied to the compacted powder can be described with the accuracy comparable to the scale of the experimental measurement error. For a known pressure applied to the powder, and for the established laws of hardening, the theory allows estimating the evolving density (and/or porosity) of the powder specimen and, consequently, selecting the optimal pressing conditions enabling the achievement of a desirable level of final porosity. The developed theoretical model allows analyzing a variety of pressing conditions, without making expensive, and sometimes unrealizable with the existing equipment, experimental studies. Although a decrease in the powder mass can slightly increase the final density of powder compacts, more promising would be the downsizing of the accelerated components of the pressing device (Fig. 9.7).

Rather high requirements to the strength characteristics of the impactor prevent such a tangible reduction in its mass. Increasing the diameter of the powder specimen needed during the transition from test lab-scale experiments to the fabrication of practically significant components, for fixed energy costs, of course, reduces the final density. The developed theoretical model enables the prediction of the results of compaction in the dies with different diameters. In particular, for dies with diameters of up to 35 mm, calculations show that a proportional increase in the charging voltage and the diameter of the die does not change the final porosity of the powder specimen. However, note that, in general, the presence of the dissipative terms in the

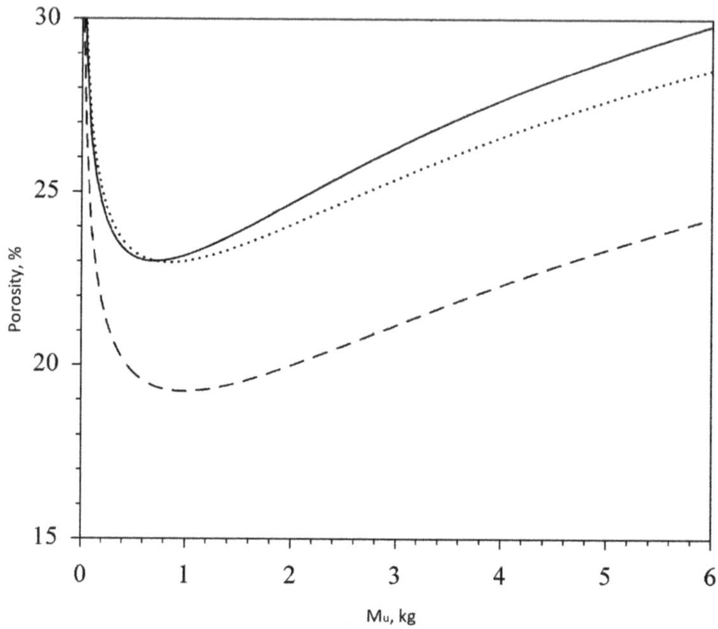

Fig. 9.7 Porosity θ of the compacted alumina as a function of the impactor mass M_u (solid line, m_p= 1.5 g (powder mass), U_0=2.0 kV; dashed line, m_p=1.5 g, U_0=2.5 kV; dotted line, m_p=1.0 g, U_0=2. kV. (Reprinted from Olevsky et al. [21], Copyright (2013) with permission of Springer)

constitutive equations of the model violates this simple way to scale. The ways to significantly increase the final density of a powder compact at a given charging voltage of the capacitive storage are particularly relevant for large-diameter compacted products. One approach consists in varying the natural period of the electric circuit. When the electric current increases the time in a circuit equal to (approximately) the time of the realization of the pressing process (time before the stop of the impactor), there is a kind of resonance – the force exerted by the impactor on the powder specimen can substantially exceed the amplitude of the magnetic field impact on the impactor. This effect is achieved through the best use of the inertia properties of the impactor.

Double-sided MPC was studied by Barbarovich [3] as a more economical and more efficient variant of MPC. In search of methods of enhancing the efficiency of MPC, double-sided MPC has been recently revisited [28]. Double-sided MPC allows achieving higher densities of the compact relative to the single-sided scheme (Fig. 9.8, see also Fig. 9.2). The double-sided method uses two impactors, which are moving toward each other and densify the powder volume located between them (Fig. 9.8). The utilized concentrators have a serial connection. A comparison of the single- and double-sided compression schemes is shown in Fig. 9.9 for an alumina nanopowder. An increase in the total inductance in the double-sided scheme reduces the amplitude of the current in the circuit and, consequently, leads to a decrease in

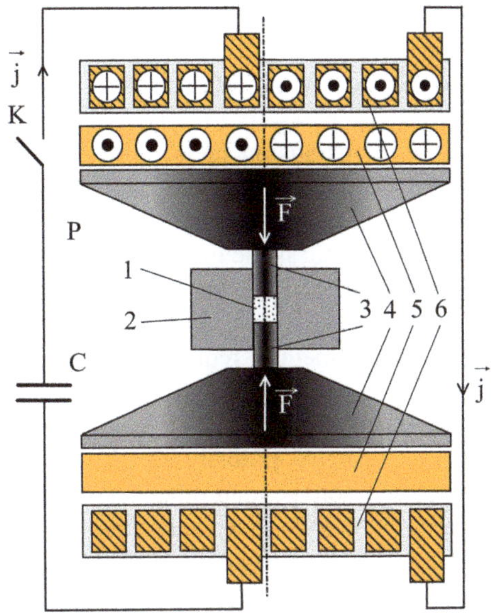

Fig. 9.8 Principle scheme of double-sided MPC: 1, powder; 2, die; 3, punches; 4, concentrators; 5, satellites; 6, inductors. (Drawn using data of Ref. [28])

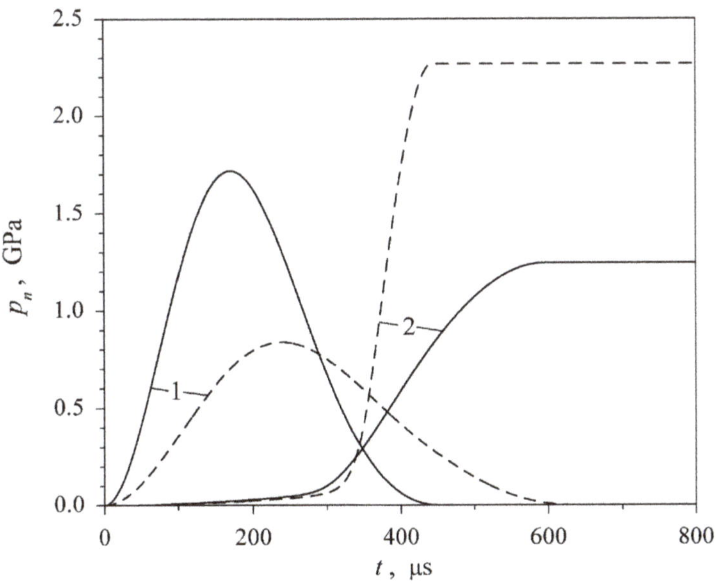

Fig. 9.9 "Magnetic pressure" normalized with respect to the specimen's cross-sectional area (lines 1) and pressure acting on the densified powder (lines 2). Solid lines, single-sided MPC; dashed lines, double-sided MPC. (Reprinted from Olevsky et al. [21], Copyright (2013) with permission of Springer)

the "magnetic" force acting on each of the impactors (Fig. 9.9, lines marked 1). However, the force on the powder is not reduced; on the contrary, as seen in Fig. 9.9, lines marked 2, it increases, leading to a decrease in the final porosity (from 34.8% in single-sided pressing to 29.6% in double-sided pressing). This occurs mainly because of exclusion of the effective elasticity of the pressing device from the process. In such a setup, all the momentum reaching the powder is spent on its pressing and not wasted on the elastic deformation of the lower part of the press. A positive factor is the growth of the natural period of the circuit oscillations due to an increase in its inductance. This shifts the mode of operation of the pressing device to the "resonant" pressing conditions. Note, however, that the double-sided compression method requires substantial modifications of the existing equipment for MPC.

The final goal of MPC is to fabricate parts with desired characteristics, which include geometry, porosity, density distribution, microhardness, fracture toughness, breakdown voltage, and other functional properties. In order to fabricate a part, an optimal choice of independent as well as dependent parameters should be made including the parameters of the powder (the material of the particles, composition of the powder mixture, particle size, particle shape), parameters of preliminary treatment (pre-pressing, degassing, preheating), and parameters of the pulse (the pressure amplitude in the pulse, waveform of the pulse, pulse duration, the number of pulses in a series or the law governing the series of successive pulses).

With many parameters inherent to MPC, a large number of experiments may be necessary for the process optimization. Knowledge of the process mechanisms at the pre-compaction stages as well as those during MPC makes it possible to develop an efficient fabrication technology. The problem becomes even easier to solve when the understanding level of the processes allows describing them mathematically and developing software capable of calculating the required technological parameters.

Poor densification of alumina nanopowders predicted theoretically was verified experimentally using powders having particles in the range of 21–82 nm [10]. It was found that the size effect is observed at a pulse pressure of 0.1 GPa, while at 1 GPa, the residual porosities of the compacts produced from different powders coincide within the limits of the measurement error. The authors suggest that this qualitatively distinguishes oxide powders from metallic powders, in which the size effects are observed up to pressures of several GPa. In metallic powders, the ability to deform increases with the particle size; in oxide powders, even coarse particles are mechanically strong. It is thus believed that the only reason for the size effect is the action of inter-particle dispersion forces of attraction. The authors of Ref. [10] have also modeled multi-cycle compaction, in which the powder is subjected to several pressing pulses. The calculations have shown that multi-cycle loading allows achieving densities close to the theoretical one. The efficiency of the proposed method was verified by the experiments with the Ti6Al4V alloy [12] and aluminum alloy powders [11].

Microstructure investigations of a copper compact obtained by uniaxial MPC showed that the porosity increased with the distance from the moving punch [29]. The same effect was observed in uniaxial MPC of titanium nitride [30].

In order to achieve a more uniform densification, it was proposed to conduct another pressing cycle after turning the compact in the die by 180° [29]. Applying two pressing cycles has given good results for an aluminum-based powder [11]. It was found that the compact density decreased as the amount of the powder in the die increased. This effect was more pronounced at low compaction pressures.

The role of preliminary pressing on the final density of the material produced by MPC was studied. Densification of an alumina powder prior to MPC resulted in a decrease in the final density of the compact [31]. However, an opposite result was described in a later publication [32]. An increase in the initial density of the powder load by static pre-pressing resulted in higher densities after MPC as no work of the external force was necessary to achieve denser packing of strong powder agglomerates, so the energy was more efficiently used for the fracturing process.

It was shown experimentally that the agglomerated state of nanopowders plays a major role in the pressing processes [32]. The presence of large and strong agglomerates makes it more challenging to compact a powder by MPC and obtain high final densities. This can be a consequence of the size effect; however, this was not taken into consideration in the modeling studies presented in Refs. [33, 34].

A certain number of aspects of MPC have not been investigated to date. Processes at the particle scale have been rather scarcely addressed. The influence of the compact size on the results of uniaxial MPC was not studied, while this was studied for radial MPC. Effects specific to compaction of shapes other than axisymmetric have not been yet investigated. The roles of pre-pressing, the presence of absorbed species, and temperature of the process have not been addressed in detail. In nanopowders, the content of absorbed species can be up to several percent, and their presence can cause the formation of cracks in the compacts. This issue was addressed in Ref. [35] for alumina and rather briefly in Ref. [36] for zirconia.

Understanding of the process mechanisms includes not only the micro- and macro-level phenomena in the powder compact but processes in the MPC tooling, such as elastic vibrations of the punches in uniaxial MPC, deformation kinetics of shells in radial MPC, and the interaction of the external magnetic field with the powder particles.

Sintering of materials compacted by MPC is not widely presented in the literature, nor is the influence of the defects and strain accumulated during MPC on their sintering behavior.

MPC offers possibilities of preserving the nanostructure of the compacted materials and making parts with a wide range of shapes – from disks and pellets to long-length rods and tubes. A significant advancement in the understanding of the processes during MPC has been made in the past decade thanks to computer modeling. The future development of the technology of MPC is thought to benefit from multiple compaction pulses, compaction in two pulses with turning of the sample before the second pulse, expansion of the shell at the end of the pulse, as well as a combination with spark plasma sintering (SPS). Using multiple pulses helps achieve high compact densities without any risk of tooling damage. The second pressing cycle applied to the sample can be helpful in reaching a uniform density in thicker samples. Process parameters favoring shell expansion would make it easy to remove the compact from the tooling. Recent studies have shown that a possibility exists to switch from conventional sintering of MPC-produced parts to SPS. Ceramic materials obtained by this combination are fully dense and possess a nanostructure inherited from the powders.

9.4 Selected Examples of Application of MPC to Different Materials

The main goals of experimental MPC studies of metals are to fabricate nanostructured metals, which presumably possess improved mechanical properties important for practical applications, and gain a deeper understanding of the mechanisms of MPC of nanopowders using simple model systems. As MPC can be successful for nanopowders, fine powders with a narrow particle size distribution are the systems that draw major attention. Figure 9.10 shows that the grains in a compact produced by MPC are much finer than in a conventionally sintered compact [37].

Figure 9.11 shows the relative densities of different materials produced by MPC including soft aluminum and brittle alumina and silicon carbide.

MPC is effective in producing metallic parts from powders. The compacts exhibit properties that make them suitable for practical application without a post-compaction annealing operation. MPC is effective for both coarse powders and nanopowders. At low compaction pressures, in order to achieve full densification and desired properties, ductile metals should be compacted by a series of pulses. A higher density and a more uniform microstructure of the compact are obtained when two pulses of compaction are applied, the part being turned by 180° before the second pulse. It is believed that for each metal, a critical pulse pressure exists that can ensure full densification in a single-pulse compaction process. If several pulses are applied to metals that lack plasticity, fragmentation of the particles occurs, and properties of the compact affected by the inter-particle interaction forces deteriorate.

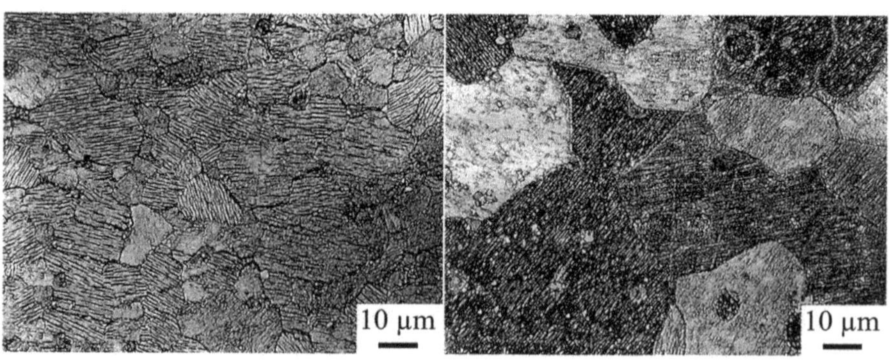

Fig. 9.10 Microstructure of a SmCo compact produced by MPC (left) and conventional sintering (right) [37]

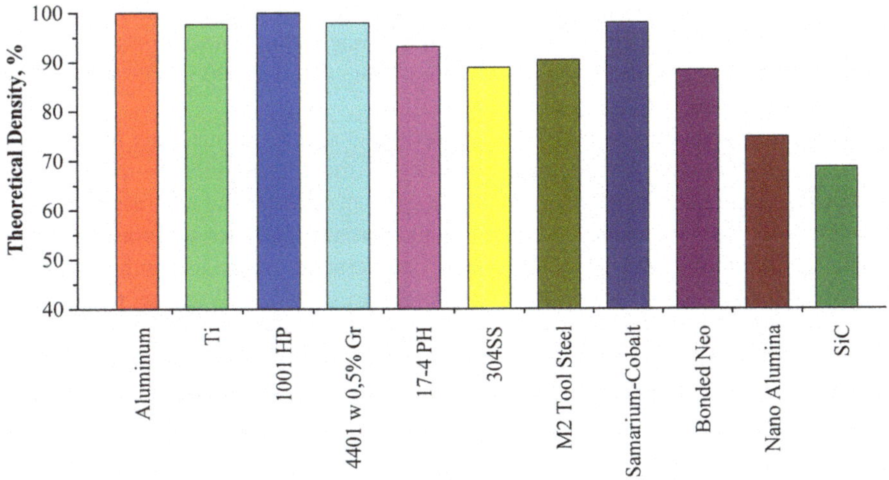

Fig. 9.11 Relative densities of materials compacted by MPC [37]

Uniaxial MPC of copper was reported in Ref. [29, 38, 39]. The following results have been obtained: uniaxial MPC of copper nanopowder at 1.5 GPa results in the formation of compacts 95% dense; the hardness of the compacts increases with the compaction temperature up to 300 °C, while their nanocrystalline structure is preserved; when the temperature increases up to 400 °C, significant grain growth occurs with simultaneous drops in density and microhardness.

The density of aluminum compacted at 1.6 GPa and 300 °C varied between 94% and 95% [40]. As the compaction temperature increased, the pore size also increased but the total volume fraction of the pores decreased. In Ref. [41], fabrication of a 97.6% dense Al compact at a pressure of 1.6 GPa and 300 °C was reported. The same evolution of pore distribution as in Ref. [40] was observed and related to fragmentation of the oxide films. Uniaxial MPC of an Al–20% Si powder with an average particle size of 106 μm at room temperature was studied [11]. The highest density (98%) was achieved when two pressing pulses were performed; the compaction results largely depending on the value of pressure applied during the second compaction cycle.

In Ref. [42], it was found that pretreatment of the powder affected the density of the iron-based compacts consolidated by MPC. The maximum density of the compact (82.1%) produced by MPC was reached when the powder was milled for 4 h (the particle size up to 50 μm) and annealed at 600 °C for 1 h. Uniaxial MPC was carried out at a pressure of 1.5 GPa. The only information on record on MPC of steels comes from publications by IAP Research (USA) [43]. Fabrication of compacts 99% dense from steel powders was reported.

Uniaxial MPC of $Mg–Zn_{4.3}Y_{0.7}$ was presented in Ref. [44]. It was shown that under the maximum value of discharging voltage, the compact density was only 96%. The authors suggest that the residual porosity and poor mechanical characteristics of the compact were due to the presence of an oxide film on the powder surface and a limited slip system in the magnesium crystalline structure. A nearly fully dense compact was obtained after compaction at 300 °C showing a 13% increase in hardness.

Nanostructured ceramic materials are attractive structural materials due to their high strength, hardness, and fracture toughness. Among consolidation methods that allow preserving the nanostructure of the powders in the consolidated material, MPC is one of the most promising. Unlike metals, ceramic particles are not capable of plastic deformation, and, therefore, MPC of ceramic materials cannot produce compacts ready for the practical use. In order to obtain fully dense materials, a post-compaction sintering step is necessary. Alumina powders are the most popular objects in the studies of dynamic compaction as these powders are inexpensive and have a broad range of applications. The uniaxial MPC of alumina nanopowders was systematically addressed in Ref. [31]. The size of the powder particles was 20 nm. The pre-pressed powder was subjected to vacuum degassing at 300 and 600 °C in a chamber up to a pressure of 1 Pa. It was found that preliminary densification of the powder leads to a decrease in the density of the MPC-produced compact. As the pulse pressure increases up to 2 GPa, the density of the compact increases. Densities ranging from 52% to 83% were obtained; the compacts did not show any inner cracks. In a continuation of this research [45], it was found that when a static pressure of 1.6 GPa is used, the optimal degassing temperature is room temperature, which indicates a positive effect of the absorbed species reducing the inter-particle friction. It was concluded that pressure in the pulse is not the only parameter affecting the compaction results.

Sintering of nanostructured alumina compacts produced by uniaxial MPC was studied in Ref. [46] as the final processing stage of the consolidated material. It was found that sintering would result in a fully dense compact, if the compacts produced by MPC were 70–80% dense, which is rather easy to achieve. Optimization of the heating regime has been performed considering the phase changes in the material. It was found that fast uncontrolled heating is not suitable for sintering of pressed parts from nanosized powders as it causes expansion of the material and results in low final densities. The result of the study was the preparation of 97% dense α-Al_2O_3 with a grain size of 100–150 nm.

A study had been carried out to compare the effectiveness of the Z-pinch and Θ-pinch methods for pressing of two types of alumina nanopowders containing magnesia [7]. An original powder holder has been designed, in which both methods can be realized under the same conditions. The holder was designed in such a way that it was possible to conduct hot degassing and seal it before placing it in the MPC facility. Better results were obtained with the Θ-pinch method, in which the same parameters of the generator resulted in a longer pulse. Pressing of powders having particles of 20 and 90 nm by radial MPC and by the static method showed that in all cases higher densities were obtained using non-agglomerated powders. However, compacts from agglomerated powders showed a more uniform density along the radius of the compact.

The dependence of the final density of the MPC-produced compact on the green density was studied for several alumina-based nanopowders containing different additives [47]. For experiments on radial MPC, the powders were poured into a thin-walled tube with an outer diameter of 12–14 mm and a central channel 2–3 mm in diameter. Radial MPC was performed at a pressure of 0.3 GPa and a pulse duration of about 100 μs. A compact with a uniform microstructure was thus obtained and was 55–60% dense. In parallel experiments, powders were processed by uniaxial MPC, which resulted in the formation of a disk 15 mm in diameter and 1 mm thick with a relative density of 56–85%. The pressed compacts were sintered at temperatures up to 1550 °C for 5 h. When sintering of pure α-Al_2O_3 was carried out at 1550 °C, densities of the pressed compacts exceeding 2 g/cm^3 led to a decrease in the final sintered density. When a nanopowder of an Al–Mg alloy was added to α-Al_2O_3, the final sintered density was practically independent on the density obtained at the compaction stage. For powders annealed at 1400 °C, the final sintered density increased as the density of the MPC-produced compact increased up to 2.8 g/cm^3. The best mechanical properties and structural characteristics were obtained using a corundum powder containing 15% of an Al–Ag alloy. The density of the sintered material was 97%, its microhardness was 18–20 GPa, and its grain size was about 1 μm.

References [48, 49] report the formation of 94% dense crack-free α-Al_2O_3 compacts after uniaxial MPC at 1.25 GPa and sintering at 1450 °C for 3 h. However, better results were produced by a combination of uniaxial MPC and SPS [50]. After uniaxial MPC at a pressure of 2.1 GPa and subsequent SPS for 10 min at 1350 °C (heating rate 100 °C/min) and a pressure of 50 MPa, bulk samples with a relative density of 99.7% were obtained. No grain growth was observed after these operations. It is interesting to note that Al_2O_3–$ZrO_2(Y_2O_3)$ nanosized composite powders compacted by MPC up to a relative density of 55% could be sintered to a density of 99% by SPS at 1350 °C [51]. This shows that a combination of MPC and SPS is very promising for the fabrication of dense ceramics.

The behavior of ZrO_2 during compaction was reported in a number of publications [52, 53]. The compaction resulted in the formation of 80% dense materials, which were further densified during sintering up to a density of 96%. Radial MPC followed by sintering at 1600 °C was used to consolidate yttria powders [54, 55]. After MPC the density was 69%; after sintering it increased to 98.6%.

The successful process of MPC depends on the plasticity of the material or the content of a binder. Therefore, composite mixtures can be anticipated to behave during MPC in a way similar to metals. The details of MPC of Al–Al_2O_3 were presented in Ref. [56]. The size of Al_2O_3 particles was 40 nm, while the size of Al particles was 154 nm. The grain size of the sintered materials was 1–2 μm. The best mechanical properties were observed in the composite containing 15% of Al compacted at 0.3 GPa and sintered at 1550 °C. The microhardness of the composite was 19.4 GPa, and its fracture toughness was 7.3 MPa·m$^{1/2}$.

MPC of Co–diamond composites was studied by Lee et al. [57]. After compaction at 4 GPa, the density of the material was 86.4%, and upon subsequent sintering it increased up to 99.6%. The hardness of the material was 99.8 HRB. Compared with conventionally processed Co–diamond composites, composites produced by MPC were denser and had higher hardness and smaller grains. The compacts produced by this technique can be used in drilling tools and offer longer service lives of the drilling tools made of these materials and higher drilling rates.

9.5 Summary

MPC is a technique based on the quasi-dynamic pressing of powders. It does not employ strain rates of the level usual for the dynamic compaction; however, its deformation rates by orders of magnitude exceed those in the conventional quasi-static powder pressing. It should be also noted that MPC does not usually involve any direct interaction of the magnetic field with powder materials. Rather, it uses the kinetic energy of the tooling rendering the conditions of the quasi-dynamic compaction. This intermediate position of MPC between quasi-static and dynamic compaction of powders provides it with some rather unique capabilities, including the preservation of the nanostructure of the obtained materials and the suitability for a wide range of shapes – from disks and tablets to rods and tubes of great length. Unlike dynamic pressing, MPC generates an acclivous leading edge of the compression wave, which excludes the emergence of the next discharge wave and avoids exfoliation of the compacts. The method is divided into uniaxial pressing and radial pressing. The understanding of the processes taking place during MPC has been advanced in the last decade thanks to computer modeling. The following are promising directions for the development of MPC technology: the use of multiple pulses, repeated pressing of tablets with overturning, expansion of the shell at the end of the pulse, and the combination of MPC with SPS.

References

1. Mironov VA (1980) Magnetic pulsed pressing of powders. Zinatie, Riga, USSR, 196 p, (in Russian)
2. Sandstrom DJ (1964) Consolidating metal powders magnetically. Met Prog 8:215–221
3. Barbarovich YK (1969) Use of the energy of a strong pulsed magnetic field for powder compaction. Soviet Powder Metall Metal Ceram 8(10):798–803
4. Knopfel G (1972) Superstrong magnetic fields. Mir, Moscow, 383 p, (in Russian)
5. Ivanov VV, Paranin SN, Vikhrev AN, Nozdrin AA (1997) Materialovedenie (Materials Science) 5:49 (in Russian)
6. Ivanov VV, Vikhrev AN (1997) Fizika I Khimiya Obrabotki Materialov. Phys Chem Mater Process 3:67 (in Russian)
7. Paranin S, Ivanov V, Nikonov A, Spirin A, Khrustov V, Ivin S, Kaygorodov A, Korolev P (2006) Densification of nano-sized alumina powders under radial magnetic pulsed compaction. Adv Sci Technol 45:899–904
8. Ivanov VV, Ivin SY, Khrustov VR, Kotov YA, Murzakaev AM, Nikonov AV, Paranin SN, Spirin AV (2005) Fabrication of nanoceramic thin-wall tubes by magnetic pulsed compaction and thermal sintering. Sci Sinter 37:55–60

9. Boltachev GS, Nagayev KA, Paranin SN, Spirin AV, Volkov NB (2010) Magnetic pulsed compaction of nanosized powders. Nova Science Publishers, New York
10. Boltachev GS, Volkov NB, Kaygorodov AS, Loznukho VP (2011) The peculiarities of uniaxial quasistatic compaction of oxide nanopowders. Nanotechnologies Russia 6(9–10):639–646
11. Park HY, Fatih Kilicaslan M, Hong SJ (2012) Effect of multiple pressures by magnetic pulsed compaction (MPC) on the density of gas-atomized Al-20Si powder. Powder Technol 224:360–364
12. Li M, Yu H, Li C (2010) Microstructure and mechanical properties of Ti6Al4V powder compacts prepared by magnetic pulse compaction. Trans Nonferrous Met Soc China 20:553–558
13. Dobrov SV, Ivanov VV (2004) Simulation of pulsed magnetic molding of long powdered products. Tech Phys 49:413–419
14. Boltachev GS, Volkov NB, Ivanov VV, Kaygorodov AS (2009) Shock-wave compaction of the granular medium initiated by magnetically pulsed accelerated striker. Acta Mech 204:37–50
15. Boltachev GS, Kaygorodov AS, Volkov NB (2009) Densification of the granular medium by the low amplitude shock waves. Acta Mech 207:223–234
16. Boltachev GS, Nagayev KA, Paranin SN, Spirin AV, Volkov NB (2010) Theory of the magnetic pulsed compaction of nanosized powders. In: Cabral V, Silva R (eds) Nanomaterials: properties, preparation and processes. Nova Science Publishers, New York, pp 1–58
17. Boltachev GS, Volkov NB, Dobrov SV, Ivanov VV, Nozdrin AA, Paranin SN (2007) Simulation of radial pulsed magnetic compaction of a granulated medium in a quasi-static approximation. Tech Phys 52(10):1306–1315
18. Boltachev GS, Volkov NB, Paranin SN, Spirin AV (2010) Dynamics of cylindrical conducting shells in a pulsed longitudinal magnetic field. Tech Phys 55:753–761
19. Boltachev GS, Volkov NB (2009) Expansion of a conducting shell by magnetic field of an external inductor. Tech Phys Lett 35:334–336
20. Boltachev GS, Volkov NB (2009) Bimetallic cylinder in a pulsed magnetic field. Tech Phys Lett 35:916–919
21. Olevsky EA, Bokov AA, Boltachev GS, Volkov NB, Zayats SZ, Ilyina AM, Nozdrin AA, Paranin SN (2013) Modeling and optimization of uniaxial magnetic pulse compaction of nanopowders. Acta Mech 224(12):3177–3195
22. Shtern MB, Serdyuk GG, Maksimenko LA, Truhan YV, Shulyakov YM (1982) Phenomenological theories of powder pressing. Naukova Dumka, Kiev (in Russian)
23. Olevsky E (1998) Theory of sintering: from discrete to continuum. Mater Sci Eng R 23:41–100
24. Olevsky EA, Molinari A (2000) Instability of sintering of porous bodies. Int J Plast 16:1–37
25. Olevsky EA, Molinari A (2006) Kinetics and stability in compressive and tensile loading of porous bodies. Mech Mater 38:340–366
26. Olevsky EA, LaSalvia JC, Ma J, Meyers MA (2007) Densification of porous bodies in a granular pressure-transmitting medium. Acta Mater 55:1351–1366
27. Shtern M, Olevsky E (2008) Plastic behavior of agglomerated powder. Comput Mater Sci 43:704–709
28. Ivashutenko AS (2010) Alumina-zirconia nanoceramics obtained with the use of high energy flows. PhD Dissertation, Tomsk Polytechnic University, Tomsk (in Russian)
29. Yu HP, Li CF (2007) Dynamic compaction of pure copper powder using pulsed magnetic force. Acta Metall Sin (English Letters) 20(4):277–283
30. Andrievskii RA, Vikhrev AN, Ivanov VV, Kuznetsov RI, Noskova NI, Sazonova VA (1996) Magnetic-pulse and high-pressure shear-strain compaction of nanocrystalline titanium nitride. Fiz Metall Metalloved 81:137–145 (in Russian)
31. Ivanov VV, Kotov YA, Samatov OH, Böhme R, Karow HU, Schumacher G (1995) Synthesis and dynamic compaction of ceramic nano powders by techniques based on electric pulsed power. Nanostruct Mater 6(1–4):287–290
32. Nozdrin AA (2007) Investigation of the possibilities of dynamic pressing of alumina-based nanopowders. Perspectivnye Materialy 6:79 (in Russian)

33. Boltachev GS, Volkov NB (2010) Size effect in compaction of nanopowders. Pis'ma v Zhurnal Technicheskoi Fiziki (Technical Physics Letters) 36(17):96 (in Russian)
34. Boltachev GS, Volkov NB (2011) Simulation of nanopowder compaction in terms of granular dynamics. Tech Phys 56(7):919–930
35. Kaygorodov AS, Ivanov VV, Paranin SN, Nozdrin AA (2007) The role of adsorbed species on pulse pressing of oxides. Rossiiskie Nanotechnologii 2(1–2):112–118 (in Russian)
36. Guzeyev VV (1995) Temperature control of zirconia ceramics sintering. Glas Ceram 10:25–29 (in Russian)
37. Chelluri B, Knoth E (2010) 4th international conference on high speed forming. Columbus
38. Lee GH, Rhee CK, Lee MK, Kim WW, Ivanov VV (2004) Nanostructures and mechanical properties of copper compacts prepared by magnetic pulsed compaction method. Mater Sci Eng A 375–377(15):604–608
39. Rhee CK, Lee GH, Kim WW, Ahn JH, Hahn YD (2003) Correlation between structure and mechanical properties for nano-crystalline copper prepared by pulsed compaction. J Metastable Nanocryst Mater 15–16:757–763
40. Lee GH, Rhee CK, Kim KH (2003) The effect of compaction temperature and pressure on the pores in nanostructured metal compacts prepared by magnetic pulsed compaction. Metals Mater Int 9(4):375–378
41. Han YS, Seong BS, Lee CH (2004) SANS study of microstructural inhomogeneities on Al nano-powder compacts. Physica B Cond Matter 350(1–3):E1015–E1018
42. Hong SJ, Lee GH, Rhee CK (2007) Magnetic pulsed compaction of ferromagnetic nano-powders for soft-magnetic core. Mater Sci Eng A 449–451(25):401–406
43. Chelluri B (1994) Dynamic magnetic consolidation (DMC) process for powder consolidation of advanced materials. Mater Manuf Proc 9(6):1127–1142
44. Chae HJ, Kim YD, Kim TS (2011) Microstructure and mechanical properties of rapidly solidified Mg alloy powders compacted by magnetic pulsed compaction (MPC) method. J Alloys Compd 509(1):S250–S253
45. Banin VE, Boehme R, Schumacher G, Vikhrev A (1995) Dynamic compaction of nanosized ceramic oxide powders. Mater Sci Forum 225–227:623–628
46. Ivanov VV, Khrustov VR (1998) Inorg Mater 34(4):39–43 (in Russian)
47. Kaygorodov A, Rhee C, Kim W, Ivanov V, Paranin S, Spirin A, Khrustov V (2007) Nozzles from alumina ceramics with submicron structure fabricated by radial pulsed compaction. Mater Sci Forum 534–536:1053–1056
48. Hong SJ, Lee JK, Lee MK, Sung JH, Lee CG, You YZ (2006) Consolidation of Al_2O_3 nanopowder by magnetic pulsed compaction and sintering. Solid State Phenom 118:615–622
49. Hong SJ, Koo JM, Lee JG (2009) Precompaction effects on density and mechanical properties of Al_2O_3 nanopowder compacts fabricated by magnetic pulsed compaction. Mater Trans 50 (12):2885–2890
50. Lee J, Hong SJ, Lee MK, Sung JH, Lee CG, You YZ (2006) Fabrication of high-density nanostructured alumina by the combined processes of magnetic pulsed compaction (MPC) and Spark Plasma Sintering (SPS). Solid State Phenom 118:597–602
51. Dyatlova YG, Rumyantsev VI, Ordan'yan SS, Osmakov AS, Zayats SV, Ivanov VV, Paranin SN (2013) Website of "Virial", Ltd. (electronic source, free access): http://www.virial.ru/upload/medialibrary/85a/Tezisy_2.pdf. (in Russian)
52. Ivanov VV, Paranin SN, Khrustov VR (2002) Nanostructured ceramics based on aluminum and zirconium oxides produced using magnetic pulsed pressing. Phys Metals Metallogr 94(1):S98–S106
53. Banin VE, Paranin S, Khrustov V, Medvedev A, Shtol'ts A (2002) Processing of nanostructured oxide ceramics with magnetic pulsed compaction technique. Key Eng Mater 206–213:377–380
54. Lee JG, Hong SJ, Park JJ (2010) Fabrication of an yttria thin-wall tube by radial magnetic pulsed compaction of powder-based tapes. Mater Trans 51(9):1689–1693
55. Lee JK, Hong SJ, Lee MK, Rhee CK (2007) Fabrication of high density Y_2O_3 ceramics by magnetic pulsed compaction. Solid State Phenom 119:175–178

56. Ivanov VV, Kaigorodov AS, Khrustov VR, Paranin SN, Spirin AV (2006) High-strength alumina-based ceramics produced by magnetic pulsed compaction of composite nanopowders. Rossiiskie Nanotechnologii 1(1–2):201–207 (in Russian)
57. Lee JG, Lee MK, Hong SJ, Lee HW, Pyun SP, Rhee CK (2010) Consolidation of mixed diamond and cobalt granule powders by magnetic pulsed compaction. Mater Lett 64(1):35–37

Chapter 10
Field Effects on Reacting Systems

10.1 Reactive Sintering: General Remarks

In modern consolidation methods, the term "reactive sintering" refers to processes, in which a bulk material forms as a result of chemical reactions occurring during heat treatment of powder mixtures. Energy from an external source is used to initiate the chemical reaction and is continuously provided until the synthesis is complete and during further sintering. A more efficient approach is pressure-assisted reactive sintering, which facilitates densification of the reaction products. Therefore, reactive sintering can be considered as a combination of two processes: chemical reaction and densification. The extent of densification is a matter of materials design; either fully dense or porous materials can be obtained after the reaction step is complete.

Densification of a reaction mixture itself can occur under applied pressure at low temperatures before the reaction starts. In an ideal situation, when the reaction is not accompanied by a volume change, a dense reaction product can be obtained [1]. However, in most cases, the volume of the product is different from the sum of the volumes of the reactants, and the reaction interferes with the initially achieved densification through the formation of cracks and other macrodefects, which makes it necessary to consolidate the product after the reaction.

The advantages of compaction and sintering of the reaction products while they are still warm after an exothermic formation reaction have been widely appreciated in the field of SHS. Substantial experience has been accumulated in the SHS accompanied by the product densification [2–6] and thermal explosion under pressure [7, 8]. The SHS–dynamic compaction of the reaction product is performed immediately after the reaction is complete using a uniaxial press specially designed to accommodate the reacting systems. Mechanical pressure in such facilities can be applied within seconds from the ignition of the reaction [6]. The remaining problems of the SHS–dynamic compaction and thermal explosion under pressure are achieving 100% density and preserving grains smaller than 1 μm.

E. A. Olevsky, D. V. Dudina, *Field-Assisted Sintering*,
https://doi.org/10.1007/978-3-319-76032-2_10

The choice of reactive sintering is often dictated by its advantages for the production of nanostructured materials. Many interesting and attractive properties found in nanostructured materials – increased mechanical strength, hardness, improved ductility and fracture toughness, increased wear resistance, superplasticity, and tailorable magnetic behavior [9] – present a constant stimulus to look for new processing routes offering flexible control over the material structure and chemical composition. In preparation of bulk nanostructured materials by powder sintering, successful consolidation of powders without sacrificing the previously tailored microstructure and phase composition of the material is the major challenge. The term "in situ processing" is usually used to refer to processes, in which the desired phase/phases of a material form during sintering or consolidation. The in situ processing is suggested to give better results in terms of a more uniform microstructure compared to the ex situ processing [10, 11]. During an in situ processing, the target phase is constructed from precursors at atomic, molecular, or crystallite levels. Better interfacial compatibility, coherent interfaces, and interfaces free from contamination are aimed by the in situ formation of inclusions in a matrix. However, choosing reactive processing rather than conventional mixing and sintering of non-reacting powders does not guarantee a fine-grained microstructure unless the processing parameters are optimized.

The following problems can be encountered during a sintering process that is accompanied by the formation of new phases by chemical reactions:

- The reaction is complete at lower temperatures than those required for efficient densification of the product because the diffusion in the reaction product is too slow.
- The product has already well-established contacts between the particles (severely agglomerated product), which are too strong to allow for the particle rearrangement [12].
- In multiphase products, the phases may have different sintering behavior (plays a significant role when the phases are not mixed at the grain scale and form agglomerates).

10.2 Driving Forces in Reactive Sintering

Savitskii [13, 14] emphasizes the importance of understanding the driving forces acting during sintering of binary and multicomponent powder mixtures. A porous body that is composed of powders of different chemical nature and capable of interacting with each other possesses, beside an excess free surface energy, an excess free energy associated with the potential of the system to form an alloy or a new compound/compounds. Therefore, in order to determine which of the processes – the action of the capillary forces or the alloy/compound formation – dominates during sintering, the associated released energy amounts should be quantitatively compared. The energy released during a chemical reaction between micron-sized powders is normally two to three orders of magnitude higher than that associated with a reduction in the surface area of the particles of a densifying porous body. An increase in porosity and an associated increase in the free surface energy are thus thermodynamically allowed during a reactive sintering process when the chemical reaction causes a release of a large amount of energy. Savitskii argues that much more powerful chemical processes of the alloy and compound formation are not to be considered as additional factors; rather, they should be dealt with as the main driving forces for sintering of reacting powder mixtures. The surface energy factor becomes dominant once the system has reached a chemical equilibrium. The difference in the partial diffusion coefficients can lead to the formation of the diffusion-caused porosity decreasing shrinkage during sintering or inducing swelling of the compact. The process of chemical homogenization can accelerate sintering and promote densification if the alloy formation during sintering decreases the material viscosity and stimulates the material flow under the action of capillary forces. So, according to Savitskii, the mechanism by which the system attains its chemical equilibrium during sintering determines whether the sintering process is enhanced or hindered.

The application of electromagnetic fields to assist the densification process is thus fully justified for systems that swell rather than shrink as a result of the chemical interactions. For the mixtures of nanopowders giving a nano-particulate reaction product, a change in the free surface energy associated with an increase in the porosity and a change in the free energy of the system due to the chemical reaction can be comparable. Therefore, no considerable increase in the compact porosity is to be expected after the chemical reaction is complete.

10.3 Modeling of Reactive Sintering

Reactive sintering is a complex process because of a large number of parameters that influence the sintering results. A unique modeling study was conducted by Olevsky et al. [15], who elaborated a phenomenological model of reactive sintering that involves chemical reactions. The basis of the model is the nonlinear viscous constitutive relationship of the continuum sintering theory discussed in Chap. 4:

$$\sigma_{ij} = \frac{\sigma(W)}{W}\left[\varphi \dot{\varepsilon}_{ij} + \left(\psi - \frac{1}{3}\varphi\right)\dot{e}\,\delta_{ij}\right] + P_L\delta_{ij}.$$

In order to describe the behavior of the system that experiences phase transformations during sintering or in which new phases form as a result of chemical reactions, additional terms P_{dc} and P_{hd} are introduced:

$$\sigma_{ij} = \frac{\sigma(W)}{W}\left[\varphi \dot{\varepsilon}_{ij} + \left(\psi - \frac{1}{3}\varphi\right)\dot{e}\,\delta_{ij}\right] + (P_L - P_{\mathrm{dc}} - P_{\mathrm{hd}})\delta_{ij}. \tag{10.1}$$

Here P_{dc} and P_{hd} take into account the change of the theoretical density caused by the formation of the reaction product and the volume Kirkendall effect due to the

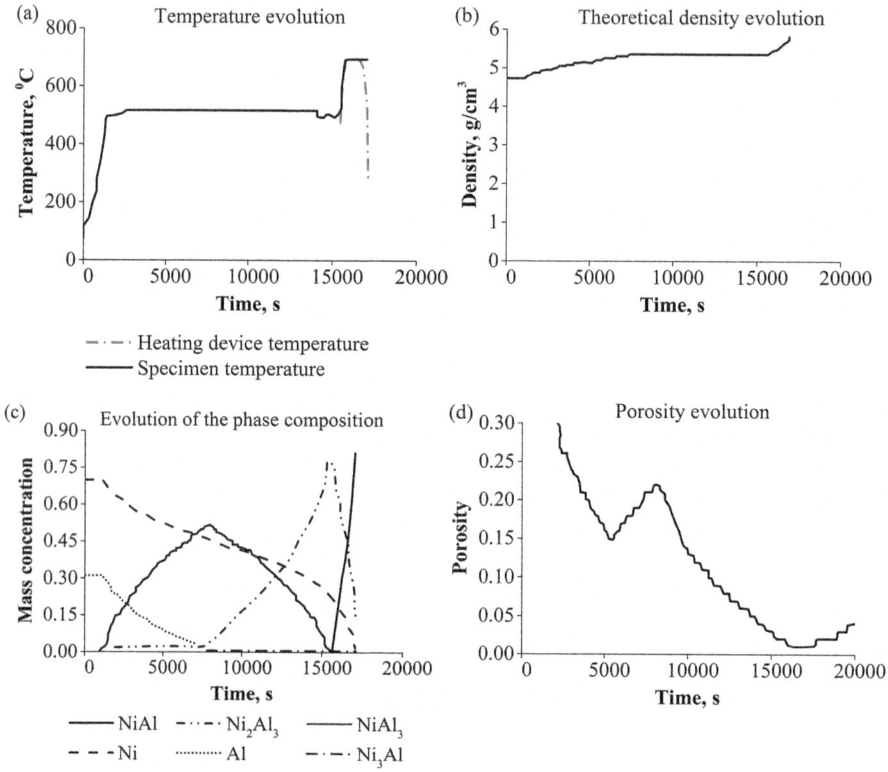

Fig. 10.1 Results of the optimization of the temperature regime for reactive sintering of Ni–Al: (**a**) temperature, (**b**) theoretical density, (**c**) phase composition, and (**d**) porosity evolution. An external hypothetical heating/cooling device is introduced to control the self-propagating high-temperature synthesis. (Reprinted from Olevsky et al. [15]. Copyright (1997), with permission from Elsevier)

difference in the intrinsic coefficients of diffusion, respectively. Modeling based on the solution of Eq. (10.1) utilizes the dependences of P_{dc} and P_{hd} on porosity. Using the results of modeling of reactive sintering of Ni–Al, it was suggested to use an external hypothetical heating device to control the self-propagating high-temperature synthesis. A heating regime that prevents the explosive progress of the reaction was determined by calculations (Fig. 10.1).

10.4 Diffusion During Heat Generation by a Contact Source and During Isothermal Annealing

The consequences of the highly localized heat generation at the inter-particle contacts in electric pulse sintering can be revealed by analyzing the diffusion-induced mass transfer in a pair of metals subjected to electric current. Intensive heat evolution occurring at the inter-particle contacts causes interdiffusion of the contacting metals. In highly nonuniform temperature field, the contribution of thermal diffusion is negligible, as was shown by Krishtal et al. for laser-treated alloys [16]. Therefore, the calculations of the concentration profiles of the metals can be performed not taking thermal diffusion into account. In a pair of metals with unlimited mutual solubility (e.g., Cu–Ni), the calculated asymmetric concentration profiles are explained by the differences in diffusion and thermal parameters of the two metals [17]. The efficiencies of the mass transfer in the case of heating from the inter-particle contact heat source and in isothermal heating can be compared using a model system of two contacting cubes of different metals [18]. The comparison can be made assuming the same thermal energy input in both cases. In the first case, the heat is conducted from the contact source according to the laws of heat conduction; in the second case, the same amount of heat is utilized to uniformly and instantly heat the contacting metals up to an average temperature T_{av}. The heat evolved at the inter-particle contact during time τ by the heat source of power density q can be expressed as follows:

$$\Delta Q_c = qs\tau,$$

while the energy required to heat the pair of the contacting metals from temperature T_0 to T_{av} is

$$\Delta Q_{is} = (c_1\rho_1 + c_2\rho_2)V(T_{av} - T_0),$$

where s is the contact area, V is the volume of each metal, c_1 and c_2 are the specific heat capacities of the metals, and ρ_1 and ρ_2 are their densities. Assuming the equal energy input $\Delta Q_c = \Delta Q_{is}$, T_{av} can be shown to depend on the $\frac{s}{V}$ ratio. Burenkov and Raichenko [18] considered a model of two contacting semi-infinite elements representing materials differing in thermal properties and solved the diffusion problem using the boundary layer method. This allowed calculating the ratio of the volume of the species diffusing through the unit area in the case of the contact heat source m_c to that diffusing during isothermal annealing m_{is}. Figure 10.2 shows that, when q is equal to 1 MW·cm^{-2} and τ is equal to $2{\cdot}10^{-5}$ s, the $\frac{m_c}{m_{is}}$ ratio increases from approximately 3 to 10^5 as the particle size increases from 45 to 330 μm. Such significant intensification of diffusion in the case of the contact source is due to the highly localized evolution of heat in the regions adjacent to the contact interface.

10.5 Initiation of Reactions by Electric Current

In conventional sintering, the thermal energy is supplied from an external source. The process of reactive sintering can use the heat of an exothermic reaction between the mixed powders [19]. In this case, only initiation of the reaction requires external

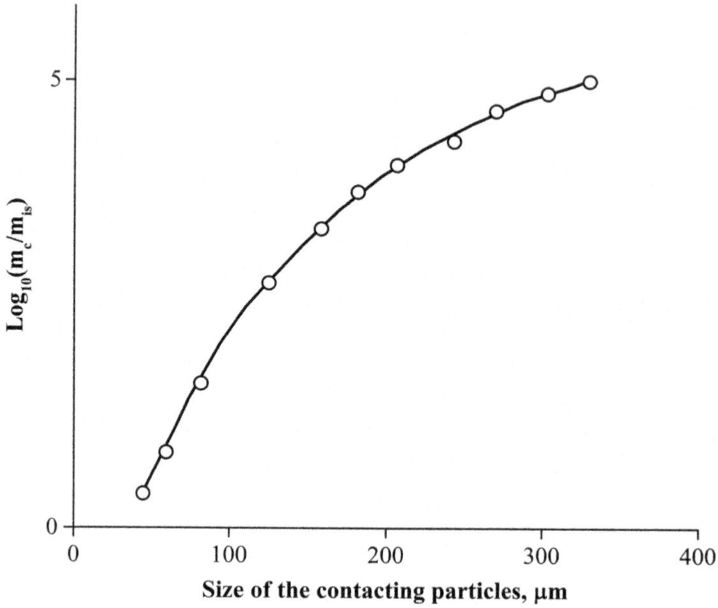

Fig. 10.2 Dependence of $\log_{10}\frac{m_c}{m_{is}}$ on the size of the contacting particles, $q = 1$ MW·cm^{-2}, $\tau = 2{\cdot}10^{-5}$ s. (Drawn using data of Ref. [18])

heating. In reacting systems with sufficient adiabatic heating, densification occurs via liquid-phase sintering [20].

A direct current of low voltage (4–24 V) was used to initiate a reaction of self-propagating high-temperature synthesis by Belousov et al. [21]. They suggested three types of possible reaction initiation: contact, breakdown, and volume initiation.

During the contact initiation, the synthesis starts at the contact surface between the powder specimen and the plunger as a higher-temperature region due to high contact resistance. The synthesis in the compacted reactants proceeds in the form of a wave.

The concept of breakdown is based on a reduction of electric resistance along one of the current lines. The breakdown initiation is caused by low-quality mixing or by nonuniformities of the powder loading in the die. Regions adjacent to the low-resistance line are the only regions that experience transformation, and the rest of the sample is left unreacted.

A complete transformation is achieved when the volume initiation takes place. It becomes possible when the whole sample experiences Joule heating and precautions are taken not to initiate the reaction according to the other scenarios mentioned above.

Morsi and Mehra [22] highlighted the challenge of quantifying the electric current passing through the sample sintered in a die in setups, in which both the sample and the die carry a certain fraction of the total current. In order to overcome this difficulty, a container-less setup utilizing a direct current of 600–1000 A was designed. Only a contacting (low) pressure was applied to the pre-pressed sample; for this reason the setup was suitable for the production of porous materials. For a mixture of aluminum and titanium powders, which were pre-pressed to form a sample of cylindrical shape, it was found that the time of the current application needed to fully engulf the sample in a combustion reaction to form Al_3Ti as the major phase decreased with increasing electric current. An interesting observation was an increase in porosity of the sintered samples with an increase in electric current indicating poor sintering of the reaction product. Diametral swelling of the samples was observed: the higher was the electric current applied, the greater was the swelling effect. The central part of the reacted cylinder was less dense than the top and the bottom regions. The role of the contact of the sample with the WC–Co electrodes was twofold: sparking was observed at the contact regions possibly resulting in higher than the average local temperatures; at the same time, heat dissipated through the electrodes. Both factors could have contributed to the differences in the sample density along the vertical axis.

10.6 Faster Reactions Under Applied Field

Experiments conducted by Bertolino et al. [23, 24] revealed the intrinsic effect of electric current on the growth of the product layer between the foils of two metals reacting with each other. A layer of a metal (Au) was sandwiched between two layers of another metal (Al) and annealed with and without a electric current at a constant temperature. Samples shown in Fig. 10.3 were annealed at 450 °C for 4 h without any current and under current densities of $0.51 \cdot 10^3$ A cm^{-2} and $1.02 \cdot 10^3$ A cm^{-2}. The micrographs of the cross sections of the annealed samples show that the growth of the intermetallic layer at the Al and Au interfaces was greatly affected by the electric current, the product layer becoming thicker with increasing current density.

From the quantitative representation of the effect (Fig. 10.4), one can see the difference in the slopes of the lines corresponding to experiments with different values of electric current and the experiment without electric current. There exits an incubation time for the nucleation of the product layer. In the absence of current, it was necessary to anneal the samples for 20 h before the product layer could be detected. The incubation time for the nucleation of the intermetallic phase was drastically reduced as the current was applied. A current density could be chosen to reduce the incubation period to essentially zero.

Anselmi-Tamburini et al. [25] investigated the solid-state reactivity under electric field by making Si and Mo foils interact under the spark plasma sintering (SPS) conditions. An assembly shown in Fig. 10.5 was used to conduct the experiments. The primary reaction product was $MoSi_2$ with Mo_5Si_3 as a minor phase. The direction of the electric current did not influence the thickness of the $MoSi_2$ layer, but the growth rate of the product phase showed a strong dependence on the current value. The activation energy for the formation of $MoSi_2$ was calculated to be

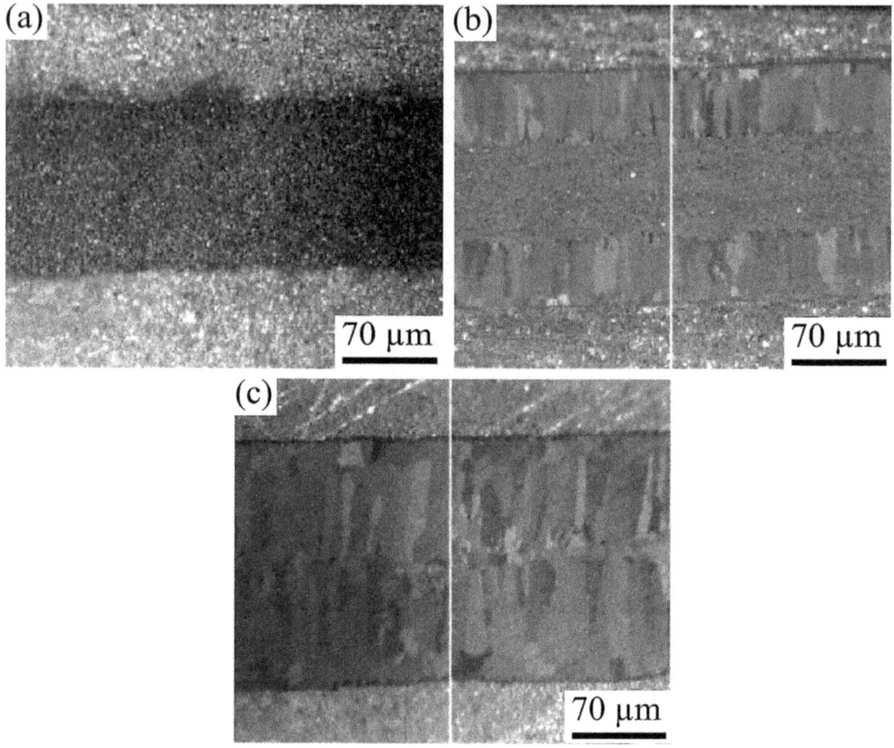

Fig. 10.3 Effect of electric current on the growth of the product layer at Al/Au interfaces in the samples (Al/Au/Al sandwich) annealed at 450 °C for 4 h: (**a**) no current, (**b**) current density $0.51 \cdot 10^{3}$ A cm^{-2}, and (**c**) current density $1.02 \cdot 10^{3}$ A cm^{-2}. (Reprinted from Bertolino et al. [23]. Copyright (2001), with permission from Elsevier)

168 kJ·mol^{-1} for the case of applied current, which is in general agreement with reported values for growth in the absence of a current. This allowed the authors to suggest that the reaction mechanism did not change when the reaction was performed under an electric current; however, the mass transport was enhanced by the applied current, very likely due to increased defect mobility and/or concentration [26]. A conclusion that the $MoSi_2$ growth rate did not depend on the direction of electric current was made based on the equal thicknesses of the product layers on both sides of the silicon wafer [25]. This result may seem to be opposite to what could be anticipated from the electromigration theory. The authors suggested that the classical electromigration studies cannot be performed using a system, in which the processes occurring during annealing under electric current are complicated by the formation of a compound/compounds. In order to form a new phase, the diffusion of both reactants may be necessary as in the case of the formation of $MoSi_2$. If diffusion of one of the reactants is enhanced by electromigration, the growth rate will still depend on the slower diffusing species. For electromigration studies, such systems as Cu–Ni should be chosen, in which only continuous solid solutions exist [27].

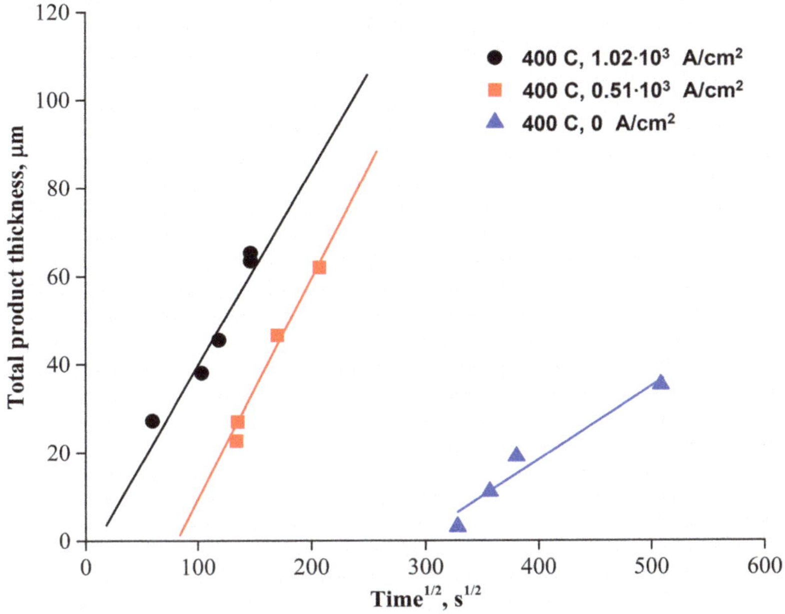

Fig. 10.4 Effect of electric current on the growth rate of the layer of an intermetallic compound between the Al and Au foils. The thickness of the layer is plotted versus the square root of the time following the parabolic growth law. The samples are disks with a diameter of 5 mm. Data are for experiments at 400 °C. (Reprinted from Bertolino et al. [24]. Copyright (2002) by permission of Taylor & Francis Ltd., www.tandfonline.com)

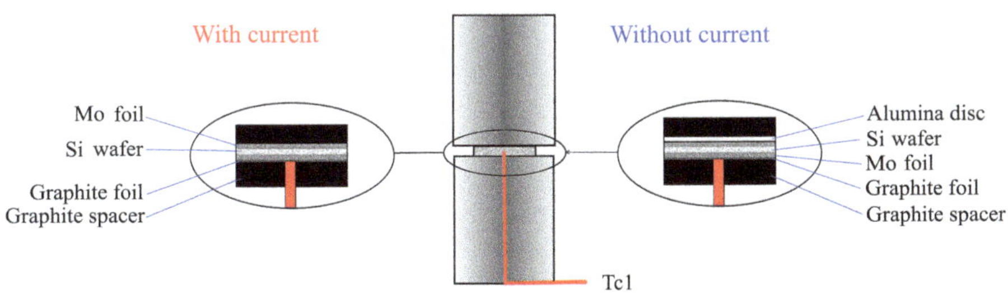

Fig. 10.5 Schematic of an assembly for conducting chemical reaction between Mo and Si in the presence and absence of electric current in the SPS facility. (Reprinted from Anselmi-Tamburini et al. [25]. Copyright (2005), with permission from Elsevier)

Unlike the Mo–Si system (Figs. 10.6, 10.7), the Mo–C and Ni–Ti systems reacting under an applied current showed a dependence of the activation energy of the reaction product formation on the current density. The product layer between carbon and molybdenum layers grows faster as the current density increases (Fig. 10.8). The dependence of the activation energy on the current density for the growth of the β-Mo_2C layer as a result of the reaction between Mo and C is shown in Fig. 10.9 [28]. In the Ni–Ti system [29], the effect was observed for all reaction

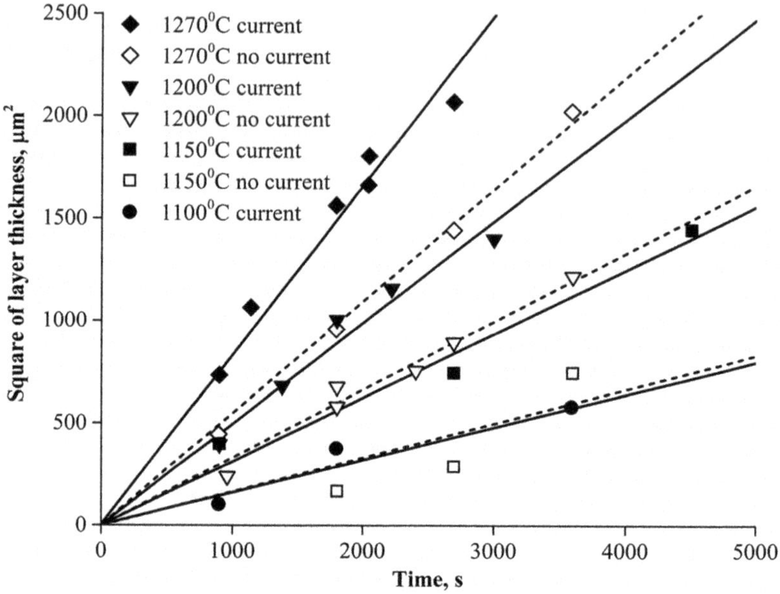

Fig. 10.6 Growth rates of the $MoSi_2$ layer between a Mo foil and at a Si wafer at different temperatures in the SPS in the presence and absence of electric current flowing through the sample. (Reprinted from Anselmi-Tamburini et al. [25]. Copyright (2005), with permission from Elsevier)

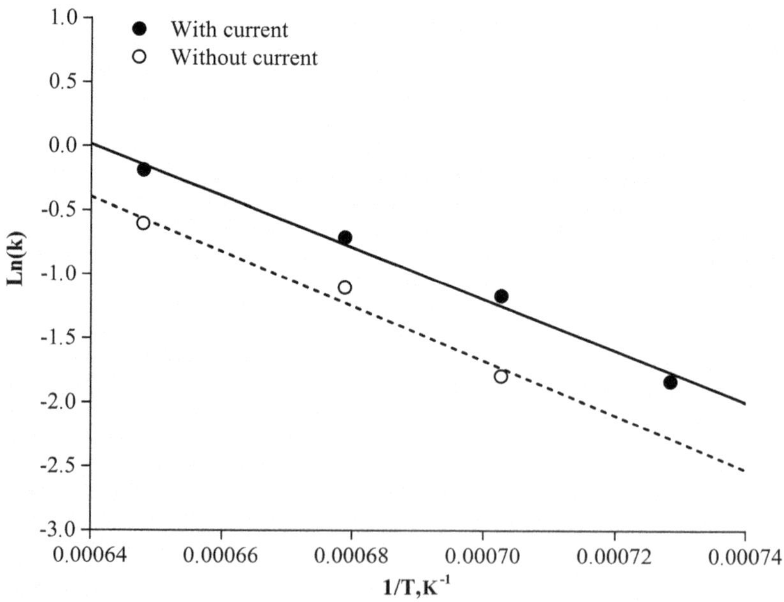

Fig. 10.7 Arrhenius plots of the temperature dependence of the growth rate of $MoSi_2$ in the presence and absence of electric current flowing through the sample. (Reprinted from Anselmi-Tamburini et al. [25]. Copyright (2005), with permission from Elsevier)

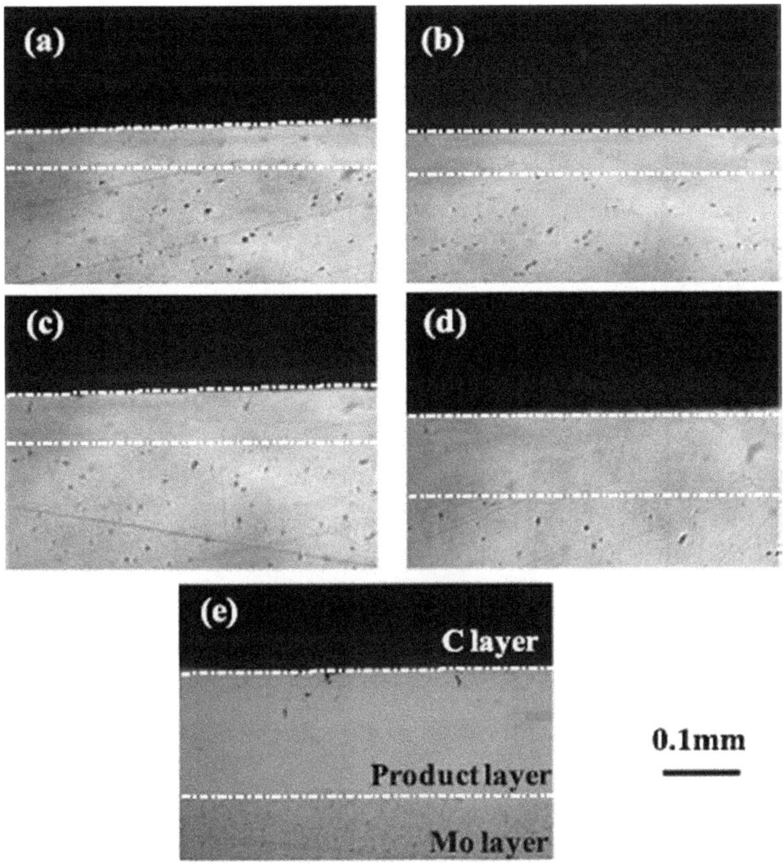

Fig. 10.8 Optical images of the Mo–C diffusion couple annealed at 1569 °C for 20 min at different current densities 0 A cm^{-2} (**a**), 521 A cm^{-2} (**b**), 732 A cm^{-2} (**c**), 1145 A cm^{-2} (**d**), and 1476 A cm^{-2} (**e**). (Reprinted from Kondo et al. [28]. Copyright (2008) with permission from Japan Society of Powder and Powder Metallurgy)

product phases; however, the most significant reduction of the activation energy was found for the Ni_3Ti intermetallic. The results were explained in terms of an altered product growth mechanism due to current-enhanced mobility of defects.

Anselmi-Tamburini et al. [30] compared the thickness of the HfB_2 product layer (Figs. 10.10, 10.11) formed on hafnium particles in a Hf-B mixture partially reacted during the SPS with that observed in the diffusion couple experiments. The thicknesses of the HfB_2 product layers agreed well with results of calculations from the diffusion couple experiments. A conclusion was made that the reactivity in this system was not changed by the applied current.

Munir et al. [31] consider three main mechanisms of enhanced mass transport under applied current, which are electromigration, increased concentrations of point defects, and reduced activation energy for the mobility of defects, which can lead to

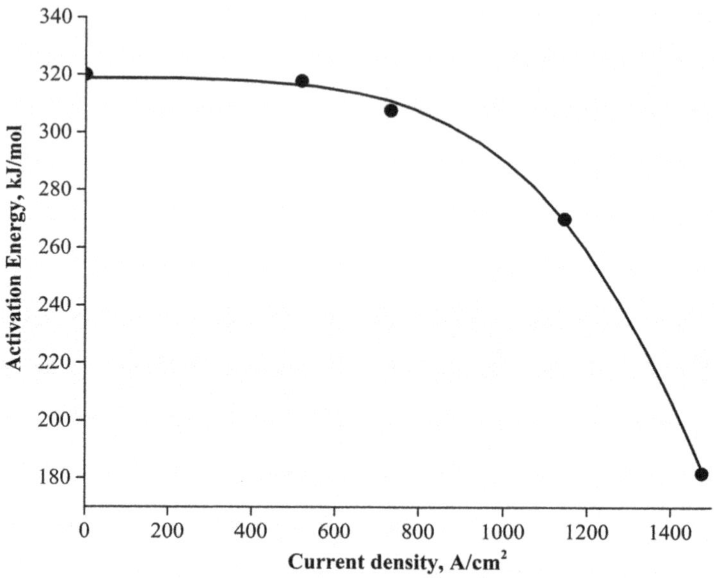

Fig. 10.9 Calculated values of the activation energy for the growth of the β-Mo_2C layer as a function of the current density. (Reprinted from Kondo et al. [28]. Copyright (2008) with permission from Japan Society of Powder and Powder Metallurgy)

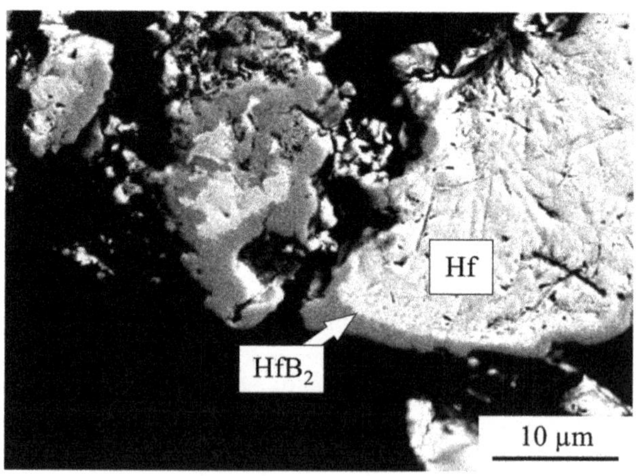

Fig. 10.10 A HfB_2 product layer formed on hafnium articles in the partially reacted Hf-B mixture during the SPS; the maximum temperature in the experiment was 1100 °C. (Reprinted from Anselmi-Tamburini et al. [30]. Copyright (2006) with permission of Springer)

their faster annealing. If defects are already present in a material, they tend to anneal at a higher rate under an imposed current, which was shown by positron annihilation spectroscopy [26]. An increased defect annealing rate was observed in Ni_3Ti under applied current and was attributed to a decrease in the activation energy of the defect mobility.

Similarly, Mackenzie et al. [32] considered DC field as an external factor influencing the mobility of the migrating species in the course of a chemical reaction.

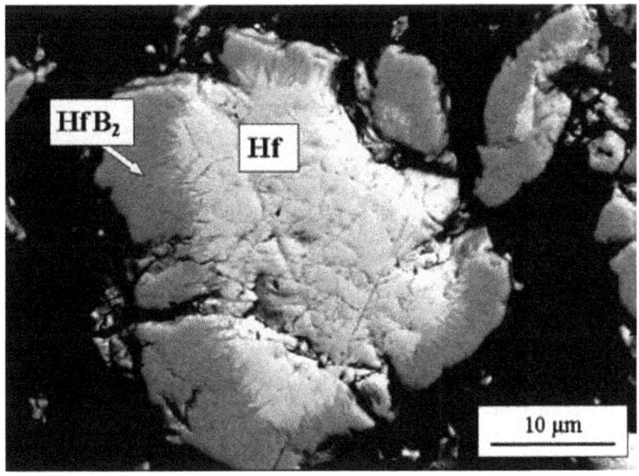

Fig. 10.11 A HfB_2 product layer formed on hafnium articles in the partially reacted Hf-B mixture during the SPS; the maximum temperature in the experiment was 1445 °C. (Reprinted from Anselmi-Tamburini et al. [30]. Copyright (2006) with permission of Springer Nature)

In the studies of the influence of a DC field on the formation of calcium aluminates in the mixtures of CaO and Al_2O_3 powders, the contents of the product phases were greater in the vicinity of the positive electrode than in the vicinity of the negative electrode. The similarity of the curves describing the phase formation dynamics with increasing temperature suggested that the application of an electric field did not change the reaction mechanism but increased the reaction rate. In a conventional solid-state reacting system, Ca^{2+} diffuses into Al_2O_3. The electric field vector acts to reinforce the diffusion of Ca^{2+} into Al_2O_3 for all grain orientations, in which the Ca-containing component is more positive than the Al-containing component. For grain orientations, in which the Al-containing component is more positive, the effect of the field will be opposite to the direction of thermally induced diffusion, so that Ca^{2+} will experience little displacement. The electric field vector will also promote some diffusion of Al^{3+} into CaO under these conditions. However, this effect is not expected to take place to any significant extent as the diffusivity of Al^{3+} is relatively low. Based on these considerations, Mackenzie et al. concluded that the net effect of the field is to promote diffusion of Ca^{2+} into Al_2O_3 grains at the positive electrode, resulting in the enhanced formation of calcium aluminates in that region. The concentrations of all reaction products were lower near the negative electrode than near the positive electrode, but still higher than in the sample, to which no electric field was applied.

Neiman et al. [33] derived an equation describing the kinetics of diffusion-controlled reaction under a constant applied field. They considered reactions of synthesizing ABO_4 compounds from the AO and BO_3 oxides, where A is Pb or Ba and B is Mo or W. In these reactions, diffusion of Mo or W determines the rate of the process. The thickness of the reaction product ABO_4 was found to be related to the applied voltage U through the following equation:

$$l^2 = 2V\tau \frac{\sigma t_B}{z_B F} \left[\frac{|\Delta G^0|}{z_B F} + U \right],$$

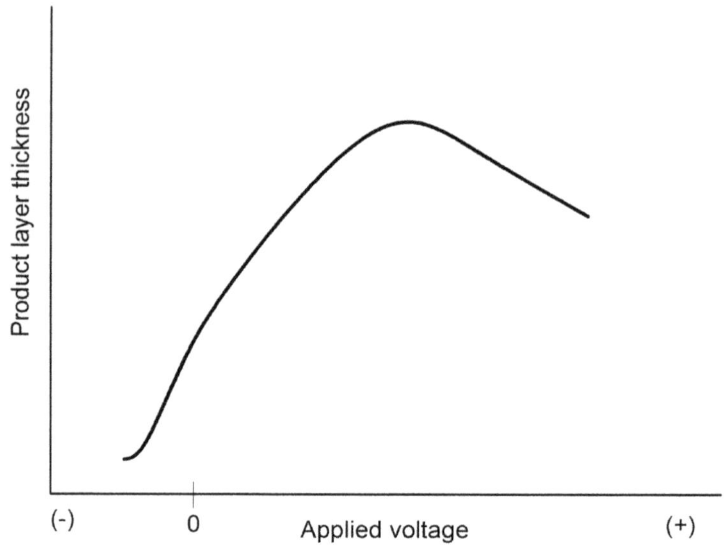

Fig. 10.12 The thickness of the product layer $PbMoO_4$ versus the applied voltage (shown schematically) (Drawn using data of Ref. [33])

where l is the thickness of the reaction product, V is the molar volume of ABO_4, τ is the time, $z_B F$ is the charge of 1 g-atom of ion B, t_B is the transport number of ion B, σ is the conductivity of ABO_4, and ΔG^0 is the standard Gibbs free energy of the reaction $AO + BO_3 = ABO_4$. When the applied field promotes transport of the rate-limiting species, which is B ion in this case, the reaction accelerates and vice versa and can be stopped by an applied voltage of $U = -|\Delta G^0|/z_B F$. Experiments were conducted using compacts of Pb_2MoO_5 and MoO_3, which were put in contact and subjected to the action of a constant field by means of Pt electrodes. The current in the circuit was of the order of 2 mA. $+U$ polarization corresponded to (−) Pt | Pb_2MoO_5 | MoO_3 | Pt(+). Indeed, the acceleration of the reaction up to a certain voltage has been proved experimentally; however, with a further increase in the applied voltage, ABO_4 experiences decomposition due to a reaction of electrolysis (Fig. 10.12). At $U < 0$, the reaction rate is low, but the reaction does not stop completely, which points to the participation of species carrying no electrical charge.

10.7 Slower Reactions Under Applied Field

An interesting effect of electric field slowing down a chemical reaction was observed when the CaO and Al_2O_3 reactants were in the form of pre-pressed pellets brought in contact by their surfaces [34]. The product layer was thicker when the CaO pellet was positive relative to the Al_2O_3 pellet than in the case of the opposite polarity.

However, in this configuration, a greater degree of interdiffusion was observed in the sample without electric field than in the samples subjected to the electric field. The observed difference was explained by a greater degree of intergranular sintering occurring in the CaO pellet under the applied field and hindering grain-boundary diffusion, which plays an important role in the interdiffusion processes between CaO and Al_2O_3. These results show that when electric current is applied with an aim of accelerating diffusion, early sintering between the reactant particles should be taken into account, as it may hinder grain-boundary diffusion and thus slow down the reaction kinetics. This example is significant, as it shows the multidirectional influence of electric current on the compacts composed of mixed powders. Sintering of particles of a reactant will, of course, not be an issue when the reactants are thoroughly mixed with each other and the number of contacts between particles of the same material is limited.

Zingel [35] conducted thermolysis of $KMnO_4$ in a constant field in the noncontact mode and found that as the field intensity increases, the kinetic constant of the intermediate stage of thermal decomposition decreases, reaches a minimum, and starts increasing again approaching the "zero-field" value. This behavior of the reaction constant was explained by the changes in the electron concentration and mobility in the applied field. The concentration and mobility of electrons determine the electron-jump mechanism of electron exchange between MnO_4^- and MnO_4^{2-}, which explains the effect of applied field on the decomposition rate of $KMnO_4$.

10.8 Chemical Reactions Involved in High-Voltage Processes

When the suitability of a field-assisted sintering technique is considered for performing the synthesis simultaneously with consolidation, the total duration of the process is the first parameter that needs to be taken into account. If the consolidation process is designed to be very fast by supplying a large amount of energy to the powder compact within a short period of time (fractions of a second), only reactions that are self-sustaining can fully convert the reaction mixture into a product. Otherwise, the conversion will proceed to give only small reaction yields.

A single electric pulse formed by a capacitor discharge was able to initiate an exothermic reaction of the formation of titanium diboride TiB_2 in the Ti–B–Cu compact [36], but it was insufficient for densification even at a current density as high as 10^9 A·m^{-2}. The compact obtained by reactive pulse sintering was a porous TiB_2–Cu composite.

Under the action of a high-voltage discharge, powders of reactive metals are prone to chemical reactions with the surrounding atmosphere. An et al. [37] have shown that environmental electro-discharge sintering of a titanium powder in an atmosphere of nitrogen results in the formation of titanium nitride TiN and titanium oxynitrides TiO_xN_y. The pre-existing oxide films of the particle surface present no barrier for the reaction to occur as they experience dielectric breakdown at high voltages and mechanical destruction under applied pressure. Electric pulse sintering can, therefore, be used as a method of surface modification of powders. If the electric pulse sintering equipment is to be used for the powder surface modification only, the pressure levels should be adjusted to avoid unnecessary particle sintering. When nanosized powders of reactive metals are selected for sintering, the fraction of the metal transformed into oxide, and nitride compounds can be quite high such that a better description of the process would be a "synthesis" rather than surface modification.

Sizonenko et al. [38] proposed high-voltage electrical discharge applied to suspensions of metal powders in hydrocarbon liquids as a promising method of the synthesis of fine particles of carbides. In their studies, suspensions of a titanium powder were subjected to high-voltage electric discharge. It was found that the yield of titanium carbide TiC increased with increasing current density in the discharge channel.

New synthesis possibilities discovered upon the development of electric discharge-assisted mechanical milling [39] are inspiring for the future applications of electric pulse sintering, as there are similarities between these two processes. In electric discharge-assisted mechanical milling, a voltage of 30 kV is used, which is applied to the milling bodies and the bottom of the milling vial. The currents are reported to be low (in the microampere range), but the current densities can be high when the current has to pass through an agglomerate of powder particles or a single particle. The powder particles experience a grinding action from the milling bodies combining impact and shear forces. The conditions of the method are suitable for performing the synthesis of compounds from powders and components of the gaseous atmosphere as reactants. Calka et al. [40] demonstrated that during electric discharge-assisted mechanical milling, the Al–Mg powder mixtures fully transform into oxides, which further react with each other to form the $MgAl_2O_4$ spinel. The $MgAl_2O_4$ spinel obtained by electric discharge-assisted mechanical milling had a larger lattice parameter compared with that reported for the stoichiometric compound. The milling is performed for several minutes, which certainly raises a question of whether or not the action of a single electric discharge in the electric pulse sintering processes would be enough to conduct a similar synthesis. Further investigations are necessary to clarify this issue. Similarly, an attractive feature of the synthesis during electric pulse sintering is a possibility of obtaining and preserving metastable crystalline states in the reaction products. It can be assumed that, even though by applying a single electric pulse to a powder mixture it may not be possible to achieve a complete transformation, metastable compounds of interesting crystalline state can form containing high concentrations of lattice defects. Such compounds are likely to show an increased chemical reactivity if used as precursors in further syntheses and interesting defect state-dependent physical properties.

10.9 Synthesis and Sintering by Microwaves

Microwave energy can be used either to enhance the reaction kinetics or to trigger chemical reactions at lower temperatures. Electromagnetic fields in the form microwaves have been used to conduct synthesis of many materials, as has been reviewed by Agrawal [41] and Rao et al. [42]. In the synthesis of oxide materials, a concept of prereduction of phases (reactants) was adopted [41]. The main idea was to create a defect structure in the oxide to produce a highly microwave absorptive precursor. This approach allowed enhancing the reaction kinetics dramatically and synthesizing complex oxide materials at temperatures several hundred degrees lower than in conventional processes.

Synthesis of different ceramic materials, such as SiC, TiC, Mo_2C, and $MoSi_2$, has been implemented from the mixtures of elemental powders [42]. The main requirement to the reaction mixtures suitable for conducting synthesis by initiating a reaction by microwave heating is the presence of at least one microwave susceptor among the reactants, which would be heated and would transfer heat to the other components by thermal conduction. If the reaction is associated with a sufficient enthalpy change, local ignition may be enough to transform all the reactants by a moving high-temperature reaction front. For the synthesis of carbides, a high microwave susceptibility of graphitic carbon is utilized. The reaction between molybdenum and silicon powders cannot be initiated by microwave heating, as both silicon and molybdenum are poor microwave susceptors. In order to synthesize $MoSi_2$ from the elements by microwave heating, carbon powders were added to the reaction mixture [43]. This led to the formation of both SiC and $MoSi_2$ phases. It was, therefore, suggested to pelletize the Mo–Si reaction mixture and cover it by a layer of amorphous carbon. The product of the microwave synthesis was a well-sintered pellet of $MoSi_2$ after a surface layer of SiC had been removed.

Microwave synthesis and sintering can be conducted as separate stages [44, 45]. In this case, the solid-state synthesis is conducted in the powder state, and the product is cooled down to room temperature and pressed into a pellet to be further sintered by the same microwave treatment [44]. Another possibility is synthesis in the pre-compacted pellet, which is further ground into a powder and then pressed again before sintering [45]. Grinding the reaction product can eliminate the problem of the formation of strong agglomerates during the high-temperature synthesis reaction that hinder further densification.

In Ref. [44], a ball-milled mixture of $Mg(OH)_2$ and silica gel was calcined at different temperatures using microwave heating. As can be seen from Fig. 10.13, the forsterite Mg_2SiO_4 was formed after calcination at 900 °C. Upon treatment at a higher temperature, the forsterite crystallites grew in size, as indicated by narrower reflections on the XRD patterns of the synthesized product. The product was cold-pressed up to a relative density of 49.5% and then sintered by conventional sintering and microwave sintering at the same temperatures. It was found that microwave-sintered specimens showed higher relative densities than the conventionally sintered ones (Fig. 10.14).

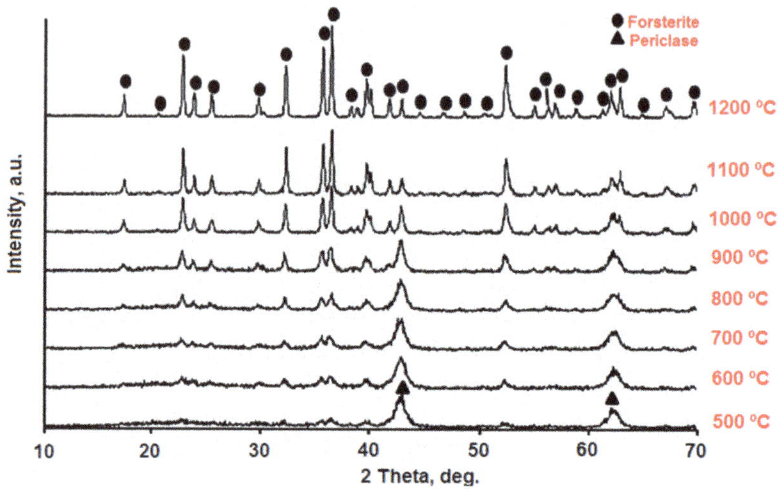

Fig. 10.13 XRD patterns of the $Mg(OH)_2$–silica mixtures ball milled and calcined at different temperatures using microwave heating. (Reprinted from Barzegar Bafrooei et al. [44]. Copyright (2013) with permission from Elsevier)

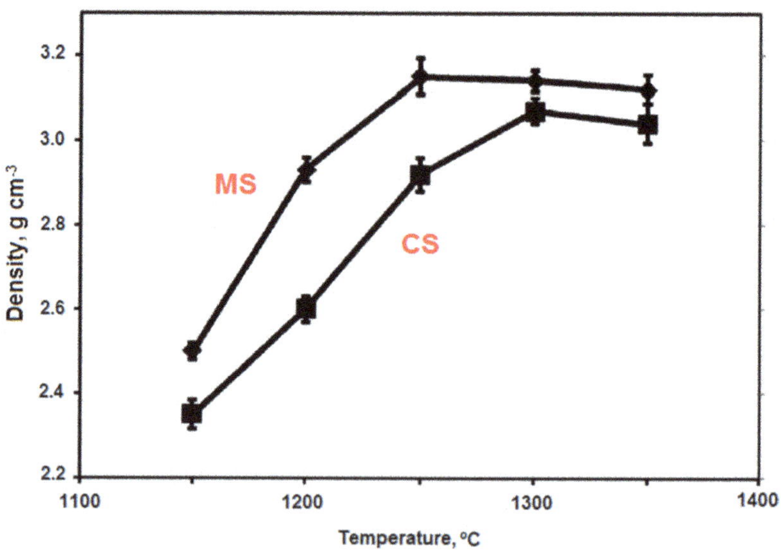

Fig. 10.14 Density of the sintered forsterite ceramics prepared by conventional sintering (CS) and microwave sintering (MS). (Reprinted from Barzegar Bafrooei et al. [44]. Copyright (2013) with permission from Elsevier)

Microwave synthesis was reported to present advantages over conventional solid-state synthesis and melting-based techniques for the rapid synthesis of TiNiSn – a promising thermoelectric material [45]. The microwave treatment of the Ti–Ni–Sn powder mixture was carried out using a SiC powder as a microwave susceptor

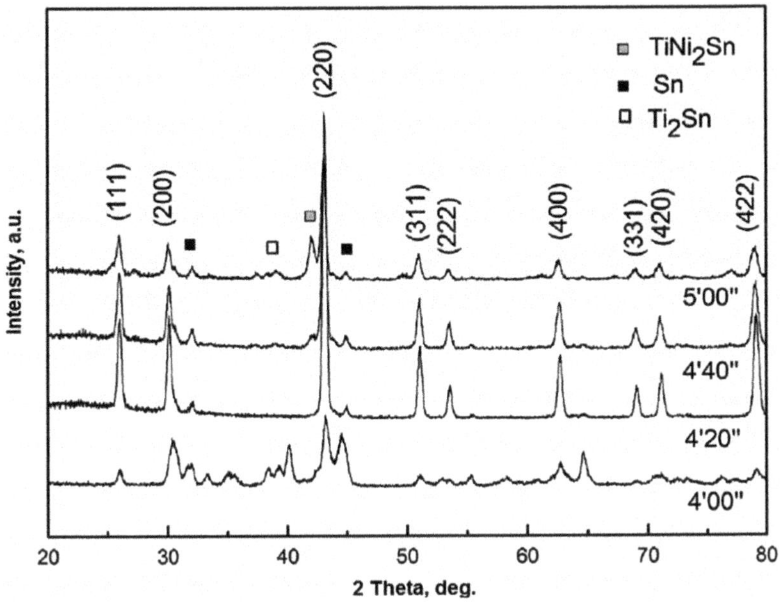

Fig. 10.15 XRD patterns of the Ti–Ni–Sn mixture microwave processed for different durations; the Miller indices mark reflections of the TiNiSn phase. (Reprinted from Lei et al. [45]. Copyright (2015) with permission from Elsevier)

placed around the sealed tube containing the reaction mixture. Alloying in the Ti–Ni–Sn powder mixture was complete in 4 min and 20 s such that the major phase was TiNiSn. Further treatment of the reaction product resulted in decomposition of the TiNiSn phase (Fig. 10.15).

Cesário et al. [46] synthesized and sintered $Ce_{0.8}Yb_{0.2}O_{1.9}$ from a mixture of CeO_2 and Yb_2O_3 powders using microwave sintering and conventional sintering; the latter was used for comparison. The final temperature reached during the microwave sintering and conventional sintering was the same and equal to 1500 °C. The synthesis was complete in both processes resulting in the formation of a single-phase material of fluorite structure. The main result of their work was the achievement of a higher relative density and a finer grain size by the microwave processing (96.54%, 2 μm) compared with conventional sintering (94.82%, 5–10 μm). The microstructures of the ceramics obtained by reactive sintering are shown in Fig. 10.16. Compacts sintered by microwave exhibited a higher bulk electrical conductivity compared with conventionally sintered compacts.

Reactive sintering by microwaves was also shown to be beneficial for the development of $CaBi_4Ti_4O_{15}$ ceramics with a high dielectric constant and low dielectric losses [47], synthesis and sintering of potassium sodium niobate lead-free piezoelectric ceramics [48], and $Gd_2Zr_2O_7$ pyrochlore waste forms promising for immobilization of nuclear waste [49]. Shorter holding times, faster heating, and promising properties of the obtained ceramics allow considering microwave reactive

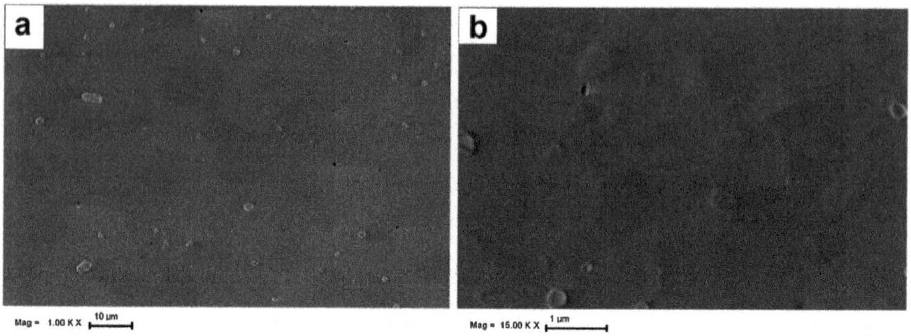

Fig. 10.16 $Ce_{0.8}Yb_{0.2}O_{1.9}$ reactively sintered from a mixture of CeO_2 and Yb_2O_3 powders by conventional furnace sintering at 1500 °C, heating rate 2 °C min^{-1}, holding time 5 h (**a**), microwave sintering 1500 °C, heating rate 120 °C min^{-1}, holding time 15 min (**b**). (Reprinted from Cesário et al. [46]. Copyright (2016) with permission from Elsevier)

processing as an energy-saving alternative to conventional sintering that also produces materials with better functional performance.

Microwaves can be used to synthesize intermetallics from the mixtures of metal powders [50, 51]. Microwave ignition of a combustion reaction presents a promising alternative to thermal ignition, as microwave heating is limited to the sample (reaction mixture) alone and does not involve its surroundings (holders, plates, etc.). Considering a certain penetration depth of microwaves into the material (which depends on the material properties), the ignition can be highly localized. Rosa et al. [51] studied the initiation of combustion synthesis of aluminide intermetallics starting from the elemental powder mixtures containing ferromagnetic metals (Fe, Co, Ni) in a microwave single-mode applicator. The samples were placed in the regions, in which either electric field or magnetic field dominated. The ignition of the combustion reactions in the predominant magnetic field allowed avoiding the generation of electric arcs and reducing the ignition times and the power required to synthesize the intermetallics. The experiments conducted in the predominant magnetic field were also more reproducible than the experiments conducted in the predominant electric field. The reactions ignited in the predominant magnetic field were more prone to occur in the self-propagating combustion mode than the reactions initiated in the electric field due to a smaller reactive volume reaching ignition conditions in the magnetic field. In the Fe–Al system, the product of synthesis in magnetic field had a more uniform microstructure and was less oxidized than the product synthesized in electric field.

10.10 Enhancement of Chemical Reactivity by Magnetic Field

An example of a higher chemical reactivity in a magnetic field was presented by Mitsui et al. [52]. Ferromagnetic MnBi was synthesized by a solid-state reaction between manganese and bismuth powders in a high magnetic field. The synthesis

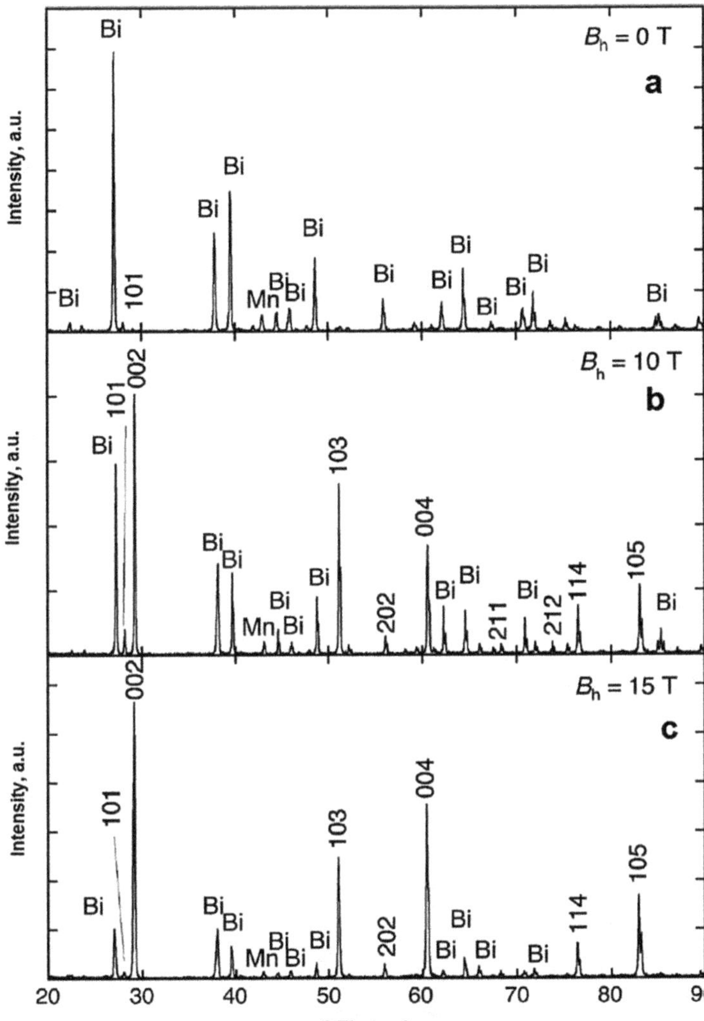

Fig. 10.17 XRD patterns of the Mn + Bi powder mixtures sintered in magnetic field of 0 T (**a**), 10 T (**b**), and 15 T (**c**); *hkl* numbers denote the Miller indices of MnBi. (Reprinted from Mitsui et al. [52]. Copyright (2014) with permission from Elsevier)

was conducted at a temperature of 523 K, which is below the eutectic temperature in the system. Therefore, the synthesis was conducted in the solid state. The concentration of the ferromagnetic MnBi phase in the sample sintered at 15 T was about 70 wt.%, which was much greater than in the mixture sintered in the absence of magnetic field (5 wt.%), as can be deduced from the XRD patterns of the products of annealing (Fig. 10.17). It was concluded that magnetic field dramatically enhances the formation of MnBi phase from Bi and Mn reactants. The authors suggested that magnetic field lowers the activation energy of the reaction between Mn and Bi. A large uniaxial magnetic anisotropy of MnBi caused the crystal orientation in the process of reaction sintering. In the synthesized product, the *c*-axis of the hexagonal structure of MnBi was oriented parallel to the magnetic field. The advantage of in-field solid-state sintering over the conventional methods, such as arc-melting, induction melting, and crystallization of the melt-spun ribbons, was in the possibility to achieve crystal orientation during the synthesis of the MnBi phase.

In Ref. [53], MnBi was synthesized from solid Mn and liquid Bi using in-field heat treatment. In the presence of liquid bismuth, the reaction was fast, and the influence of magnetic field on the synthesis was not detected using the experimental design selected by the authors. Reactive liquid-phase sintering in the presence of magnetic field produced the reaction product of the same composition as sintering in the absence of field. However, the magnetic field did influence the crystallite size of the synthesized MnBi: the crystallite size of the samples obtained at 0 and 15 T was 36 and 63 nm, respectively. So, the crystallites coarsened under conditions of in-field reactive sintering. The ferromagnetic MnBi crystals in the in-field sintered material were oriented along the direction of the external magnetic field.

10.11 SPS Dies as Chemical Reactors with Controlled Temperature and Atmosphere

The introduction of SPS facilities into laboratory practice in many research institutes and universities has significantly broadened the research audience interested in this powder processing technique. Thanks to the possibilities of heating with a controlled rate under dynamic vacuum, SPS facilities can be used as convenient and user-friendly high-temperature reactors for conducting solid-state synthesis. Multiple electric pulses used in the SPS technique have a positive effect on the reaction completion and the structural uniformity of the synthesized products. Orrù et al. [54] have thoroughly reviewed the accumulated data on the consolidation of materials by electric current-assisted sintering using a systematic approach based on the metals–ceramics–composites classification. From that review, it is clearly seen that a great variety of materials – ceramics, composites, nanostructured materials, and porous materials – have been synthesized using SPS. Indeed, SPS dies possess several advantages as chemical reactors (Fig. 10.18):

- High-temperature synthesis under protective conditions of dynamic vacuum
- Rapid heating and cooling
- Enhanced reactivity and lower reaction onset temperatures for certain systems
- Reducing atmosphere, a possibility of reduction of contaminating oxides present on metal powders
- In situ reduction of oxides that can be used to form oxygen-deficient oxide phases with promising properties

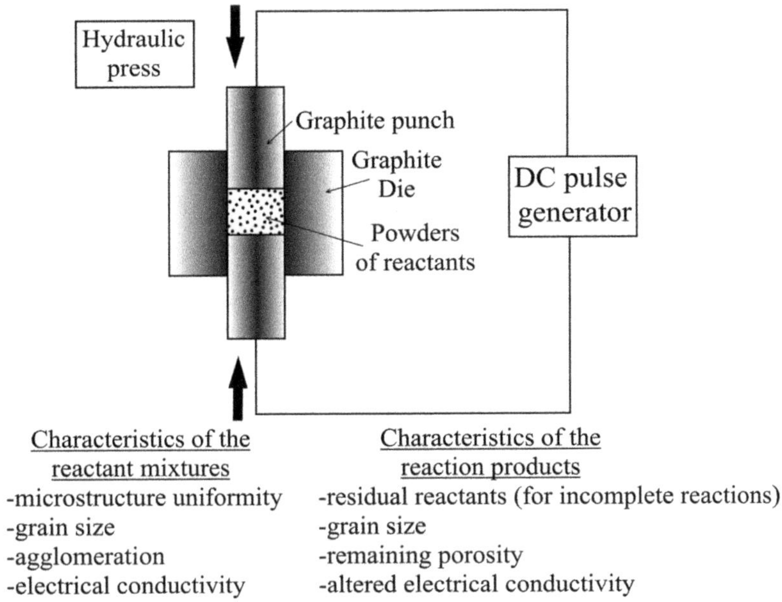

Fig. 10.18 SPS die as a high-temperature chemical reactor. (Reprinted from Dudina et al. [55]. Copyright (2013) D. V. Dudina, A. K. Mukherjee. This is an open access article distributed under the Creative Commons Attribution License, https://creativecommons.org/licenses/by/3.0/)

- Syntheses can result in the formation of products in different states (unsintered powders, porous compacts, or fully consolidated materials)

The following types of reactions are possible to conduct in the SPS chamber:

- Targeted synthesis of new compounds from the mixtures of solid reactants
- Decomposition reactions
- Interfacial reactions/partial alloying between the phases in composites to modify their properties and control adhesion between the phases

Characteristic features of reactive sintering in comparison with sintering of non-reacting powders are related to additional factors, such as uniformity of distribution of the reactants in the mixture, heat release during exothermic reactions, specific volume change, and the presence of reaction by-products or remaining reactants due to incomplete reactions. Table 10.1 summarizes the characteristic features of reactive sintering and their consequences for the SPS process.

The formation of extensive amounts of liquid phases should be avoided in order to keep the structural integrity of the SPS die. The formation of liquid phases is not desired when products with fine grains need to be synthesized. When reactions are performed in the SPS dies, certain issues can be addressed as allowed by the capabilities of the SPS facilities. The applied pressure is instrumental in rearranging the reactant particles during early stages of the synthesis as well as in densification of the porous reaction product.

Table 10.1 Characteristic features of reactive sintering and their consequences for the SPS process

Characteristic features of reactive sintering	The consequences for the SPS process
Distribution uniformity of the reactants in the reaction mixtures	Nonuniform distribution of zones of high electrical conductivity can result in local temperature-induced processes (chemical reactions, melting, phase redistribution, decomposition)
Initiation of the reaction locally in certain preferred zones	The reaction initiates in the vicinity of zones of higher electrical conductivity; additives of high electric conductivity can be used to initiate the reaction
Heat release in exothermic reactions	The programmed temperature schedule can still be followed (little temperature overshooting)
The formation of reaction by-products	Gaseous by-products are removed under conditions of dynamic vacuum

Reactive SPS can be aimed at the synthesis of one phase of a composite; the other phase can be added to the reaction mixture as a previously synthesized compound [56]. For exothermic reactions, such ex situ introduced compounds serve as diluents decreasing the adiabatic temperature of the system and making it possible to control the grain size of the compound synthesized directly during the SPS.

Syntheses of refractory materials and composites are often based on highly exothermic reactions. The initiation of a reaction in a self-propagating mode and a thermal explosion mode relies on heating of the system locally and through the entire volume, respectively. When the reaction mixture is placed in a SPS die, the initiation of reactions occurs throughout the sample volume when the reaction mixture is electrically conductive and uniformly packed, while the reactants are uniformly intermixed.

During the SPS processes involving highly exothermic reactions, a controlled power input allows maintaining the temperature schedule with only small deviations of the temperatures from the programmed values. This is achieved by a controlled drop of current through the sample once the reaction is initiated in the system. Even when the reaction mixture transforms with the evolution of a large amount of heat, a strong control of the temperature in the SPS facilities still allows keeping the selected heating schedule. Experimentally, a stability of the programmed temperature schedule has been proven during the synthesis of titanium diboride [57] and rhenium diboride [58] from the elements in the SPS. During the synthesis of rhenium diboride from the elemental powders, only a minor hump can be noticed on the temperature evolution curve (Fig. 10.19a). A programmed sintering schedule in the SPS apparatus does not allow excessive heat of the reaction to significantly influence the process temperature as the current and voltage drop the very same moment the ignition of a combustion reaction takes place (Fig. 10.19b), which is also accompanied by shrinkage of the compact (Fig. 10.19c).

Contact surfaces between reactant particles are potential reaction initiation sites in pre-mixed powder mixtures. Under pulsed electric current, initiation of the reaction is more likely to occur in the high-temperature regions near contact points between

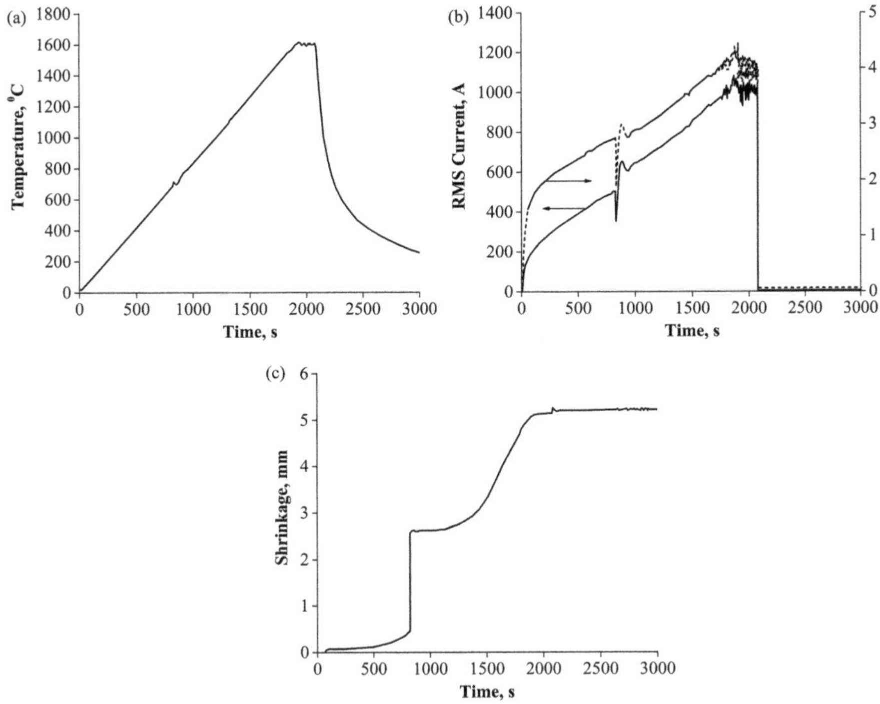

Fig. 10.19 Temperature (**a**), current and voltage profiles (**b**), and shrinkage (**c**) during the reactive SPS of ReB_2. (Reprinted from Locci et al. [58]. Copyright (2008) with permission from Elsevier)

the powder agglomerates. These regions possess higher resistance than the volume of the agglomerates at the early stages of synthesis/sintering. The local high-temperature regions forming in the vicinity of necking points within a sample can be the spots of the reaction initiation. If an additive with a high electrical conductivity is introduced into a powder with a low electrical conductivity, local high-temperature spots will form within the sample in the regions adjacent to the additive particles. In order to avoid confusion between the influence of the high-resistance regions, such as inter-particle contacts, and that of the added particles of a lower-resistivity material, let us consider a scheme below (Fig. 10.20). The strings of white circles represent particles of the major component of the sintered material, while gray circles represent particles of a lower-resistivity material. According to $P = \frac{U^2}{R}$, where P is the power, U is the voltage, and R is the resistance, the power evolved in the string of gray circles will be higher than in the string of white circles.

It was shown that when a pulsed electric current passes directly through the sample, Ti–B powder mixtures with additions of Mg react at a lower temperature to form TiB_2 in comparison with the case when no electric current passes directly through the sample [57]. The reaction is ignited in the hot spots generated within the

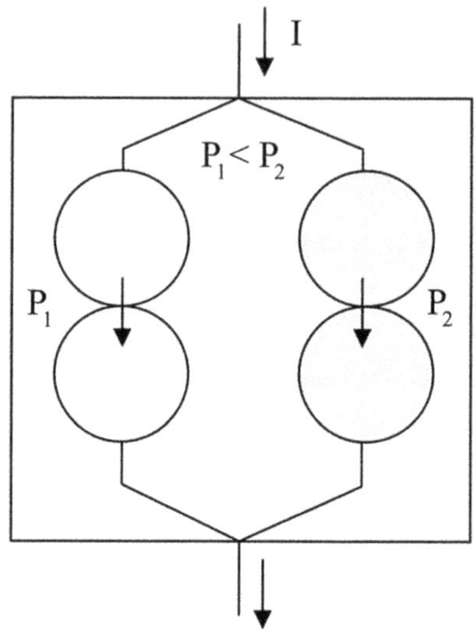

Fig. 10.20 Introduction of particles of a lower-resistivity material into the powder compacts creates higher-temperature regions within the compacts. The strings of white circles represent particles of the major component of the sintered material, while gray circles represent particles of the added lower-resistivity material

sample subjected to the electric current. The presence of hot spots as reaction initiation centers is assumed based on the influence that the Mg additions exert on the reaction between titanium and boron. Mg additions increase the conductivity of the mixture of titanium and boron (the electrical conductivities of titanium and magnesium are $2.38 \cdot 10^6$ and $2.26 \cdot 10^7$ $S \cdot m^{-1}$, respectively). The observed effects of the Mg additions are a reduced ignition temperature and a more uniform sintered product. When two samples of the same composition were sintered using the same time–temperature schedule, the sample containing Mg additions showed a more uniform microstructure along with a slightly lower density. The Ti–B–Mg reactive system transforming under pulsed electric current provided evidence of its beneficial influence on the microstructure uniformity of the sintered material. Mixtures of titanium and boron powders containing Mg additions were sintered in a normal SPS run in a graphite die and in a die insulated by means of BN. When the products of the reaction in these two specimens were compared, the one that experienced the action of electric current had a more uniform microstructure than the specimen processed in the presence of the insulating layer (Fig. 10.21a, b). A more uniform microstructure of the compact was obtained when the heating rate was increased from 20 to 100 $°C \cdot min^{-1}$ due to a more uniform distribution of the ignition points during the synthesis of TiB_2 in the mixture of Ti and B containing the additions of Mg (Fig. 10.21b, c).

Similar to conventional SHS, the reaction mode in reactive SPS depends on the heating rate with a certain minimum heating rate required to ignite the reaction and carry it out in a self-sustaining or explosion mode. Under slow heating, titanium diboride forms from the coarse-grained elemental powders as a result of a sequence of processes starting from the allotropic phase transition in titanium followed by the formation of titanium monoboride TiB, a Ti-rich liquid phase and a further reaction of the liquid with the remaining TiB and B to form the final product TiB_2 [59].

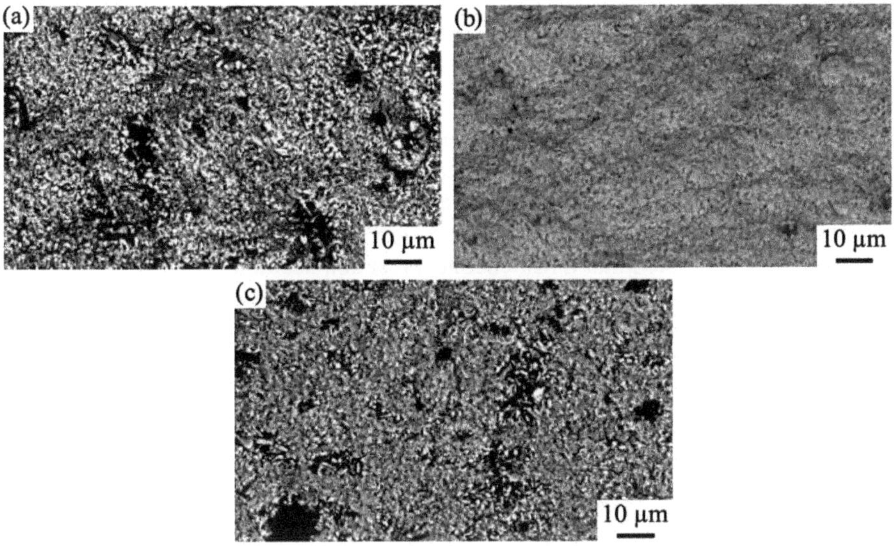

Fig. 10.21 Microstructure of the reaction product in the mixture of Ti–B–Mg processed by SPS: (**a**) reacted in the die with a BN-insulating layer, 100 °C·min^{-1}; (**b**) in a graphite die without insulation, 100 °C·min^{-1}; and (**c**) in a graphite die without insulation, 20 °C·min^{-1}. (Reprinted from Salamon et al. [57]. Copyright (2007) with permission of John Wiley & Sons)

The reactants remain in the compact along with the product phase after SPS at 1150 °C, while the reaction is complete after SPS at 1250 °C. During further densification at increased temperatures, the synthesized needlelike TiB_2 grains with micron and submicron cross-sections retain their shape and do not show any significant growth.

The chemical properties and microstructural characteristics of the reaction mixtures, such as the mixing uniformity, grain size, and degree of agglomeration, influence the microstructure of the product synthesized during SPS. This is also true for other processes of reactive sintering. Important for the synthesis in the SPS is a change in the electrical conductivity of the material in the sintering die with the reaction advancement and a possibility of controlling the synthesis by varying the electrical conductivity of the reaction mixtures. In order to initiate decomposition of magnesium hydride MgH_2, which is an insulator [60], the electrical conductivity of the powder sample was increased by adding graphite. Magnesium – the product of decomposition – also increases the electrical conductivity of the material in the SPS die. This example shows that, as the reaction product accumulates, the conductivity of the material in the die can change. If the reaction product has a higher electrical conductivity than the initial compound or the reaction mixture, the reaction is self-accelerated due to the in situ formation of conductive particles, which induce the formation of hot spots in the remaining, not yet fully reacted, powder mixture or not yet fully decomposed powder of a compound.

Partial reduction of oxides during the SPS presents another way of increasing electrical conductivity of the material in the sintering die. Interestingly enough, it can be used to advantage to facilitate densification of materials that would otherwise have been heated through thermal conduction from the die only. During the SPS of nanocrystalline rutile TiO_2, its partial reduction led to the formation of oxygen vacancies and an increase in its electrical conductivity, which, in turn, made it possible for the current to pass through the sample ultimately resulting in successful densification of the material at unexpectedly low temperatures and pressures [61].

Munir [62] suggested that a fast chemical reaction accompanied by shrinkage of the sample is the best scenario for the formation of a dense nanostructured or sub-micron-grained material. However, the reaction and densification steps often do not coincide during reactive SPS processing. When reaction is complete before densification starts, then, in order to obtain a fully dense product, one has to resort to sintering at higher temperatures. Such a situation develops when boron carbide forms in the mixtures of amorphous boron and carbon black at temperatures as low as 1200 °C, while temperatures as high as 1900 °C are required to produce a dense compact from the reaction product, as was shown by Anselmi-Tamburini et al. (Fig. 10.22) [63]. When chemical reaction and densification do not coincide, a further increase in temperature is required to obtain a fully dense product for compounds, in which self-diffusion is retarded. In certain cases, full-density sintered products could only be obtained at the expense of excessive grain growth [64].

The interrelation between the grain size and the density of boron carbide B_4C synthesized from the mixtures of amorphous boron and carbon black nanopowders was demonstrated by Hulbert et al. [65, 66], who used an off-set SPS die. Modeling of the process suggested that the difference between the temperatures of the top and bottom parts of the sample was 400 K. The sintered material had a structure with gradients in the grain size and porosity. In the sample, the reaction product changed from porous B_4C with grains 200 nm in size to dense boron carbide with grains about 2 μm in size forming in the region of the sample that experienced higher sintering temperatures.

An important factor in the SPS synthesis is the temperature range of the reaction. The occurrence of the reaction at a low temperature is beneficial for making a fine-grained product. If the reaction does occur within a wide temperature range during heating of the reaction mixture in a SPS chamber, the upper temperatures of the range are likely to destroy the nanostructure of the synthesized product formed at the early heating stages. These considerations were illustrated by the synthesis of $AlMgB_{14}$ from a mechanically milled mixture of Al, Mg, and B during SPS (Fig. 10.23) [67]. The first boride phase to form was the lower boride $AlMgB_4$, which appeared in the sintered sample at 600 °C. The $AlMgB_{14}$ phase was found after SPS at 800 °C, and the synthesis continued until the reaction was complete at 1325 °C. A noticeable shrinkage of the sample was observed starting from 1250 °C, which is much higher than the reaction onset temperature. A fine-grained reaction product was not favored as grains that nucleated in the beginning of the process coalesced and continued to grow as the SPS temperature increased.

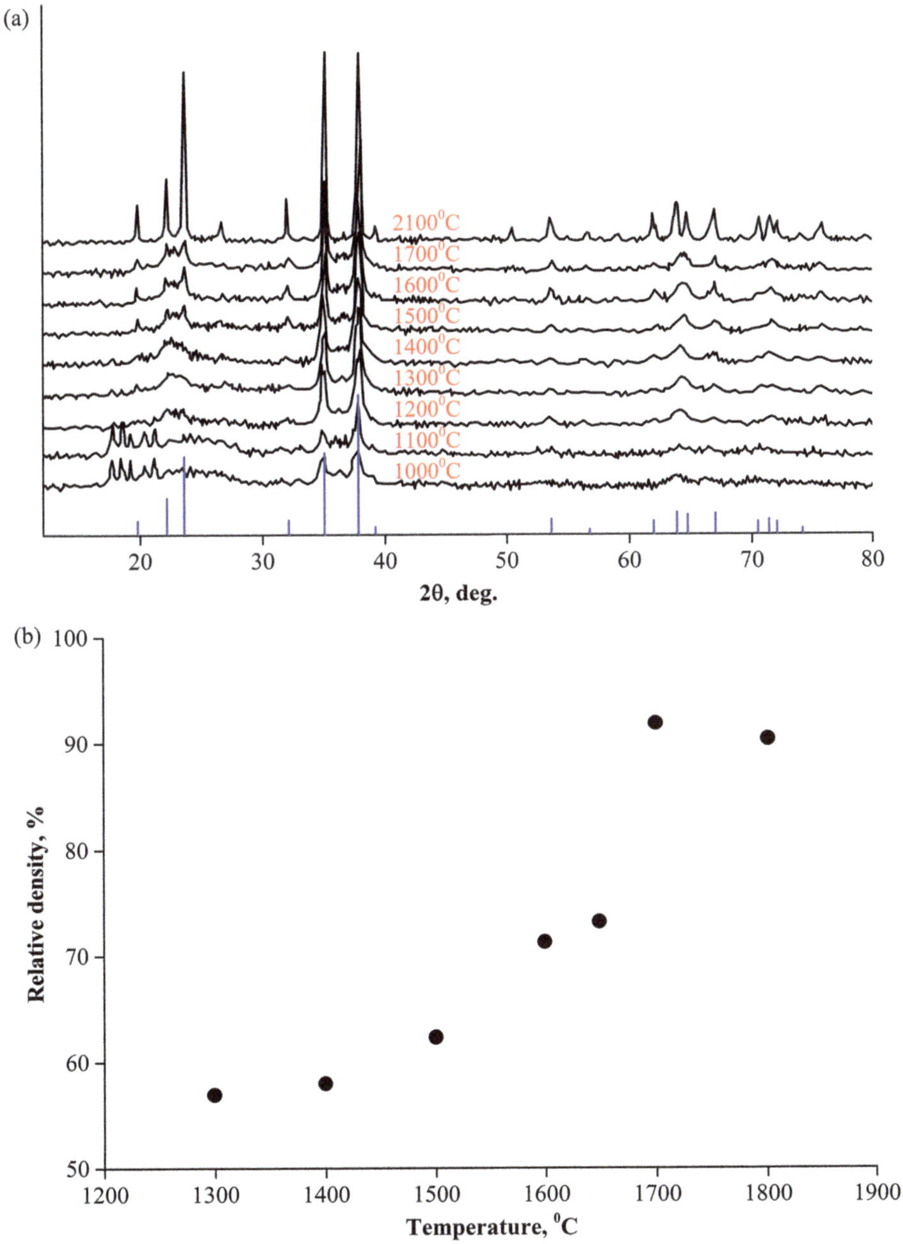

Fig. 10.22 XRD patterns of B–C mixtures heated and reacted in the SPS (**a**) and relative density of the B_4C reaction product as a function of the sintering temperature (**b**). (Reprinted from Anselmi-Tamburini et al. [63]. Copyright (2005) with permission of John Wiley & Sons)

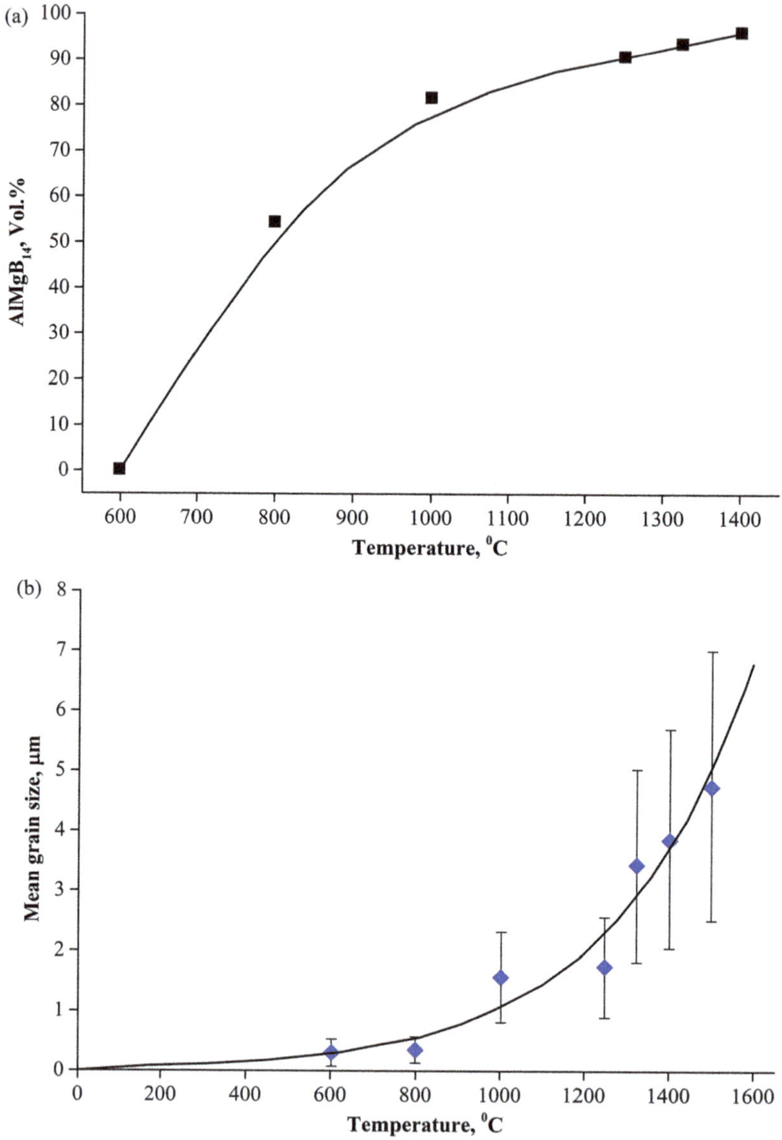

Fig. 10.23 Volume fraction (**a**) and grain size (**b**) of the $AlMgB_{14}$ phase as functions of the SPS temperature. (Reprinted from Roberts et al. [67]. Copyright (2008) with permission from Elsevier)

In single-phase materials synthesized by reactive SPS, the microstructure evolution can be described using the grain size and porosity characteristics. For single-phase ceramics formed by highly exothermic reactions, it is very difficult to obtain a fully dense material with nanosized grains because of local heat evolution and the concomitant temperature rise. If the reaction is moderately exothermic, fine grains form, and further microstructural development depends on the densification behavior of the reaction product. Thus, nanostructured molybdenum disilicide $MoSi_2$ with 140 nm grains [62] and iron aluminides with grains between 30 and 90 nm [68, 69] were successfully obtained from the corresponding mechanically milled mixtures, in which the reactants were refined to the nanoscale and thoroughly intermixed.

In composite materials, the grain size may not be the same for all constituent phases depending on the mechanism of the synthesis reactions and the tendency of each material to coarsen during sintering. As a result, the porosity distribution may not be uniform throughout the volume of the composite. When dealing with a multiphase material, one has to take into account the differences in the properties of the phases, such as characteristic grain growth and densification rates. An illustration of the above considerations is the in situ synthesized B_4C-23 vol.%–TiB_2 composites obtained by reactive SPS of the Ti–B–C mechanically milled mixtures [56, 70]. The agglomeration of titanium diboride is seen in the sintered B_4C–TiB_2 composites that were milled for durations insufficient for a uniform distribution of titanium. The agglomerates of titanium diboride appear as areas consisting of poorly sintered faceted grains 1–2 μm in size or larger, while boron carbide-rich areas show submicron grains. In the areas where large titanium diboride agglomerates are present, the fracture occurred in the intergranular mode. In the same material, the boron carbide phase showed much smaller grains. The presence of a pore inside a TiB_2 agglomerate can be explained by a decrease in the specific volume that accompanies the reaction between metallic titanium and boron. Boron carbide, whose content in the composite was 77 vol.% and which played a the role of a matrix, did not allow the TiB_2 grains to rearrange and better sinter between themselves (Fig. 10.24a). Due to a higher melting temperature of titanium diboride in comparison with boron carbide, TiB_2 agglomerates comprised the major part of the total porosity of the composites. In order to eliminate this porosity, the distribution of titanium should be more uniform, which was achieved by a longer mechanical milling of the reaction mixtures (Fig. 10.24b). These results demonstrated that

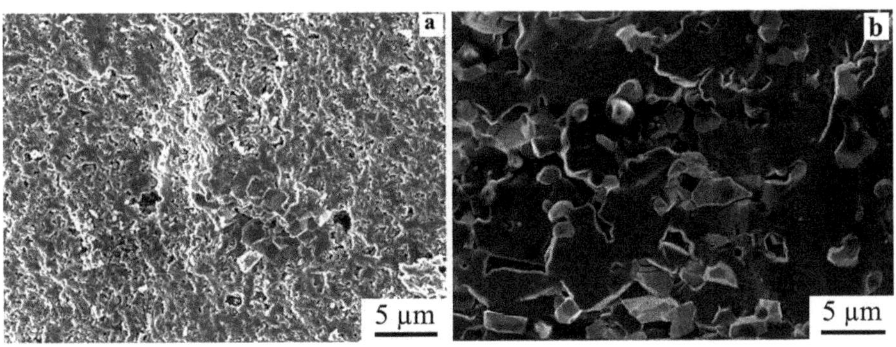

Fig. 10.24 Fracture surface of the TiB_2–B_4C composite showing an agglomerate of TiB_2 grains, reaction mixture milled for 4 h, SPS at 1600 °C (**a**) and a nearly fully dense TiB_2–B_4C, reaction mixture milled for 16 h, SPS at 1700 °C (**b**). (Reprinted from Hulbert et al. [56]. Copyright (2008), with permission from Elsevier)

agglomerates of a phase with a higher melting temperature compared with other phases of a composite may comprise a disproportionate part of the total porosity. In order to eliminate this type of porosity, the distribution of particles of this phase should be as uniform as possible. The formation of agglomerates should be avoided, which is achieved through a uniform distribution of the reactant participating in the formation of this phase in the reaction mixture.

10.12 Comparison of Reactive SPS and SPS of the Products of Self-Propagating High-Temperature Synthesis (SHS)

Licheri et al. [71] and Musa et al. [72] performed comparative studies of reactive SPS and SPS of the products of SHS as applied to the production of ultrahigh-temperature ceramics. The results of those studies were analyzed by Orrù and Cao [73]. Two synthesis routes are schematically described in Fig. 10.25 [72]. In the synthesis of a single-phase ceramic, the reaction during SPS occurs in a combustion mode. HfB_2 samples 95.4% dense could be produced by reactive SPS from the elemental powder mixtures at 1800 °C and 50 MPa [72]. This pressure was applied after the combustion reaction was complete; the initial pressure was 20 MPa. This was the only difference in terms of the SPS parameters from the other processing

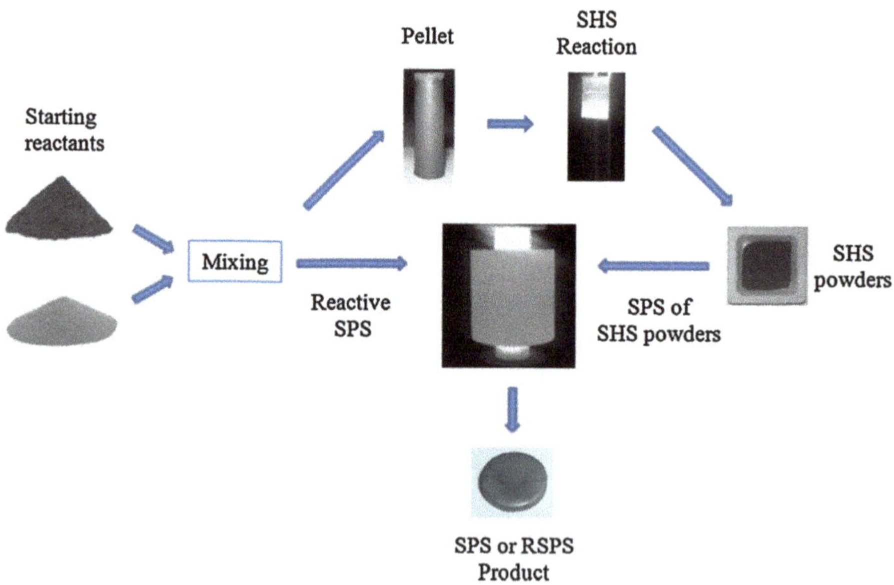

Fig. 10.25 Synthesis routes for producing ultrahigh-temperature ceramics: reactive SPS (reaction and consolidation in a single process) and SPS of the products of SHS (RSPS – reactive SPS). (Reprinted from Musa et al. [72]. Copyright (2012) with permission from Elsevier)

route – SPS of the SHS product – in which a pressure of 50 MPa was kept constant throughout the sintering cycle. When the product of SHS was sintered by SPS, temperatures as high as 1850–1900 °C were needed to produce a 93% dense sample. The composite system studied by Licheri et al. [71] was ZrB_2–SiC; this system was not formed by a combustion reaction upon heating in the SPS; rather, the B_4C, Si, and Zr reactants transformed into the reaction products in a gradual manner. In order to achieve the same relative densities using the same SPS pressure, the SPS temperature that was needed to consolidate the product of the SHS was 100 degrees lower than that used in the reactive SPS of the mixture of B_4C, Si, and Zr reactants. Orrù and Cao [73] concluded on a better suitability of the reactive SPS route for dense single-phase ceramics and the two-step processing (SHS followed by SPS) for dense multiphase (composite) ceramics.

The benefits of the reactive SPS route for dense single-phase ceramics may be related to the differences in the particle size and particle aggregate structure of the reaction products obtained in the SPS die, in which heat dissipation is allowed, and of those obtained via the SHS. With the temperature control during the SPS processing, the temperatures in the sample reacting in the SPS die in a combustion mode did not rise to the level of the conventional SHS. However, for reactions that proceed gradually upon heating, the temperature-related considerations would be hardly suitable to explain the differences between the densification behavior of the SHS product and the material reacted in the SPS die. Licheri et al. [71] did not report experiments with a two-step SPS pressure cycle for the ZrB_2–SiC system, as was done for the synthesis of HfB_2 [72]. Therefore, there remains a question to be answered in the future studies of whether completing the reaction under very low pressure or in a pressureless experiment could serve as a good starting point for manufacturing dense composite ceramics by means of reactive SPS with subsequent higher-pressure densification stages. Sintering of composite powders synthesized by the SHS is a promising alternative to sintering of mixtures obtained by blending of the ceramic constituents [74, 75].

10.13 Preparation of Reaction Mixtures for Reactive Sintering

Considering the availability of a wide variety of commercial nanopowders, one can choose between sintering without a reaction step and reactive sintering in order to produce a bulk nanostructured material. When selecting the state of reactants or the type of precursors to be used in the synthesis, the quality of the available powders such as purity, particle size, the presence of aggregates, as well as possible chemical and phase changes during pre-sintering processing should be taken into account. In order to prepare reaction mixtures by mechanical milling and mechanical alloying, planetary ball mills, attritors, or vibratory mills are frequently used [76]. The reaction between the components of the mechanically milled mixture occurs under conditions dramatically different from those typical of simple mixtures: the composite agglomerates with an increased interface area and a hierarchical structure form during high-energy mechanical milling. The intensity and duration of mechanical milling are important parameters, which influence the mixing uniformity, the structure of the composite agglomerates, as well as the nature and extent of the chemical interaction in the milled mixtures during milling and further sintering.

In order to obtain a nanocrystalline or a nanocomposite product by a chemical reaction, a careful choice of reactants and their preparation procedures should be made. In order to initiate the reaction at many nucleation sites at the same time, a large number of contact points between the solid reactants should be established, which is favored by well-developed interfaces between the reactants. The most visible solution along this line is to have particles of reactants as small as possible and mix them thoroughly avoiding the formation of aggregates of the particles of the same type. However, as will be discussed later, a mixture of fine particles or grains is not the only way to create conditions for the formation of nanosized reaction products.

If two nanopowders are selected as the reactants and the aggregates of each component are retained in the mixture, then the availability of the reactants to each other is greatly reduced. In such cases, sintering of the particles of the same material can occur before the reaction starts. Kim et al. [77] applied dry mixing and ultrasonic treatment in liquid in an attempt to improve the mixing uniformity of nickel and aluminum nanopowders and proved the former to be more efficient in achieving the mixing uniformity and breaking the nanoparticle aggregates.

In order to obtain a nanograined product, it may be sufficient to have only some reactants as nanopowders and others as coarser-grained powders. An example of this possibility is the synthesis of nano-HfB_2 through the reaction between a B_4C micron-sized powder, a HfO_2 nanopowder, and phenolic resin [78].

The influence of the preparation procedure of a two-phase mixture of nanopowders on the reaction onset temperature and the resultant microstructure of the reaction product is clearly demonstrated in a comparative study conducted by Stanciu et al. [79], who used three types of precursors for reactive sintering of Al_2TiO_5: co-gelified Al_2O_3 and TiO_2 powders, mechanical mixtures of Al_2O_3 and TiO_2 individual sol–gel powders, and powders synthesized by coprecipitation. As can be seen from the XRD patterns, the reaction between the co-gelified Al_2O_3 and TiO_2 powders is complete after the SPS at 1150 °C and occurs only partially in the other two precursors (Fig. 10.26). As can be seen from Fig. 10.27, the synthesis from the co-gelified Al_2O_3 and TiO_2 powders produces a much finer microstructure than the synthesis from the powders obtained by coprecipitation.

The presence of the reaction products in the mechanically milled mixtures can influence the subsequent SPS behavior of the system and the microstructure development [62]. The partial formation of the reaction product reduces the driving force needed for the reaction to be complete; at the same time, the product grains serve as the nucleation centers. The presence of the reaction products in the mechanically milled mixture of Mo–Si had a positive effect on the development of the nanostructured bulk product. On the other hand, during the synthesis of TiB_2–TiN composites,

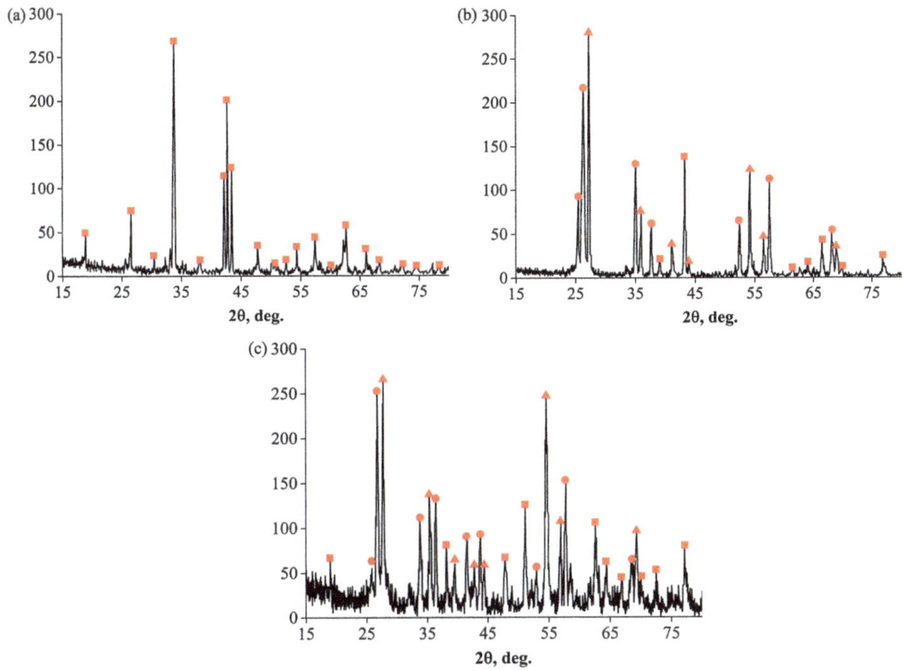

Fig. 10.26 XRD patterns of the compacts produced from Al_2O_3 and TiO_2 nanopowder mixtures of different processing history by SPS at 1100 °C: (**a**) co-gelified Al_2O_3 and TiO_2 powders, (**b**) mechanical mixtures of Al_2O_3 and TiO_2 individual sol–gel powders, and (**c**) powders synthesized by coprecipitation (squares, Al_2TiO_5; triangles, TiO_2; circles, Al_2O_3). (Reprinted from Stanciu et al. [79]. Copyright (2003) with permission from Elsevier)

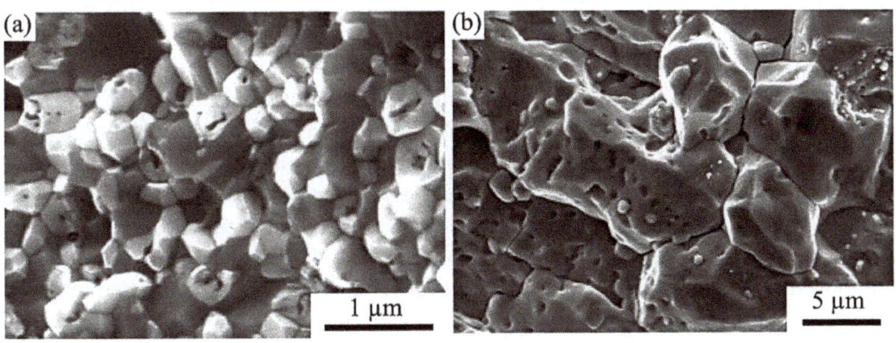

Fig. 10.27 Fracture surface of the compacts produced from Al_2O_3 and TiO_2 nanopowder mixtures of different processing history by SPS at 1100 °C: (**a**) co-gelified Al_2O_3 and TiO_2 powders and (**b**) powders synthesized by coprecipitation. (Reprinted from Stanciu et al. [79]. Copyright (2003) with permission from Elsevier)

the opposite effect was observed, and coarse grains were found in the bulk material produced from the mixtures containing certain amounts of the reaction products after milling. Handtrack et al. [80] performed mechanical milling of Ti–Si mixtures before applying SPS to produce Ti–Ti_5Si_3 nanocomposites. The dispersion of silicon during milling was important for the formation of fine particles of the reinforcing Ti_5Si_3 phase and full consumption of Si in the chemical reaction. The partial interaction of the finest particles of Si with the Ti matrix during milling did not show any detrimental effect on the formation of a nanocomposite material during SPS.

Mechanical milling of the powder mixtures prior to SPS enhances the kinetics of solid-state reactions. Shorter SPS time was needed to fully convert Ti, B_4C, and C into the TiC–TiB_2 composites when the mixture was mechanically milled [81, 82].

In reactive sintering of boron carbide from the elemental constituents by applying pressure and AC current, the improvement in the final density with increasing preliminary mechanical milling time was observed by Heian et al. [83]. However, iron contamination was introduced during milling at levels detectable by the XRD phase analysis.

Process control agents (PCA), which are additives of organic nature, are frequently used in the mechanical milling practice to avoid excessive cold welding of powders to vial walls, milling balls, and powder particles between themselves. PCAs can also be a source of reacting species thus altering the phase composition and influencing the microstructure development of the synthesized material. Thus, methanol served as a PCA during milling of 3Fe–Al mixtures and was a source of carbon for the in situ formation of a carbide reinforcing phase [84]. During the SPS, κ-carbide Fe_3AlC formed rather than Fe_3Al, and the resultant material was a two-phase composite Fe–Fe_3AlC.

SPS can provide conditions for crystallization of amorphous alloy powders resulting in the formation of nanocomposites [85, 86]. Such situations are not usually classified as reactive sintering and are dealt with in the field of devitrification and nanocrystallization of amorphous alloys. However, an amorphous solid can play the role of a reactant capable of transforming into new phases by reaction with another solid. Duan et al. [87, 88] successfully synthesized nanograined Si_3N_4–SiC ceramics during SPS by crystallization of an amorphous silicon carbonitride obtained by cross-linking of a polymer precursor followed by pyrolysis. The nanostructured composite was not fully dense. However, when a nanocrystalline TiO_2 powder was added to the amorphous Si–C–N precursor, improved densification was achieved. In the synthesized composite, the $TiC_{0.3}N_{0.7}$ phase formed a network composed of 100–300 nm grains (Fig. 10.28). This network penetrated through the regions of nanosized grains (20–30 nm) of the Si_3N_4–SiC composite. The XRD patterns of the composites are shown in Fig. 10.29. This chemical modification resulted in a fully dense material with an attractive combination of high fracture toughness and electrical conductivity.

A phase with grains as small as 100 nm can form as a result of a solid-state reaction between micron-sized powder reactants carried out in a SPS chamber [89, 90]. Indeed, a fine-grained microstructure of the reaction products is favored

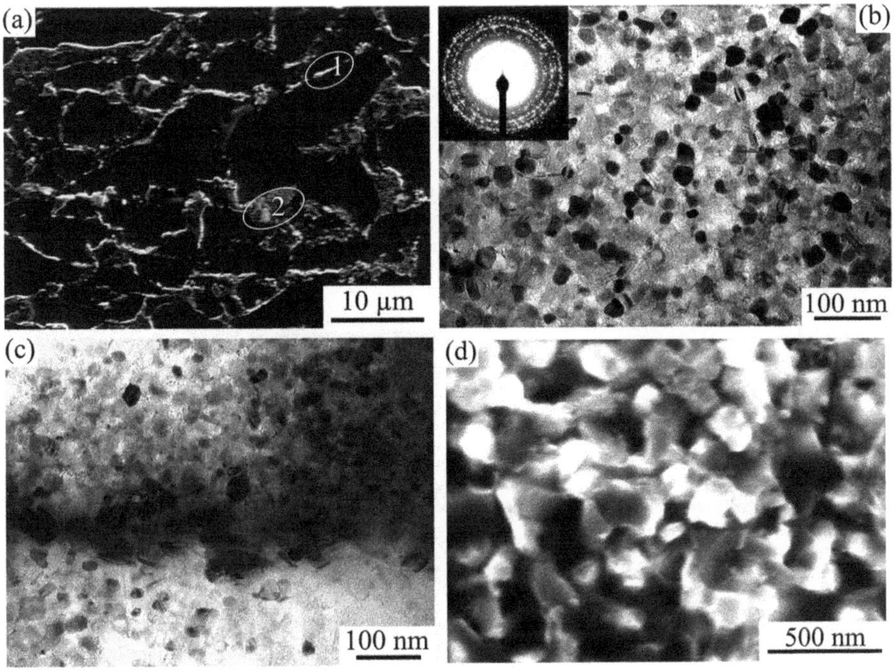

Fig. 10.28 Si_3N_4–SiC–$TiC_{0.3}N_{0.7}$ nanocomposite: (**a**) fracture surface, scanning electron microscopy (SEM); (**b**) microstructure of dark areas in (**a**) corresponding to Si_3N_4–SiC, TEM; (**c**) region 1 in (**a**), TEM; (**d**) region 2 in (**a**) corresponding to the $TiC_{0.3}N_{0.7}$ phase, SEM. (Reprinted from Duan et al. [87]. Copyright (2004) with permission from Elsevier)

when the reactants are mixed at the nanolevel or the process involves multiple steps. Zhang et al. [89] implemented reactive SPS in the mixture of Ti, C, and Si powders. The intermediate crystalline phases TiC_x and $Ti_5Si_3C_y$ formed first from Ti, Si, and C and then participated in the following reaction:

$$(4/3 - 3z)TiC_x + 1/3Ti_5Si_3C_y + (2 - 4/3x - (1 - 3x)z - 1/3y)C \rightarrow (1 - z)Ti_3SiC_2 + zSiC.$$

The SiC phase had a grain size of about 100 nm, while the other phase – Ti_3SiC_2 – had grains up to 5 μm (Fig. 10.30). The formation of SiC through intermediate solid phases is crucial for the microstructure development of the Ti_3SiC_2–SiC nanocomposite from the mixture of coarse-grained powders. It can be assumed that the mixture of the synthesized intermediate phases possessed a refined structure compared to that of the initial reaction mixture.

Wang et al. [91] reported the formation of submicron-grained TiN–Al_2O_3 from AlN, TiO_2, and Ti powders, all three having particles from several to several tens of microns. Mixing of the powders was performed by milling in ethanol and grain refinement was not favored. The fine-grained structure of the composite could have

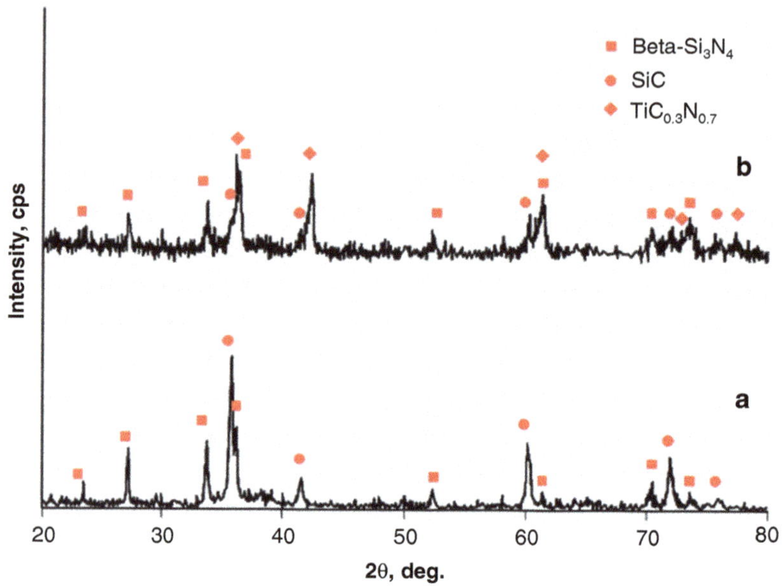

Fig. 10.29 XRD patterns of the Si_3N_4–SiC ceramic nanocomposite obtained by SPS of a Si–C–N amorphous precursor (**a**) and the Si_3N_4–SiC–$TiC_{0.3}N_{0.7}$ ceramic nanocomposite obtained by reactive SPS of a mixture of nanocrystalline TiO_2 with the amorphous Si–C–N (**b**). (Reprinted from Duan et al. [87]. Copyright (2004) with permission from Elsevier)

Fig. 10.30 Fracture surface of Ti_3SiC_2–SiC nanocomposite sintered by SPS at 1280 °C and showing grains of SiC about 100 nm in size (**a**) lower magnification and (**b**) higher magnification. (Reprinted from Zhang et al. [89]. Copyright (2006) with permission from Elsevier)

been the result of solid-state interactions of complex mechanisms with three solid phases AlN, TiO_2, and Ti participating. However, the detailed mechanism of the reactions and the possible nature of the intermediate products have not been elucidated for this system.

10.14 Decomposition Reactions During SPS

The majority of studies on reactive SPS use chemical reactions between pre-mixed powders to synthesize ceramic phases. At the same time, very little research has been done on decomposition reactions during SPS. Decomposition of a compound is another possible pathway to nanocrystalline materials. Thermal decomposition of intercalated layered hydroxides [92] and salts [93] resulting in the formation of metal nanoparticles may encourage further investigations of reactions of these types under pulsed electric current. A decomposition reaction can be used as a step in the complex solid-state synthesis. This approach helps overcome the storage and handling problems of highly reactive powders. Nanosized tungsten formed as a result of decomposition of W_2N served as a reactant in the synthesis of nanograined tungsten carbide WC [94]. When highly reactive metals, such as zirconium, are needed for the synthesis, fine powders available commercially may contain oxide phases. In order to work with metallic reactants of better quality and higher reactivity, metal hydrides can be taken as raw materials forming pure metals upon decomposition during heating in the SPS [95]. The in situ silicon formed upon decomposition of Si_3N_4 was used as a reactant to form SiC-bonded carbon composites [96]. The evolution of gaseous products upon decomposition of solid compounds can be used to advantage when pore-forming agents are required. Thus, decomposition of TiH_2 was used to form pores in spark plasma sintered NiTi [97]. In a distinct work of Schmidt et al. [60], decomposition of magnesium hydride MgH_2 has been studied with a goal of determining the effect of electrical current on the onset temperature of decomposition, which was found to decrease when electric current was passing through the powder. However, no microstructural investigation was presented in that publication.

Partial decomposition upon sintering can have a deleterious effect on the properties of some materials and, therefore, should be brought to minimum by choosing the optimal time–temperature sintering schedule. Decomposition of hydroxyapatite, a promising biocompatible material for medical applications, is highly undesirable, as the product of decomposition – α-tricalcium phosphate – is soluble in body liquids. Miao et al. [98] reported successful SPS of hydroxyapatite–yttria-stabilized zirconia composites accompanied by only slight decomposition of hydroxyapatite.

10.15 Evolution of C–C Bonds Under Electric Current

The catalyst-free graphitization of amorphous carbon can be conducted by applying electric current to the sample. Honda et al. [99] reported graphitization of amorphous carbon under electric discharge and pressure applied simultaneously to the powder specimen. In that work, both current and temperature effects could be responsible for the graphitization phenomenon. Asaka et al. [100] reported the catalyst-free graphitization of amorphous carbon on the surface of carbon nanotubes, which experienced Joule heating by electric current with a current density of the order of 10^8 A cm^{-2}. Despite high local current densities at the inter-particle contacts during the early stages of the SPS process, the conditions of the SPS alone are not sufficient to induce graphitization of amorphous carbon, as can be concluded from a study by Toyofuku et al. [101], who sintered amorphous carbon in the SPS. It was found that the sample remains amorphous even after sintering at 2200 °C and is poorly densified.

High currents passing through carbon nanotubes can break C–C bonds, as was reported by Kim et al. [102]. Placed in a die of a standard SPS facility and heated up to 2000 °C, single-walled carbon nanotubes experienced high currents, which caused their unzipping and transformation into graphene stacks consisting of two to three layers. Similarly, Sribalaji et al. [103] observed unzipping of nanotubes during SPS of TiC–WC–carbon nanotube composites leading to the in situ formation of graphene nanoribbons and increased levels of toughening in the composites.

Huang et al. [104] suggested that in alumina–carbon nanotube composites subjected to standard SPS conditions at 1300 °C, the nanotubes experienced high current densities, and the local temperature near the nanotubes was much higher than the temperature inside the alumina grains. Softened and thermally activated, the nanotubes become reactive and either burn out or react with alumina. Raman spectroscopy was used to gain evidence of the destruction of the nanotube hexagonal units after SPS of alumina-nanotube composites. Some alumina grains showed "footprints" left by the nanotubes, which indicated the diffusion of carbon into alumina. The nanotubes acted as grain growth inhibitors, so their destruction led to significant grain coarsening of alumina. When hexagonal boron nitride BN layers were introduced between the sample and the graphite punches, the nanotubes in the composites sintered in the same SPS regime were preserved. The hexagonal boron nitride has a high electrical resistivity but is thermally conductive, so the heat flux from the punches to the sample was not obstructed.

The transformation of carbon nanotubes into submicrometer- and micrometer-sized diamonds during SPS at 1500 °C and 80 MPa was reported by Zhang et al. [105], who suggested that the effects of high temperatures and sparking were similar to those of the electron or ion beams or high pressures in terms of the influence on the carbon nanotubes. The formation of diamonds from carbon nanotubes during SPS proceeded through the intermediate products – carbon onion structures. It was found that fullerenes also transformed into diamond under the SPS conditions, while graphite remained stable [106].

10.16 Interaction of the Materials Sintered Using Graphite Foil and Graphite Tooling with Carbon

The influence of carbon of the graphite tooling and graphite foil on the elemental and phase composition of the sintered materials is an important issue in the SPS processing. When only a thin subsurface layer of the sintered compact is affected by carbon, this layer can be removed by polishing. However, in many cases, carbon contamination and partial reduction of oxides caused by the presence of carbon occur within volumes comparable to the total volume of the sintered part and deteriorate the properties of the sintered materials. Until now, a number of papers have been published that tackle this problem and look into possible mechanisms of carbon accumulation in the sintered specimens (solid-state diffusion and evaporation–condensation at high temperatures). As will be discussed below, during SPS, carbon can accumulate at grain boundaries of the sintered material not forming a separate phase or participate in the formation of new phases in the subsurface regions of the compact, and, under certain conditions, throughout the compact volume. Carbon uptake by the sintered specimens becomes more significant at high sintering temperatures.

Zapata-Solvas et al. [107] observed carbon enrichment of grain boundaries of zirconium diboride ZrB_2 sintered in the presence of graphite foil. The formation of carbon-containing phases caused by diffusion of carbon from graphite foil was reported by Solodkyi et al. [108], who observed the presence of boron carbide B_4C in the compact sintered from boron suboxide B_6O. The B_4C phase was not found in the compact when tantalum foil was used to prevent carbon diffusion into the B_6O compact. When a molybdenum powder was sintered in contact with graphite foil, β-Mo_2C formed on the surface of the specimen [109]. The growth of the molybdenum carbide layer followed the parabolic law, and the determined activation energy was close to that of the activation energy of carbon diffusion in molybdenum. The carbide layer was found on the samples sintered at temperatures higher than 1550 °C. At lower temperatures, the layer either did not form or was so thin that detached easily from the specimen's surface together with the graphite foil.

Surface carburization of carbide-forming metals during SPS can be used to advantage to produce surface-strengthened parts. An example of such a treatment was demonstrated by Hayashi et al. [110], who used SPS to form a TiC coating on titanium substrates. For that, a graphite powder was placed into the die cavity, and a titanium substrate was embedded into the powder. The surface carburization was carried out at a pressure of 10.5 MPa at 770–970 °C. After holding the assembly at 970 °C for 30 min, a continuous TiC coating 7 μm thick was formed on the surface of the titanium substrate. Similar to the growth of Mo_2C, the growth of the TiC layer was described by the parabolic law. The use of a graphite powder as a reactant in such coating deposition processes helps protect the punches from being damaged due to reaction with the substrate.

Grasso et al. [111] showed that an increase in the temperature during the SPS of a WC-based nanopowder conducted in contact with graphite foil in the 1200–1400 °C range leads to an increase in the concentration of carbon in the sintered material. The authors attributed this effect to diffusion of carbon from the graphite foil into the specimen. As the losses of carbon from the nanopowder were due to its undesirable oxidation prior to sintering, the restoration of the carbon content during sintering had a positive effect on the material properties.

Experiments conducted by Lee et al. [112] showed that carbon uptake during SPS of a tungsten powder in contact with graphite foil results in the formation of a carbide phase in the surface layers of the sintered compacts. A tungsten carbide $W_2C_{0.85}$-

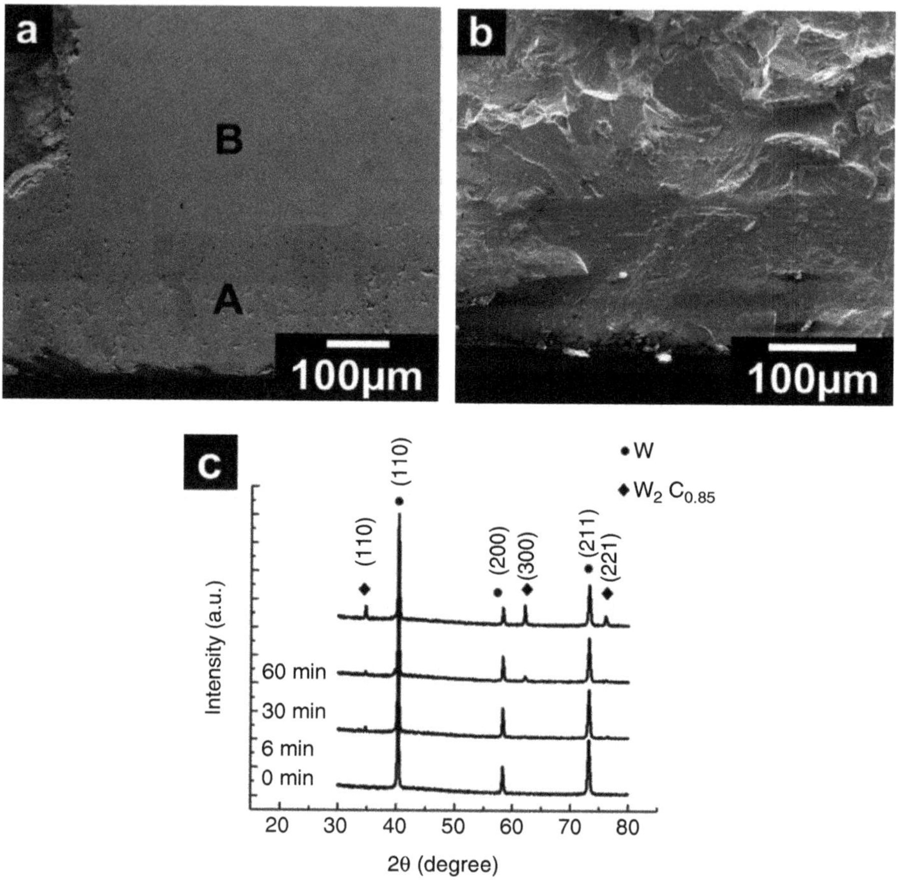

Fig. 10.31 (**a**) Cross section of the tungsten compact sintered in contact with graphite foil showing a carbide layer (no boron nitride coating was applied to the foil), (**b**) fracture surface of the tungsten compact sintered in contact with graphite foil showing a carbide layer (a boron nitride coating was applied to the foil), and (**c**) XRD patterns of the tungsten compacts sintered with different holding times (0–60 min); the amount of $W_2C_{0.85}$ on the specimen's surface increases with the holding time. (Reprinted from Lee et al. [112]. Copyright (2016) with permission from Elsevier)

containing external shell was generated on the specimen's surface. In order to prevent the interaction of the sintered material with carbon, boron nitride coatings are usually applied on the internal surface of the die, punches, and the surface of the graphite foil. A cross-sectional view of the compact sintered without a BN coating on the graphite foil is shown in Fig. 10.31a. The material in zone A corresponds to the carbon-affected layer and contains 2.2 wt.% of C, while zone B is free from carbon. In the case of SPS of tungsten, a boron nitride coating on the graphite foil did not block the carbon uptake completely but slowed down the inward diffusion of carbon. The carbide-containing layer was also detectable on the fracture surface of

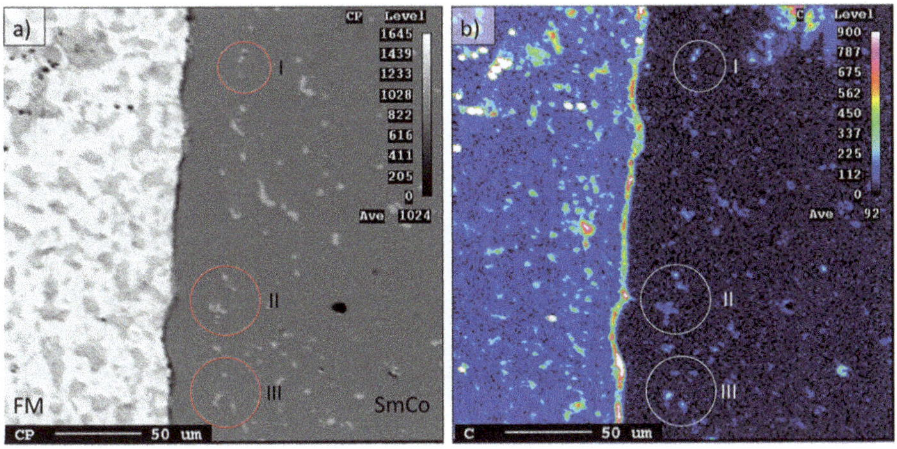

Fig. 10.32 Back-scattered electron image (**a**) and carbon map (**b**) of the boundary between the mounting material (Field's metal (FM) – Bi–In–Sn) and the SPS-processed Sm(Co, Fe, Cu, Zr)$_z$ material. I, II, and III circles mark Sm-rich regions, which coincide with C-rich regions. (Reprinted from Mackie et al. [113], this article is available under the terms of the Creative Commons Attribution License (CC BY), https://creativecommons.org/licenses/by/4.0/)

the compacts. Figure 10.31b shows such a layer formed in the experiment, in which a BN coating was sprayed over the surface of the graphite foil. The concentration of the $W_2Co_{0.85}$ phase increased with the holding time (Fig. 10.31c). Assuming a parabolic growth of the carbide layer, the diffusion coefficient of carbon in tungsten was calculated, and the obtained value was close to that reported in the literature.

Mackie et al. [113] studied the distribution of carbon in the surface layers of Sm (Co, Fe, Cu, Zr)$_z$ sintered by SPS using graphite tooling and graphite foil. The sintered Sm(Co, Fe, Cu, Zr)$_z$ material contained Sm-rich regions. These regions coincided with C-rich regions (Fig. 10.32), which allowed the authors to suggest a possibility of the formation of the carbide phase Sm_3C.

Due to carbon inward diffusion and the formation of chromium carbides, Boulnat et al. [114] observed strengthening of the surface layers of oxide dispersion-strengthened ferritic steel samples sintered by SPS. Increased carbon concentrations and increased hardness were observed within thicknesses of about 200 μm from the sample's edge in the samples sintered at 1150 °C (Fig. 10.33).

The thickness of the carbon-affected layer on the surface of the compact obtained by SPS of a glassy Fe-based powder of the $Fe_{75}Si_{20}B_5$ (at.%) composition at 900 °C was only 2–3 μm (Fig. 10.34) [115]. The first phase to form upon heating of the glassy powder was α-Fe(Si). This phase formed at 490 °C, and then the crystallization continued by the formation of the Fe_2B and Fe_3Si phases. A small thickness of the carbon-affected layer may be related to low solubility of carbon in the boride and silicide phases.

The W_2C phase was observed by Rodriguez-Suarez et al. [116] in both the surface layers and the interior of the compacts spark plasma sintered from tungsten

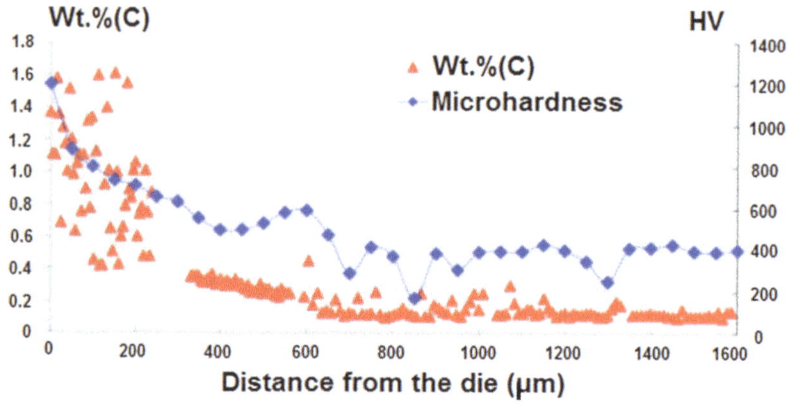

Fig. 10.33 Microhardness and carbon concentration profiles starting from the edge of the sample of the oxide dispersion-strengthened steel spark plasma sintered at 1150 °C for 20 min. (Reprinted from Boulnat et al. [114]. Copyright (2014), The Institute of Materials, Minerals and Mining, by permission of Taylor & Francis Ltd., www.tandfonline.com on behalf of The Institute of Materials, Minerals and Mining)

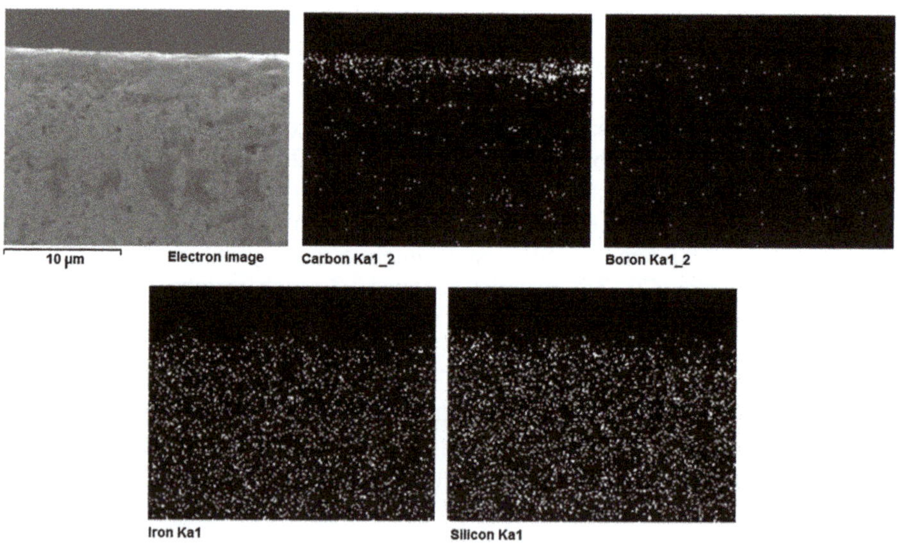

Fig. 10.34 Cross-sectional view of the compact consolidated by SPS of the glassy $Fe_{75}Si_{20}B_5$ powder obtained by mechanical alloying and the corresponding elemental maps. The SPS temperature was 900 °C; the holding time was 1 min. (Reprinted from Neamţu et al. [115]. Copyright (2014) with permission from Elsevier)

nanoparticle-coated Al_2O_3 powders using graphite tooling. Carbon had diffused deep into the volume of the specimen. It was found that the carbide formation depended on the pretreatment procedures applied to the initial powder (calcination, milling), as the structure of the Al_2O_3 agglomerates influenced the accessibility of

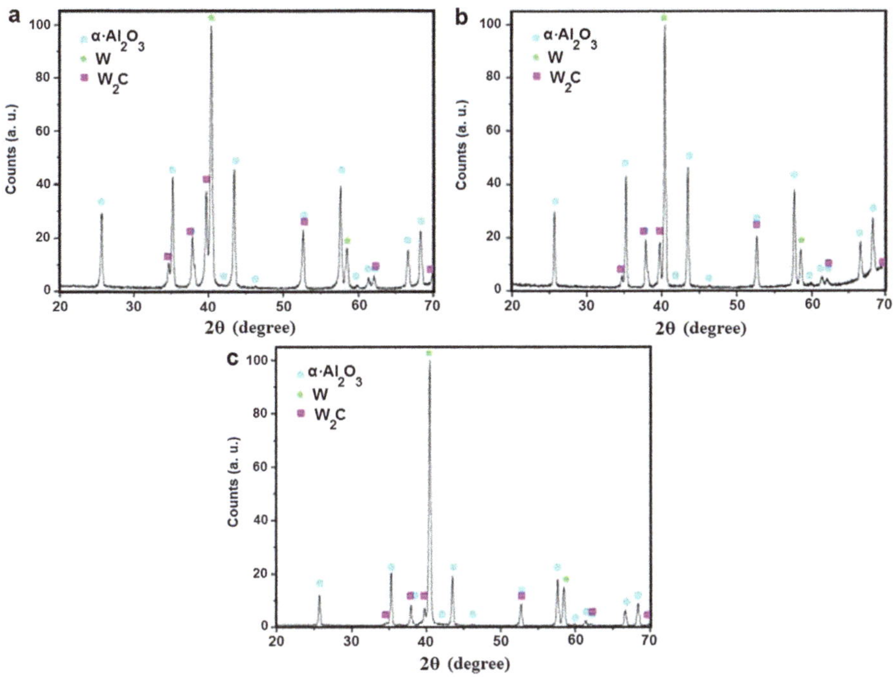

Fig. 10.35 XRD patterns of the compacts spark plasma sintered from tungsten nanoparticle-coated Al_2O_3 powders: (**a**) the surface of the compact, Al_2O_3-2.5 vol.%W powder, Al_2O_3 as-received; (**b**) cross section of the compact, Al_2O_3-2 vol.%W, Al_2O_3 calcined; (**c**) cross section of the compact, Al_2O_3-4 vol.%W, Al_2O_3 calcined and milled. (Reprinted from Rodriguez-Suarez et al. [116]. Copyright (2009) with permission from Elsevier)

tungsten to carbon diffusing into the sample during sintering (Fig. 10.35). The authors assumed that the in situ formed W_2C phase influenced the sintering behavior of the composite powders by hindering grain-boundary sliding.

In Refs. [117, 118], the reactivity of Ni-15 at.%W powders prepared by mechanical alloying or simple mixing toward carbon of graphite foil during SPS has been addressed. In the Ni-15 at.%W mixtures, tungsten as a dispersed phase or a solute in a Ni-based matrix acted as a "tracer" of carbon diffusion into the sample observed through the formation of binary or ternary carbides. The mixture of Ni and W powders was quite uniform (Fig. 10.36a) with smaller particles of tungsten (brighter particles in the image) distributed between nickel particles (darker particles in the image). Mechanical milling resulted in alloying between the metals and dissolution of tungsten in nickel. The mixture transformed into composite agglomerates (Fig. 10.36b), in which the main phase was a Ni(W) solid solution.

XRD patterns taken from the flat ends and cross sections of the Ni–W compacts are presented in Fig. 10.37 [118]. Tungsten carbide WC was found at the flat ends of

Fig. 10.36 Morphology of Ni-15 at.%W powders: (**a**) mixed and (**b**) mechanically alloyed in a high-energy planetary ball mill. (Reprinted from Dudina et al. [118]. Copyright (2016) with permission from Elsevier)

compacts obtained from both powder mixtures. In the case of the mechanically alloyed powders, carbon diffusion into the sintered compact resulted in the formation of Ni_2W_4C particles in a Ni-based matrix throughout the compact volume. In contrast to the compact obtained from the mechanically alloyed powders, no carbon-containing phases were detected in the interior of the compacts produced from the mixed powders by sintering under the same conditions. No significant differences were found in the microstructures of the subsurface regions and the interior of the compact sintered from the mixed powders (Fig. 10.38a, c). The black, gray, and light-gray areas were assigned to particles of nickel, Ni(W) solid solutions, and residual tungsten, respectively.

When a Ni(W) solid solution obtained by mechanical alloying decomposed upon carbon attack through the reaction

$$\mathrm{Ni(W)} + \mathrm{C} = \mathrm{WC} + \mathrm{Ni},$$

the WC crystals nucleated and grew from the Ni-based matrix; they could be easily distinguished in the subsurface regions of the specimens sintered from the mechanically alloyed powders (Fig. 10.38b). The subsurface layer of the compact produced from the mechanically alloyed powders had a distinct boundary with the interior of the compact (Fig. 10.38d). In the inner part of the compact, submicron particles of Ni_2W_4C were distributed in a Ni-based matrix. In order to reveal the details of distribution of WC particles in the subsurface layers of the compact, the polished cross section was etched with HCl solution. Submicron WC particles were organized in chains clearly visible in the microstructure (Fig. 10.39a). The correspondence of the bright particles in the images to the WC phase was confirmed by the energy-dispersive spectroscopy (EDS) point analysis (Fig. 10.39b, c). Chains formed by the submicron particles were a feature structurally “inherited” from the powder state of the Ni-15 at.%W alloy, as the size of regions encircled by the chains was comparable to the size of the powder agglomerates in the mechanically milled mixture. Paths for

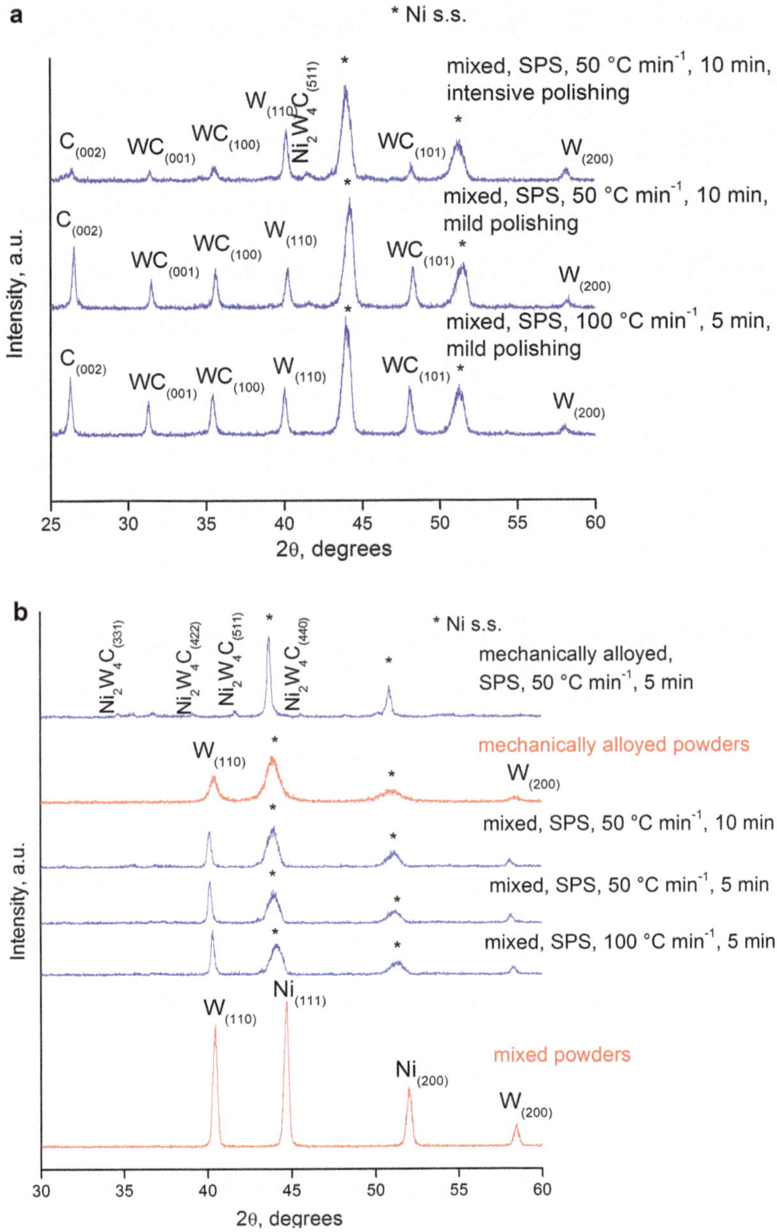

Fig. 10.37 XRD patterns of the Ni-15 at.%W powders and compacts spark plasma sintered at 900 °C: (**a**) flat ends of the compacts sintered from the mixed powders after partial removal of the graphite foil and (**b**) mixed and mechanically alloyed powders and cross section of the sintered compacts. (Reprinted from Dudina et al. [118]. Copyright (2016) with permission from Elsevier)

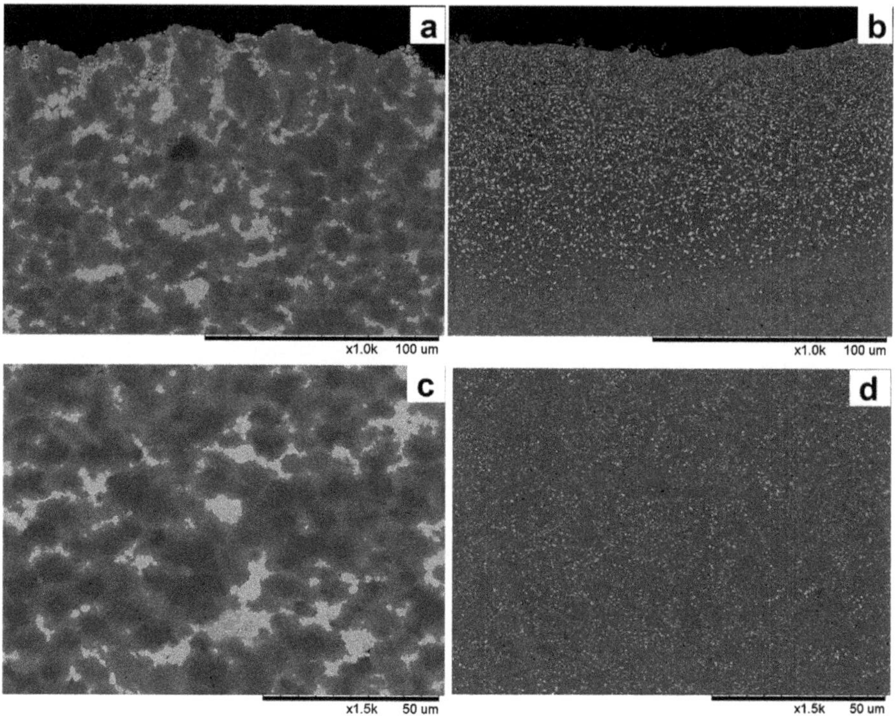

Fig. 10.38 Microstructure of the Ni-15 at.%W compacts spark plasma sintered at 900 °C, heating rate 50 °C min^{-1}, holding time 5 min from mixed (**a**, **c**) and mechanically milled (**b**, **d**) powders; (**a** and **b**) subsurface layers of the compacts adjacent to the graphite foil during SPS, (**c** and **d**) interior of the compacts. (Reprinted from Dudina et al. [118]. Copyright (2016) with permission from Elsevier)

faster diffusion were offered by boundaries between the initial powder agglomerates. The uptake of carbon caused by its inward diffusion was facilitated by a much higher concentration of defects and a greater volume of grain boundaries in the mechanically alloyed powder compared with the mixed powders. The distribution pattern of the Ni_2W_4C particles in the compact's interior (Fig. 10.39d) was similar to that of the submicron WC particles in the subsurface layers of the compact. The character of distribution of the carbon-containing compounds – WC and Ni_2W_4C – was to a great extent predetermined by the particle size and morphology of the mechanically milled powder. It is possible that diffusion of carbon occurred faster along the surface of the powder particles not yet fully consolidated into a bulk material upon heating during the SPS.

The thickness of the carburized layer x was estimated as

$$x \sim (Dt)^{1/2},$$

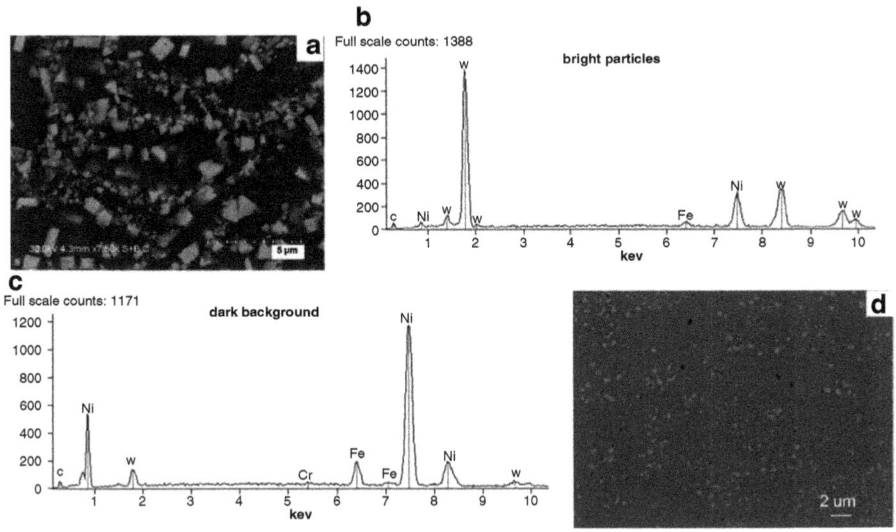

Fig. 10.39 (**a**) Subsurface (WC-containing) layer of the compact spark plasma sintered from a Ni-15 at.%W mechanically alloyed powder mixture in contact with graphite foil, EDS taken from the bright particles (**b**) and dark background (**c**) in the image shown in (**a**), (**d**) the interior of the compact. Etching in 10% HCl solution. (Reprinted from Bokhonov et al. [117])

where *D* is the diffusion coefficient and *t* is the time, and was found to be of the order of 100 μm at 1000 °C, which agreed well with the observed thickness of the WC-containing subsurface layer in the sintered sample [117]. However, the presence of carbon in the sintered material should also be taken into account at greater distances from the interface with the graphite foil due to much faster (orders of magnitude) diffusion along the surface of pores and grain boundaries during sintering. The hardness of the compacts spark plasma sintered from the mechanically alloyed Ni-15 at.%W powders measured on the cross section in the specimen center was significantly higher than that of the compacts produced from the mixed powders (Tables 10.2, 10.3). This difference was due to the presence of submicron Ni_2W_4C particles distributed in a Ni-based solid solution-strengthened matrix and a low porosity of the compact.

No tungsten carbide-containing layer formed when a mechanically milled Cu-15 at.%W powder mixture was spark plasma sintered under the same conditions (Fig. 10.40). This difference can be explained by a very low solubility of carbon in copper. Instead of a carbide-containing layer adjacent to the graphite foil, a uniform microstructure through the thickness of the sintered compact was observed. It was concluded that it is not the presence of the carbide-forming element that determines the character of interaction of the graphite foil with the sintered material but the solubility of carbon in the component that plays the role of a matrix in the sintered material. In nickel, carbon has appreciable solubility, which is 1.3 at.% at 1000 °C [119].

Table 10.2 Vickers hardness of different areas of the compact spark plasma sintered from a mechanically alloyed Ni-15 at.%W powder mixture; SPS was conducted in contact with graphite foil [117]

Area of the cross section	Distance from the interface with the graphite foil to the measurement point (μm)	Vickers hardness (HV)
Flat ends of the compact	20–50	470 ± 10
	50–100	610 ± 10
	100–150	620 ± 10
Cylindrical surface	20–30	510 ± 10
	30–70	620 ± 10
	70–120	650 ± 10
Interior of the compact	500–1000	620 ± 10

Table 10.3 Vickers hardness of the central part of the compacts spark plasma sintered at 900 °C from mixed Ni-15 at.%W powders; SPS was conducted in contact with graphite foil

Heating rate (°C min^{-1})	Holding time (min)	Hardness (HV)
100	5	315 ± 10
50	5	340 ± 10
50	10	370 ± 15

Fig. 10.40 Cross-sectional view of the compact spark plasma sintered from a mechanically milled Cu-15 at.%W powder mixture; SPS was conducted in contact with graphite foil at 900 °C (Reprinted from Bokhonov et al. [117])

In Ref. [118], the evidence of carbon uptake from graphite foil by metallic nickel was obtained. For this purpose, a carbonyl nickel powder was spark plasma sintered at 1000 °C in the presence of graphite foil. Figure 10.41 shows enlarged views of the Ni(111) and Ni(220) reflections of the XRD patterns taken from the flat ends of the compact (the adhered graphite foil was partially removed by mild polishing). The reflections shift to lower angles relative to their positions in the XRD patterns taken from the initial nickel powder and cross section of the compact. This shift points to

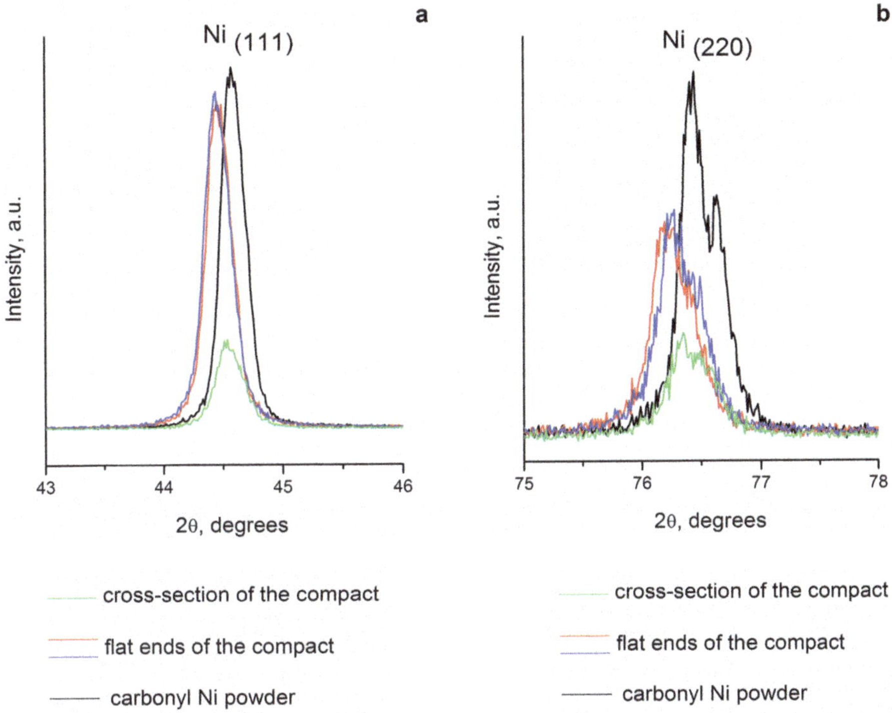

Fig. 10.41 (**a**) Ni(111) and (**b**) Ni(220) reflections of the XRD patterns of the carbonyl nickel powder and the compact spark plasma sintered from this powder at 1000 °C in contact with graphite foil. (Reprinted from Dudina et al. [118]. Copyright (2016) with permission from Elsevier)

an increased lattice parameter of nickel, which is caused by dissolution of carbon in the subsurface layers of the compact as a result of inward diffusion during SPS.

Collet et al. [120] conducted SPS of an oxidized copper powder at different temperatures and a pressure of 28 MPa and found that the reducing capacity of the SPS chamber with graphite tooling is not sufficient to fully reduce oxides contained in the partially oxidized copper powder during consolidation. Figure 10.42 shows the fracture surface and the polished cross section of the compact spark plasma sintered from the oxidized copper powder in vacuum. Although the sintered compact shows the fracture surface typical of a ductile metal, the presence of pores and oxide inclusions is unambiguously confirmed by an image taken from the polished cross section. Collet et al. had to resort to conducting SPS in hydrogen to eliminate the oxide phase from the compact. During SPS in a hydrogen atmosphere, the oxide concentration in the sintered compact decreased by an order of magnitude.

Microstructural indications of the chemical reduction of oxide films present on the surface of metals during treatment in a SPS chamber were found in Refs. [121–124]. Toyofuku et al. [122] sintered tungsten wires to tungsten plates in a SPS chamber and observed the formation of tungsten deposits on the surface of the initial

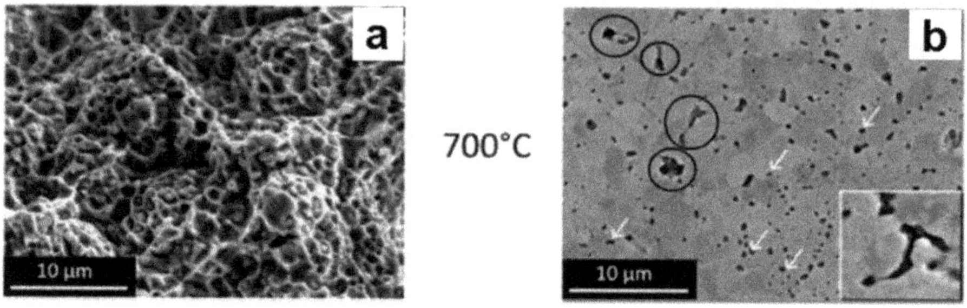

Fig. 10.42 Fracture surface (**a**) and polished cross section (**b**) of the compact spark plasma sintered from an oxidized copper powder containing 3.5 wt.% of Cu_2O. SPS was conducted at 700 °C, 28 MPa, a heating rate of 50 °C min^{-1}, and without the dwell time at the maximum temperature in contact with graphite foil in vacuum. In (**b**), the circles indicate pores, the arrows – particles of Cu_2O. (Reprinted from Collet et al. [120]. Copyright (2016) with permission from Elsevier)

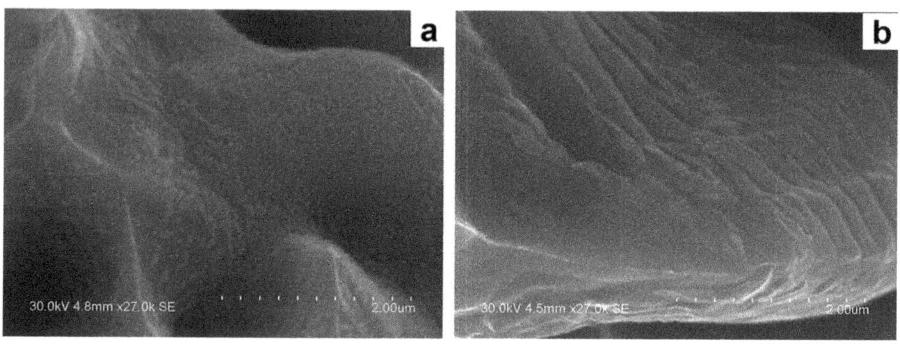

Fig. 10.43 The surface of ligaments of porous copper obtained by SPS of an electrolytic copper powder in contact with graphite foil (12 MPa, 650 °C) in a layer adjacent to the foil (**a**) and the surface of ligaments of porous copper obtained by cold-pressing of the powder and annealing in a flow of argon at 750 °C (**b**)

plates as well as supporting SiC rods. The latter were not in direct contact with any of the tungsten parts. The process of deposition was explained by evaporation of tungsten oxides present on the initial tungsten wires and plates and their reduction by carbon monoxide CO present in the SPS chamber resulting in the formation of particles of metallic tungsten.

Very fine crystallites were observed on the surface of the ligaments of spark plasma sintered porous copper in the areas of the compact that had been in contact with graphite foil during consolidation and were assigned to metallic copper formed by reduction of the surface oxide (Fig. 10.43a) [123]. Comparative annealing experiments conducted in argon, during which no carbon was introduced into the furnace environment, did not reveal any precipitates on the copper ligaments (Fig. 10.43b). In a dense Cu-based compact, oxide reduction during SPS occurred in a very thin layer adjacent to the graphite foil. When a partially oxidized copper

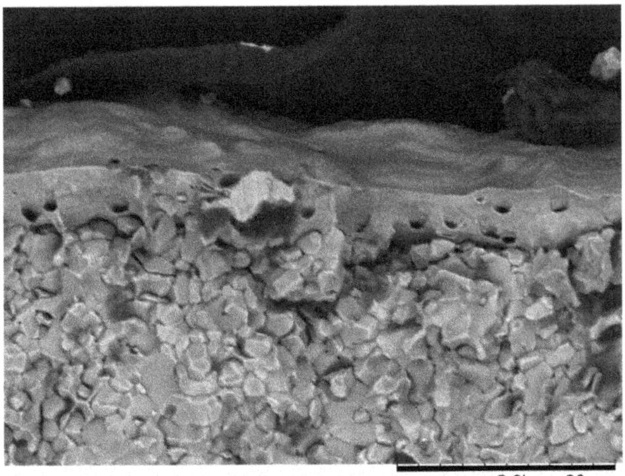

Fig. 10.44 Fracture surface of the compact spark plasma sintered from a partially oxidized copper powder of the Cu-50 vol.%Cu_2O composition at 900 °C and 40 MPa in contact with graphite foil. A layer of the metal reduced from the oxide is seen adjacent to the graphite foil

powder of the Cu-50 vol.%Cu_2O composition was spark plasma sintered at 40 MPa and 900 °C, a continuous layer of the metal reduced from the oxide formed on the surface of the specimen (Fig. 10.44).

Dudina and Bokhonov [124] addressed the issue of cleaning of metallic powders from oxide films during SPS on the scale of the size of the sintered compact, i.e., in the material volumes comparable in size to the sintered object. Using SPS experiments with a partially oxidized nickel powder and different configurations of the SPS assemblies providing the contact of the powder sample with either graphite or copper foil or excluding any direct contact with graphite (annealing of the powder placed on the flat end of a graphite punch covered with a layer of hexagonal boron nitride BN), it was shown that a direct contact with a source of carbon is critical for elimination of NiO from the sintered material on the sample size scale. The O/Ni atomic ratio in the initial (non-oxidized) nickel powder was 0.05, and in the oxidized powder, this ratio was increased to 0.20 (Table 10.4). The non-oxidized carbonyl nickel powder is gray. After partial oxidation, the powder acquired a black color. After the partially oxidized nickel powder had been annealed in the SPS chamber at 800 °C using an assembly without the upper punch (Fig. 10.45a), a change in color was observed: the surface of the powder layer turned gray. In that experiment, a layer of the powder about 1 mm thick was placed on the flat end of the punch coated with a boron nitride layer. The O/Ni ratio measured on the gray surface was 0.02, which is lower than in the initial non-oxidized nickel powder. When the powder bed was stirred with a glass rod, it was found that the interior of the sample was still black. The O/Ni measured on the black powder was 0.07 (Table 10.4). The surface morphology of the particles in the black areas was different from that of the particles found in the gray areas: the former showed fine grains of the oxide phase covering the surface (Fig. 10.46a, b), while the latter had smooth surfaces (Fig. 10.46c, d). This experiment showed that when the powder is heated in the SPS chamber but is

Table 10.4 O/Ni atomic ratio in the powders and sintered compacts determined by energy-dispersive spectroscopy

Powder/compact	O/Ni	Area analyzed by EDS
Carbonyl nickel powder	0.05	Surface of a single particle
Partially oxidized carbonyl nickel powder	0.20	Surface of a single particle
Partially oxidized carbonyl nickel powder annealed in the SPS chamber at 800 °C using an assembly without the upper punch (the powder was placed on the flat end of the lower punch coated with a BN layer)		
• Gray "undisturbed" surface of the powder bed	0.02	Surface of a single particle
• Inner part of the powder bed (black)	0.07	Surface of a single particle
Compact sintered by SPS from the partially oxidized carbonyl nickel powder (sintering in contact with graphite foil over the surface of the compact, 10 MPa, 800 °C)	0.02	30 μm × 40 μm (fracture surface of the compact)
Compact sintered by SPS from the partially oxidized carbonyl nickel powder (sintering in contact with circles of copper foil at the flat ends and graphite foil over the cylindrical surface, 10 MPa, 800 °C)		
• Center of the cross section of the compact	0.15	30 μm × 40 μm (fracture surface of the compact)
• Rim areas (near the cylindrical surface of the sintered disk)	0.04	30 μm × 40 μm (fracture surface of the compact)

Reprinted from Dudina and Bokhonov [124]. Copyright (2016) with permission from Elsevier
The spectra were recorded from powder layers or fracture surfaces of the compacts

not in direct contact with graphite, the reduction of the oxides covering the powder particles is limited to the surface layers of the powder bed. Reduction was obviously due to the reducing character of the atmosphere of the SPS chamber – the presence of carbon monoxide CO forming upon heating of graphite tooling and spacers in forevacuum.

The O/Ni ratio in a 2-mm-thick compact sintered from the partially oxidized nickel powder at 800 °C for 3 min in contact with graphite foil over the surface was 0.02, which is lower than in the initial non-oxidized carbonyl nickel powder (Table 10.4). Comparative experiments were conducted with graphite and copper foils using a configuration shown in Fig. 10.45b. The fracture surface of the compact sintered in contact with graphite foil was gray, while that of the compact sintered under the same conditions but in contact with copper foil at the flat ends was black except for the rim areas that were in contact with the graphite foil lining the wall of the die (Fig. 10.47). The following chemical reactions take place as carbon diffuses into the compact:

$$\mathrm{NiO} + \mathrm{C} = \mathrm{Ni} + \mathrm{CO(gas)};$$
$$2\mathrm{NiO} + \mathrm{C} = 2\mathrm{Ni} + \mathrm{CO_2(gas)}.$$

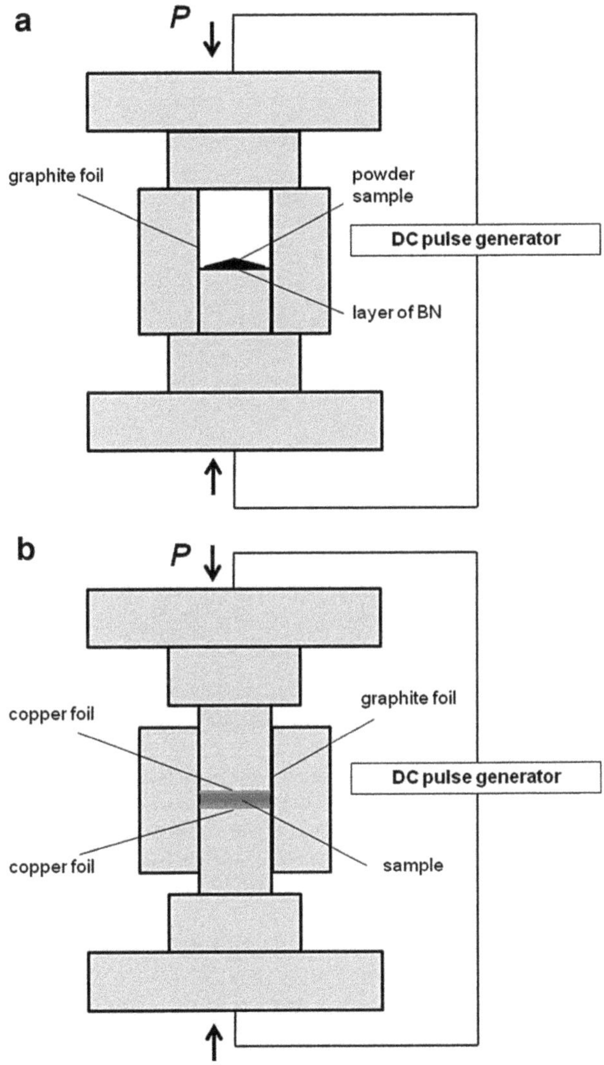

Fig. 10.45 Sintering and annealing of the partially oxidized nickel powder using a SPS assembly: (**a**) annealing on the flat end of the punch and (**b**) sintering under pressure in contact with copper foil at the flat ends of the compact and graphite foil along the cylindrical surface. (Reprinted from Dudina and Bokhonov [124]. Copyright (2016) with permission from Elsevier)

Calculations showed that the amount of carbon contained in the circles of graphite foil 250 μm thick placed between the flat end of the punch and the sample at both sides of the sample is approximately twice the amount that is needed to fully reduce NiO contained in the partially oxidized nickel powder having O/Ni(at.) = 0.2 for a 2-mm-thick sample. In the compact sintered in contact with graphite foil, interparticle necking was more pronounced (Fig. 10.48a, b) than in the compact sintered in contact with copper foil at the flat ends (Fig. 10.48c, d). In the former, the fracture surface of the necks showed dimples characteristic of ductile fracture. These microstructural features were also found in the rim areas of the compact sintered in contact

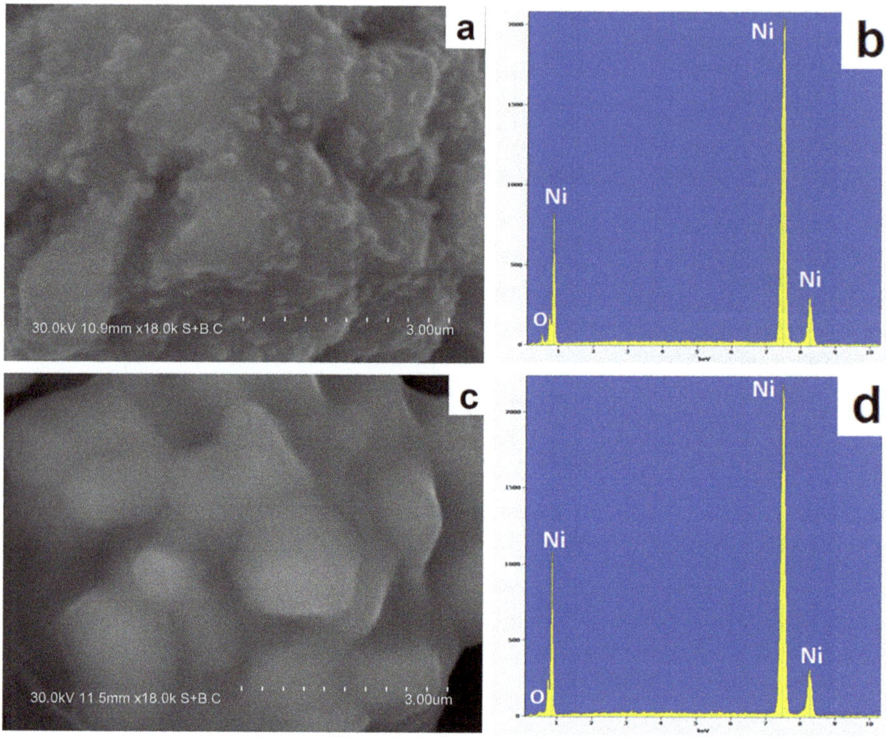

Fig. 10.46 (**a**) The surface of the powder particles located in the interior of the powder bed that was placed on the flat end of the lower punch coated with a boron nitride layer; (**b**) corresponding EDS spectrum from the particle surface; (**c**) the surface of the powder particles located in the upper layer of the powder bed; (**d**) corresponding EDS spectrum from the particle surface. (Reprinted from Dudina and Bokhonov [124]. Copyright (2016) with permission from Elsevier)

Fig. 10.47 Fractured compact spark plasma sintered from the partially oxidized carbonyl nickel powder, sintering in contact with circles of copper foil at the flat ends and graphite foil over the cylindrical surface of the compact, sintering at 10 MPa and 800 °C. (Reprinted from Dudina and Bokhonov [124]. Copyright (2016) with permission from Elsevier)

Fig. 10.48 (**a**) Fracture surface of the compact sintered from the partially oxidized nickel powder in contact with graphite foil fully covering the surface of the compact, (**b**) corresponding EDS spectrum; (**c**) fracture surface of the compact shown in Fig. 10.47, center of the compact, (**d**) corresponding EDS spectrum; (**e**) fracture surface of the compact shown in Fig. 10.47, rim area, (**f**) corresponding EDS spectrum. The EDS spectra were taken from the areas shown in the images. (Reprinted from Dudina and Bokhonov [124]. Copyright (2016) with permission from Elsevier)

with copper foil (Fig. 10.48e, f), as those were in contact with graphite foil lining the wall of the die.

When metal powders contain oxide surface films and the purpose of consolidation is obtaining oxygen-free metallic material, the reduction of the oxide films by carbon diffusing into the sample during sintering is an advantage. The results obtained in Ref. [124] show that when using graphite foil during SPS, it is possible to start from an oxidized powder of a metal, in which carbon has appreciable solubility, and produce oxide-free metallic compacts provided the sintering temperature and time are sufficient for the carbon to diffuse into the sample at a desired depth and the amount of carbon as a reactant is enough to reduce the oxides contained in the powder. These results have important implications for the powder metallurgy industry: one can ease the handling requirements or synthesis/processing conditions of the powders and lower the associated costs. In a single SPS operation, a partially oxidized powder can be reduced and consolidated into a compact.

The influence of carbon on oxide materials during SPS is detrimental, as the interaction of oxides with carbon leads to deterioration of optical transparency [125, 126]. Bertrand et al. [125] spark plasma sintered an amorphous powder of the $85TeO_2$–$15WO_2$ composition with and without carbon diffusion barriers and with and without a pressureless sintering operation prior to SPS. Even with the carbon diffusion barriers, black spots were found in the glassy matrix (Fig. 10.49a), which were proven to correspond to carbon. The formation of carbon was explained by the following reaction:

$$2CO(g) = CO_2(g) + C(s)$$

taking place inside the pores during densification of the compact. It was found that pressureless sintering prior to SPS can reduce carbon contamination during the SPS step. The Raman spectrum of the sample obtained without the pressureless sintering step showed characteristic peaks of carbon – D-band and G-band – while these were absent in the spectrum of the sample pre-sintered before the SPS (Fig. 10.49b). Experiments with fine and coarse particles showed that carbon contamination was greater in the compacts sintered from the former.

Graphite particles were observed by Bernard-Granger et al. [126] in the volume of the $MgAl_2O_4$ compacts formed by slip casting and further processed by SPS. Pre-compaction of the samples excluded incorporation of carbon pieces into the sample during the procedure of placing the specimen into the graphite die. However, graphite particles were found in the compact's interior, based on which the authors concluded that the source of carbon was carbon monoxide contained in the SPS chamber.

Morita et al. [127] found that the carbon contamination of $MgAl_2O_4$ processed by SPS is independent of temperature. This allowed suggesting evaporation of carbon from the graphite foil or the graphite die during the heating process rather than the solid-state diffusion of carbon as the mechanism of carbon incorporation into the specimens. The evaporated carbon was encapsulated in the closed pores during sintering. Evaporation was enhanced by rapid heating during SPS. Morita et al. [128] found that in the $MgAl_2O_4$ specimen produced by SPS at a high heating rate (100 °C min^{-1}), the carbon contamination originating from the graphite foil and graphite die was present throughout the specimen volume, while in the specimen produced at a low heating rate (10 °C min^{-1}), carbon contamination of this origin

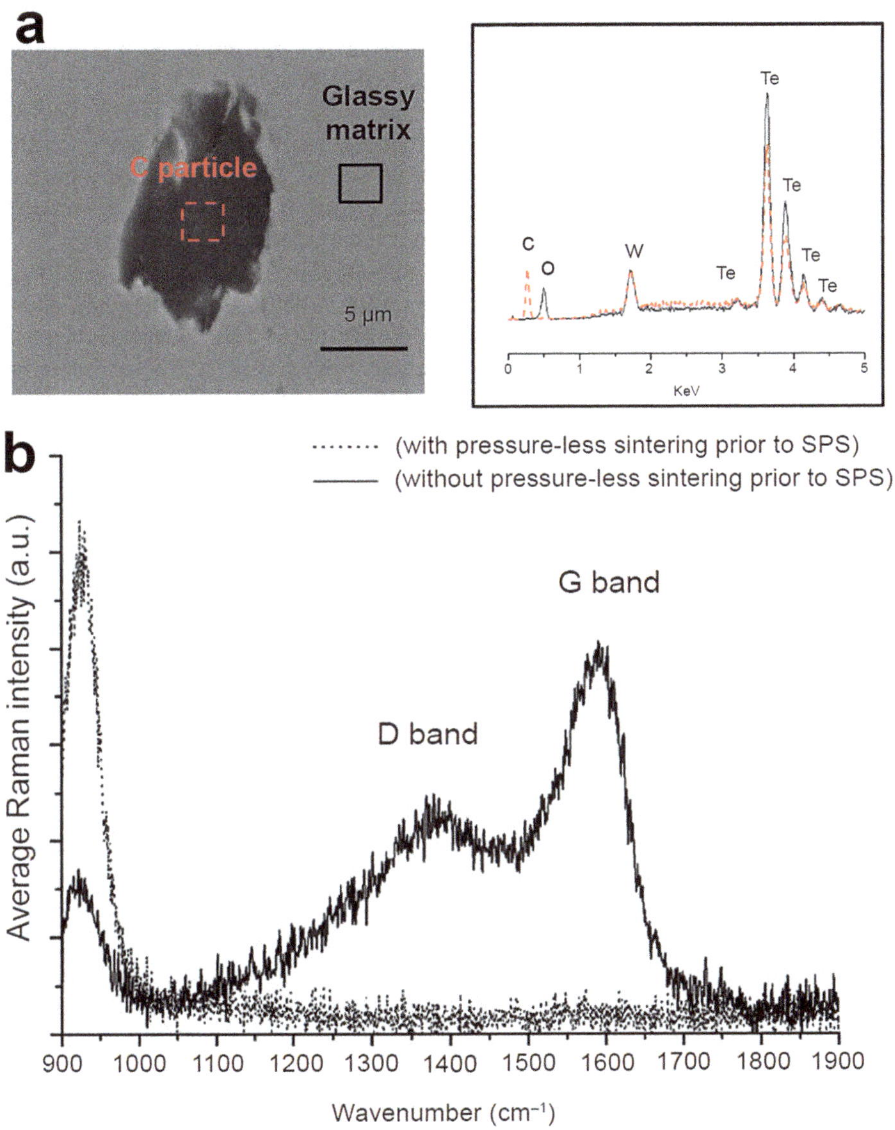

Fig. 10.49 SEM image and EDS analysis of a carbon particle embedded in the glassy $85TeO_2$–$15WO_2$ matrix (**a**) and Raman spectra of the $85TeO_2$–$15WO_2$ compacts spark plasma sintered with and without a pressureless sintering step before SPS (**b**). (Reprinted from Bertrand et al. [125]. Copyright (2013) with permission of John Wiley & Sons)

was present only on the surface of the compact. They also emphasized that post-sintering annealing of oxide ceramics after SPS may not fully eliminate the carbon contamination, if carbon just oxidizes and stays in the form of CO and CO_2 in the closed pores. Therefore, it is vital to reduce the carbon contamination during the SPS process itself.

Partial reduction of oxides in the SPS environment can lead to the formation of oxygen vacancies. Jiang and Mukherjee sintered sol–gel produced Y_2O_3–MgO nanocomposite powders and obtained a nanograined compact [129]. Due to oxygen losses, the crystalline lattices of both phases experienced contraction, as indicated by decreased lattice parameters determined from the shift of the positions of the corresponding peaks on the XRD patterns (Fig. 10.50). Low infrared transmission of the sintered Y_2O_3–MgO nanocomposite was greatly improved when the oxygen content was restored by annealing the compact in air after the SPS (Fig. 10.51).

The reducing environment of the SPS is favorable for conducting reduction reactions to form new phases and nanocomposite structures and tailor the phase composition and microstructure of the surface of the sintered specimens. Isobe et al. [130] showed that nickel nanoparticle-toughened alumina (Fig. 10.52) can be in situ formed from co-precipitated Al_2O_3 and NiO oxides by reduction of NiO during SPS. The XRD patterns of the composites presented in Fig. 10.53 confirm the presence of metallic nickel that was reduced from its oxide. An increase in the fracture toughness of the composites was observed with increasing Ni content (Fig. 10.54).

Magnetic FePt/Fe_3Pt nanocomposites were produced by SPS of solution synthesized FePt and Fe_3O_4 nanoparticles [131]. The Fe_3Pt phase formed as a result of the in situ reduction of Fe_3O_4 during SPS. With increasing SPS temperature, the oxygen content in the composites decreased. Consequently, the reducing action of the SPS environment can be enhanced by increasing the SPS temperature.

Another example of an advantageous in situ reduction of oxides occurring simultaneously with densification during SPS was demonstrated in Ref. [132]: the in situ reduction was used as a basis for developing graphene–ceramic composites using graphene oxide as a starting material.

Lia et al. [133] proposed an approach to solving the problem of carbon losses (decarburization) in thermally sprayed WC–Co coatings using the potential of SPS as a process of thermal treatment in the presence of carbon and carbon monoxide. When the WC–Co coatings suffer from carbon losses, their abrasive wear resistance decreases. Post-spray SPS treatment was conducted on plasma-sprayed WC–Co coatings and resulted in the restoration of tungsten carbide WC through the reaction of W_2C or W with carbon. After 6 min of SPS treatment at 800 °C, WC was the dominant phase in the coatings. The restoration of carbon in the SPS-treated coatings resulted in a 40% increase in the microhardness relative to the as-sprayed coatings. Under the selected conditions, carbon restoration was limited to a thickness of less than 10 μm.

The surface of the spark plasma sintered specimen may differ in composition from that of the core if the material is sensitive to the reducing environment of the SPS chamber. The development of the zoned structure is due to the easy removal of gaseous reaction products from the surface layer of the sintered specimen and a direct contact of the sintered material with the graphite die. During SPS of the α-$Al_{1.86}Fe_{0.14}O_3$ solid solution of corundum structure, in the surface layer of the specimen, the $FeAl_2O_4$ spinel formed along with small amounts of Fe according to the following reactions [134]:

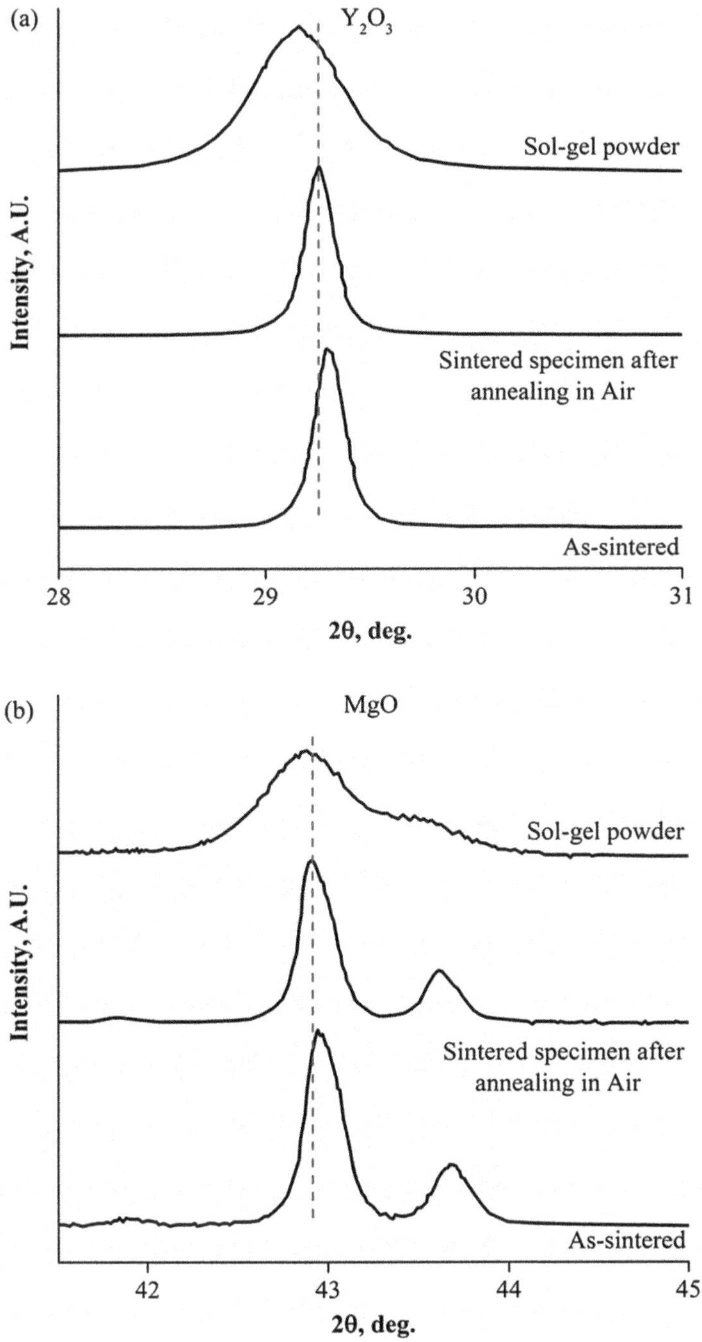

Fig. 10.50 XRD peaks of the oxide phases contained in the Y_2O_3–MgO nanocomposites: synthesized powder, compact sintered by the SPS and compact annealed in air after the SPS. Note the shift of peak positions of both oxides. Y_2O_3 peak (**a**), MgO peak (**b**) (Reprinted from Jiang and Mukherjee [129]. Copyright (2011) with permission from Elsevier)

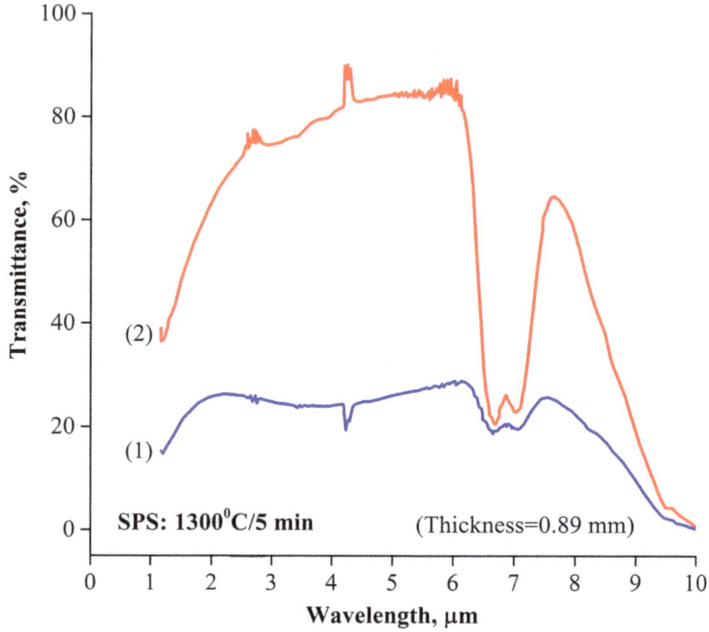

Fig. 10.51 Transmission spectrum of spark plasma sintered Y_2O_3–MgO nanocomposite (1) and after annealing in air (2). (Reprinted from Jiang and Mukherjee [129]. Copyright (2011) with permission from Elsevier)

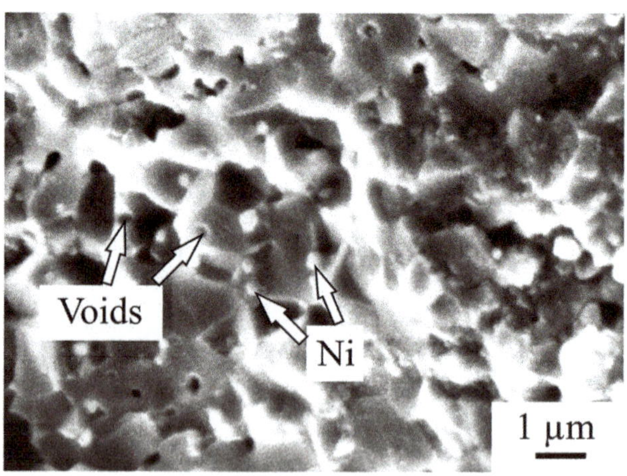

Fig. 10.52 Microstructure of the Al_2O_3–Ni composite produced by SPS of Al_2O_3–NiO powders (Ni nanoparticles formed in situ as a result of reduction of NiO during SPS). (Reprinted from Isobe et al. [130]. Copyright (2006) with permission from Elsevier)

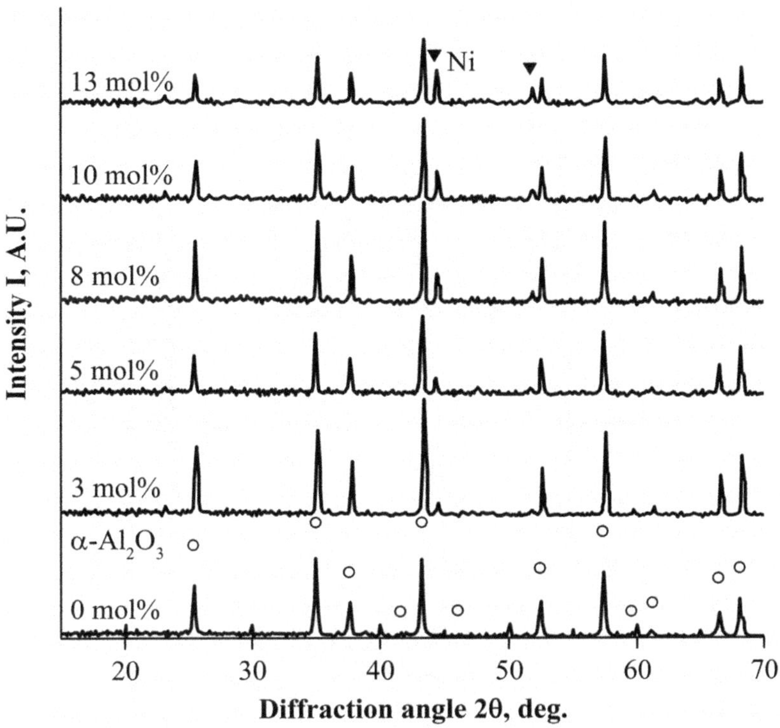

Fig. 10.53 XRD patterns of Al_2O_3–Ni composites produced by SPS of Al_2O_3–NiO powders, 1350 °C, 6 min (Ni nanoparticles formed in situ as a result of reduction of NiO during SPS). (Reprinted from Isobe et al. [130]. Copyright (2006) with permission from Elsevier)

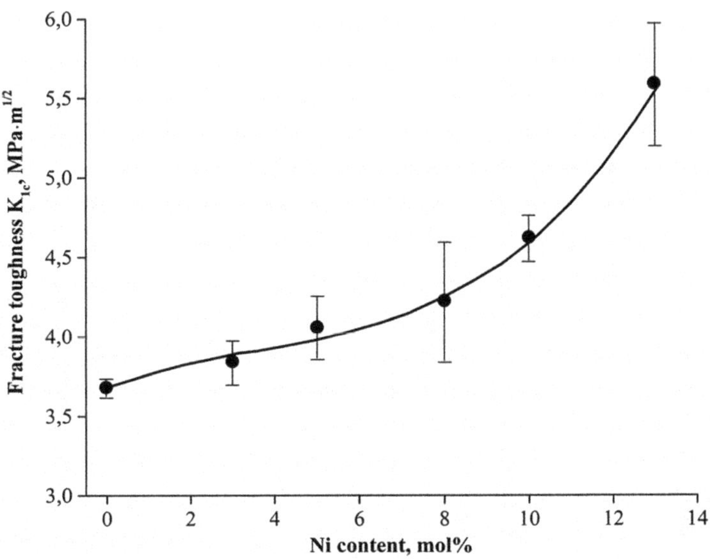

Fig. 10.54 Fracture toughness of Al_2O_3–Ni composites (Ni nanoparticles formed in situ as a result of reduction of NiO during SPS). (Reprinted from Isobe et al. [130]. Copyright (2006) with permission from Elsevier)

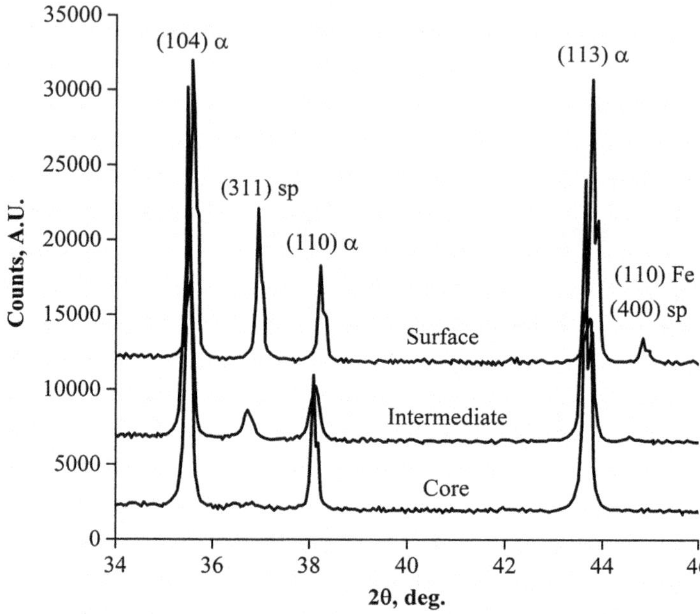

Fig. 10.55 XRD patterns taken from different part of the specimen obtained by SPS of $Al_{1.86}Fe_{0.14}O_3$ powder: α, corundum structure; sp, spinel. (Reprinted from Gurt Santanach et al. [134]. Copyright (2008) with permission from Elsevier)

$$Al_{2-2x}Fe_{2x}O_3 \rightarrow 2xFeAl_2O_4 + (1-3x)Al_2O_3 + x/2O_2$$

and

$$FeAl_2O_4 \rightarrow Fe + Al_2O_3 + 1/2O_2.$$

In the intermediate zone of the specimen, the spinel content was lower than in the surface layer. In the core of the specimen, the spinel phase was not detected by the XRD (Fig. 10.55). The surface layer having a nanocomposite structure was separated from the rest of the specimen by a sharp boundary. It was suggested that this structure is a result of electric current more easily passing along the surface of the specimen. By partial reduction of the α-$Al_{1.86}Fe_{0.14}O_3$ solid solution prior to SPS, the reduction yields could be increased. By forming a nanocomposite with significant concentrations of Fe and $FeAl_2O_4$ during the SPS, it was possible to obtain a material with high hardness and fracture strength. Further studies of SPS of the solid solutions of this type revealed that a Fe-depleted layer can form between the core of the sample and its surface layer due to electric current-induced diffusion of Fe^{3+} toward the upper punch, which is cathode in a SPS setup [135].

Alternative materials for the SPS dies are being sought. One of the possible solutions is conducting SPS in SiC dies, as was suggested in Refs. [136, 137]. The experiments proved to be successful provided additional heating is imposed on a SiC

Fig. 10.56 Flat end of a porous iron compact spark plasma sintered in contact with copper foil (pressureless SPS) after the foil has been torn off: (**a**) 600 °C and (**b**) 800 °C

die (from external heaters) in the beginning of a pulsed current sintering process to increase the electrical conductivity of SiC. As far as substitutes for graphite foil are concerned, none of the so far attempted materials has proven satisfactory for the needs of SPS. The protective foil should be electrically conductive not to cause disruption of electric current. At the same time, it should be easy removable from the surface of the compact after the sintering operation is complete. Metallic foils tend to adhere to the sintered compacts. In Refs. [138, 139], tantalum and copper foils were used during the SPS processing of metallic materials and metal-containing composites. Severe sticking of tantalum foil to Ni–diamond compacts [138] and copper foil to iron compacts [139] was observed. In contact with Ni–diamond composites, tantalum foil reacted with carbon to form TaC. With increasing sintering temperature, the contacts developed between the copper foil and the iron particles grew in size (Fig. 10.56). Copper remained on the surface of iron particles in the form of islands after the foil had been removed from the compact by tearing (Fig. 10.57), which indicated alloying at the Fe/Cu interface.

10.17 Selected Examples of Materials with Improved Properties Achieved by Reactive SPS. Syntheses in Non-conventional Assemblies and from Reactants of Unusual Morphology

When nanocomposites are synthesized, the in situ route is often claimed to be a better choice to obtain a fine-grained product. However, only a few reports are available on the direct comparison of the sintered materials of the same composition produced by in situ and ex situ processes. Such a comparison has been made by Wang et al. [91] for spark plasma sintered TiN–Al_2O_3 composites. Submicron-

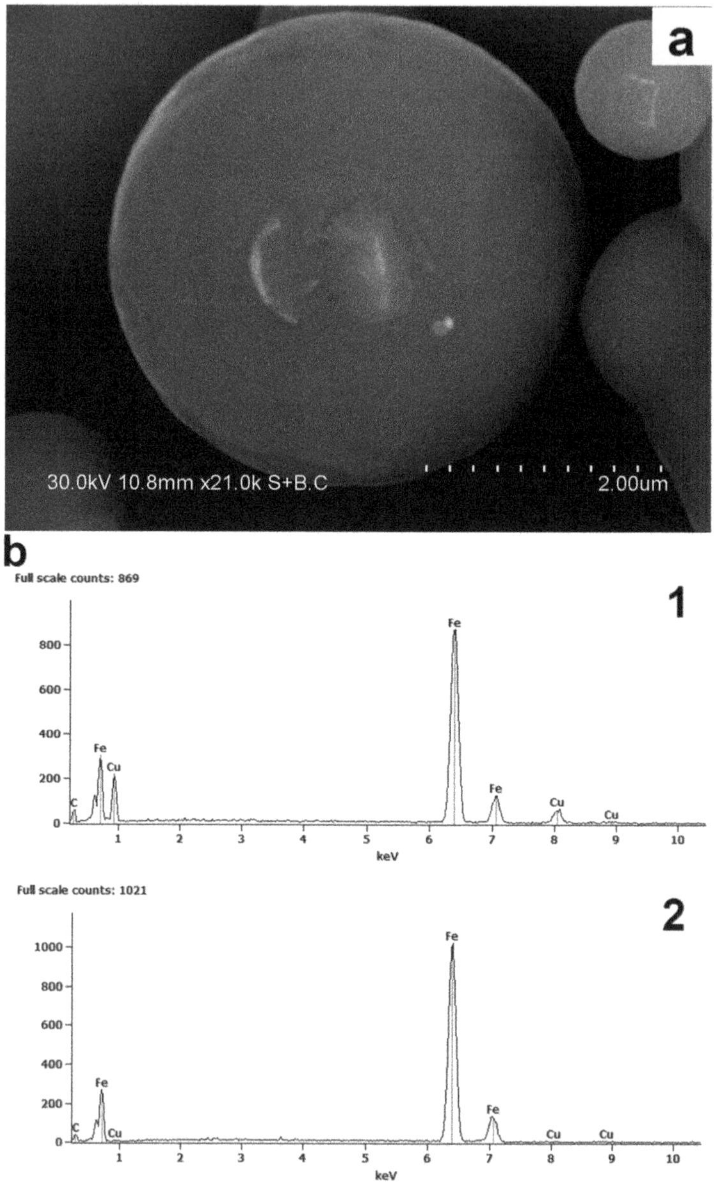

Fig. 10.57 (**a**) The surface of an iron particle of the porous compact spark plasma sintered in contact with copper foil at 800 °C (pressureless SPS) and (**b**) the elemental analysis of the points marked 1 and 2. Pieces of copper on the iron particles indicate adherence of the foil to the compact

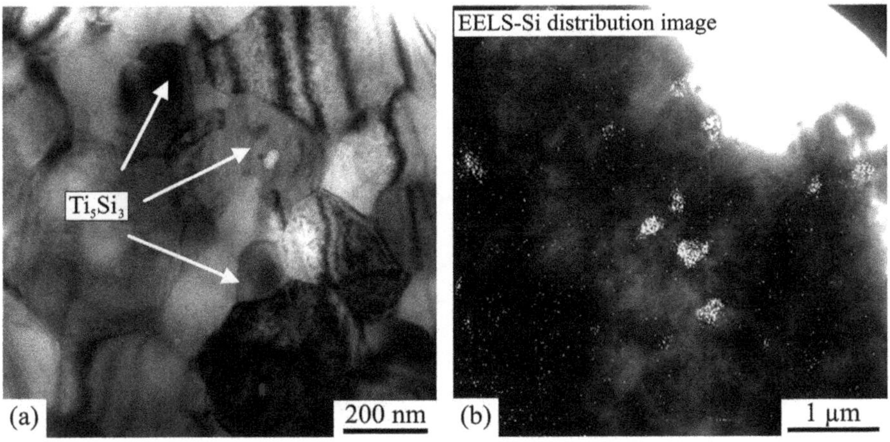

Fig. 10.58 (**a**) Transmission electron microscopy of Ti–Ti_5Si_3 composites produced by reactive SPS, (**b**) distribution of Si from the electron energy loss spectra (EELS). (Reprinted from Handtrack et al. [80]. Copyright (2006) with permission from Elsevier)

grained TiN–Al_2O_3 were synthesized from AlN, TiO_2, and Ti reactants, all three having particles from several micrometers to several tens of micrometers; in parallel, composites of the same composition were fabricated by sintering of TiN and Al_2O_3 submicron powders. The spark plasma sintered composites had similar microstructures. In terms of bending strength, the in situ-produced composite was superior, but the ex situ composite was harder and tougher.

In metal matrix composites, it is advantageous to have the same metal playing the roles of a matrix and a reactant to form a ceramic reinforcement; otherwise there remains a concern to establish a direct contact between the reactants distributed in a matrix. Examples of the synthesis of reinforcements using the matrix metal are the synthesis of titanium monoboride TiB in a titanium matrix [140, 141], cementite Fe_3C dispersoids in an iron matrix [142], and Ti_5Si_3 nanoparticles in a titanium matrix [80]. Rapid sintering by SPS allowed keeping fine Ti_5Si_3 dispersoids distributed in the titanium matrix [80]. The distribution uniformity of silicon was key to the formation of fine particles of Ti_5Si_3 in the sintered composites (Fig. 10.58). These particles stabilized the ultrafine-grained structure of the titanium matrix, as smaller grains were found in composites with higher silicon content (Fig. 10.59). Under optimized conditions of SPS, the reinforcement particles were in the range of 100–200 nm, while the Ti matrix grains remained in the submicron range of 200–400 nm. This unique microstructure of the composite is responsible for attractive mechanical properties of the composite comparable in terms of strength and even better in terms of wear resistance to the widely known Ti-6Al-4V alloy.

In a number of studies, attractive properties of single-phase ceramics and ceramic nanocomposites synthesized by reactive SPS were reported. TiB monolithic nearly fully dense compacts with needle-shape grains showed a fracture toughness of 5.9 ± 0.4 MPa·m$^{1/2}$ [143]. Ceramic nanocomposites and composites with

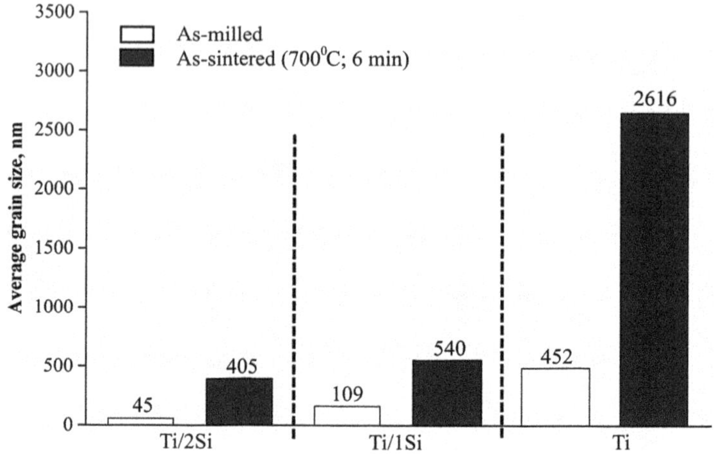

Fig. 10.59 The dependence of the grain size of the Ti matrix in the Ti–Ti_5Si_3 nanocomposites on Si content (0, 1, 2 wt.%) in the initial mixtures, SPS temperature 700 °C, and holding time 6 min. (Reprinted from Handtrack et al. [80]. Copyright (2006) with permission from Elsevier)

submicron-sized grains demonstrated excellent combinations of properties: a fracture toughness of 6.5 MPa·m$^{1/2}$ and a microhardness of 20.6 GPa in TiN–TiB_2 [144], a flexural strength of 865 MPa and a hardness of 39.3 GPa in B_4C–TiB_2 [145], a fracture toughness of 6.7 MPa·m$^{1/2}$, and metallike electrical conductivity in Si_3N_4/SiC/$TiC_{0.3}N_{0.7}$ [87, 88]. Nano-organized $MoSi_2$ synthesized and consolidated in the SPS exhibited better oxidation resistance than its microcrystalline analogue and a fracture toughness of 5.8 ± 0.4 MPa·m$^{1/2}$ [146].

Due to the potential of low-temperature fast sintering, SPS has become a method of choice for dealing with materials sensitive to temperature. Campayo et al. [147] and Le Gallet et al. [148] suggested SPS processing for iodine-containing apatite. A high chemical stability of iodine-containing apatite can be used for immobilizing radioactive iodine. Lead phosphovanadate is a reactant for the synthesis of iodoapatite. The synthesis and consolidation of iodine-containing compounds should be performed at temperatures low enough to avoid volatilization of iodine. Le Gallet et al. [148] examined the applicability of SPS for both sintering of iodine-containing apatite powders and in situ synthesis of apatite from a mixture of lead iodide and lead phosphovanadate and concluded on the advantages of the reactive process in terms of a more uniform and finer structure of the densified material. Figure 10.60 shows the cross-sections of the iodoapatite compacts obtained via nonreactive and reactive sintering routes. The reactive route was based on the following reaction:

$$3Pb_3(VO_4)_{2(1-x)}(PO4)_{2x} + PbI_2 \rightarrow Pb_{10}(VO_4)_{6(1-x)}(PO4)_{6x}I_2.$$

As sintering has to be performed at low temperatures due to decomposition of $Pb_{10}(VO_4)_{6(1-x)}(PO4)_{6x}I_2$ at 500 °C, there is a density gradient along the axis of the compact with denser regions located closer to the surface and more porous regions in

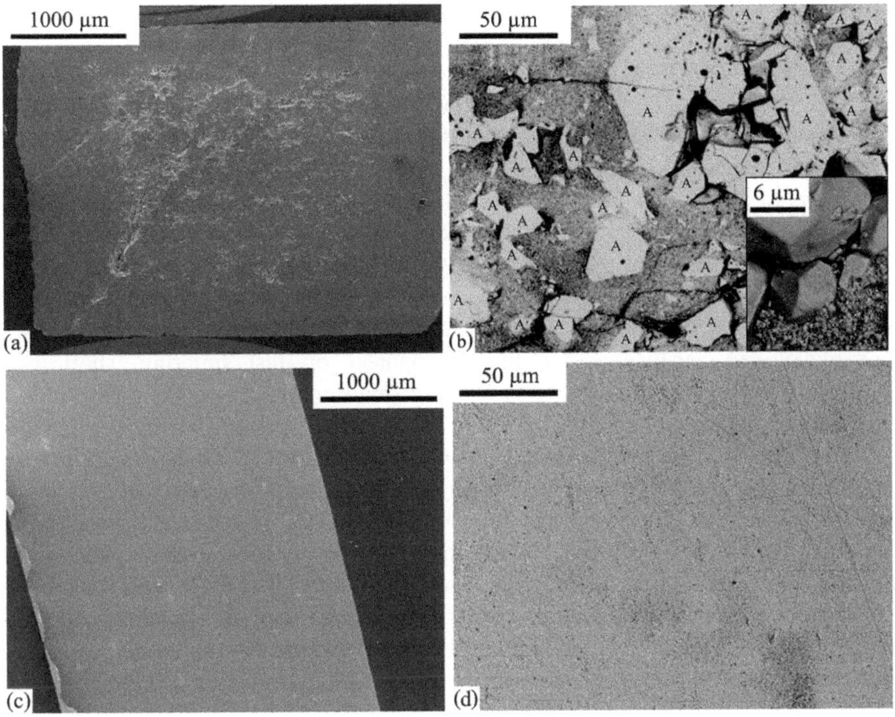

Fig. 10.60 Microstructure of the spark plasma sintered $Pb_{10}(VO_4)_{6(1-x)}(PO4)_{6x}I_2$ produced by nonreactive sintering (**a**, **b**) and by performing the reaction during the SPS (**c**, **d**). Grains formed via abnormal grain growth are marked by letter "A" in (**b**). (Reprinted from Le Gallet et al. [148]. Copyright (2010) with permission from Elsevier)

the core of the compact. Abnormal grain growth was characteristic of the nonreactive processing, which hindered the formation of a dense, uniform, and fine-grained structure. The highest density achieved was 92%. Unlike nonreactive sintering, the reaction performed in situ during the SPS yielded densities higher than 96%, which implied closed porosity needed for the practical applications to reduce the surface area of the consolidated $Pb_{10}(VO_4)_{6(1-x)}(PO4)_{6x}I_2$ in contact with the environment. The dramatic difference in the consolidation behavior of the two material systems was explained by participation of liquid phases formed on the basis of PbI_2 (and probably PbO forming a eutectic with PbI_2) in sintering in the case of the reactive processing. This example shows that although liquid-phase sintering is usually associated with conventional processing and enhanced grain growth, conditions can be found, under which the liquid phase facilitates fast densification but does not lead to undesired grain growth. These conditions can be reached when the liquid phase is formed by one of the reactants and not by the product.

In the synthesis of compounds, there is always a challenge to use reactants with large differences in their melting points or vapor pressures. Because of these differences, the synthesis has to be conducted at low temperatures and, consequently, under conditions of slow diffusion. The synthesis can still be successful if nonthermal enhancement of the reaction is enabled. Beekman et al. [149] proposed the SPS technique as a means of dealing with such challenging synthesis situations and demonstrated the applicability of a SPS apparatus for the synthesis and crystal growth of intermetallic clathrates, such as $Na_{24}Si_{136}$. The grown crystals were 200–300 μm in size and did not contain defects typical for the compound produced by thermal decomposition of Na_4Si_4. The role of the applied field was in oxidation of Si_4^{4-} cluster anions at the anode followed by the formation the clathrate framework. This study pointed to a high potential of SPS facilities for the crystal growth processes. The crystal growth-oriented area of the SPS research remains significantly underdeveloped because of the initial emphasis on fast sintering, short total processing time, and a possibility of keeping small grain sizes as the major advantages of this technique.

A decomposition reaction was used by Chakravarty et al. [150] to produce porous alumina by the SPS. Decomposition of $Al(OH)_3$ occurred in situ at the early stages of heating during the SPS cycle and was followed by the phase transformations in alumina. Figure 10.61 shows that the in situ route gave better combinations of porosity and strength than the route used by Oh et al. [151], who sintered a commercially available submicron α-alumina powder without any chemical reactions involved. This comparison should be, however, considered with care, as the authors of these publications used different mechanical property testing methods.

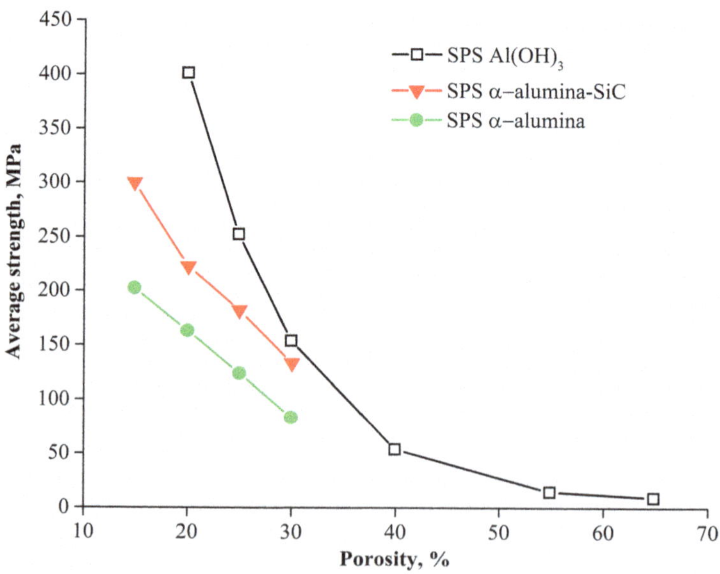

Fig. 10.61 The dependence of strength of the porous alumina compacts on the porosity according to Chakravarty et al. [150]; alumina is synthesized in situ during SPS from $Al(OH)_3$ (comparison is made with the ex situ produced porous α-alumina and Al_2O_3–SiC composites obtained by Oh et al. [151]). (Reprinted from Chakravarty et al. [150]. Copyright (2008) with permission from Elsevier)

The main result of Chakravarty et al. [150] is the porosity levels of 20–30% combined with high strength of the compacts, which are suitable for practical applications of porous alumina. High strength of the porous alumina compacts was explained by extensive necking during sintering accelerated by the exothermic transformation of θ-Al_2O_3 to α-Al_2O_3, and, therefore, it could not have been achieved without the use of the in situ reactive technique.

Initially designed for solid-state sintering, SPS facilities have been proven to be suitable for conducting solid-state chemical reactions [54, 55, 152]. SPS has been developed into a method of solid-state synthesis under applied electric current and mechanical pressure. Chemists consider SPS as a new tool, full capabilities of which are yet to be discovered [153, 154]. The successes of reactive SPS in synthesizing bulk nanostructured materials can be further extended to the simultaneous synthesis and joining of materials [155] as well as manufacturing of coatings [156, 157]. Joining of materials requires heat input to facilitate mutual diffusion. Joining of materials differing in the chemical nature becomes even more challenging when the grain size of the joint and the adjacent areas has to be preserved in the nano- or submicron range. In order to solve these tasks, fast cooling has to be applied to minimize grain growth. A joint of two materials can be obtained by sintering a pellet consisting of two layers. In this case, two porous layers are stacked together and sintered. The key to the formation of a strong bond between the layers is interdiffusion at the interface or chemical interaction resulting in the formation of compounds. Two layers to be joined can be formed by in situ chemical reactions, in which case the synthesis and joining occur simultaneously [155] (Fig. 10.62). The challenge in such situations is to perform two reactions simultaneously, for which a temperature and pressure window should be found for both reactions to be complete.

Reactive synthesis presents opportunities for making gradient microstructures [65, 66] including those of tubular geometries (Fig. 10.63) [158] by varying the temperature distribution in the reacting sample. The features of the obtained microstructures are gradients in the grain size and porosity (Fig. 10.64). Developed for the synthesis of boron carbide and boron carbide-based composites produced through

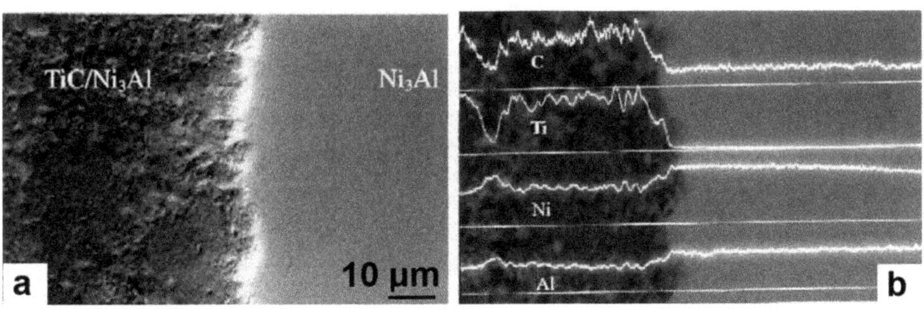

Fig. 10.62 Interface between Ni_3Al and TiC/Ni_3Al layers produced by the reactive SPS (**a**) and corresponding concentration profiles (**b**). (Reprinted from Liu and Naka [155]. Copyright (2003) with permission from Elsevier)

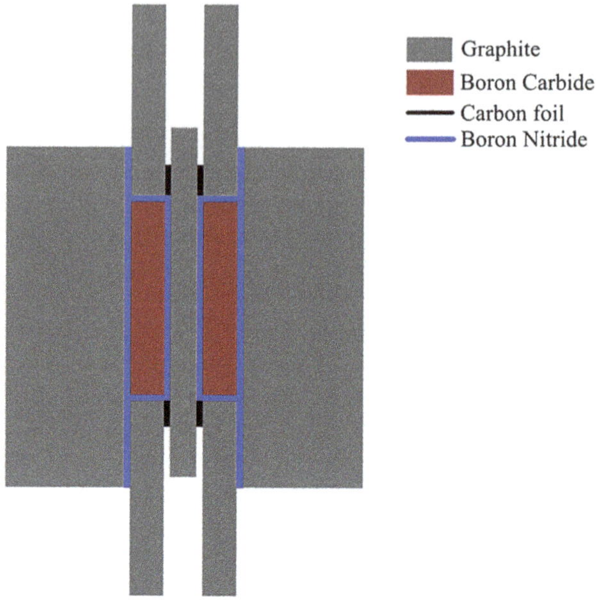

Fig. 10.63 A setup for producing tube–geometry B_4C specimens by SPS. (Reprinted from Holland et al. [158]. Copyright (2010) with permission of Elsevier)

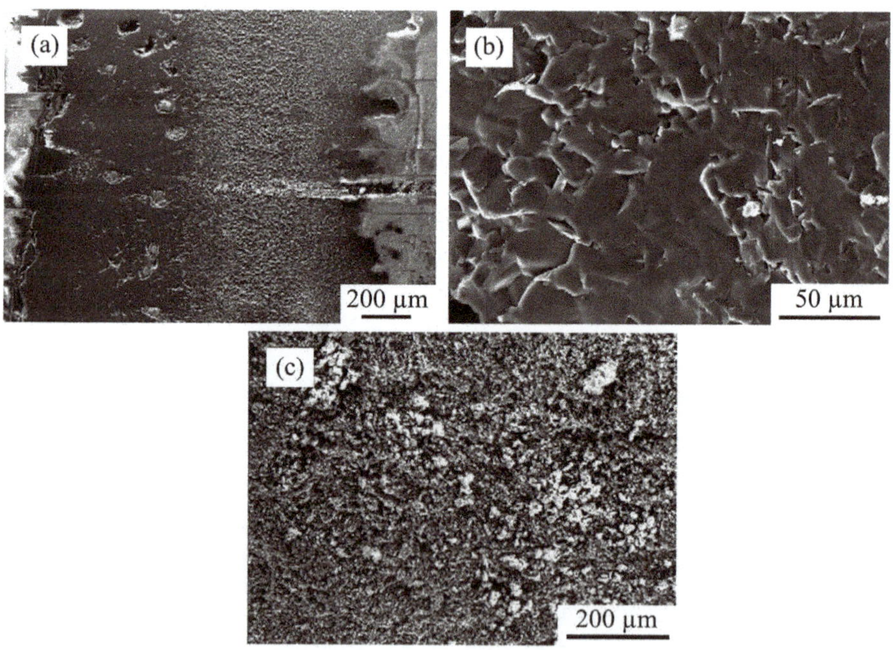

Fig. 10.64 Microstructure of different zones of the spark plasma sintered B_4C tube of a gradient structure: (**a**) general view of the cross section of the tube wall, (**b**) dense zone adjacent to the central rod (Fig. 10.63) during the SPS, and (**c**) porous zone adjacent to the die wall. (Reprinted from Holland et al. [158]. Copyright (2010) with permission of Elsevier)

Fig. 10.65 Porous FeAl synthesized by pressureless SPS at 900 °C from a mixture of iron and aluminum powders (open porosity 46%). (Reprinted from Dudina et al. [161]. Copyright (2017) with permission from Elsevier)

infiltration of the ceramic performs, the approach can be used for different ceramic and composite materials.

SPS treatment is suitable for the synthesis of products that are not densified; rather, they present porous and weakly bonded agglomerates [78]. When a low pressure or no pressure is applied to the mixture of reactants, porous ceramic [159] and intermetallic materials [160, 161] can also be obtained (in the form of robust compacts of desired shape). Dudina et al. [161] reported the synthesis of porous iron aluminide FeAl with high open porosities – 41–51% – by pressureless SPS of the mixtures of iron and aluminum powders at 800–900 °C. Highly porous compacts were obtained without any pore-forming agents. Figure 10.65 shows a porous structure of FeAl synthesized by pressureless SPS at 900 °C. Single-phase FeAl with an open porosity of 46% was obtained using a heating rate of 70 °C·min^{-1} and a holding time of 3 min at 900 °C. When the heating rate was lowered to 30 °C·min^{-1}, compacts with a higher open porosity (51%) were formed [162]. SPS conducted with insulating layers of boron nitride BN between the sample and the graphite punches showed that, even without electric current directly passing through the sample, high open porosities of FeAl could be achieved. Sintering in a hot press conducted in an argon atmosphere produced compacts with an open porosity lower than that of compacts obtained in the SPS facility. It was concluded that treatment of a loosely packed Fe–Al mixture in the SPS chamber with or without an electric current passing directly through the sample ensures a fast transformation of the reaction mixture into a highly porous FeAl product, the passage of electric current through the sample not being responsible for the preservation of high open porosities.

In recent years, particles of unusual morphology have attracted attention as interesting objects to be consolidated by SPS [163–165]. Scheele et al. [163] spark plasma sintered PbTe–PbSe core–shell nanostars preserving their morphology in the partially densified compacts, which, however, maintained mechanical integrity and stability. Bokhonov and Dudina [164] reported the results of pressureless SPS of

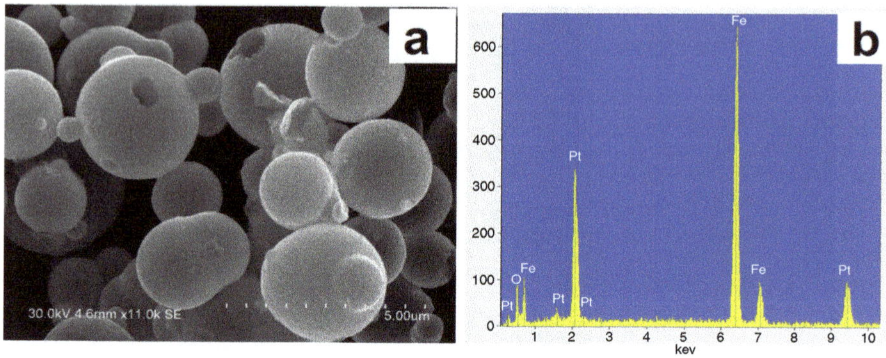

Fig. 10.66 SEM image (**a**) and EDS (**b**) of the Fe@Pt core–shell particles. (Reprinted from Bokhonov and Dudina [164]. Copyright (2016) with permission from Elsevier)

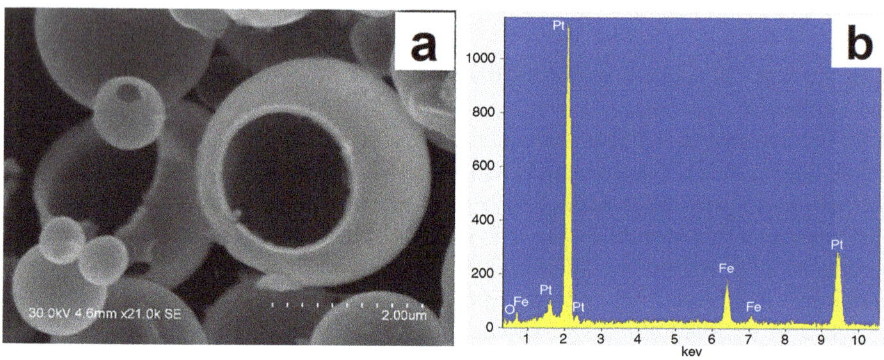

Fig. 10.67 SEM image (**a**) and EDS (**b**) of the Pt(Fe) hollow particles obtained by HCl treatment. (Reprinted from Bokhonov and Dudina [164]. Copyright (2016) with permission from Elsevier)

Fe@Pt core–shell particles obtained by a galvanic replacement reaction (Fig. 10.66) and Pt(Fe) hollow particles synthesized by dissolving iron from the core–shell particles in HCl solution (Fig. 10.67). SPS led to the formation of porous compacts and alloying between Fe and Pt. The porous material produced from the Fe@Pt via SPS followed by HCl treatment was composed of a FCC Fe-based solid solution, while the porous material produced via HCl treatment followed SPS was the Pt_3Fe intermetallic (Fig. 10.68). During pressureless SPS, a network formed from the core–shell (Fig. 10.69a) and hollow particles (Fig. 10.69b). In the porous compact obtained by sintering of the core–shell Fe@Pt particles followed by HCl treatment (Fig. 10.69c), the spherical morphology of the elements of the porous structure was better preserved than in the compact obtained by sintering of the hollow particles. This study demonstrated that acid treatment and SPS can be used together to tailor the elemental and phase composition and morphological characteristics of the sintered materials obtained from core–shell metallic particles.

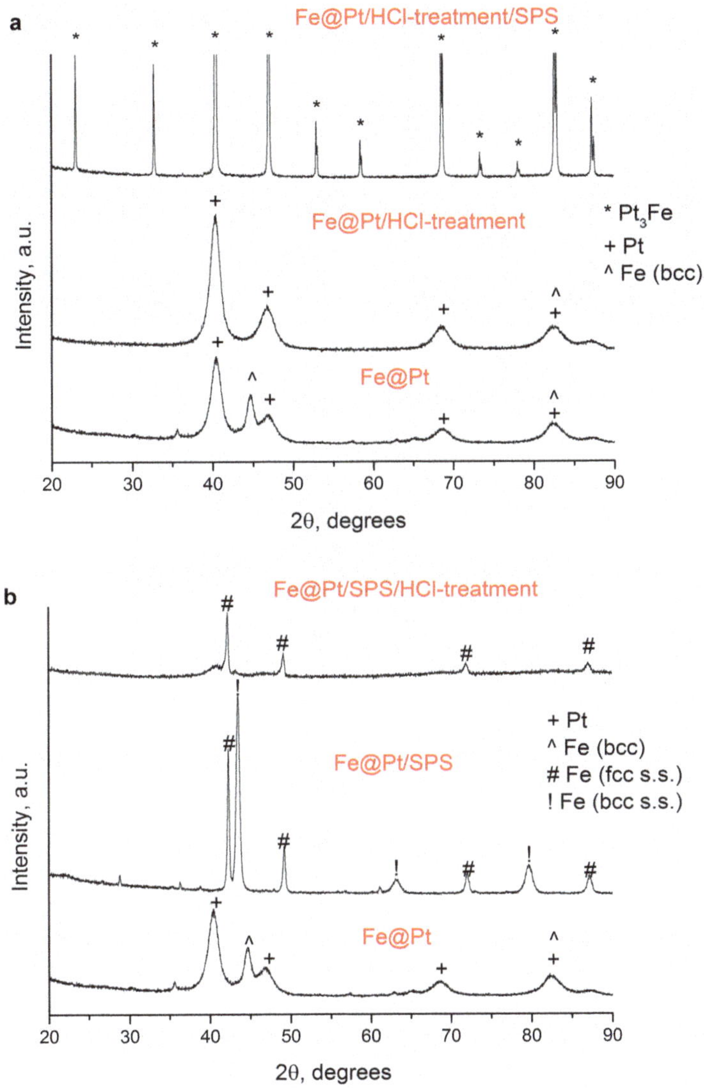

Fig. 10.68 XRD patterns showing the phase changes accompanying the processing of the Fe@Pt core–shell particles through the "Fe@Pt/HCl treatment/SPS" (**a**) and "Fe@Pt/SPS/HCl treatment" (**b**) routes. The pattern marked "Fe@Pt/HCl treatment" corresponds to the hollow Pt(Fe) particles. (Reprinted from Bokhonov and Dudina [164]. Copyright (2016) with permission from Elsevier)

A promising application of controlled reactions during SPS is the design of composite materials with local graded structure within nano- or submicron grains. Periodic fluctuations of the chemical composition in dielectric materials offer possibilities to flexibly modify their electrical properties. An interesting microstructure was formed when core–shell particles were sintered by SPS [165]: by careful

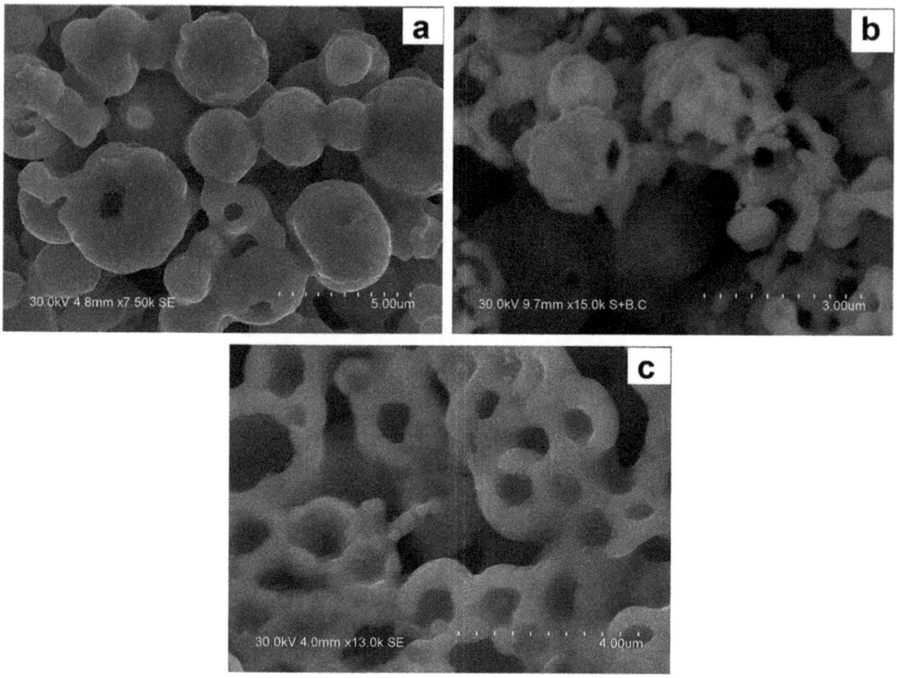

Fig. 10.69 Fracture surface of the compacts obtained by (**a**) SPS of the Fe@Pt core–shell particles ("Fe@Pt/SPS"), (**b**) SPS of the Pt(Fe) hollow particles ("Fe@Pt/HCl treatment/SPS"), and (**c**) SPS of the Fe@Pt core–shell particles followed by treatment in HCl ("Fe@Pt/SPS/HCl treatment"). (Reprinted from Bokhonov and Dudina [164]. Copyright (2016) with permission from Elsevier)

adjustment of the sintering parameters, it was possible to limit the interdiffusion between the surface layer (a shell of $SrTiO_3$ or $BaZrO_3$) and the particle core ($BaTiO_3$) and tune the dielectric properties of the core–shell material, which were different from those of the single-phase solid solution and parent perovskite phases.

10.18 Summary

Reactive sintering is a process of sintering that is accompanied by chemical reactions between the powder components, the reaction products forming the major phases of the sintered material. During reactive sintering, the energy is supplied from an external source and is used to initiate the chemical reaction and to complete sintering of the particles of the synthesized products. In the case of micrometer-sized powder reactants, the energy released during a chemical reaction is much greater than the energy associated with a reduction in the surface area during sintering. Therefore, the energy release due to chemical transformations should be considered as the main driving force for sintering of reaction mixtures composed of micrometer-sized powders. Reactive sintering is a more complex process than sintering that is not associated with chemical transformations. Under applied electromagnetic fields, the behavior of the reaction system will be determined by different responses of the components, which will depend on their physical properties. Furthermore, the phase composition of the system will evolve during the process, and its response to the electromagnetic field may also change.

Chemical reactions can be initiated by electric current, including high-voltage electric pulses. In pulsed electric current-assisted sintering processes of powders of electrically conductive materials, the contacts between the particles are regions of a higher temperature compared with the volume of the particles. When a mixture of powders of two metals soluble in each other is placed under conditions of highly localized heat generation at the inter-particle contacts, significant intensification of diffusion between the metals should be expected.

Experimental evidence of both acceleration and deceleration of chemical reactions under applied electric fields has been obtained. Electric field can cause acceleration of a chemical reaction due to alteration of the reaction mechanism. For some systems, under imposed electric current, the reaction mechanism may remain unchanged, but the mass transport may be enhanced due to increased defect mobility and/or concentration. Deceleration of chemical reactions under electric field is caused by enhanced intergranular sintering in the reactant layers.

The energy for reactive sintering can be supplied in the form of microwaves. The microwave energy is used to enhance the reaction kinetics and to trigger chemical reactions at lower temperatures compared with conventional processing. The main requirement to the reaction mixtures suitable for conducting the microwave synthesis is the presence of at least one microwave susceptor among the reactants, which would be heated and would transfer heat to the other components by thermal conduction. Shorter holding times and faster heating allow considering the microwave reactive processing as an energy-saving alternative to conventional sintering.

An advantage of solid-state sintering in a constant magnetic field is a possibility to achieve crystal orientation during the synthesis of ferromagnetic phases.

In recent years, SPS has become a popular synthesis method in solid-state chemistry and a materials design tool at nano-, micro-, and macroscales. Comparative studies of reactive SPS and SPS of the products of SHS showed that reactive SPS is preferable for single-phase materials, which acquire a higher relative density when processed by the reactive sintering route.

Transformations of carbon allotropes under applied current have been reviewed. Under electric current, amorphous carbon experiences graphitization. Under pulsed current during SPS, carbon nanotubes transform into diamond. In mixtures with particles of insulating materials, carbon nanotubes carry the electric current and experience overheating relative to the particles of the insulating material and become more chemically unstable. During SPS, materials being sintered interact with carbon of graphite tooling and graphite foil and with carbon monoxide formed by oxidation of graphite at high temperatures under conditions of lack of oxygen in the SPS chamber. The mechanisms of carbon incorporation into materials during SPS are inward diffusion of carbon and carbon evaporation and deposition inside the sample's pores. The inward diffusion of carbon is facilitated by a high solubility of carbon and the presence of defects and grain boundaries in the material being sintered.

Reactive SPS has allowed obtaining bulk materials with attractive mechanical and functional properties. The successes of reactive SPS in synthesizing bulk materials can be utilized to develop the simultaneous synthesis and joining of different materials and manufacture coatings by sintering of powder layers to substrates. Recent studies have shown that particles with core–shell morphology can be processed by SPS into bulk porous or dense solids with unique microstructures.

Modeling of reactive SPS should involve the simulation of electromagnetic–thermal–mechanical phenomena coupled with chemical and phase transformation thermodynamics and kinetics. This is an extremely challenging problem, which will have to be addressed in the future studies to provide predictions of the conditions of a *controlled non-equilibrium* needed for the positive outcome of most field-assisted sintering techniques.

References

1. Basu B, Balani K (2011) Advanced structural ceramics. Wiley, Hoboken, NJ 475 p
2. Meyers MA, Olevsky EA, Ma J, Janet M (2001) Combustion synthesis/densification of an Al_2O_3-TiB_2 composite. Mater Sci Eng A 311(1–2):83–99
3. Zhang X, He X, Han J, Qu W, Kvanin VL (2002) Combustion synthesis and densification of large-scale TiC-*x*Ni cermets. Mater Lett 56(3):183–187
4. Xu Q, Zhang X, Han J, He X, Kvanin VL (2003) Combustion synthesis and densification of titanium diboride-copper matrix composite. Mater Lett 57(28):4439–4444
5. Dargar SR, Groven LJ, Swiatkiewicz JJ, Puszynski JA (2007) In situ densification of SHS composites from nanoreactants. Int J Self-Propag High Temp Synth 16(3):125–132
6. Mishra SK, Das SK, Sherbacov V (2007) Fabrication of Al_2O_3-ZrB_2 in situ composite by SHS dynamic compaction: a novel approach. Compos Sci Technol 67(11):2447–2453
7. Gutmanas EY, Gotman I (1999) Dense high temperature ceramics by thermal explosion under pressure. J Eur Ceram Soc 19(13–14):2381–2393
8. Horvitz D, Gotman I, Gutmanas EY, Claussen N (2002) In situ processing of dense Al_2O_3–Ti aluminide interpenetrating phase composites. J Eur Ceram Soc 22(6):947–954
9. Zehetbauer MJ, Zhu YT (eds) (2009) Bulk nanostructured materials. Wiley, Hoboken, NJ 736 p
10. Greil P (2002) Advanced engineering ceramics. Adv Mater 14(10):709–716
11. Tjong SC, Ma ZY (2000) Microstructural and mechanical characteristics of in situ metal matrix composites. Mater Sci Eng R 29:49–113
12. Olevsky E, Bogachev I, Maximenko A (2013) Spark-plasma sintering efficiency control by inter-particle contact area growth: a viewpoint. Scr Mater 69(2):112–116
13. Savitskii AP (1991) Liquid-phase sintering of systems with interacting components. Nauka, Novosibirsk (in Russian)
14. Savitskii AP (2005) Scientific approaches to problems of mixtures sintering. Sci Sinter 37:3–17
15. Olevsky E, Skorohod V, Petzow G (1997) Densification by sintering incorporating phase transformations. Scr Mater 37(5):635–643

16. Krishtal MA, Zakharov PN, Kokora AN (1976) On the contribution of diffusion processes to re-distribution effects in solids under laser treatment. Fiz Khim Obrab Mater (Phys Chem Mater Process) 4:24–28 (in Russian)
17. Raichenko AI (1987) Basics of electric current-assisted sintering. Metallurgiya, Moscow 128 p. (in Russian)
18. Burenkov GL, Raichenko AI (1980) On diffusion during heat evolution at the contact of the diffusion pair components. Ukr Fiz Zh (Ukr J Phys) 25(12):2037–2045 (in Russian)
19. German RM (1996) Sintering theory and practice. Wiley, New York, NY 568 p
20. Misiolek W, German RM (1991) Reactive sintering and reactive hot isostatic compaction of aluminide matrix composites. Mater Sci Eng A 144(1–2):1–10
21. Belousov VY, Pilipchenko AV, Lutsak LD (1988) Some relationships governing initiation of self-propagating synthesis in direct electric heating. Sov Powder Metall Met Ceram 27 (10):813–816
22. Morsi K, Mehra P (2014) Effect of mechanical and electrical activation on the combustion synthesis of Al_3Ti. J Mater Sci 49(15):5271–5278
23. Bertolino N, Garay J, Anselmi-Tamburini U, Munir ZA (2001) Electromigration effects in Al-Au multilayers. Scr Mater 44:737–742
24. Bertolino N, Garay J, Anselmi-Tamburini U, Munir ZA (2002) High-flux current effects in interfacial reactions in Au–Al multilayers. Philos Mag B 82:969–985
25. Anselmi-Tamburini U, Garay JE, Munir ZA (2005) Fundamental investigations on the spark-plasma sintering/synthesis process. III. Current effect on reactivity. Mater Sci Eng A 407 (1–2):24–30
26. Garay JE, Glade SC, Anselmi-Tamburini U, Asoka-Kumar P, Munir ZA (2004) Electric current enhanced defect mobility in Ni_3Ti intermetallics. Appl Phys Lett 85:573
27. Zhao J, Garay JE, Anselmi-Tamburini U, Munir ZA (2007) Directional electromigration-enhanced interdiffusion in the Cu–Ni system. J Appl Phys 102(11):114902 7 p
28. Kondo T, Kuramoto T, Kodera Y, Ohyanagi M, Munir ZA (2008) Enhanced growth of Mo_2C formed in Mo–C diffusion couple by pulsed dc current. J Jpn Soc Powder Powder Metall 55:643–650
29. Garay JE, Anselmi-Tamburini U, Munir ZA (2003) Enhanced growth of intermetallic phases in the Ni–Ti system by current effects. Acta Mater 51:4487–4495
30. Anselmi-Tamburini U, Kodera Y, Gasch M, Unuvar C, Munir ZA, Ohyanagi M, Johnson SM (2006) Synthesis and characterization of dense ultra-high temperature thermal protection materials produced by field activation through spark plasma sintering (SPS): I. Hafnium diboride. J Mater Sci 41(10):3097–3104
31. Munir ZA, Anselmi-Tamburini U, Ohyanagi M (2006) The effect of electric field and pressure on the synthesis and consolidation of materials: a review of the spark plasma sintering method. J Mater Sci 41(3):763–777
32. Mackenzie KJD, Banerjee RK, Kasaai MR (1979) Effect of electric fields on solid-state reactions between oxides. Part 1. Reaction between calcium and aluminum oxides. J Mater Sci 14:333–338
33. Neiman AY, Krylov AO, Kuznetsov VA (1985) The influence of electric field on solid-state reactions between oxides. Russ J Phys Chem A 59(9):2360–2361 (in Russian)
34. Mackenzie KJD, Banerjee RK (1979) Effect of electric fields on solid-state reactions between oxides. Part 2. Interdiffusion studies in polycrystalline calcium and aluminium oxide pellets. J Mater Sci 14:339–344
35. Zingel EM (1982) The influence of electric field on the thermolysis rate of $KMnO_4$. Russ J Phys Chem A 57(3):766–768 (in Russian)
36. Anisimov AG, Mali VI (2010) Possibility of electric-pulse sintering of powder nanostructural composites. Combust Explos Shock Waves 46(2):237–241
37. An YB, Oh NH, Chun YW, Kim DK, Park JS, Choi KO, Eom TG, Byun TH, Kim JY, Byun CS, Hyun CY, Reucroft PJ, Lee WH (2006) One–step process for the fabrication of Ti porous compact and its surface modification by environmental electro–discharge sintering of spherical Ti powders. Surf Coat Technol 200(14–15):4300–4304

38. Sizonenko ON, Baglyuk GA, Taftai EI, Zaichenko AD, Lipyan EV, Torpakov AS, Zhdanov AA, Pristash NS (2013) Dispersion and carburization of titanium powders by electric discharge. Powder Metall Met Ceram 52(5–6):247–253
39. Calka A, Wexler D (2002) Mechanical milling assisted by electrical discharge. Nature 419:147–151
40. Calka A, Chowdhury AA, Konstantinov K (2012) Rapid synthesis of functional oxides by electric discharge assisted mechanical method. J Alloys Compd 536:3–8
41. Agrawal DK (1998) Microwave processing of ceramics. Curr Opin Solid State Mater Sci 3 (5):480–485
42. Rao KJ, Vaidhyanathan B, Ganguli M, Ramakrishnan PA (1999) Synthesis of inorganic solids using microwaves. Chem Mater 11:882–895
43. Rao KJ, Vaidhyanathan B (1995) A process of preparing molybdenum disilicide using microwaves. Indian Patent No. 788/MAS/95
44. Barzegar Bafrooei H, Ebadzadeh T, Majidian H (2014) Microwave synthesis and sintering of forsterite nanopowder produced by high-energy ball milling. Ceram Int 40:2869–2876
45. Lei Y, Lia Y, Xu L, Yang J, Wan R, Long H (2016) Microwave synthesis and sintering of TiNiSn thermoelectric bulk. J Alloys Compd 660:166–170
46. Cesário MR, Savary E, Marinel S, Raveau B, Caignaert V (2016) Synthesis and electrochemical performance of $Ce_{1-x}Yb_xO_{2-x/2}$ solid electrolytes: the potential of microwave sintering. Solid State Ionics 294:67–72
47. Sivanagi Reddy E, Sukumaran S, James Raju KC (2016) Microwave assisted synthesis and sintering of lead-free ferroelectric $CaBi_4Ti_4O_{15}$ ceramics. Mater Today Proc 3:2213–2219
48. Feizpour M, Barzegar Bafrooei H, Hayati R, Ebadzadeh T (2014) Microwave-assisted synthesis and sintering of potassium sodium niobate lead-free piezoelectric ceramics. Ceram Int 40:871–877
49. Lu X, Ding Y, Dan H, Yuan S, Mao X, Fan L, Wu Y (2014) Rapid synthesis of single phase $Gd_2Zr_2O_7$ pyrochlore waste forms by microwave sintering. Ceram Int 40:13191–13194
50. Lekse JW, Stagger TJ, Aitken JA (2007) Microwave metallurgy: synthesis of intermetallic compounds via microwave irradiation. Chem Mater 19(15):3601–3603
51. Rosa R, Veronesi P, Casagrande A, Leonelli C (2016) Microwave ignition of the combustion synthesis of aluminides and field-related effects. J Alloys Compd 657:59–67
52. Mitsui Y, Umetsu RY, Koyama K, Watanabe K (2014) Magnetic-field-induced enhancement for synthesizing ferromagnetic MnBi phase by solid-state reaction sintering. J Alloys Compd 615:131–134
53. Mitsui Y, Abematsu K, Umetsu RY, Takahashi K, Koyama K (2016) Magnetic field effects on liquid-phase reactive sintering of MnBi. J Magn Magn Mater 400:304–306
54. Orrù R, Licheri R, Locci AM, Cincotti A, Cao G (2009) Consolidation/synthesis of materials by electric current activated/assisted sintering. Mater Sci Eng R 63(4–6):127–287
55. Dudina DV, Mukherjee AK (2013) Reactive spark plasma sintering: successes and challenges of nanomaterial synthesis. J Nanomater 2013 article ID 625218, 12 p
56. Hulbert DM, Jiang D, Dudina DV, Mukherjee AK (2009) The synthesis and consolidation of hard materials by spark plasma sintering. Int J Refract Metals Hard Mater 27(2):367–375
57. Salamon D, Eriksson M, Nygren M, Shen Z (2007) Homogeneous TiB_2 ceramic achieved by electric current-assisted self-propagating reaction sintering. J Am Ceram Soc 90 (10):3303–3306
58. Locci AM, Licheri R, Orrù R, Cao G (2009) Reactive spark plasma sintering of rhenium diboride. Ceram Int 35(1):397–400
59. Schmidt J, Boehling M, Burkhardt U, Grin Y (2007) Preparation of titanium diboride TiB_2 by spark plasma sintering at slow heating rate. Sci Technol Adv Mater 8(5):376–382
60. Schmidt J, Niewa R, Schmidt M, Grin Y (2005) Spark plasma sintering effect on the decomposition of MgH_2. J Am Ceram Soc 88(7):1870–1874

61. Noh JH, Jung HS, Cho IS, An JS, Cho CM, Han HS, Hong KS (2010) Enhancing the densification of nanocrystalline TiO_2 by reduction in spark plasma sintering. J Am Ceram Soc 93(4):993–997
62. Munir ZA (2000) Synthesis and densification of nanomaterials by mechanical and field activation. J Mater Synth Process 8(3–4):189–196
63. Anselmi-Tamburini U, Munir Z, Kodera Y, Imai T, Ohyanagi M (2005) Influence of synthesis temperature on the defect structure of boron carbide: experimental and modeling studies. J Am Ceram Soc 88(6):1382–1387
64. Propescu B, Enache S, Ghica C, Valeanu M (2011) Solid-state synthesis and spark plasma sintering of $SrZrO_3$ ceramics. J Alloys Compd 509(22):6395–6399
65. Hulbert DM, Jiang D, Anselmi-Tamburini U, Unuvar C, Mukherjee AK (2008) Experiments and modeling of spark plasma sintered functionally graded boron-carbide-aluminum composites. Mater Sci Eng A 488(1–2):333–338
66. Hulbert DM, Jiang D, Anselmi-Tamburini U, Unuvar C, Mukherjee AK (2008) Continuous functionally graded boron carbide-aluminum nanocomposites by spark plasma sintering. Mater Sci Eng A 493(1–2):251–255
67. Roberts DJ, Zhao J, Munir ZA (2009) Mechanism of reactive sintering of $MgAlB_{14}$ by pulse electric current. Int J Refract Metals Hard Mater 27(3):556–563
68. Paris S, Gaffet E, Bernard F, Munir ZA (2004) Spark plasma synthesis from mechanically activated powders: a versatile route for producing dense nanostructured iron aluminides. Scr Mater 50(5):691–696
69. Bernard F, Le Gallet S, Spinassou N, Paris S, Gaffet E, Woolman JN, Munir ZA (2004) Dense nanostructured materials obtained by spark plasma sintering and field activated pressure assisted synthesis starting from mechanically activated powder mixtures. Sci Sinter 36 (3):155–164
70. Dudina DV, Hulbert DM, Jiang D, Unuvar C, Cytron SJ, Mukherjee AK (2008) In situ boron carbide-titanium diboride composites prepared by mechanical milling and subsequent spark plasma sintering. J Mate Sci 43(10):3569–3576
71. Licheri R, Orrù R, Locci AM, Cao G (2007) Efficient synthesis/sintering routes to obtain fully dense ZrB_2–SiC Ultra-High-Temperature Ceramics (UHTCs). Ind Eng Chem Res 46:9087–9096
72. Musa C, Orrù R, Sciti D, Silvestroni L, Cao G (2013) Synthesis, consolidation and characterization of monolithic and SiC whiskers reinforced HfB_2 ceramics. J Eur Ceram Soc 33:603–614
73. Orrù R, Cao G (2013) Comparison of reactive and non-reactive spark plasma sintering routes for the fabrication of monolithic and composite ultra high temperature ceramics (UHTC) materials. Materials 6:1566–1583
74. Licheri R, Orrù R, Musa C, Cao G (2008) Combination of SHS and SPS techniques for fabrication of fully dense ZrB_2-ZrC-SiC composites. Mater Lett 62(3):432–435
75. Licheri R, Orrù R, Musa C, Cao G (2010) Efficient technologies for the fabrication of dense TaB_2-based ultra-high-temperature ceramics. Appl Mater Interfaces 2(8):2206–2212
76. Suryanarayana C (2001) Mechanical alloying and milling. Prog Mater Sci 46(1–2):1–184
77. Kim JS, Choi HS, Dudina D, Lee JK, Kwon YS (2007) Spark plasma sintering of nanoscale (Ni+Al) powder mixture. Solid State Phenom 119:35–38
78. Wang H, Lee SH, Kim HD (2012) Nano-hafnium diboride powders synthesized using a spark plasma sintering apparatus. J Am Ceram Soc 95(5):1493–1496
79. Stanciu L, Groza JR, Stoica L, Plapcianu C (2004) Influence of powder precursors on reaction sintering of Al_2TiO_5. Scr Mater 50(9):1259–1262
80. Handtrack D, Despang F, Sauer C, Kieback B, Reinfried N, Grin Y (2006) Fabrication of ultra-fine grained and dispersion-strengthened titanium materials by spark plasma sintering. Mater Sci Eng A 437(2):423–429
81. Locci AM, Orrù R, Cao G, Munir ZA (2006) Effect of ball milling on simultaneous sprak plasma synthesis and densification of TiC-TiB_2 composites. Mater Sci Eng A 434(1–2):23–29

82. Locci AM, Licheri R, Orrù R, Cincotti A, Cao G (2007) Mechanical and electric current activation of solid-state reactions for the synthesis of fully dense advanced materials. Chem Eng Sci 62(18–20):4885–4890
83. Heian EM, Khalsa SK, Lee JW, Munir ZA, Yamamoto T, Ohyanagi M (2004) Synthesis of dense, high-defect-concentration B_4C through mechanical activation and field-assisted combustion. J Am Ceram Soc 87(5):779–783
84. Koizumi Y, Tanaka T, Minamino Y, Tsuji N, Mizuuchi K, Ohkanda Y (2003) Densification and structural evolution in spark plasma sintering process of mechanically alloyed nanocrystalline Fe-23Al-6C powder. Mater Trans 44(8):1604–1612
85. Ishihara S, Zhang W, Kimura H, Omori M, Inoue A (2003) Consolidation of Fe–Co–Nd–Dy–B glassy powders by spark-plasma sintering and magnetic properties of the consolidated alloys. Mater Trans 44(1):138–143
86. Perrière L, Thai MT, Tusseau-Nenez S, Blétry M, Champion Y (2011) Spark plasma sintering of a Zr-based metallic glass. Adv Mater Eng 13(7):581–586
87. Duan RG, Kuntz JD, Garay JE, Mukherjee AK (2004) Metal-like electrical conductivity in ceramic nano-composite. Scr Mater 50(10):1309–1313
88. Duan RG, Garay JE, Kuntz JD, Mukherjee AK (2005) Electrically conductive *in situ* formed nano-$Si_3N_4/SiC/TiC_xN_{1-x}$ ceramic composite consolidated by pulse electric current sintering (PECS). J Am Ceram Soc 88(1):66–70
89. Zhang J, Wang L, Shi L, Jiang W, Chen L (2007) Rapid fabrication of Ti_3SiC_2–SiC nanocomposite using the spark plasma sintering-reactive synthesis (SPS-RS) method. Scr Mater 56(3):241–244
90. Wang L, Zhang J, Jiang W (2013) Recent development in reactive synthesis of nanostructured bulk materials by spark plasma sintering. Int J Refract Metals Hard Mater 39:103–112
91. Wang L, Wu T, Jiang W, Li J, Chen L (2006) Novel fabrication route to Al_2O_3–TiN nanocomposites via spark plasma sintering. J Am Ceram Soc 89(5):1540–1543
92. Isupov VP, Chupakhina LE, Mitrofanova RP, Tarasov KA, Rogachev AY, Boldyrev VV (1997) The use of intercalation compounds of aluminium hydroxide for the preparation of nanoscale systems. Solid State Ionics 101–103(1):265–270
93. Bokhonov BB, Burleva LP, Whitcomb DR, Usanov YE (2001) Formation of nano-sized silver particles during thermal and photochemical decomposition of silver carboxylates. J Imaging Sci Technol 45(3):259–266
94. Sun SK, Kan YM, Zhang GJ (2011) Fabrication of nanosized tungsten carbide ceramics by reactive spark plasma sintering. J Am Ceram Soc 94(10):3230–3233
95. Ran S, Van der Biest O, Vleugels J (2010) ZrB_2–SiC composites prepared by reactive pulsed electric current sintering. J Eur Ceram Soc 30(12):2633–2642
96. Chen W, Tojo T, Miyamoto Y (2012) SiC ceramic-bonded carbon fabricated with Si_3N_4 and carbon powders. Int J Appl Ceram Technol 9(2):313–321
97. Zhao Y, Taya M (2006) Processing of porous NiTi by spark plasma sintering method. Proc SPIE 6170:313–318
98. Miao X, Chen Y, Guo H, Khor KA (2004) Spark plasma sintered hydroxyapatite-yttria stabilized zirconia composites. Ceram Int 30(7):1793–1796
99. Honda H, Kobayashi K, Inoue K, Ishiyama M (1967) Electrical discharge sintering and graphitization of carbon powders. Carbon 5(5):545–546
100. Asaka K, Karita M, Saito Y (2011) Graphitization of amorphous carbon on a multiwall carbon nanotube surface by catalyst-free heating. Appl Phys Lett 99:091907
101. Toyofuku N, Nishimoto M, Arayama K, Kodera Y, Ohyanagi M, Munir Z (2010) Consolidation of carbon with amorphous graphite transformation by SPS. In: Munir ZA, Ohji T, Hotta Y, Singh M (eds) Innovative processing and manufacturing of advanced ceramics and composites: ceramic transactions 2010. Wiley, Hoboken, NJ, pp 32–40
102. Kim WS, Moon SY, Park NH, Huh H, Shim KB, Ham H (2011) Electrical and structural feature of monolayer graphene produced by pulse current unzipping and microwave exfoliation of carbon nanotubes. Chem Mater 23:940–944

103. Sribalajia M, Mukherjee B, Rao Bakshi S, Arunkumar P, Suresh Babu K, Kumar Keshri A (2017) In-situ formed graphene nanoribbon induced toughening and thermal shock resistance of spark plasma sintered carbon nanotube reinforced titanium carbide composite. Compos Part B 123:227–240
104. Huang Q, Jiang D, Ovid'ko IA, Mukherjee A (2010) High-current-induced damage on carbon nanotubes: the case during spark plasma sintering. Scr Mater 63:1181–1184
105. Zhang F, Shen J, Sun J, Zhu YQ, Wang G, McCartney G (2005) Conversion of carbon nanotubes to diamond by spark plasma sintering. Carbon 43(6):1254–1258
106. Zhang F, Mihoc C, Ahmed F, Lathe C, Burkel E (2011) Thermal stability of carbon nanotubes, fullerene and graphite under spark plasma sintering. Chem Phys Lett 510:109–114
107. Zapata-Solvas E, Gómez-García D, Domínguez-Rodríguez A, Todd RI (2015) Ultra-fast and energy-efficient sintering of ceramics by electric current concentration. Sci Rep 5:8513
108. Solodkyi I, Xie SS, Zhao T, Borodianska H, Sakka Y, Vasylkiv O (2013) Synthesis of B_6O powder and spark plasma sintering of B_6O and B_6O–B_4C ceramics. J Ceram Soc Jpn 121 (11):950–955
109. Mouawad B, Soueidan M, Fabrègue D, Buttay C, Bley V, Allard B, Morel H (2012) Full densification of molybdenum powders using spark plasma sintering. Metall Mater Trans A 43 (9):3402–3409
110. Hayashi T, Matsuura K, Ohno M (2013) TiC coating on titanium by carbonization reaction using spark plasma sintering. Mater Trans 54(11):2098–2101
111. Grasso S, Poetschke J, Richter V, Maizza G, Sakka Y, Reece MJ (2013) Low-temperature spark plasma sintering of pure nano WC powder. J Am Ceram Soc 96(6):1702–1705
112. Lee G, McKittrick J, Ivanov E, Olevsky EA (2016) Densification mechanism and mechanical properties of tungsten powder consolidated by spark plasma sintering. Int J Refract Metals Hard Mater 61:22–29
113. Mackie AJ, Hatton GD, Hamilton HGC, Dean JS, Goodall R (2016) Carbon uptake and distribution in spark plasma sintering (SPS) processed Sm(Co, Fe, Cu, Zr)z. Mater Lett 171:14–17
114. Boulnat X, Fabrègue D, Perez M, Urvoy S, Hamon D, de Carlan Y (2014) Assessment of consolidation of oxide dispersion strengthened ferritic steels by spark plasma sintering: from laboratory scale to industrial products. Powder Metall 57(3):204–211
115. Neamţu BV, Marinca TF, Chicinaş I, Isnard O, Popa F, Păşcuţă P (2014) Preparation and soft magnetic properties of spark plasma sintered compacts based on Fe–Si–B glassy powder. J Alloys Compd 600:1–7
116. Rodriguez-Suarez T, Díaz LA, Torrecillas R, Lopez-Esteban S, Tuan WH, Nygren M, Moya JS (2009) Alumina/tungsten nanocomposites obtained by spark plasma sintering. Compos Sci Technol 69:2467–2473
117. Bokhonov BB, Ukhina AV, Dudina DV, Anisimov AG, Mali VI, Batraev IS (2015) Carbon uptake during spark plasma sintering: investigation through the analysis of the carbide "footprint" in a Ni–W alloy. RSC Adv 5:80228–80237
118. Dudina DV, Bokhonov BB, Ukhina AV, Anisimov AG, Mali VI, Esikov MA, Batraev IS, Kuznechik OO, Pilinevich LP (2016) Reactivity of materials towards carbon of graphite foil during spark plasma sintering: a case study using Ni–W powders. Mater Lett 168:62–67
119. Singleton M, Nash P (1989) The C–Ni (carbon–nickel) system. Bull Alloy Phase Diagrams 10:121–126
120. Collet R, le Gallet S, Charlot F, Lay S, Chaix JM, Bernard F (2016) Oxide reduction effects in SPS processing of Cu atomized powder containing oxide inclusions. Mater Chem Phys 173:498–507
121. Shearwood C, Ng HB (2007) Spark plasma sintering of wire exploded tungsten nano-powder. Proc SPIE 6798:67981B
122. Toyofuku N, Kuramoto T, Imai T, Ohyanagi M, Munir ZA (2012) Effect of pulsed DC current on neck growth between tungsten wires and tungsten plates during the initial stage of sintering by the spark plasma sintering method. J Mater Sci 47:2201–2205

123. Dudina DV, Anisimov AG, Mali VI, Bulina NV, Bokhonov BB (2015) Smaller crystallites in sintered materials? A discussion of the possible mechanisms of crystallite size refinement during pulsed electric current-assisted sintering. Mater Lett 144:168–172
124. Dudina DV, Bokhonov BB (2017) Elimination of oxide films during spark plasma sintering of metallic powders: a case study using partially oxidized nickel. Adv Powder Technol 28:641–647
125. Bertrand A, Carreaud J, Delaizir G, Duclère JR, Colas M, Cornette J, Vandenhende M, Couderc V, Thomas P (2013) A comprehensive study of the carbon contamination in tellurite glasses and glass–ceramics sintered by spark plasma sintering (SPS). J Am Ceram Soc 97:163–172
126. Bernard-Granger G, Benameur N, Guizard C, Nygren M (2009) Influence of graphite contamination on the optical properties of transparent spinel obtained by spark plasma sintering. Scr Mater 60:164–167
127. Morita K, Kim BN, Yoshida H, Hiraga K, Sakka Y (2015) Spectroscopic study of the discoloration of transparent $MgAl_2O_4$ spinel fabricated by spark-plasma-sintering (SPS) processing. Acta Mater 84:9–19
128. Morita K, Kim BN, Yoshida H, Hiraga K, Sakka Y (2016) Influence of pre- and post-annealing on discoloration of $MgAl_2O_4$ spinel fabricated by spark-plasma-sintering (SPS). J Eur Ceram Soc 36(12):2961–2968
129. Jiang D, Mukherjee AK (2011) The influence of oxygen vacancy on the optical transmission of an yttria–magnesia nanocomposite. Scr Mater 64(12):1095–1097
130. Isobe T, Daimon K, Sato T, Matsubara T, Hikichi Y, Ota T (2008) Spark plasma sintering technique for reaction sintering of Al_2O_3/Ni nanocomposite and its mechanical properties. Ceram Int 34(1):213–217
131. Rong CB, Nandwana V, Poudyal N, Liu JP, Saito T, Wu Y, Kramer MJ (2007) Bulk FePt/ Fe_3Pt nanocomposite magnets prepared by spark plasma sintering. J Appl Phys 101:09K515 3 p
132. Ramírez C, Vega-Diaz SM, Morelos-Gómez A, Figueiredo FM, Terrones M, Isabel Osendi M, Belmonte M, Miranzo P (2013) Synthesis of conducting graphene/Si_3N_4 composites by spark plasma sintering. Carbon 57:425–432
133. Li H, Khor KA, Yu LG, Cheang P (2005) Microstructure modifications and phase transformation in plasma-sprayed WC-Co coatings following post-spray spark plasma sintering. Surf Coat Technol 194(1):96–102
134. Gurt Santanach J, Estournès C, Weibel A, Peigney A, Chevallier G, Laurent C (2009) Spark plasma sintering as a reactive sintering tool for the preparation of surface-tailored Fe-$FeAl_2O_4$-Al_2O_3 nanocomposites. Scr Mater 60(4):195–198
135. Gurt Santanach J, Estournès C, Weibel A, Chevallier G, Bley V, Laurent C, Peigney A (2011) Influence of pulse current during spark plasma sintering evidenced on reactive alumina-hematite powder. J Eur Ceram Soc 31(13):2247–2254
136. Kakegawa K, Wen CM, Uekawa N, Kojima T (2014) SPS using SiC die. Key Eng Mater 617:72–77
137. Byon C, Li MH, Kakegawa K, Han YH, Lee DY (2015) Numerical study of a SiC mould subjected to a spark plasma sintering process. Scr Mater 96:49–52
138. Bokhonov BB, Ukhina AV, Dudina DV, Gerasimov KB, Anisimov AG, Mali VI (2015) Towards a better understanding of nickel/diamond interactions: the interface formation at low temperature. RSC Adv 5:51799–51806
139. Dudina DV, Mali VI, Ukhina AV, Anisimov AG, Brester AE, Bokhonov B (2016) Interparticle interactions in partially densified compacts of electrically conductive materials during spark plasma sintering. In: Proceedings of the 11th International Forum on Strategic Technology, IFOST 2016, Article number 7884067, pp 139–143
140. Feng H, Zhou Y, Jia D, Meng Q (2005) Rapid synthesis of Ti alloy with B addition by spark plasma sintering. Mater Sci Eng A 390(1–2):344–349

141. Feng H, Jia D, Zhou Y (2005) Spark plasma sintering reaction synthesized TiB reinforced titanium matrix composites. Compos Part A 36(5):558–563
142. Zhang HW, Gopalan R, Mukai T, Hono K (2005) Fabrication of bulk nanocrystalline Fe-C alloy by spark plasma sintering of mechanically milled powder. Scr Mater 53(7):863–868
143. Zhang Z, Shen X, Wang F, Lee S (2011) A new rapid route for in situ synthesizing monolithic TiB ceramic. J Am Ceram Soc 94(9):2754–2756
144. Lee JW, Munir ZA, Shbuya M, Ohyanagi M (2001) Synthesis of dense TiB_2-TiN nanocrystalline composites through mechanical and field activation. J Am Ceram Soc 84(6):1209–1216
145. Huang SG, Vanmeensel K, Van der Biest O, Vleugels J (2011) In situ synthesis and densification of submicrometer-grained B_4C–TiB_2 composites by pulsed electric current sintering. J Eur Ceram Soc 31(4):637–644
146. Cabouro G, Chevalier S, Gaffet E, Grin Y, Bernard F (2008) Reactive sintering of molybdenum disilicide by spark plasma sintering from mechanically activated powder mixtures: processing parameters and properties. J Alloys Compd 465(1–2):344–355
147. Campayo L, Le Gallet S, Grin Y, Courtois E, Bernard F, Bart F (2009) Spark plasma sintering of lead phosphovanadate $Pb_3(VO_4)_{1.6}(PO_4)_{0.4}$. J Eur Ceram Soc 29(8):1477–1484
148. Le Gallet S, Campayo L, Courtois E, Hoffmann S, Grin Y, Bernard F, Bart F (2010) Spark plasma sintering of iodine-bearing apatite. J Nucl Mater 400(3):251–256
149. Beekman M, Baitinger M, Borrmann H, Schnelle W, Meier K, Nolas GS, Grin Y (2009) Preparation and crystal growth of $Na_{24}Si_{136}$. J Am Chem Soc 131:9642–9643
150. Chakravarty D, Ramesh H, Rao TN (2009) High strength porous alumina by spark plasma sintering. J Eur Ceram Soc 29:1361–1369
151. Oh ST, Tajima K, Ando M, Ohji T (2000) Strengthening of porous alumina by pulse electric current sintering and nanocomposite processing. J Am Ceram Soc 83(5):1314–1316
152. Dudina DV, Mukherjee AK (2013) Reactive spark plasma sintering for the production of nanostructured materials. In: Sinha S, Navani NK (eds) Nanotechnology series, vol. 4: Nanomaterials and nanostructures. Studium Press LLC, Houston, TX, pp 237–264
153. Galy J, Dolle M, Hungria T, Rozier P, Monchoux JP (2008) A new way to make solid state chemistry: spark plasma synthesis of copper or silver vanadium oxide bronzes. Solid State Sci 10(8):976–981
154. Dumont-Botto E, Bourbon C, Patoux S, Rozier P, Dolle M (2011) Synthesis by spark plasma sintering: a new way to obtain electrode materials for lithium ion batteries. J Power Sources 196(4):2274–2278
155. Liu W, Naka M (2003) In situ joining of dissimilar nanocrystalline materials by spark plasma sintering. Scr Mater 48(9):1225–1230
156. Matsubara T, Shibutani T, Uenishi K, Kobayashi KF (2002) Fabrication of TiB_2 reinforced Al_3Ti composite layer on Ti substrate by reactive-pulsed electric current sintering. Mater Sci Eng A 329–331:84–91
157. Mulukutla M, Singh A, Harimkar S (2010) Spark plasma sintering for multi-scale surface engineering of materials. JOM 62(6):65–71
158. Holland T, Hulbert D, Anselmi-Tamburini U, Mukherjee AK (2010) Functionally graded boron-carbide and aluminum composites with tubular geometries using pulsed electric current sintering. Mater Sci Eng A 527(18–19):4543–4545
159. Yuan H, Li J, Shen Q, Zhang L (2012) *In situ* synthesis and sintering of ZrB_2 porous ceramics by the spark plasma sintering–reactive synthesis (SPS–RS) method. Int J Refract Metals Hard Mater 34:3–7
160. Dudina DV, Bokhonov BB, Mukherjee AK (2016) Formation of aluminum particles with shell morphology during pressureless spark plasma sintering of Fe-Al mixtures: current-related or Kirkendall effect? Materials 9:375 10 p
161. Dudina DV, Legan MA, Fedorova NV, Novoselov AN, Anisimov AG, Esikov MA (2017) Structural and mechanical characterization of porous iron aluminide FeAl obtained by pressureless spark plasma sintering. Mater Sci Eng A 695:309–314

162. Dudina DV, Bokhonov BB, Legan MA, Novoselov AN, Skovorodin IN, Bulina NV, Esikov MA, Mali VI (2017) Analysis of the formation of FeAl with a high open porosity during electric current-assisted sintering of loosely packed Fe-Al powder. Vacuum 146:74–78
163. Scheele M, Oeschler N, Veremchuk I, Peters S, Littig A, Kornowski A, Klinke C, Weller H (2011) Thermoelectric properties of lead chalcogenide core-shell nanostructures. ACS Nano 5:8541–8551
164. Bokhonov BB, Dudina DV (2017) Preparation of porous materials by spark plasma sintering: peculiarities of alloy formation during consolidation of Fe@Pt core-shell and hollow Pt (Fe) particles. J Alloys Compd 707:233–237
165. Buscaglia MT, Vivani M, Zhao Z, Buscaglia V, Nanni P (2006) Synthesis of $BaTiO_3$ core-shell particles and fabrication of dielectric ceramics with local graded structure. Chem Mater 18(17):4002–4010

Chapter 11
Other Field-Assisted Sintering Techniques

A number of sintering methods use electromagnetic radiation to rapidly heat the powder samples. The character and mechanisms of interaction of electromagnetic waves with materials depend on the wavelength of radiation. Microwave sintering has been discussed in Chap. 7. Other types of radiation can also be used to deliver energy to particle assemblies and induce inter-particle sintering. Infrared (IR) radiation is one of the means of energy transfer in furnaces. Recently, IR radiation-producing modules have been designed allowing for rapid heating and sintering of powder layers into films. IR radiation can also be produced by lasers and is constantly produced by the sun. Metallic particles meeting certain size requirements can be heated by intense light from a xenon lamp, while ultraviolet (UV) radiation can assist in conducting chemical reactions that facilitate sintering. In this chapter, principles and possibilities of powder sintering using IR radiation emitted by specially designed modules and laser-assisted, solar, photonic, and ultraviolet sintering are briefly discussed.

11.1 IR Radiation-Assisted Sintering

This section describes sintering assisted by IR radiation produced by specially designed modules. In the manufacturing of flexible electronics, conventional sintering limits the feasibility of the roll-to-roll processing [1] due to long times required for sintering. Sowade et al. [2] reported IR radiation drying and sintering of inkjet-printed silver layers on a nonabsorbent polyethylene naphthalate substrate. Drying and sintering were conducted using an IR module equipped with tube emitters made of quartz glass (Fig. 11.1). Each emitter consisted of two tungsten filaments. A U-shaped aluminum reflector was placed above the tube emitters to increase the efficiency of the IR system. Another reflector was installed under the substrate to intensify the IR irradiation due to reflection. At a higher filament

E. A. Olevsky, D. V. Dudina, *Field-Assisted Sintering*,
https://doi.org/10.1007/978-3-319-76032-2_11

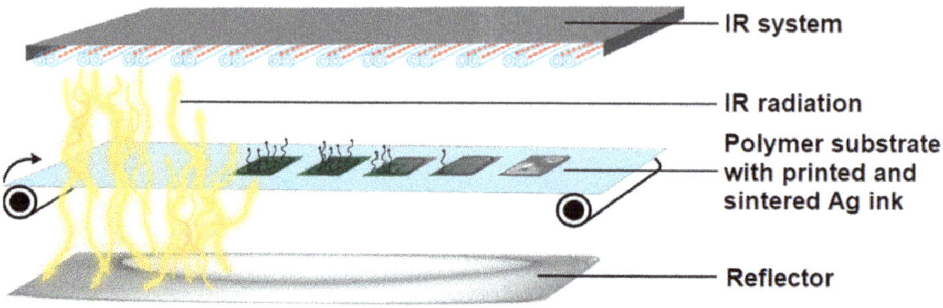

Fig. 11.1 IR radiation module for sintering of inkjet-printed layers of silver on a polymer substrate. (Reproduced from Ref. [2] with permission of The Royal Society of Chemistry)

temperature, drying and sintering was more efficient, as the silver layer has a higher absorbance in the shorter wavelength range.

Georgiadis et al. [3] emphasized the benefit of IR radiation heating over convection and conduction, which lies in the energy efficiency. They reported the formation of smooth and crack-free films from a hard latex using IR radiation sintering of dried layers deposited from emulsion. Using IR radiation-assisted sintering, the film could be obtained without adding plasticizers, which is a great benefit over the conventional processing, as during the film formation, plasticizers cause the evolution of volatile organic compounds – substances harmful from the environmental and health standpoints. In order to assist absorption of the IR radiation by the film, an additive – a near-IR-absorbing polymer – was used at a concentration of 1 wt.%. The non-sintered film consisted of spherical particles with voids between them (Fig. 11.2a). After treatment under a 250 W near-IR lamp for 4 min, the particles flattened, and the size of the voids decreased (Fig. 11.2b). After 30 min of treatment (Fig. 11.2c), the separate particles were more difficult to distinguish, which indicated sintering. The particle flattening was quantitatively analyzed by measuring the topographic profiles of the films after IR radiation-assisted sintering and heating in an oven. Figure 11.3 shows the dependence of the peak-to-valley height of the films on the time of exposure making it obvious that direct IR treatment of the films leads to a more rapid flattening (and sintering) of the particles in comparison with conventional treatment in an oven.

11.2 Solar Sintering

Solar radiation is a powerful and ecological energy source. A series of recent laboratory studies have been conducted using a solar furnace of the Plataforma Solar de Almería, Spain [4]. This solar furnace consists of the following elements: a continuous solar-tracking flat heliostat, a parabolic concentrator mirror (collector), an attenuator (shutter), and a test zone located in the concentrator focus center (Fig. 11.4). The shutter is used to regulate the amount of the incident light. When

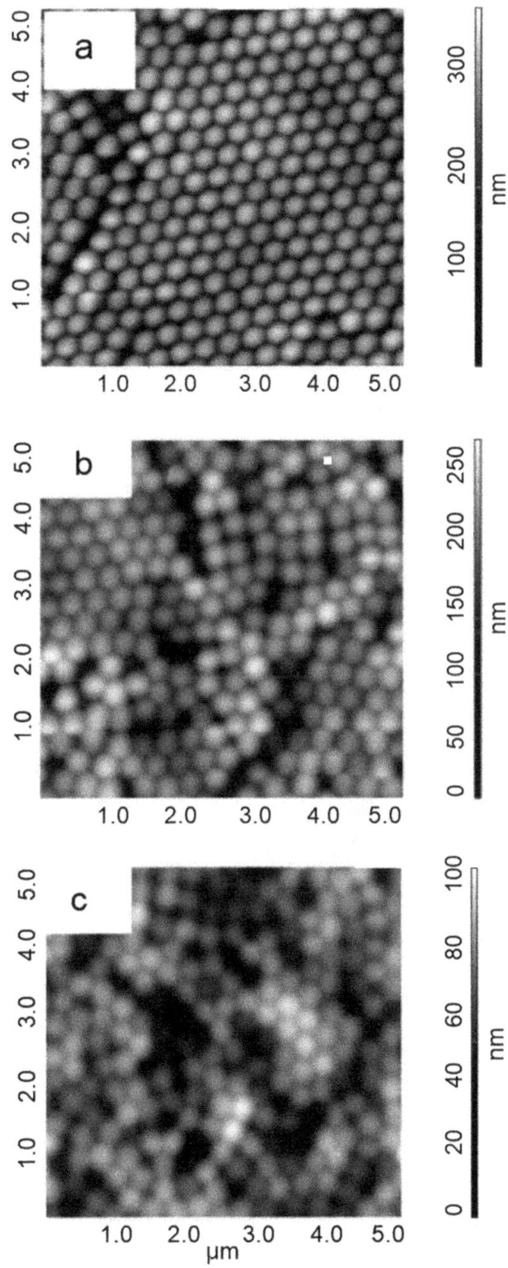

Fig. 11.2 Atomic force microscopy height images of the latex films deposited at room temperature (**a**), after 4 min (**b**), and 30 min (**c**) of IR radiation treatment. The images size is 5 µm × 5 µm. (Reprinted with permission from Georgiadis et al. [3]. Copyright (2011) American Chemical Society)

it is 100% open and the direct solar irradiance is 1000 W·m^{-2}, the focus has an irradiance peak of 3051 W·m^{-2}, a total power of 70 kW, and a focal diameter of 26 cm. A possibility of sintering of alumina at 1780 °C was reported [4]. Technical issues associated with solar furnaces include overshoot of temperature at the onset of

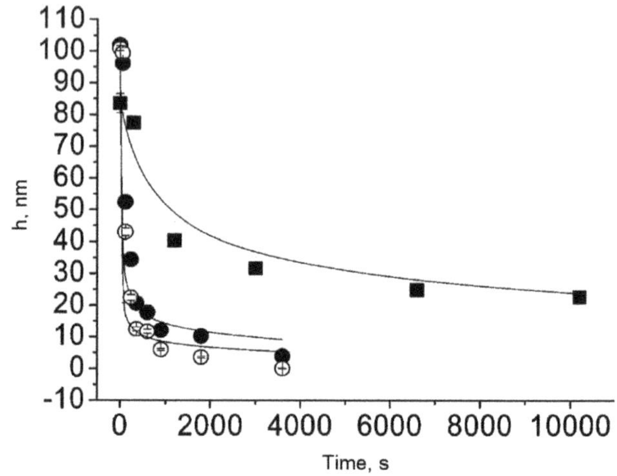

Fig. 11.3 Dependences of the peak-to-valley height on the time of continuous exposure to near-IR radiation for the acrylic additive-free latex (filled circles) and latex with 1 wt.% of IR-absorbing polymer (empty circles) and treatment in an oven at 60 °C (squares). Error bars span the size of the symbols. During the first 10 min in the oven, there was not significant sintering due to slow heating, so $t = 0$ corresponds to 10 min of holding. (Reprinted with permission from Georgiadis et al. [3]. Copyright (2011) American Chemical Society)

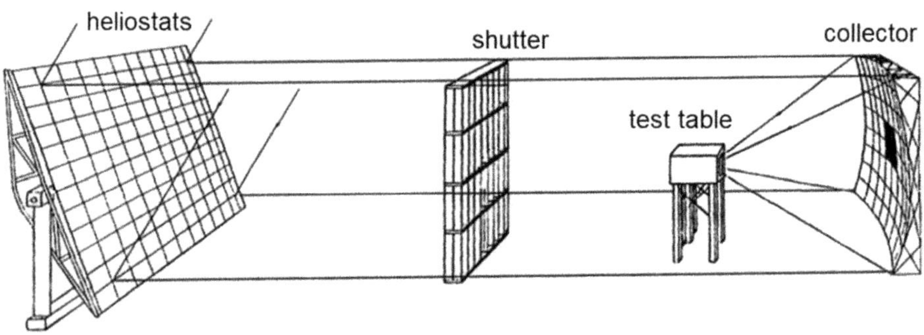

Fig. 11.4 Solar furnace of the Plataforma Solar de Almería, Spain. (Reprinted from Román et al. [4]. Copyright (2008), with permission from Elsevier)

solar heating and difficulties in the precise temperature control of the sintered part [5]. The temperature overshoot may result in the formation of a liquid phase and undesirable shape distortions. Due to large investment costs, solar sintering facilities are still expensive tools having a potential of cost reduction in the medium term [4].

A comparison between the microstructure of alumina consolidated by solar sintering and that of alumina sintered in an electric furnace was made (Fig. 11.5) [4]. The compacts were sintered by the two methods at the same maximum temperature of 1600 °C but with different heating rates and dwell times. The compact

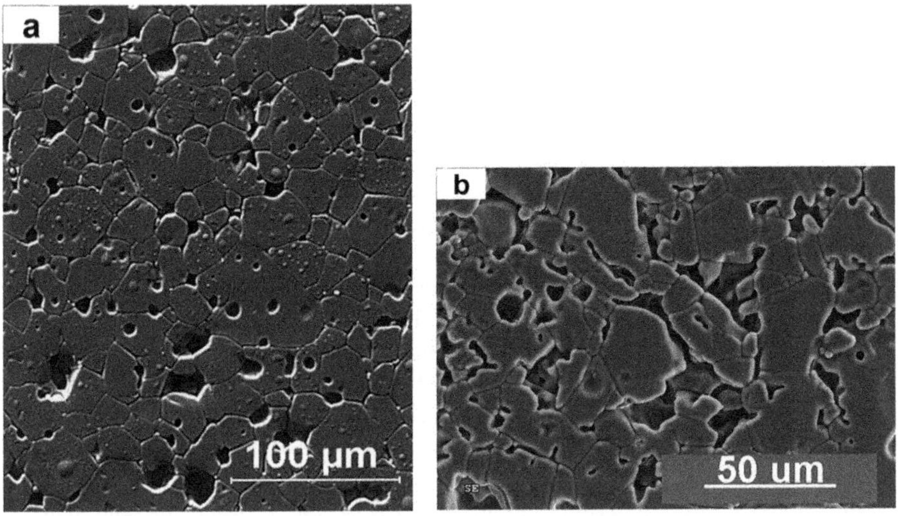

Fig. 11.5 Microstructure of alumina consolidated by solar sintering, heating rate 50 °C·min^{-1}, dwell time 60 min, relative density 95% (**a**), and sintered in a resistance furnace, heating rate 5 °C·min^{-1}, dwell time 240 min, relative density 89% (**b**). Maximum sintering temperature 1600 °C. (Reprinted from Román et al. [4]. Copyright (2008), with permission from Elsevier)

consolidated by solar sintering was denser than that sintered in a furnace (95% vs. 89% relative density). The degree of sintering was also reflected in the microstructure. The solar sintered compact had well-defined polygonal grains: the 300 nm particles of the powder experienced growth such that the compact consisted of grains as large as 20 µm. A significant number of voids of spherical shape are seen in the microstructure of the solar sintered material; these voids are located within the matrix grains (Fig. 11.5a). The voids were observed on a polished surface. However, using a different sample preparation technique, it was proved that these voids were initially (before polishing) filled with particles, which indicates incomplete sintering. In the compact sintered from the same powder in an electric furnace, only a few areas that experienced sintering could be found (Fig. 11.5b), the voids were located at grain boundaries, and the sintered material consisted of 50 µm grains. The microstructures of the compacts sintered by the two methods show that an improvement of densification was achieved by using solar sintering instead of sintering in a furnace. With the use of solar sintering, a significantly reduced duration of sintering was achieved for other materials: e.g., cordierite ceramics was sintered by solar sintering within 60 min, while in conventional sintering, about 24 h is needed for the process completion [5].

11.3 Laser-Assisted Sintering

Laser-assisted sintering is used for the fabrication of films in printed electronics and 3D objects. A detailed description of the interactions of laser radiation with materials and selective laser sintering is beyond the scope of this book and can be found in the literature on laser processing of materials [6] and 3D manufacturing [7–9]. The use of different types of lasers for different materials has been described by Lee et al. [10]. Selective laser sintering of metals is also referred to as direct metal laser sintering. In this process, Nd:YAG lasers (output wavelength 1064 nm) or Yb-fiber lasers (output wavelengths 1030–1070 nm) exhibit a higher throughput than CO_2 lasers with an operating wavelength of 10.6 μm. Nd:YAG and Yb-fiber lasers are also used for sintering of carbide ceramics. CO_2 lasers are suitable for sintering of polymers as the latter have high absorptivity at the operating wavelength of these lasers. Sintering of oxide ceramics is also possible with the use of CO_2 lasers.

In selective laser sintering, partial melting of the powder material can be achieved, and in this way, selective laser sintering bears similarities with electric current-assisted sintering, in which the inter-particle contacts in an electric current-carrying specimen can experience melting, while the particle volumes will continue to remain solid. Another similarity of electric current-assisted sintering and selective laser sintering is the dynamic nature of the processes, as the responses of the materials to the external field change with time. Normally, in the processes of electric current-assisted sintering, the contribution of inter-particle contacts to the total resistance of the sample decreases as sintering progresses due to the formation of well-established inter-particle contacts. At the same time, the resistivity of materials is a function of temperature and changes upon heating. During laser treatment, the absorptivity and thermal conductivity of the material are the important properties; these properties change with the temperature and the structure of the powder layer as the powder particles sinter between themselves [11, 12]. Absorption of laser radiation is favored by porosity of the powder layer and non-flat surfaces. Olakanmi et al. [8] indicate the relevance of the information on the sintering behavior of a powder during pulsed electric current-assisted sintering to selective laser sintering and suggest using the accumulated knowledge in the field of pulsed current sintering for gaining insights into the mechanisms of laser-assisted sintering.

11.4 Photonic Sintering

Sintering of nanoparticles using irradiation by light – photonic sintering in a continuous [13] or flash mode [14–22] – is attracting attention due to possibilities of fast processing of printed electronics and flexible dye-sensitized solar cells under ambient conditions as well as using large-area and temperature-sensitive substrate materials. In photonic sintering, films formed by nanoparticle inks or pastes are heated by irradiation. The method uses the differences in the absorption properties of the substrate and the film composed of nanoparticles. The solvent is evaporated and sintering between the nanoparticles occurs. By choosing radiation with a desired emission spectrum, energy can be "pumped" selectively into the printed ink structures without directly affecting the substrate. As the electromagnetic radiation directly affects the printed structures and not the film/substrate system as a whole, sintering is accelerated compared with the conventional process of heating in an oven.

Photonic sintering is conducted without the application of pressure making temperature the key factor influencing the film densification [23]. Govorov and Richardson [24] analyzed the generation of heat in metal nanoparticles irradiated by light. Heating is strongly enhanced under the conditions of the plasmon resonance. It was found that the temperature increase of the nanoparticles $\Delta T_{\max}$ is proportional to the second power of the nanoparticle radius R_{np}:

$$\Delta T_{\max} \propto R_{\mathrm{np}}^2$$

This size dependence of the temperature increase is governed by the total rate of heat generation and by heat transfer through the nanoparticle surface. The heating process also depends on the shape and organization of nanoparticles due to collective effects.

The results of photonic sintering depend on the size of metallic nanoparticles. Park and Kim [19] studied flash light sintering of nickel nanoparticles and found that while particles with a broad size distribution (5–500 nm) were successfully sintered by the flash light irradiation (xenon lamp, wavelengths 380 nm–1 μm), the films consisting of 50 nm particles were poorly sintered. The sheet resistance of the sintered films decreased with increasing sintering energy, which indicates the progress of sintering. The sheet resistance of the films obtained from the 50 nm particles was 2–3 orders of magnitude greater than that of the films obtained from the ink containing nickel nanoparticles with a broad size distribution. Better sinterability of the films obtained from the ink containing 5–500 nm nickel particles was explained by a better absorption of light (absorption of light with a wide range of wavelengths) by particles of different sizes, as shown in Fig. 11.6, and the presence of very fine nanoparticles, whose melting temperatures are much lower than that of bulk nickel. Early melting of very fine particles was suggested as a crucial step in flash light sintering, as the molten material can connect larger particles and facilitate the sintering process as a whole.

Park and Kim [19] utilized preheating before the main sintering step to improve the quality of the sintered films. As can be seen from images of the films sintered at 17.5 J·cm^{-2} at the main sintering step, the film preheated at 12.5 J·cm^{-2} had a more densely connected necking structure (Fig. 11.7a) than the film preheated at 7.5 J·cm^{-2} (Fig. 11.7b).

MacNeill et al. [13] analyzed photonic sintering of silver nanoparticles both experimentally and theoretically and found that it is an inherently self-damping process. The progress of densification reduces the magnitude of subsequent photonic

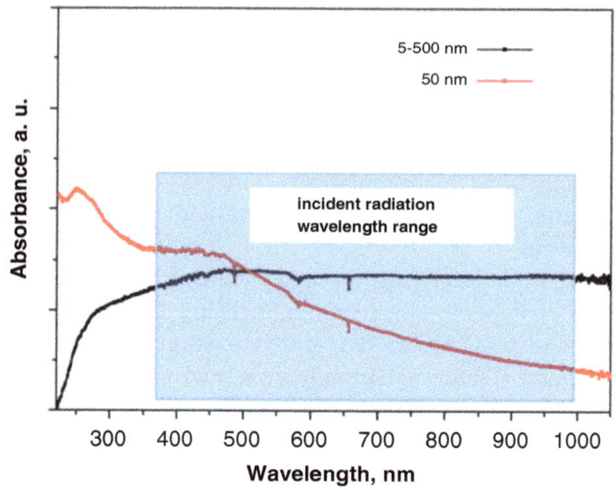

Fig. 11.6 Absorbance of light by nickel nanoparticles with different sizes and size distributions. (Reprinted from Park et al. [19]. Copyright (2013), with permission from Elsevier)

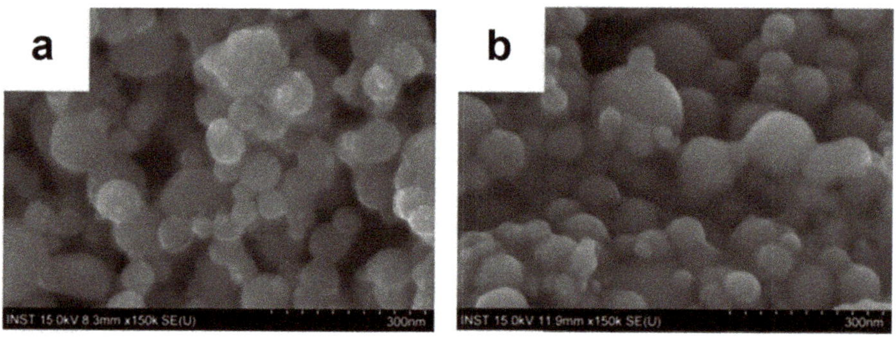

Fig. 11.7 Morphology of the films obtained from nickel nanoparticles 5–500 nm in size: (**a**) preheating at 12.5 J·cm^{-2}, (**b**) 7.5 J·cm^{-2} (the main sintering step at 17.5 J·cm^{-2}). (Reprinted from Park and Kim [19]. Copyright (2013), with permission from Elsevier)

heating. Bansal and Malhotra [23] measured the temperature of the silver nanoparticle films during their sintering by intense pulsed light. Measurements were done using a thermal camera. Under high fluencies, a turning point in the evolution of the film temperature during sintering was observed: with increasing the number of pulses, the temperature dropped down after the initial rise. The film densification leveled off beyond a critical pulse fluence and a critical number of pulses. A computational model that is able to capture the experimentally observed turning point in the temperature during sintering was developed. This model links the electromagnetic finite element analysis of optical energy absorption and semi-analytical models of inter-particle neck growth to mesoscale transient heat transfer and densification of the films. It was found that the temperature turning occurs due to a coupling between optical absorption and densification in the nanoparticle film: a reduction of optical absorption was caused by the inter-particle neck growth and changes in the particle shape.

11.5 UV-Assisted Sintering

UV radiation is useful in the processes that require conducting chemical reactions along with particle sintering. UV light was used by Oh et al. [25] to sinter TiO_2 nanoparticles for the fabrication of dye-sensitized solar cells. An organic compound-containing titanium was introduced into the TiO_2-based ink. This compound decomposed under UV radiation forming TiO_2 particles, which bonded the pre-existing TiO_2 particles together. In a study by Hwang et al. [26], deep UV light caused decomposition of a poly(N-vinylpyrrolidone) coating on copper nanoparticles so that the oxide shells on the particles were efficiently reduced.

11.6 Selected Examples of Materials Obtained Using Infrared, Solar, and Photonic Sintering

IR radiation-assisted sintering has been used for nanoparticles of metals [2], polymers [3], and oxides [27]. A limiting step to the roll-to-roll production of dye-sensitized solar cells on metals is sintering of nanoparticles of TiO_2. Watson et al. [27] suggested a near-IR heating method, in which titanium substrates are heated directly causing the removal of the binder and sintering of the TiO_2 particles. Sintering of the same assemblies in a conventional oven at temperatures close to those achieved by IR heating and for much longer time yielded cell efficiencies equal to or lower than those manufactured by using IR heating.

Solar sintering is suitable for consolidating metals [28] and ceramic materials [4, 5, 29, 30] as well as metal–ceramic composites [31]. Rosa et al. [31] showed that WC–10wt.%Co materials with properties comparable to those of the conventionally processed cemented carbides can be obtained by fast solar sintering. The differences in the heating schedule in these two processes can be seen in Fig. 11.8, from which it follows that solar sintering allowed reaching a significant reduction of the sintering time. As hardness of the solar sintered WC–10 wt.% Co materials was as high as that of the conventionally processed cemented carbide (13 GPa) and the fracture toughness was 12 $MPa{\cdot}m^{1/2}$, which is only 10% lower than the value of the fracture toughness of the conventionally processed cemented carbide, solar sintering was recommended as a cost-effective and environmentally friendly sintering method of cemented carbides. Close levels of hardness and fracture toughness could be expected from the similarity of the microstructures of these two materials (Fig. 11.9).

Photonic sintering was shown to be suitable for sintering of layers composed of metals [14–20], metal-containing composites [32], as well as conducting oxides [22] and semiconductors [21, 22]. Using photonic flash sintering, current collecting grids for solar cell applications were obtained starting from a silver nanoparticle ink, and the sintering results were compared with those produced by thermal treatment of the printed structures [20]. Grids obtained by photonic sintering

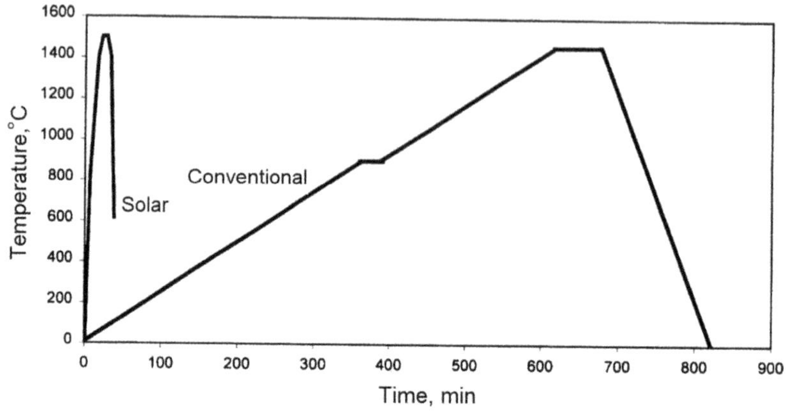

Fig. 11.8 Heating schedules in solar and conventional sintering of WC–10wt.%Co. (Reprinted from Rosa et al. [31]. Copyright (2002), with permission from Elsevier)

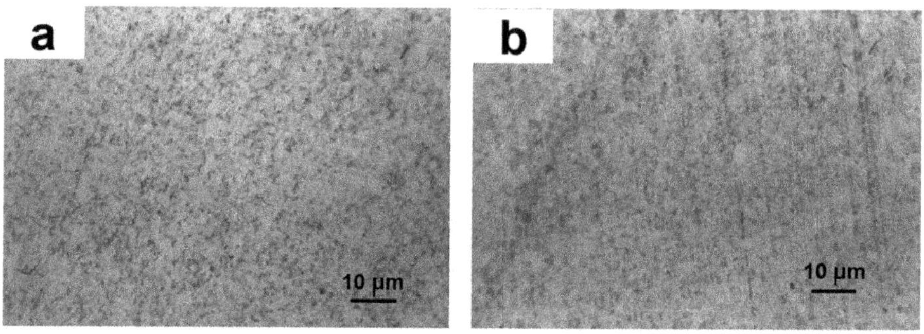

Fig. 11.9 Microstructure of solar (**a**) and conventionally sintered (**b**) WC–10wt.%Co. (Reprinted from Rosa et al. [31]. Copyright (2002), with permission from Elsevier)

exhibited advantages over thermally sintered ones in terms of geometry and conductivity. Worth mentioning are similar conductivities obtained after 5 s of photonic flash sintering and 6 h of thermal sintering. While a large number of studies on the feasibility of radiation-assisted methods of sintering of printed electronics have been published, direct assessment of different electromagnetic wave-assisted methods using the same object is rarely conducted. From a practical perspective, worth particular attention is study by Niittynen et al. [17], who compared the structure and properties of inkjet-printed samples of silver dispersion on polymer substrates sintered by different techniques – sintering in a convection oven (thermal sintering) and plasma, laser, and photonic sintering – and concluded on a better suitability of photonic sintering realized using a xenon flash lamp with an emission spectrum of 350–900 nm for scalable processing of films with good adhesion, high electrical conductivity, and a uniform and dense nanostructure. The temperature during photonic sintering could be high enough to cause substrate damage due to dissipation of heat from the metallic structure. The authors emphasized the importance of tuning the pulse length, intensity, as well as sintering time to bring substrate damage to a minimum. An advantage of selectivity of sintering enabled by the laser sintering technique due to digital control was also highlighted.

Intense pulsed light can be used to initiate chemical reactions. In this regard, photonic treatment of printed CuO accompanied by its reduction to Cu_2O and further to metallic copper presents a viable route to the fabrication of elements of printed electronics [33]. Reduction of CuO nanoparticles contained in the ink was due to a process of photoreduction as well as the presence of reducing agents in the ink. Morphological transformations from the isolated copper nanoparticles to fully sintered Cu films could be controlled by changing the light intensity.

Joo et al. [34] tested the sintering behavior of films deposited from inks containing both copper nano- and microparticles. The resistivity of the film sintered from the microparticle ink (246 μΩ·cm) was greater than that of film obtained from the nanoparticle ink (210 μΩ·cm); these films were sintered under the same flash light sintering conditions. An interesting result was a lower resistivity of the film obtained from the ink containing both nano- and microparticles in the 50:50 volume ratio (80 μΩ·cm). The authors saw the reason for an advantage of inks with particles of mixed sizes in the capability of the nanoparticles to fill the pores between the microparticles. In the film fabricated from the nanoparticle ink, nanopores remained, while in the film sintered from the microparticles, the remaining pores had sizes comparable to the size of the microparticles. The problem of residual pores was successfully solved by a proper choice of the composition of the initial ink in terms of the size of nanoparticles. In order to increase the relative density of the flash light sintered films obtained from copper nanoparticle inks, Chung et al. [35] suggested adding copper-containing precursors that undergo chemical transformations resulting in the in situ formation of copper during flash light treatment of the films. Following this strategy, films with a low resistivity (27.3 μΩ·cm) were obtained from a mixed ink with copper (II) nitrate trihydrate as a copper-containing precursor.

In addition to single metal features, alloys and composite films can be obtained by radiation-assisted sintering. Zhao et al. [36] sintered copper features by using a pulsed green light laser and a pulsed UV laser and, in order to increase the resistance of copper to oxidation, suggested alloying it with gold. Alloying was also beneficial in reducing the melting point of the material. The Cu_xAu_{100-x} alloys when printed and laser-sintered can be used as a basis for flexible chemical sensors for the detection of environmental pollutants and breath biomarkers. Composite films consisting of copper and multiwalled carbon nanotubes were obtained by flash light sintering [32]. For that, carbon nanotubes were dispersed in the copper nanoparticle ink. In the films containing the same concentration of nanotubes, the size of the pores decreased with increasing nanotube length. The presence of carbon nanotubes improved the environmental stability of the films by slowing down the process of oxidation of copper to Cu_2O.

In addition to sintering of nanoparticles, welding of nanowires can occur under flash light treatment [37–39]. In order to improve the optical efficiency of a flash lamp and light absorbing yield, silver nanowires deposited on a substrate were encapsulated by a graphene layer [38]. In the presence of graphene, ultrafast welding of the nanowires (in less than 20 ms) was observed under the intense pulsed light treatment, while pristine (uncoated) nanowires remained intact. The graphene layer is thought to effectively absorb the energy of radiation and transfer it to the silver nanowires. Mixed inks containing copper nanoparticles and copper nanowires in various proportions were tested by subjecting them to sintering under the same conditions [39]. The film flash light sintered from the ink with 5 wt.% of copper nanowires had a lower resistivity (22.77 μΩ·cm) than those obtained from the inks containing copper nanoparticles or nanowires only (94.01 μΩ·cm and 104.15 μΩ·cm, respectively).

Intense white light-assisted sintering was shown to be suitable for sintering of TiO_2 photoelectrodes [21]. The heating of the TiO_2 layer on tin-doped indium oxide-polyethylene terephthalate substrates was suggested to be caused by the sub-band absorption. The treatment of the TiO_2 film with intense pulsed light helped achieve interconnection between the TiO_2 particles and led an increase in the mechanical stability of the photoelectrode and an improvement of the power conversion efficiency over the entire spectral range.

11.7 Summary

This chapter reviewed the principles of sintering techniques based on the application of electromagnetic radiation to powder materials – IR radiation-assisted, laser, solar, photonic, and UV sintering. To enable rapid heating and sintering of powders, IR radiation-producing modules have been designed. IR radiation can also be generated by lasers and is contained in the solar radiation spectrum. Laser sintering is widely used in the layer-by-layer 3D manufacturing of parts from metallic, ceramic, and polymer materials. Solar energy is a clean and ecological type of energy. However, it requires large initial investments into the facilities. Solar sintering is suitable for metals, ceramics, and composite materials and offers faster heating and more rapid densification than conventional furnace sintering. Photonic sintering is conducted mainly by intense pulsed light radiation in the 380 nm–1 μm wavelength range. It is suitable for sintering of metallic nanoparticles, metal-containing composites, as well as semiconductors. A growing interest to IR radiation-assisted and photonic sintering is due to their compatibility with the roll-to-roll fabrication of flexible electronics. UV light is instrumental in initiating chemical reactions and can be used to facilitate sintering when the products of the in situ conducted reactions have advantages in terms of composition and particle size for the progress of sintering. For almost all of the mentioned sintering techniques, the published studies are predominantly descriptive and provide mainly qualitative analyses. An in-depth consideration of these sintering approaches' underlying physical phenomena and the related predictive modeling represent the future research directions.

References

1. Morrison NA (2016) Roll-to-roll processing of flexible devices and components: utilization in wearable and mobile electronics and the coming IOT era. Vakuum in Forschung und Praxis 28 (4):30–35
2. Sowade E, Kang H, Mitra KY, Weiß OJ, Weber J, Baumann RR (2015) Roll-to-roll infrared (IR) drying and sintering of an inkjet-printed silver nanoparticle ink within 1 second. J Mater Chem C 3:11815–11826
3. Georgiadis A, Bryant PA, Murray M, Beharrell P, Keddie JL (2011) Resolving the film-formation dilemma with infrared radiation-assisted sintering. Langmuir 27:2176–2180
4. Román R, Cañadas I, Rodríguez J, Hernández MT, González M (2017) Solar sintering of alumina ceramics: microstructural development. Sol Energy 82:893–902
5. Oliveira FAC, Rosa LG, Fernandes JC, Rodríguez J, Cañadas I, Martínez D, Shohoji N (2009) Mechanical properties of dense cordierite discs sintered by solar radiation heating. Mater Trans 50:2221–2228
6. Gladush GG, Smurov I (2011) Physics of laser materials processing: theory and experiment, Springer Series in Materials Science (SSMATERIALS, V. 146), USA, 534 p
7. Gibson I, Rosen DW (2015) Stucker B (2010) additive manufacturing technologies: 3D printing, rapid prototyping, and direct digital manufacturing, 2nd edn. Springer Science+Business Media, New York, 498 p
8. Olakanmi EO, Cochrane RF, Dalgarno KW (2015) A review on selective laser sintering/melting (SLS/SLM) of aluminium alloy powders: processing, microstructure, and properties. Prog Mater Sci 74:401–477
9. Shishkovsky I, Yadroitsev I, Bertrand P, Smurov I (2007) Alumina–zirconium ceramics synthesis by selective laser sintering/melting. Appl Surf Sci 254(4):966–970
10. Lee H, Huat Joel Lim C, Ji Low M, Tham N, Matham Murukeshan V, Kim YJ (2017) Lasers in additive manufacturing: a review. Int J Prec Eng Manuf Green Technol 4(3):307–322
11. West C, Wang X (2017) Modeling of selective laser sintering/selective laser melting. In: Gu B, Helvajian H, Piqué A, Dunsky CM, Liu J (eds) Laser 3D manufacturing IV. Proc SPIE 10095, USA, p 1009506
12. Boley CD, Khairallah SA, Rubenchik AM (2014) Calculation of laser absorption by metal powders in additive manufacturing. LLNL-JRNL-665313, USA
13. MacNeill W, Choi CH, Chang CH, Malhotra R (2015) On the self-damping nature of densification in photonic sintering of nanoparticles. Sci Rep 5:14845
14. Kim HS, Dhage SR, Shim DE, Hahn HT (2009) Intense pulsed light sintering of copper nanoink for printed electronics. Appl Phys A Mater Sci Process 97:791–798
15. Hwang HJ, Chung WH, Kim HS (2012) In situ monitoring of flash-light sintering of copper nanoparticle ink for printed electronics. Nanotechnology 23:485205
16. Hösel M, Krebs FC (2012) Large-scale roll-to-roll photonic sintering of flexo printed silver nanoparticle electrodes. J Mater Chem 22:15683–15688
17. Niittynen J, Abbel R, Mäntysalo M, Perelaer J, Schubert US, Lupo D (2014) Alternative sintering methods compared to conventional thermal sintering for inkjet printed silver nanoparticle ink. Thin Solid Films 556:452–459
18. Park SH, Chung WH, Kim HS (2014) Temperature changes of copper nanoparticle ink during flash light sintering. J Mater Proc Technol 214:2730–2738
19. Park SH, Kim HS (2014) Flash light sintering of nickel nanoparticles for printed electronics. Thin Solid Films 550:575–5811
20. Galagan Y, Coenen EWC, Abbel R, van Lammeren TJ, Sabik S, Barink M, Meinders ER, Andriessen R, Blom PWM (2013) Photonic sintering of inkjet printed current collecting grids for organic solar cell applications. Org Electron 14:38–46
21. Jin HY, Kim JY, Lee JA, Lee K, Yoo K, Lee DK, Kim B, Kim JY, Kim H, Son HJ, Kim J, Lim JA, Ko MJ (2014) Rapid sintering of TiO_2 photoelectrodes using intense pulsed white light for flexible dye-sensitized solar cells. Appl Phys Lett 104:1439027

22. Marjanovic N, Hammerschmidt J, Perelaer J, Farnsworth S, Rawson I, Kus M, Yenel E, Tilki S, Schubert US, Baumann RR (2011) Inkjet printing and low temperature sintering of CuO and CdS as functional electronic layers and Schottky diodes. J Mater Chem 21:13634–13639
23. Bansal S, Malhotra R (2016) Nanoscale-shape-mediated coupling between temperature and densification in intense pulsed light sintering. Nanotechnology 27(49):495602
24. Govorov AO, Richardson HH (2007) Generating heat with metal nanoparticles. NanoToday 2 (1):30–38
25. Oh Y, Lee SN, Kim HK, Kim J (2012) UV-assisted chemical sintering of inkjet-printed TiO_2 photoelectrodes for low-temperature flexible dye-sensitized solar cells. J Electrochem Soc 159: H777–H781
26. Hwang HJ, Oh KH, Kim HS (2016) All-photonic drying and sintering process via flash white light combined with deep-UV and near-infrared irradiation for highly conductive copper nano-ink. Sci Rep 6:19696
27. Watson T, Mabbett I, Wang H, Peter L, Worsley D (2011) Ultrafast near infrared sintering of TiO_2 layers on metal substrates for dye-sensitized solar cells. Prog Photovolt Res Appl 19 (4):482–486
28. Cañadas I, Martínez D, Rodríguez J, Gallardo JM (2004) Sintering of multilayered copper wires in a solar furnace. Proc. European Congress and Exhibition on Powder Metallurgy, The European Powder Metallurgy Association, UK
29. Oliveira FAC, Shohoji N, Fernandes JC, Rosa LG (2005) Solar sintering of cordierite-based ceramics at low temperatures. Sol Energy 78:351–361
30. Fernandes JC, Amaral PM, Rosa LG, Shohoji N (2000) Weibull statistical analysis of flexure breaking performance for alumina ceramic disks sintered by solar radiation heating. Ceram Int 26:203–206
31. Rosa LG, Amaral PM, Anjinho C, Fernandes JC, Shohoji N (2002) Fracture toughness of solar-sintered WC with Co additive. Ceram Int 28:345–348
32. Hwang HJ, Joo SJ, Kim HS (2015) Copper nanoparticle/multiwalled carbon nanotube composite films with high electrical conductivity and fatigue resistance fabricated via flash light sintering. ACS Appl Mater Interf 7:25413–25423
33. Paglia F, Vak D, van Embden J, Chesman ASR, Martucci A, Jasieniak JJ, Gaspera ED (2015) Photonic sintering of copper through the controlled reduction of printed CuO nanocrystals. ACS Appl Mater Interf 7:25473–25478
34. Joo SJ, Hwang HJ, Kim HS (2014) Highly conductive copper nano/microparticles ink via flash light sintering for printed electronics. Nanotechnology 252:656014
35. Chung WH, Hwang HJ, Kim HS (2015) Flash light sintered copper precursor/nanoparticle pattern with high electrical conductivity and low porosity for printed. Thin Solid Films 580:61–70
36. Zhao W, Rovere T, Weerawarne D, Osterhoudt G, Kang N, Joseph P, Luo J, Shim B, Poliks M, Zhong CJ (2015) Nanoalloy printed and pulse-laser sintered flexible sensor devices with enhanced stability and materials compatibility. ACS Nano 9(6):6168–6177
37. Mallikarjuna K, Hwang HJ, Chung WH, Kim HS (2016) Photonic welding of ultra-long copper nanowire network for flexible transparent electrodes using white flash light sintering. RSC Adv 6:4770–4779
38. Yang SB, Choi H, Lee DS, Choi CG, Choi SY, Kim ID (2015) Improved optical sintering efficiency at the contacts of silver nanowires encapsulated by a graphene layer. Small 11:1293–1300
39. Joo SJ, Park SH, Moon CJ, Kim HS (2015) A highly reliable copper nanowire/nanoparticle ink pattern with high conductivity on flexible substrate prepared via a flash light-sintering technique. ACS Appl Mater Interf 7:5674–5684

Concluding Remarks

This book describes the scientific and technological concepts of field-assisted sintering that have evolved over the past 120 years. Starting in the late nineteenth century with the consolidation of the heating filaments of the incandescent lights by passing high-voltage electric current discharges and up to the latest successes of ultra-high-speed production of complex net-shape metallic and ceramic products, field-assisted sintering attracts an increasing number of researchers by a broad variety of its technological capabilities and scientific problems.

The rich diversity of technical solutions is possible in this area due to a significant number of different factors involved in the control of field-assisted sintering processes. The use of electromagnetic factors in addition to the traditionally applied in materials processing heating and pressure remarkably increases the possibility of precision control of consolidation processes. At the same time, the complexity of the processing also increases significantly, reaching even higher degrees with the involvement of chemical reactions or phase transformations.

In fact, a successful field-assisted sintering process utilizes the conditions of *controlled non-equilibrium*. The high rate of field-assisted processing is based on various transient thermal and nonthermal phenomena, which, if properly managed, provide unique environment for densification and microstructure retention. Therefore, one of the most important challenges of field-assisted sintering techniques is the process stability and the possibility of the process control.

The complexity of the technological equipment for field-assisted sintering reached its highest degree in industrial devices for spark plasma sintering. These devices include, on the one hand, the possibility of the computerized control of the heating, pressure, and electric current, and on the other hand, they provide an in situ recording of all the process parameters. From this viewpoint, even aside from its field-assisted capabilities, this equipment represents an embodied dream of the previous generations of researchers in the area of materials processing. It should be admitted that the high interest of the materials processing community toward field-assisted sintering, and especially spark plasma sintering, is partially caused by

E. A. Olevsky, D. V. Dudina, *Field-Assisted Sintering*,
https://doi.org/10.1007/978-3-319-76032-2

the abovementioned comprehensive equipment capabilities and the rapid turnover of the experimental results.

At the same time, field-assisted sintering gains particular prominence in connection with its unique capabilities of processing very hard-to-deform materials, which would typically require lengthy consolidation times at significantly elevated temperatures under conditions of conventional powder pressing or sintering. Practical implementations of the field-assisted sintering's bright potential, however, are limited by the lack of theoretical concepts enabling the respective processes' predictiveness and optimization. Indeed, despite all the noticeable advantages, traditionally, the use of field-assisted sintering in many cases is still shrouded in mystery. Overwhelming majority of studies on field-assisted sintering describe empiric trial-and-error attempts to consolidate various powder material systems. Generic physically based modeling concepts are currently in strong demand to enable the understanding and control of the "field effect" – a distinguishing set of factors rendering different field-assisted sintering vs. conventional hot pressing and sintering results.

In this connection, this monograph is an attempt to systematically describe various field-assisted sintering techniques with an emphasis on the fundamental principles of these processes. The solutions of the fundamental and applied problems require the in-depth understanding of the field-assisted sintering's underlying fundamental physics and the respective predictive process modeling and modeling-based optimization. Obviously, for the further development of this field, process modeling should not be a scholastic addition to the experimental data, but a direct part of the technological process.

Index

E. Olevsky, D. Dudina, *Field-Assisted Sintering*,
https://doi.org/10.1007/978-3-319-76032-2

G

H

I

GPSR Compliance
The European Union's (EU) General Product Safety Regulation (GPSR) is a set of rules that requires consumer products to be safe and our obligations to ensure this.

If you have any concerns about our products, you can contact us on

ProductSafety@springernature.com

In case Publisher is established outside the EU, the EU authorized representative is:

Springer Nature Customer Service Center GmbH
Europaplatz 3
69115 Heidelberg, Germany

www.ingramcontent.com/pod-product-compliance
Ingram Content Group UK Ltd.
Pitfield, Milton Keynes, MK11 3LW, UK
UKHW020705041026
15590UKWH00001B/478
* 9 7 8 3 0 3 0 0 9 3 8 5 3 *